# Unified Transform for Boundary Value Problems

# Unified Transform for Boundary Value Problems

## *Applications and Advances*


Edited by

**A. S. Fokas**
University of Cambridge
Cambridge, United Kingdom

**B. Pelloni**
University of Reading
Reading, United Kingdom



siam
Society for Industrial and Applied Mathematics
Philadelphia

10 9 8 7 6 5 4 3 2 1



| | |
|---|---|
| *Publisher* | David Marshall |
| *Acquisitions Editors* | Sara Murphy and Elizabeth Greenspan |
| *Developmental Editor* | Gina Rinelli |
| *Managing Editor* | Kelly Thomas |
| *Production Editor* | David Riegelhaupt |
| *Copy Editor* | Susan Fleshman |
| *Production Manager* | Donna Witzleben |
| *Production Coordinator* | Cally Shrader |
| *Compositor* | Techsetters, Inc. |
| *Graphic Designer* | Lois Sellers |

**Library of Congress Cataloging-in-Publication Data**

Unified transform for boundary value problems : applications and advances / edited by A.S. Fokas, University of Cambridge, Cambridge, United Kingdom, B. Pelloni, University of Reading, Reading, United Kingdom.

pages cm. — (Other titles in applied mathematics ; 141)

Includes bibliographical references and index.

ISBN 978-1-611973-81-5

1. Boundary value problems. I. Fokas, A. S., 1952, editor. II. Pelloni, Beatrice, editor.

QA379.U55 2015

515'.35—dc23

2014035698

# List of Contributors

**A. C. L. Ashton**
DAMTP, University of Cambridge, Cambridge, UK, *a.c.l.ashton@damtp.cam.ac.uk*

**G. Biondini**
Department of Mathematics, State University of New York at Buffalo, Buffalo, NY, *biondini@buffalo.edu*

**S. N. Chandler-Wilde**
Department of Mathematics and Statistics, University of Reading, Reading UK, *s.n.chandler-wilde@reading.ac.uk*

**X. Claeys**
LJLL, UPMC Paris, France

**A. S. Fokas**
Department of Applied Mathematics and Theoretical Physics, Cambridge University, Cambridge, UK and Research Centre of Mathematics, Academy of Athens, Athens, 11527 Greece *t.fokas@damtp.cam.ac.uk*

**R. Hiptmair**
SAM, ETH Zurich, Switzerland, *hiptmair@sam.math.ethz.ch*

**A. Iserles**
Department of Applied Mathematics and Theoretical Physics, Cambridge University, Cambridge, UK, *t.fokas@damtp.cam.ac.uk*

**C. Jerez-Hancke**
Escuela de Ingeniería, Pontificia Universidad Católica de Chile, Santiago, Chile

**S. Langdon**
Department of Mathematics and Statistics, University of Reading, Reading, UK, *s.langdon@reading.ac.uk*

**J. Lenells**
KTM Royal Institute Technology, 10044 Stockholm, Sweden, *jlenells@kth.se*

**D. Mantzavinos**
Department of Mathematics, State University of New York at Buffalo, Buffalo, NY, *dionyssi@buffalo.edu*

**B. Pelloni**
Department of Mathematics and Statistics, University of Reading, Reading, UK, *b.pelloni@reading.ac.uk*

**S. Pintarelli**
SAM, ETH Zurich, Switzerland

**D. Shepelsky**
Verkin Institute for Low Temperature Physics and Engineering, Kharkiv, Ukraine

**D. A. Smith**
Department of Mathematical Sciences, University of Cincinnati, Cincinnati, OH, *david.smith2@uc.edu*

**S. A. Smitheman**
Department of Applied Mathematics and Theoretical Physics, Cambridge University, Cambridge, UK, *t.fokas@damtp.cam.ac.uk*

**E. A. Spence**
Department of Mathematical Sciences, University of Bath, Bath, UK, *E.A.Spence@bath.ac.uk*

**T. Trogdon**
Courant Institute of Mathematical Sciences, New York University, New York, *trogdon@cims.nyu.edu*

# Contents

# Preface

In May 2012, the International Centre for Mathematics Sciences (ICMS) in Edinburgh hosted the workshop *Boundary Value Problems for Linear Elliptic and Integrable PDEs: Theory and Computation*. This workshop brought together a small group of mathematicians interested, from several different perspectives, in the solution of boundary value problems, namely the type of problems modeled by partial differential equations (PDEs) that are most commonly arising in applications.

This meeting focused in particular on the applications of the so-called *unified transform* (also referred to as the *Fokas transform* or the *method of Fokas*) to the analysis and numerical modeling of boundary value problems for linear and integrable nonlinear PDEs and on the closely related *boundary element method*, a well-established numerical approach for solving linear elliptic PDEs. The latter method can be viewed as the counterpart in the physical space of the numerical implementation of the unified transform, which is formulated in the spectral (Fourier) space.

This book was conceived during the workshop mentioned above and collects the results of the exchanges of ideas fostered by the meeting. The chapters are closely related and, when put together, paint a picture of the state of the art in the advances and applications of the unified transform as well as its relation with the boundary element method.

It is divided in three main parts. Part I contains new theoretical results on evolution and elliptic problems, linear and nonlinear. New explicit solution representations for several classes of boundary value problems are constructed and rigorously analyzed.

Part II, at the center of the book, is a detailed overview of variational formulations for elliptic problems, building up to placing the unified transform approach in this classical context, alongside the boundary element method, and stressing its novelty.

Part III presents recent numerical applications based on the boundary element method and on the unified transform.

# Chapter 1

# Introduction

*A. S. Fokas and B. Pelloni*

## 1.1 ▪ Boundary Value Problems

In 1747, d'Alembert derived the wave equation, which was the first partial differential equation (PDE) in the history of mathematics. Soon after, d'Alembert and Euler discovered a general method for constructing large classes of solutions, namely the method of *separation of variables*. Bernoulli introduced the infinite sine series and Euler discovered the standard formula for the coefficients of a Fourier series. Fourier inaugurated in 1807 the era of linearization. In 1814 Cauchy wrote as essay on using complex variables for the evaluation of certain integrals. In 1828 Green introduced the powerful approach of integral representations that can be obtained via Green's functions. Separation of variables led to the *spectral analysis* of ordinary differential operators and to the solution of PDEs via a *transform pair*. The prototypical such pair is the Fourier transform; variations include the sine, the cosine, and the Laplace and Mellin transforms.

In the second half of the 20th century it was realized that certain nonlinear evolution PDEs, called *integrable*, can be formulated as the compatibility condition of two linear eigenvalue equations called a *Lax pair* and that this formulation gives rise to a method for solving the initial value problem for these equations. This method is now known as the *inverse scattering transform* method. One of us has emphasized that this method is based on a deeper form of separation of variables [1]. Indeed, the spectral analysis of the spatial part of the Lax pair yields an appropriate *nonlinear Fourier transform pair*, whereas the temporal part of the Lax pair yields the time evolution of the nonlinear Fourier data. In this sense, in spite of the fact that the inverse scattering transform is applicable to nonlinear PDEs, this method still follows the logic of separation of variables.

After the emergence of a method for solving the *initial value problem* for nonlinear integrable evolution equations, the most significant outstanding open problem in the analysis of these equations became the solution of *initial-boundary value problems*. A general approach for solving such problems for evolution equations was introduced in [2] and developed in the work of more than 70 researchers [3]. It is remarkable that these results have motivated the discovery of a new transform method for solving *linear* PDEs in two variables. This method, which is referred to in the current literature as the *unified transform*, is based on two novel ideas (steps): (i) *perform the simultaneous spectral analysis of both equations defining the Lax pair of the given PDE—or equivalently of a certain closed differential one-form* (this is to be contrasted with the case of initial value problems, where the spectral analysis of only the $t$-independent part of the Lax pair is performed), and (ii) *analyze a certain global relation between all initial and boundary values*. The unified transform goes *beyond* separation of variables. Indeed, since it is based on the *simultaneous* spectral analysis of both parts of the Lax pair, it corresponds to the *synthesis* as opposed to separation of variables. As a consequence of this fundamental difference, even in the case of linear PDEs the form of the solution obtained by the unified transform differs drastically from the classical representations. It should be noted that the integral representations obtained via Green's functions retain global features. Actually, it was shown in [4, 5] that in the case of linear PDEs an alternative way to construct the novel integral representations obtained by the unified transform is to use appropriate contour deformations and Cauchy's theorem starting from the integral representations obtained via Green's functions (instead of performing the simultaneous spectral analysis of the associated Lax pair). In this sense, the unified transform reveals a deep relationship between the seminal contributions of Fourier, Cauchy, and Green and furthermore extends these contributions to integrable *nonlinear* PDEs. Indeed, it is shown in [6] that for linear PDEs this method provides a unification as well as a significant extension of the classical transforms, of the method of images, of the Green's function representations and of the Wiener–Hopf technique. (The latter technique through a series of ingenious steps gives rise to a Wiener–Hopf factorization problem, which is actually equivalent to a Riemann–Hilbert (RH) problem; in the new method, such RH problems can be *immediately* obtained using the global relation.) Furthermore, the new approach provides an appropriate "nonlinearization" of some of the above concepts.

For the case of linear elliptic PDEs, a case of great significance in many different areas of application, one of the most powerful numerical approaches is the so-called *boundary element method*. This method, which is well developed and is an established area of expertise within the numerical analysis community, can be viewed as the counterpart of the numerical implementation of the unified transform for linear elliptic PDEs. In particular, both methods are based on variational formulations. For second-order linear elliptic PDEs, such variational formulations are derived after multiplying by a test function and integrating by parts or, equivalently, are based on Green's identities. However, while the boundary element method is formulated in the physical space, the unified transform is formulated in the spectral (Fourier) space.

This book summarizes some recent developments and applications of the unified transform method and presents modern applications of the boundary element method. The first part of the book is devoted to recent advances in the applications of the unified transform, for linear and integrable nonlinear PDEs of both evolution and elliptic types. The second part consists of an important chapter by E. A. Spence that explicitly sets the stage for variational formulations of both the boundary element and unified transform methods. The last part of the book presents numerical strategies for elliptic PDEs, based on either the unified transform or the boundary element method, or the numerical implementation of the unified transform for evolution PDEs.

### 1.1.1 ▪ Evolution PDEs in One and Two Spatial Dimensions

Linear PDEs with *second*-order spatial derivatives formulated in one spatial dimension, with Dirichlet, Neumann, or Robin boundary conditions, can be solved by employing an appropriate transform pair in the spatial variable. For example, the heat equation on the half-line, with either Dirichlet or Neumann boundary conditions, can be solved by either the sine or the cosine transform.

Indeed, the Dirichlet problem for the heat equation on the half-line

$$u_t = u_{xx}, \quad 0<x<\infty, \quad 0<t<T, \quad T>0, \tag{1.1}$$

$$u(x,0) = u_0(x), \quad 0<x<\infty, \tag{1.2}$$

$$u(0,t) = f_0(t), \quad 0<t<T, \tag{1.3}$$

where $u_0(x)$ and $f_0(t)$ have appropriate decay and smoothness, can be solved by the sine transform:

$$u(x,t) = \frac{2}{\pi}\int_0^\infty e^{-\lambda^2 t}\sin(\lambda x)\left[\int_0^\infty u_0(\xi)\sin(\lambda\xi)d\xi + \lambda\int_0^t e^{\lambda^2 s} f_0(s)ds\right]d\lambda. \tag{1.4}$$

However, the $\lambda$-integral on the right-hand side of the above expression is *not* uniformly convergent with respect to the parameter $x$. This lack of uniformity, in addition to rendering such expressions problematic for numerical computations, also makes it difficult to verify that the solution to an initial-boundary value problem obtained using such a transform is indeed a solution. In this respect we note that the construction of the solution via any transform method is usually carried out *assuming* that a solution exists. Thus, unless one can appeal to PDE existence results, one must verify that the final formula obtained in this way does indeed satisfy the given PDE and the prescribed initial and boundary conditions. For initial-boundary value problems, this is straightforward, but for initial-boundary value problems one must overcome the lack of uniform convergence in this representation of the solution. It is interesting to note that this problem is not addressed in any of the standard applied texts on boundary value problems for linear PDEs.

It should be emphasized that the problem of lack of uniform convergence occurs in *every* nonhomogeneous boundary value problem solved by standard transform methods.

The situation is even less satisfactory for linear PDEs involving third-order spatial derivatives. A typical example is the Stokes equation on the half-line:

$$u_t + u_x + u_{xxx} = 0, \quad 0<x<\infty, \quad 0<t<T. \tag{1.5}$$

It can be shown that this problem, supplemented with the initial and boundary conditions (1.2) and (1.3), is well-posed. However, there does not exist an analogue of the sine or cosine transforms. One may attempt to solve the above problem by a Laplace transform in $t$, but this approach has several drawbacks. In particular, if $\tilde{u}(x,s)$ denotes the Laplace transform of $u(x,t)$, then $\tilde{u}$ satisfies

$$\tilde{u}_{xxx}(x,s) + \tilde{u}_x(x,s) + s\tilde{u}(x,s) = u_0(x), \quad 0<x<\infty. \tag{1.6}$$

Thus, in order to construct an appropriate Green's function for this ODE, one must solve the homogeneous version of (1.6), that is, the cubic equation

$$\lambda(s)^3 + \lambda(s) + s = 0.$$

In contrast to the above difficulties, the unified transform (a) can be applied to linear evolution PDEs involving spatial derivatives of any order and (b) yields expressions which are always uniformly convergent at the boundary.

In particular, for the Dirichlet problem for the heat equation on the half-line, instead of (1.4), the unified transform method yields

$$u(x,t) = \frac{1}{2\pi}\left\{\int_{-\infty}^\infty e^{i\lambda x-\lambda^2 t}\hat{u}_0(\lambda)d\lambda - \int_{\partial D^+} e^{i\lambda x-\lambda^2 t}[\hat{u}_0(-\lambda) + 2i\lambda\tilde{f}_0(\lambda^2)]d\lambda\right\}, \tag{1.7}$$

where $\partial D^+$ is the boundary of the region

$$D^+ = \{\lambda \in \mathbb{C} : \mathrm{Im}\lambda \geq 0, \mathrm{Re}(\lambda^2) \leq 0\},$$

$\hat{u}_0(\lambda)$ is the Fourier transform of the initial condition, and $\tilde{f}_0(\lambda)$ is the following transform of the boundary condition:

$$\tilde{f}_0(\lambda) = \int_0^T e^{\lambda s} f_0(s) ds.$$

In contrast to (1.4), it is straightforward to verify that the function defined by (1.7) satisfies the boundary condition at $x = 0$.

Using Cauchy's theorem, the contours in (1.7) can be deformed to obtain (1.4). However, in general it is *not* possible to deform the contours in the expression obtained by the new method to obtain an expression involving integrals on the real axis. For example, such a deformation does *not* exist for the analogous expression for equation (1.5), which is consistent with the fact that in this case there does not exist a classical $x$-transform. Furthermore, even when such a deformation is possible, it appears that the expression formulated in the complex $\lambda$-plane has both analytical and numerical advantages. Regarding the latter advantage, we note that at least for those cases that the Fourier transform of the initial and boundary data can be computed explicitly, the computation of $u(x,t)$ reduces simply to the computation of an integral in the complex $\lambda$-plane. This integral contains the term $\exp(i\lambda x)$, which decays exponentially for large $\lambda$ in the upper-half $\lambda$ plane. In addition, by an appropriate contour deformation it is possible to map $\partial D^+$ to a contour in $\mathbb{C}^+$ with the property that that term involving the temporal dependence $\exp(\omega(\lambda)t)$ also decays exponentially for large $t$. Thus, the numerical computation of this integral is most efficient; see [7]. The implementation of the unified transform for linear evolution PDEs in one and two dimensions was recently reviewed in [8].

In Section 2.1, Mantzavinos uses the heat equation in one and two dimensions as an illustrative example to show that, using the unified transform, it is possible to construct analytic solutions for problems involving nonseparable boundary conditions as well as nonlocal constraints. It should be noted that even for the particular cases that the boundary conditions are separable, the unified transform gives a construction of the relevant spectral representations in a much simpler way than the classical spectral analysis. Actually, the unified transform provides a new approach to some open problems in spectral analysis. In Section 2.2, Smith illustrates the spectral interpretation of the unified transform, describing how this method can be viewed as the natural extension of Fourier transform techniques for non-self-adjoint operators. This section also discusses the spectral meaning of the transform pair defined by the unified transform, using the recent definition of a new class of spectral functionals that essentially diagonalize the non-self-adjoint spatial differential operators associated with linear evolution boundary value problems.

A great advantage of the unified transform for linear evolution initial-boundary value problems is that it provides an algebraic way for eliminating the unknown boundary values or, more precisely, appropriate transforms of the unknown boundary values. However, for integrable nonlinear PDEs this elimination is in general *not* possible. Thus, for such nonlinear PDEs one has to characterize the so-called *generalized Dirichlet-to-Neumann map*, namely characterize the unknown boundary values in terms of the given initial and boundary data. However, it turns out that for a particular class of boundary conditions, called *linearizable*, the algebraic elimination *is* possible, and hence for the corresponding problems the unified transform method is as effective as the classical inverse scattering transform. Recall that the sine and cosine transforms can be obtained from the Fourier transform by considering either an odd or an even symmetry. Thus, if an initial boundary value problem can be solved by a sine or cosine transform, it can also be solved by an odd or even extension to the full line via the Fourier transform. The situation is similar for linearizable integrable PDEs posed on the half-line. For example, linearizable initial boundary value problems for the nonlinear Schrödinger (NLS) equation (which is a second-order PDE whose linear version can therefore be solved by an appropriate sine or cosine transform) can be solved either by the unified transform or by the inverse scattering transform after an appropriate extension to the full line. However, linearizable initial-boundary value problems for the Korteweg–deVries (KdV) equation (whose linear version *cannot* be solved by an appropriate $x$-transform) so far can only be solved by the unified transform. These issues are further discussed in Chapter 3, which discusses in detail the relation between the two above approaches to the solution of initial-boundary value problems for the NLS equation on the half-line with linearizable boundary conditions.

### 1.1.2 ▪ Elliptic PDEs

It is well known that the main difficulty with boundary value problems stems from the fact that, although the solution representation requires the knowledge of all boundary values, some of them are *not* prescribed as boundary conditions. In the theory of elliptic PDEs, the determination of the unknown boundary values is known as the problem of characterizing the Dirichlet-to-Neumann map. The global relation provides the starting point for analytical, rigorous, and numerical characterizations of this map. Regarding analytical results, we note that for certain simple domains, the unified transform yields analytical expressions for the unknown boundary values [9], [10], [11], [12], [13], [14], [15], [16], [17], [18], [19], [20]. Regarding rigorous results, we note that in a series of important papers, Ashton has shown that the analysis of the global relation provides a novel approach to obtaining a plethora of rigorous results for linear elliptic PDEs. This rigorous approach is summarized in Chapter 4. Regarding numerical results, we note that the unified transform has inspired a novel numerical technique for the determination of the unknown boundary values [21], [22], [23], [24], [25], [26], [27], [28], [29], [30]. For elliptic PDEs formulated in the interior of a polygon, the above technique provides the analogue of the boundary integral method, but now the analysis takes place in the spectral (Fourier) space instead of the physical space. This technique is summarized in Section 7.1 by Fokas, Iserles, and Smitheman. Furthermore, the first steps toward extending this technique to the *exterior* of a polygon are presented in Section 7.2, by Fokas and Lenells.

Part II of the book, by Spence, presents an overview of variational formulations of second-order linear elliptic PDEs that are based on integration by parts or, equivalently, on Green's identities.

The difficulty of treating effectively boundary value problems in the exterior of a bounded domain is at the heart of Section 7.3, where Chandler-Wilde and Langdon discuss the applications of the boundary element method to high-frequency problems arising in acoustics and make connections to the unified transform by analyzing particular instances of this method for problems of acoustic scattering by diffraction gratings. The practical problem is the scattering of acoustic waves by an obstacle is naturally formulated as a boundary value problem in the domain exterior to the obstacle. The efficient solution of such problems, in particular the resolution of the high-frequency components, is still a difficult problem, and what is discussed is a state-of-the-art hybrid boundary element method for it, as well as connections to the unified transform.

The same difficulty is again the main motivation for Section 7.4, where Claeys et al. consider scalar second-order problems in the exterior of a bounded domain with Dirichlet boundary conditions on the boundary and formulate two novel well-posed *multi-trace boundary integral equations* for such problems. Compared to conventional single-trace formulations they offer the advantage of being amenable to operator preconditioning, which allows greater numerical effectiveness.

The first applications of the unified transform to nonlinear elliptic PDEs were presented in [31, 32]. In Section 5.1 the results to date on the case of the elliptic sine-Gordon equations are summarized by Pelloni, who also gives a complete characterization of linearizable boundary conditions for the case of boundary value problems posed in a semi-infinite strip.

A crucial step in the development of the unified transform was the observation by Fokas and Gelfand that linear PDEs always admit a Lax pair formulation [33]. This observation also played a crucial role in the development of a new technique for deriving transforms, which finally led to the construction by Novikov of the inverse attenuated Radon transform [34]. It is remarkable that as shown by Trogdon, this formulation is also useful for the numerical integration of the initial value problem of linear evolution PDEs. In addition to this result, Trogdon presents in section 8.1 a powerful new technique for the numerical solution of integrable PDEs based on the numerical integration of the associated RH formulation.

## Conclusions and Open Problems

Following its introduction in [2], the unified transform has been extensively developed, and it has found applications in surprisingly many and varied mathematical fields. This book touches upon recent developments and applications.

Several important problems remain open, including the following:

(a) *$t$-periodic boundary conditions.* It was noted earlier that initial-boundary value problems for evolution PDEs on the half-line with *linearizable* boundary conditions can be solved via the unified transform as efficiently as initial value problems. Furthermore, for *nonlinearizable* boundary conditions which decay for large $t$, the unified transform yields useful asymptotic information on the large $t$ behavior of the solution. However, for the physically important case of *$t$-periodic* boundary conditions, it is necessary to characterize the Dirichlet-to-Neumann map. Pioneering results in this direction for the NLS have been obtained in [35, 36, 37]. However, several problems, including the extension of these results for the case of KdV, remain open.

(b) *$x$-periodic initial conditions.* The initial value problem for nonlinear integrable PDEs with *periodic* initial conditions can be analyzed via the elegant algebro-geometric formalism. This involves finite-genus Riemann surfaces and yields finite-dimensional solutions sets [38, 39, 40]. Even if these solutions could be made computationally accessible [41, 42], one still has to answer the important question of the density of these solutions in an appropriate function space. It has been shown in [43] that this problem belongs to the linearizable class. However, the relevant implementation of the unified transform in this case remains open.

(c) *Three-dimensional problems.* The implementation of the unified transform to linear evolution PDEs in two spatial dimensions is discussed in the article of Mantzavinos. Furthermore, the first steps toward the analysis of integrable nonlinear evolution PDEs in two spatial dimensions have been taken in [44]. However, the associated analysis of the Dirichlet-to-Neumann map, as well as the implementation of the unified transform to elliptic PDEs in three dimensions, remain open.

The approach described and applied in the various sections contained in this book need to be developed further. The articles presented here, together with the extensive bibliography presented after each section, constitute an essential map for understanding the modern developments and the several ramifications of this method and for appreciating its applicability.

# Bibliography


[1] A. S. Fokas, Lax pairs: A novel type of separability, Inverse Problems **25** (2009), 123007.

[2] A. S. Fokas, A unified transform method for solving linear and certain nonlinear PDEs, Proc. R. Soc. Lond. Ser. A **453** (1997), 1411–1443.

[3] A. S. Fokas, A Unified Approach to Boundary Value Problems, SIAM (2008).

[4] A. S. Fokas and M. Zyskin, The fundamental differential form and boundary value problems, Quart. J. Mech. Appl. Math. **55** (2002), 457–479.

[5] E. A. Spence and A. S. Fokas, A new transform method I: Domain-dependent fundamental solutions and integral representations, Proc. Roy. Soc. **466** (2010), 2259–2281.

[6] A. S. Fokas and E. A. Spence, Synthesis, as opposed to separation, of variables, SIAM Rev. **54** (2012), 291–324.

[7] A. S. Fokas and N Flyer, A hybrid analytical-numerical method for solving evolution partial differential equations. I: The half-line, Proc R. Soc. Lond. Ser. A, **464** (2008), 1823–1849.

[8] B. Deconinck, T. Trogdon and V. Vasan, The method of Fokas for solving linear partial differential equations, SIAM Rev. **56** (2014), 159–186.

[9] D. G. Crowdy and A. S. Fokas, Explicit integral solutions for the plane elastostatic semi-strip, Proc. Roy. Soc. Lond. Ser. A, **460** (2004), 1285–1309.

[10] D. ben-Avraham and A. S. Fokas, Solution of the modified Helmholtz equation in a triangular domain and an application to diffusion-limited coalescence, Phys. Rev. E **64** (2001), 016114.

[11] A. S. Fokas and A. A. Kapaev, A Riemann-Hilbert approach to the Laplace equation, J. Math. Anal. Appl. **251** (2000), 770–804.

[12] A. S. Fokas, Two dimensional linear PDE's in a convex polygon, Proc. R. Soc. Lond. Ser. A **457** (2001), 371–393.

[13] A. S. Fokas and D. A. Pinotsis, Quaternions, evaluation of integrals and boundary value problems, Comp. Meth. Funct. Th. **7** (2007), 443–476.

[14] G. Dassios and A. S. Fokas, The basic elliptic equations in an equilateral triangle, Proc. R. Soc. Ser. A **461** (2005), 2721–2748.

[15] G. Dassios and A. S. Fokas, Methods for solving elliptic PDEs in spherical coordinates, SIAM J. Appl. Math. **68** (2008), 1080–1096.

[16] Y. Antipov and A. S. Fokas, A transform method for the modified Helmholtz equation on the semi-strip, Math. Proc. Cambridge Philos. Soc. **137** (2004), 339–365.

[17] A. S. Fokas and A. A. Kapaev, On a transform method for the Laplace equation in a polygon, IMA J. Appl. Math. **68** (2003), 355–408.

[18] G. Dassios and M. Doschoris, On the global relation and the Dirichlet-to-Neumann correspondence, Stud. Appl. Math. **126** (2011), 75–102.

[19] G. Dassios and M. Doschoris, Axisymmetric Stoke's flow in a spherical shell revisited via the Fokas method. Part I: Irrotational flow, Math. Methods Appl. Sci. **34** (2011), 850–868.

[20] D. G. Crowdy and A. M. J. Davis, Stokes flow singularities in a two-dimensional channel: A novel transform approach with application to microswimming, Proc. R. Soc. Lond. Ser. A **469** (2013), 20130198.

[21] B. Fornberg and N. Flyer, A numerical implementation of Fokas boundary integral approach: Laplace's equation on a polygonal domain, Proc. R. Soc. Ser. A **467** (2011), 2983–3003.

[22] C-I. R. Davis and B. Fornberg, A spectrally accurate numerical implementation of the Fokas transform method for Helmholtz-type PDEs, Complex Var. Elliptic Equ. **59** (2013), 1–14.

[23] C.-I. Davis, Numerical Tests of the Fokas Method for Helmholtz-Type Partial Differential Equations: Dirichlet to Neumann Maps, Masters thesis, University of Colorado at Boulder (2008).

[24] A. S. Fokas, N. Flyer, S. A. Smitheman, and E. A. Spence, A semianalytical numerical method for solving evolution and elliptic partial differential equations, J. Comput. Appl. Math. **227** (2009), 59–74.

[25] S. R. Fulton, A. S. Fokas, and C. A. Xenophontos, An analytical method for linear elliptic PDEs and its numerical implementation. J. Comput. Appl. Math. **167** (2004), 465–483.

[26] Y. G. Saridakis, A. G. Sifalakis, and E. P. Papadopoulou, Efficient numerical solution of the generalized Dirichlet-Neumann map for linear elliptic PDEs in regular polygon domains, J. Comput. Appl. Math. **236** (2012), 2515–2528.

[27] A. G. Sifalakis, S. R. Fulton, E. P. Papadopoulou, and Y. G. Saridakis, Direct and iterative solution of the generalized Dirichlet-Neumann map for elliptic PDEs on square domains, J. Comput. Appl. Math. **227** (2009), 171–184.

[28] A. G. Sifalakis, E. P. Papadopoulou, and Y. G. Saridakis, Numerical study of iterative methods for the solution of the Dirichlet-Neumann map for linear elliptic PDEs on regular polygon domains, Int. J. Appl. Math. Comput. Sci. **4** (2007), 173–178.

[29] A. G. Sifalakis, A. S. Fokas, S. Fulton, and Y. G. Saridakis, The generalised Dirichlet to Neumann map for linear elliptic PDEs and its numerical implementation, J. Comput. Appl. Math. **219** (2008), 9–34.

[30] S. A. Smitheman, E. A. Spence, and A. S. Fokas, A spectral collocation method for the Laplace and modified Helmholtz equations in a convex polygon, IMA J. Numer. Anal. **30** (2010), 1184–1205.

[31] J. Lenells and A. S. Fokas, Boundary-value problems for the stationary axisymmetric Einstein equations: A rotating disc, Nonlinearity **24** (2011), 177–206.

[32] A. S. Fokas, J. Lenells, and B. Pelloni, Boundary value problems for the elliptic sine-Gordon equation in a semi-strip, J. Nonlinear Sci. **23** (2013), 241–282.

[33] A. S. Fokas and I. M. Gelfand, Integrability of linear and nonlinear evolution equations and the associated nonlinear Fourier transforms, Lett. Math. Phys. **32** (1994), 189–210.

[34] R. G. Novikov, An inversion formula for the attenuated -ray transformation, Ark. Mat. **40** (2002), 145–167.

[35] A. Boutet de Monvel, A. R. Its, and V. Kotlyarov, Long-time asymptotics for the focusing NLS equation with time-periodic boundary condition, C. R. Math. **345** (2007), 615–620.

[36] A. Boutet de Monvel, A. R. Its, and V. Kotlyarov, Long-time asymptotics for the focusing NLS equation with time-periodic boundary condition on the half-line, Comm. Math. Phys. **290** (2009), 479–522.

[37] A. Boutet de Monvel, A. R. Its, and V. Kotlyarov, Decaying long-time asymptotics for the focusing NLS equation with periodic boundary condition, Int. Math. Res. Not. IMRN **3** (2009), 547–577.

[38] S. P. Novikov, S. V. Manakov, L. U. Pitaevski, and V. E. Zakharov, Soliton Theory: The Inverse Scattering Method, Plenum, New York (1984).

[39] B. A. Dubrovin, Theta functions and non-linear equations, Russian Math. Surveys **36** (1981), 11–92.

[40] I. M. Krichever and S. P. Novikov. The inverse problem method and holomorphic bundles on Riemann surfaces, Math. Phys. Rev. **3** (1980), 5–26.

[41] B. Deconinck, M. Heil, A. Bobenko, M. Van Hoeij, and M. Schmies, Computing Riemann theta functions, Math. Comput. **73** (2004), 1417–1442.

[42] B. Deconinck and M. S. Patterson, Computing the Abel map, Phys. D **237** (2008), 3214–3232.

[43] A. S. Fokas and A. R. Its, The nonlinear Schrödinger equation on the interval, J. Phys. A **37** (2004), 6091–6114.

[44] D. Mantzavinos and A. S. Fokas. The Kadomtsev-Petviashvili II equation on the half-plane, Phys. D **240** (2011), 477–511.

# Part I

# Theoretical Results

# Chapter 2

# Evolution Problems: Linear

## 2.1 ▪ Linear Initial-Boundary Value Problems via the Fokas Method

*D. Mantzavinos*

### 2.1.1 ▪ Introduction

Self-adjoint boundary value problems for linear PDEs formulated in a separable domain and involving separable boundary conditions can be solved via the usual transforms. The unified transform method introduced in 1997 by Fokas [1, 2], although originally developed for integrable *nonlinear* evolution PDEs in one spatial dimension, has had important implications for *linear* PDEs [3, 4]. In particular, it is capable of constructing analytical solutions to certain linear boundary value problems which either are non-self-adjoint or involve nonseparable boundary conditions. See [5, 6, 7] for applications of the unified method to the classical spectral theory of non-self-adjoint linear differential operators, and see [8, 9] for recent reviews comparing the unified method with the standard approaches.

Even for the simple cases where the standard transform methods can be applied, the implementation of the classical techniques for constructing the relevant classical spectral representations is quite cumbersome. For example, D. S. Cohen, in order to solve a particular boundary value problem for the heat equation formulated on the half-line, had to construct the spectral representation

$$q(x) = \frac{2}{\pi}\int_0^\infty \frac{\cos kx + k\sin kx}{k^2+1}\left[\int_0^\infty (\cos k\xi + k\sin k\xi)q(\xi)d\xi - q(0)\right]dk \\ + 2e^{-x}\left[q(0) - \int_0^\infty e^{-\xi}q(\xi)d\xi\right]. \tag{2.1}$$

The derivation of (2.26) is rather elaborate; see [10] and also the relevant discussion in [11]. Analogously, the authors of [12], in order to study a particular boundary value problem for the heat equation formulated on the finite interval involving certain nonlocal constraints, had to perform a rather involved spectral analysis; in this case, the relevant eigenvalues, denoted by $k$, satisfy a *transcendental* equation.

The advantages of the unified transform method regarding the aforementioned issues are the following: (i) it can be implemented in a straightforward way *without* the need to derive a spectral representation, and (ii) it yields an integral expression which is uniformly convergent at the boundary. Concerning (ii), in particular, we recall that the solution obtained via the standard methods has the serious disadvantage that it is *not* uniformly convergent at the boundary (unless the boundary conditions are homogeneous). Thus, since the unified transform method is easier to implement, and since it yields uniformly convergent solutions, it is clear that it is preferable to the standard methods even in the case that the latter ones can actually be applied. More specifically, the unified transform method (a) yields an effective approach for the numerical evaluation of typical boundary value problems (see [13, 14]) and (b) provides a straightforward approach to obtaining rigorous results.

What is more, we note that most of the existing results using the unified transform method have been restricted to two dimensions. Regarding linear PDEs, the only works in higher than two dimensions are (i) the analytical solution of evolution PDEs in two spatial dimensions involving either second- or third-order spatial derivatives formulated in the quarter plane and satisfying separable boundary conditions, and (ii) the analytical solution of the two-dimensional heat equation formulated in the interior of an equilateral triangle with Dirichlet boundary conditions [15].

Our goal in this article is two-fold: (1) using as an example the heat equation in one and two spatial dimensions, we illustrate how the unified transform method can be implemented in a straightforward way for problems involving nonseparable boundary conditions, as well as nonlocal constraints. (2) We emphasize a new application of the unified transform method, namely that it can be used to construct classical spectral representations (such as (2.26) and the spectral representations that appear in [12]) in an easier way than the standard approaches.

**The Unified Transform Method.** The implementation of the method for an evolution equation requires the followings steps:

1. *Integration of the divergence forms.* Rewrite the equation in a suitable divergence form. Then, integrate this form over the given domain to obtain the so-called global relation, namely an equation coupling the Fourier transform of the solution $u$, denoted by $\hat{u}$, with appropriate transforms of the initial data and of the boundary values.

2. *Inversion of the global relation.* Solve the global relation for $\hat{u}$ and use the inverse Fourier transform to obtain $u$ in terms of transforms of the initial data and of the boundary values. Deform the contour of integration from the real line to an appropriate contour in the complex spectral plane to obtain an integral representation of $u$. This representation is not yet effective because it involves transforms of unknown boundary values.

3. *Elimination of the unknowns.* Use the global relation, as well as the equation obtained from the global relation under certain symmetry transformations, to eliminate the transforms of the unknown boundary values in order to derive an effective solution expression (effective, in the sense that it only involves transforms of the given data).

**Organization of the Article.** In Section 2.1.2, we present the solution of the heat equation on the half-line with given initial data and an oblique Robin boundary condition at $x = 0$:

$$u_t(0,t) + \alpha u_x(0,t) + \beta u(0,t) = \gamma(t), \quad t > 0, \tag{2.2}$$

where $\alpha, \beta$ are real constants and $\gamma(t)$ is a given function. The particular case of $\beta = \gamma = 0$

is the one considered by Cohen in [10]; using the general solution formula obtained by the unified method, we also derive equation (2.26) in a straightforward way.

Section 2.1.3 is devoted to the heat equation on the finite interval satisfying the nonlocal constraints

$$\int_0^l u(x,t)dx = \mathcal{F}(t), \quad \int_0^l (l-x)\,u(x,t)dx = \mathcal{R}(t), \tag{2.3}$$

where $\mathcal{F}(t)$ and $\mathcal{R}(t)$ are given functions. Using the general solution formula obtained by the unified transform method, we derive in a straightforward way the classical spectral representation that appears in [12]. A detailed analysis of the problems contained in Sections 2.1.2 and 2.1.3 is available in [16].

In Section 2.1.4, we present an analysis of what may be regarded as a two-dimensional analogue of the oblique Robin initial-boundary value problem (IBVP) of Section 2.1.2; namely, we consider the two-dimensional heat equation posed on the quarter plane with given initial data and oblique Neumann boundary conditions:

$$u_t = u_{xx} + u_{yy}, \qquad 0 < x < \infty,\ 0 < y < \infty,\ t > 0, \tag{2.4a}$$

$$u(x,y,0) = u_0(x,y), \quad 0 < x < \infty,\ 0 < y < \infty, \tag{2.4b}$$

$$\cos\beta_1\, u_x(x,0,t) - \sin\beta_1\, u_y(x,0,t) = F_1(x,t), \quad 0 < x < \infty,\ t > 0, \tag{2.4c}$$

$$\cos\beta_2\, u_y(0,y,t) + \sin\beta_2\, u_x(0,y,t) = F_2(y,t), \quad 0 < y < \infty,\ t > 0, \tag{2.4d}$$

where $\beta_1, \beta_2$ are given angles and $F_1, F_2$ are known functions, satisfying appropriate compatibility conditions. The complete treatment of the IBVP (2.4) is available in [17].

Finally, one of the analytical advantages of the unified transform method, namely the uniform convergence of the solution formulae at the boundary, is demonstrated in the appendix, where the solution to the oblique Robin IBVP of Section 2.1.2 is verified via purely algebraic manipulations.

## 2.1.2 ▪ Oblique Robin Conditions on the Half-Line

In a 1932 article by Langer [18], we encounter the following problem from heat conduction. An open tank with free surface area $A_l$ is filled with a certain liquid of arbitrary initial temperature, which absorbs heat from its surrounding medium of temperature $T(t)$. Furthermore, the liquid is kept well-stirred so that its temperature $r(t)$ is uniform throughout its volume. A semi-infinite solid rod is placed in direct contact with the tank, as shown in Figure 2.1; hence, heat is transferred from the liquid to the rod. The lateral surface of the rod is insulated so that there is no heat flow through it. Also, since the length of the rod is infinite, the temperature $u$ of the rod can be regarded as uniform throughout the cross-section $A_s$. Therefore, $u = u(x,t)$, i.e., $u$ depends only on the distance $x$ from the point of contact with the tank and on the time $t$.

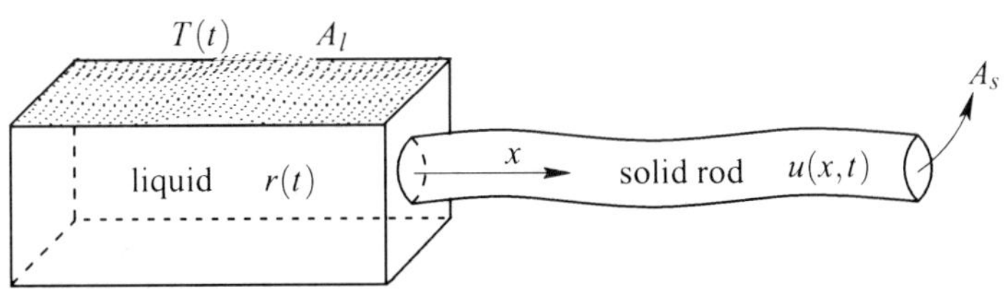


**Figure 2.1.** *Heat conduction along a solid rod.*

The above suggest that the diffusion of heat along the rod can be modeled by the following IBVP for the one-dimensional heat equation:

$$
\begin{aligned}
&u_t = u_{xx}, && 0 < x < \infty, \quad t > 0, && \text{(2.5a)}\\
&u(x,0) = u_0(x), && 0 < x < \infty, && \text{(2.5b)}\\
&u(0,t) = r(t), && t > 0, && \text{(2.5c)}
\end{aligned}
$$

where $u_0(x)$ is the initial temperature distribution of the rod, which is known, and $r(t)$ is the temperature of the tank, which is unknown.

The rate of accumulation of heat in the liquid can be expressed in terms of the incoming heat from the surrounding medium and of the heat flowing from the tank to the rod by the equation

$$\sigma \frac{dr}{dt}(t) = -u_x(0,t) + \lambda[T(t) - r(t)], \tag{2.6}$$

where the real constants $\sigma$ and $\lambda$ depend on the specific heat, the mass, and the conductivities of the rod and the liquid. Combining equations (2.5c) and (2.6), we obtain the following so-called *oblique Robin* boundary condition at $x = 0$:

$$\sigma u_t(0,t) + u_x(0,t) + \lambda[u(0,t) - T(t)] = 0. \tag{2.7}$$

Letting for convenience of notation $\alpha = 1/\sigma$, $\beta = \lambda/\sigma$, and $\gamma(t) = \lambda T(t)/\sigma$, the IBVP (2.5) becomes

$$
\begin{aligned}
&u_t = u_{xx}, && 0 < x < \infty, \quad t > 0, && \text{(2.8a)}\\
&u(x,0) = u_0(x), && 0 < x < \infty, && \text{(2.8b)}\\
&u_t(0,t) + \alpha u_x(0,t) + \beta u(0,t) = \gamma(t), && t > 0. && \text{(2.8c)}
\end{aligned}
$$

We will now derive the solution to the IBVP (2.8) via the unified transform, following steps 1–3 of the unified transform method. The complete derivation can be found in [16].

**1. Integration of the divergence form.** It is straightforward to verify that equation (2.8a) can be written in divergence form as

$$\left(e^{-ikx+k^2t}u(x,t)\right)_t = \left(e^{-ikx+k^2t}[u_x(x,t) + ik\,u(x,t)]\right)_x, \quad k \in \mathbb{C}. \tag{2.9}$$

Green's theorem in the domain $\Omega \doteqdot \{0 < x < \infty,\ 0 < s < t\}$, depicted in Figure 2.2, yields

$$\int_{\partial\Omega} e^{-ikx+k^2s}\left\{u(x,s)dx + \left[u_x(x,s) + ik\,u(x,s)\right]ds\right\} = 0, \quad k \in \mathbb{C}^-, \tag{2.10}$$

where the restriction of $k$ to $\mathbb{C}^- \doteqdot \{k \in \mathbb{C} : \mathrm{Im}(k) \le 0\}$ ensures that the exponential $e^{-ikx}$

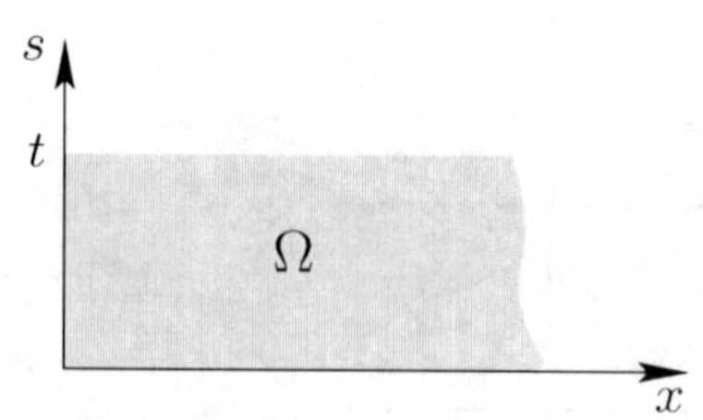


**Figure 2.2.** *The domain Ω.*

is bounded for all $x \in [0,\infty)$, in order for the corresponding integral to make sense. No such restriction is needed for $e^{k^2 s}$, since $s \in [0,t]$ is finite.

Define the Fourier transform pair on the half-line by

$$\hat{u}(k,t) = \int_0^\infty e^{-ikx} u(x,t)dx, \qquad k \in \mathbb{C}^-, \tag{2.11a}$$

$$u(x,t) = \frac{1}{2\pi}\int_{-\infty}^\infty e^{ikx}\hat{u}(k,t)dk, \quad 0 < x < \infty. \tag{2.11b}$$

**Remark 1.** *In contrast to the Fourier transform along the real line, the domain of definition of $\hat{u}(k,t)$ is now extended from the real line to $\mathbb{C}^-$, since $e^{-ikx}$ is analytic and bounded there for all $x \in [0,\infty)$.*

Under (2.11a), equation (2.9) becomes the global relation

$$e^{k^2 t}\hat{u}(k,t) = \hat{u}_0(k) - \left[g_1(k^2,t) + ikg_0(k^2,t)\right], \quad k \in \mathbb{C}^-, \tag{2.12}$$

where $\hat{u}_0(k)$ is the Fourier transform of the initial data $u_0(x)$ and

$$g_j(k^2,t) \doteqdot \int_0^t e^{k^2 s}\partial_x^j u(0,s)ds, \quad j = 0,1. \tag{2.13}$$

**2. Inversion of the global relation.** Inverting (2.12) by means of (2.11b) yields

$$u(x,t) = \frac{1}{2\pi}\int_{-\infty}^\infty e^{ikx-k^2 t}\hat{u}_0(k)dk - \frac{1}{2\pi}\int_{-\infty}^\infty e^{ikx-k^2 t}\left[g_1(k^2,t) + ikg_0(k^2,t)\right]dk. \tag{2.14}$$

Note that the exponential $e^{ikx-k^2(t-s)}$ is bounded as $|k| \to \infty$ in the region $\mathbb{C}^+ \setminus D^+$, where the region $D^+$ of Figure 2.3 is defined by

$$D^+ = \left\{k \in \mathbb{C}^+ : (\operatorname{Im} k)^2 > (\operatorname{Re} k)^2\right\}. \tag{2.15}$$

Thus, by Cauchy's theorem and Jordan's lemma we can deform the contour of the second integral in (2.14) from the real line to the contour $\partial D^+$ to obtain the integral representation

$$u(x,t) = \frac{1}{2\pi}\int_{-\infty}^\infty e^{ikx-k^2 t}\hat{u}_0(k)dk - \frac{1}{2\pi}\int_{\partial D^+} e^{ikx-k^2 t}\left[g_1(k^2,t) + ikg_0(k^2,t)\right]dk. \tag{2.16}$$

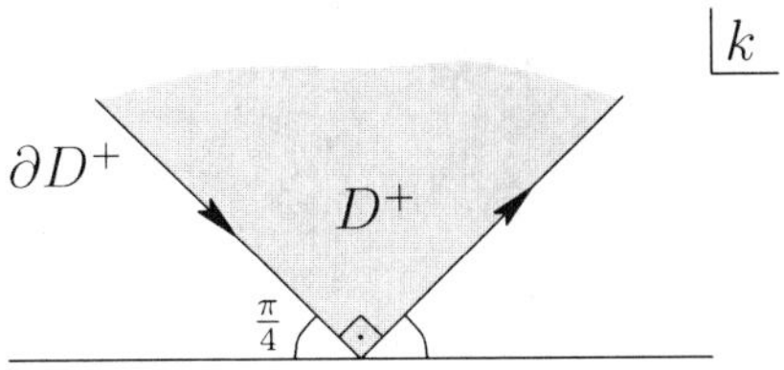


**Figure 2.3.** *The region $D^+$.*

**3. Elimination of the unknowns.** The integral representation (2.16) does not provide an effective solution expression, since the functions $g_0, g_1$ are unknown. These functions, however, depend on $k$ only through $k^2$ and, therefore, they are *invariant* under the transformation $k \mapsto -k$. Under this transformation, the global relation (2.12) assumes the form

$$e^{k^2 t}\hat{u}(-k,t) = \hat{u}_0(-k) - \left[g_1(k^2,t) - ik\, g_0(k^2,t)\right], \quad k \in \mathbb{C}^+. \tag{2.17}$$

What is more, multiplying the boundary condition (2.8c) by $e^{k^2 s}$ and integrating with respect to $s$ from 0 to $t$ yields

$$e^{k^2 t}u(0,t) - u_0(0) - k^2 g_0(k^2,t) + \alpha g_1(k^2,t) + \beta g_0(k^2,t) = \tilde{\gamma}(k^2,t), \tag{2.18}$$

where

$$\tilde{\gamma}(k^2,t) \doteqdot \int_0^t e^{k^2 s}\gamma(s)ds. \tag{2.19}$$

Solving (2.18) for $g_1$ and substituting the resulting expression in equation (2.17) gives

$$g_0(k^2,t) = \frac{1}{k^2 - i\alpha k - \beta}\left\{\alpha\left[\hat{u}_0(-k) - e^{k^2 t}\hat{u}(-k,t)\right] - \left[\tilde{\delta}(k^2,t) - e^{k^2 t}u(0,t)\right]\right\}, \ k \in \mathbb{C}^+,$$

where

$$\tilde{\delta}(k^2,t) \doteqdot \tilde{\gamma}(k^2,t) + u_0(0). \tag{2.20}$$

Thus, returning to (2.18) we find

$$\begin{aligned} g_1(k^2,t) + ik\, g_0(k^2,t) = &-\frac{k^2 + i\alpha k - \beta}{k^2 - i\alpha k - \beta}\left[e^{k^2 t}\hat{u}(-k,t) - \hat{u}_0(-k)\right] \\ &- \frac{2ik}{k^2 - i\alpha k - \beta}\left[\tilde{\delta}(k^2,t) - e^{k^2 t}u(0,t)\right], \quad k \in \mathbb{C}^+, \end{aligned} \tag{2.21}$$

so that the integral representation (2.16) becomes

$$\begin{aligned} u(x,t) = &\frac{1}{2\pi}\int_{-\infty}^{\infty} e^{ikx - k^2 t}\hat{u}_0(k)dk \\ &+ \frac{1}{2\pi}\int_{\partial D^+} \frac{e^{ikx-k^2 t}}{k^2 - i\alpha k - \beta}\Big\{(k^2 + i\alpha k - \beta)\left[e^{k^2 t}\hat{u}(-k,t) - \hat{u}_0(-k)\right] \\ &\qquad + 2ik\left[\tilde{\delta}(k^2,t) - e^{k^2 t}u(0,t)\right]\Big\}dk. \end{aligned} \tag{2.22}$$

While at first glance expression (2.22) may look no better than representation (2.16), since it involves the unknown functions $\hat{u}(-k,t)$ and $u(0,t)$, the contributions of the integrals corresponding to these unknowns can be computed explicitly. For this purpose, we need to study the roots of the polynomial $k^2 - i\alpha k - \beta$, which are given by

$$k_1, k_2 = \frac{1}{2}\left(i\alpha \pm \sqrt{-\alpha^2 + 4\beta}\right). \tag{2.23}$$

Indeed, considering the cases of 0, 1, or 2 roots lying inside the region $D^+$, we arrive at the following proposition.

**Proposition 2.1 (Solution).**

1. *Suppose that the roots $k_1, k_2$ of the polynomial $k^2 - i\alpha k - \beta$, given by equation (2.23), do not lie in the region $D^+$ depicted in Figure 2.3. Then, the oblique Robin IBVP (2.8) admits*

*the solution*

$$u(x,t)=\frac{1}{2\pi}\int_{-\infty}^{\infty} e^{ikx-k^2t}\hat{u}_0(k)dk$$
$$+\frac{1}{2\pi}\int_{\partial D^+} e^{ikx-k^2t}\frac{2ik\,\tilde{\delta}(k^2,t)-(k^2+i\alpha k-\beta)\,\hat{u}_0(-k)}{k^2-i\alpha k-\beta}dk, \quad (2.24a)$$

*where $\hat{u}_0(k)$ and $\tilde{\delta}(k^2,t)$ are defined according to equations* (2.11a) *and* (2.20), *respectively, and the contour $\partial D^+$ is shown in Figure* 2.3.

2. *If $k_1 \in D^+$ and $k_2 \notin D^+$, then the solution is given by the expression* (2.24a) *plus the function $u_1(x,t)$ defined by*

$$u_1(x,t)=-\frac{2k_1}{2k_1-i\alpha}e^{ik_1x-k_1^2t}\left[\alpha\,\hat{u}_0(-k_1)-\tilde{\delta}(k_1^2,t)\right]. \quad (2.24b)$$

3. *If $k_2 \in D^+$ and $k_1 \notin D^+$, then the solution is given by the expression* (2.24a) *plus the function $u_2(x,t)$ defined by*

$$u_2(x,t)=-\frac{2k_2}{2k_2-i\alpha}e^{ik_2x-k_2^2t}\left[\alpha\,\hat{u}_0(-k_2)-\tilde{\delta}(k_2^2,t)\right]. \quad (2.24c)$$

4. *If both $k_1$ and $k_2$ lie in the region $D^+$, then the solution is given by the expression* (2.24a) *plus the functions $u_1(x,t)$ and $u_2(x,t)$.*

**Return to the Real Line and the Classical Spectral Representation.** If $\alpha=1$ and $\beta=\gamma=0$, then the oblique Robin condition (2.8c) assumes the *separable* form

$$u_t(0,t)+u_x(0,t)=0. \quad (2.25)$$

In this case, the method of separation of variables yields a problem where the eigenvalue occurs in both the equation and the boundary condition. Using the elaborate spectral representation

$$q(x)=\frac{2}{\pi}\int_0^{\infty}\frac{\cos kx+k\sin kx}{k^2+1}\left[\int_0^{\infty}(\cos k\xi+k\sin k\xi)q(\xi)d\xi-q(0)\right]dk$$
$$+2e^{-x}\left[q(0)-\int_0^{\infty}e^{-\xi}q(\xi)d\xi\right], \quad (2.26)$$

Cohen [10] obtained the solution to the IBVP (2.8) with the boundary condition (2.8c) replaced by (2.25) in the form

$$u(x,t)=\frac{2}{\pi}\int_0^{\infty}e^{-k^2t}\frac{\cos kx+k\sin kx}{k^2+1}\left[\int_0^{\infty}(\cos k\xi+k\sin k\xi)u_0(\xi)d\xi-u_0(0)\right]dk$$
$$+2e^{-x+t}\left[u_0(0)-\int_0^{\infty}e^{-\xi}u_0(\xi)d\xi\right]. \quad (2.27)$$

Next, we derive Cohen's representation from our solution representation. Letting $\alpha=1$ and $\beta=\gamma=0$ in (2.24a) and deforming the contour $\partial D^+$ to the real line gives

$$u(x,t)=\frac{1}{2\pi}\int_{-\infty}^{\infty}e^{ikx-k^2t}\hat{u}_0(k)dk+\frac{1}{2\pi}\int_{-\infty}^{\infty}e^{ikx-k^2t}\left[\frac{2i}{k-i}u_0(0)-\frac{k+i}{k-i}\hat{u}_0(-k)\right]dk$$
$$+2e^{-x+t}\left[u_0(0)-\hat{u}_0(-i)\right]. \quad (2.28)$$

By the definition of the Fourier transform, equation (2.28) becomes

$$u(x,t)=\frac{1}{2\pi}\int_0^\infty e^{-k^2t}\left[e^{ikx}\int_0^\infty e^{-ik\xi}u_0(\xi)d\xi+e^{-ikx}\int_0^\infty e^{ik\xi}u_0(\xi)d\xi\right]dk$$
$$-\frac{1}{2\pi}\int_0^\infty e^{-k^2t}\left[e^{ikx}\frac{k+i}{k-i}\int_0^\infty e^{ik\xi}u_0(\xi)d\xi+e^{-ikx}\frac{k-i}{k+i}\int_0^\infty e^{-ik\xi}u_0(\xi)d\xi\right]dk$$
$$+\frac{1}{i\pi}\int_0^\infty e^{-k^2t}\left[\frac{e^{-ikx}}{k+i}-\frac{e^{ikx}}{k-i}\right]u_0(0)dk+2e^{-x+t}\left[u_0(0)-\int_0^\infty e^{-\xi}u_0(\xi)d\xi\right],$$

which, upon rearrangement, is Cohen's representation (2.27).

**Remark 2.** *Note that for general boundary value problems the contour deformation from $\partial D^+$ to the real line may not be possible. In this particular case, however, the relevant exponentials are bounded and analytic for $k\in\mathbb{C}^+\setminus D^+$, and thus such a deformation is allowed.*

**Remark 3.** *It is interesting to note that the spectral representation* (2.26) *can be obtained directly via the unified transform method in a much simpler way. Indeed, evaluating equation* (2.27) *at $t=0$ and replacing $u(x,0)$ and $u_0(x)$ by $q(x)$ yields equation* (2.26).

### 2.1.3 ▪ Nonlocal Constraints on the Interval

The treatment via the unified transform method of IBVPs for linear evolution equations posed on the interval, instead of the half-line, is not much different. In particular, the heat equation posed on the interval with oblique Robin boundary conditions at both endpoints can be analyzed similarly to the half-line problem of Section 2.1.2; see [16].

In the current section, we illustrate how the unified transform method applies in the case of the interval via the following IBVP with *nonlocal* constraints:

$$u_t=u_{xx},\qquad x\in(0,l),\quad t>0, \tag{2.29a}$$
$$u(x,0)=u_0(x),\qquad x\in(0,l), \tag{2.29b}$$
$$\int_0^l u(x,t)dx=\mathscr{F}(t),\qquad t>0, \tag{2.29c}$$
$$\int_0^l (l-x)\,u(x,t)dx=\mathscr{R}(t),\qquad t>0, \tag{2.29d}$$

where $\mathscr{F}$ and $\mathscr{R}$ are given functions.

Similarly to Section 2.1.2, steps 1 and 2 of the unified transform method yield the global relation

$$e^{k^2t}\hat{u}(k,t)=\hat{u}_0(k)+e^{-ikl}\left[h_1(k^2,t)+ik\,h_0(k^2,t)\right]-\left[g_1(k^2,t)+ik\,g_0(k^2,t)\right],\quad k\in\mathbb{C}, \tag{2.30}$$

and the integral representation

$$u(x,t)=\frac{1}{2\pi}\int_{-\infty}^\infty e^{ikx-k^2t}\hat{u}_0(k)dk+\frac{1}{2\pi}\int_{\partial D^-}e^{ik(x-l)-k^2t}\left[h_1(k^2,t)+ik\,h_0(k^2,t)\right]dk$$
$$-\frac{1}{2\pi}\int_{\partial D^+}e^{ikx-k^2t}\left[g_1(k^2,t)+ik\,g_0(k^2,t)\right]dk, \tag{2.31}$$

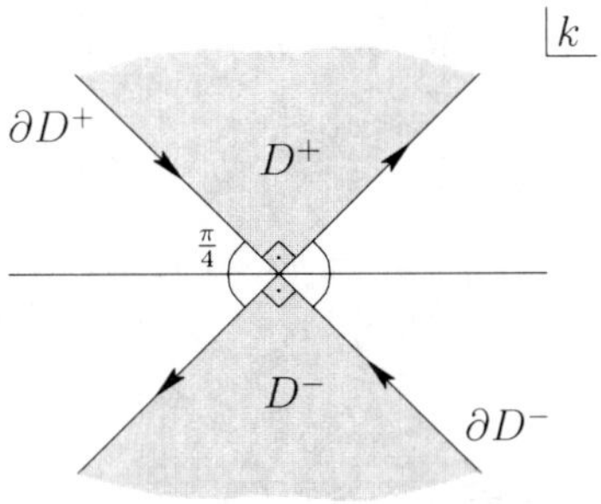


**Figure 2.4.** *The regions $D^+$ and $D^-$.*

where

$$\hat{u}_0(k) \doteqdot \int_0^l e^{-ikx} u_0(x)dx, \quad k \in \mathbb{C}, \tag{2.32}$$

the functions $g_j$ are defined by equation (2.13), the functions $h_j$ are defined by

$$h_j(k^2,t) \doteqdot \int_0^t e^{k^2 s} \partial_x^j u(l,s)ds, \quad j = 0,1, \tag{2.33}$$

and the contours $\partial D^{\pm}$ are depicted in Figure 2.4.

**Remark 4.** *Note that, in contrast to the half-line case, the global relation* (2.30) *is defined for all $k \in \mathbb{C}$, since the Fourier transform* (2.32) *makes sense for any complex $k$ due to $x$ being finite.*

**3. Elimination of the unknowns.** Letting

$$\frac{d\mathcal{F}}{dt}(t) = F(t), \quad \frac{d\mathcal{R}}{dt}(t) = R(t) \tag{2.34}$$

and performing some straightforward manipulations on the nonlocal constraints (2.29c) and (2.29d) gives

$$u_x(l,t) - u_x(0,t) = F(t), \tag{2.35a}$$
$$l\,u_x(0,t) = u(l,t) - u(0,t) - R(t). \tag{2.35b}$$

In turn, these equations imply the expressions

$$g_1\left(k^2,t\right) = \frac{1}{l}\left[h_0\left(k^2,t\right) - g_0\left(k^2,t\right) - r\left(k^2,t\right)\right], \tag{2.36a}$$
$$h_1\left(k^2,t\right) = h_0\left(k^2,t\right) - g_0\left(k^2,t\right) + f\left(k^2,t\right) - r\left(k^2,t\right), \tag{2.36b}$$

where

$$f\left(k^2,t\right) \doteqdot \int_0^t e^{k^2 s} F(s)ds, \quad r\left(k^2,t\right) \doteqdot \int_0^t e^{k^2 s} R(s)ds. \tag{2.37}$$

Combining equations (2.36) with the global relation (2.30) yields

$$e^{k^2 t}\hat{u}(k,t) = \hat{u}_0(k) + \left[\left(\frac{1}{l} + ik\right)e^{-ikl} - \frac{1}{l}\right]h_0\left(k^2,t\right) + \left[\left(\frac{1}{l} - ik\right) - \frac{e^{-ikl}}{l}\right]g_0\left(k^2,t\right)$$
$$+ e^{-ikl} f\left(k^2,t\right) + \frac{1}{l}\left(1 - e^{-ikl}\right) r\left(k^2,t\right), \quad k \in \mathbb{C}, \tag{2.38a}$$

which, under the transformation $k \mapsto -k$, additionally implies the relation

$$e^{k^2 t}\hat{u}(-k,t) = \hat{u}_0(-k) + \left[\left(\frac{1}{l} - ik\right)e^{ikl} - \frac{1}{l}\right]h_0(k^2,t) + \left[\left(\frac{1}{l} + ik\right) - \frac{e^{ikl}}{l}\right]g_0(k^2,t)$$

$$+ e^{ikl} f(k^2,t) + \frac{1}{l}\left(1 - e^{ikl}\right) r(k^2,t), \quad k \in \mathbb{C}. \tag{2.38b}$$

Equations (2.38) constitute a $2\times 2$ linear system for the unknown functions $g_0, h_0$. Writing this system in matrix form and using Cauchy's theorem and Jordan's lemma, as in the case of the half-line, we are able to obtain an explicit expression for these unknowns provided that the relevant matrix involved is invertible. This is a direct consequence of the roots of the determinant $P(k)$ of this matrix, which is given by

$$P(k) \doteq \frac{k}{l}\left[(lk - 2i)e^{-ikl} + 4i - (lk + 2i)e^{ikl}\right]. \tag{2.39}$$

Analyzing the behavior of these roots leads to the following proposition.

**Proposition 2.2.** *The initial-boundary value problem* (2.29) *admits the solution*

$$u(x,t) = \frac{1}{2\pi}\int_{-\infty}^{\infty} e^{ikx - k^2 t}\hat{u}_0(k)dk \tag{2.40}$$

$$-\frac{1}{2\pi}\int_{\partial D^-} e^{ik(x-l) - k^2 t}\frac{k}{lP(k)}\Big\{\left[2i - lk - 2i\,e^{ikl}\right]\hat{u}_0(k) - lk\,\hat{u}_0(-k)$$

$$+ 2i\left[1 + (ilk - 1)e^{ikl}\right]f(k^2,t) + 2k\left(e^{ikl} - 1\right)r(k^2,t)\Big\}\,dk$$

$$-\frac{1}{2\pi}\int_{\partial D^+} e^{ikx - k^2 t}\frac{k}{lP(k)}\Big\{\left[2i - (2i + lk)e^{ilk}\right]\hat{u}_0(k) - lk\,e^{-ikl}\hat{u}_0(-k)$$

$$+ 2i\left[e^{-ikl} + ilk - 1\right]f(k^2,t) + 2k\left(1 - e^{-ikl}\right)r(k^2,t)\Big\}\,dk,$$

*where* $\hat{u}_0(k)$ *denotes the Fourier transform* (2.32) *of the initial data* (2.29b), *the functions* $f(k^2,t)$ *and* $r(k^2,t)$ *are defined by equations* (2.37), *the determinant* $P(k)$ *is defined by equation* (2.39), *and the contours* $\partial D^{\pm}$ *are depicted in Figure* 2.4.

**Return to the Real Line and the Spectral Representation.** As in Section 2.1.2, we may obtain the "classical" representation for the solution of the IBVP (2.29) by collapsing the contours of integration of (2.40) to the real line. Indeed, such a deformation gives

$$u(x,t) = \sum_{\substack{k_m \\ k_m \neq 0}} e^{ik_m x - k_m^2 t}\frac{ik_m}{lP'(k_m)}\Big\{\left[(2i - lk_m)e^{-ik_m l} - 2i\right]\hat{u}_0(k_m) - lk_m\,e^{-ik_m l}\hat{u}_0(-k_m)$$

$$+ 2i\left(ilk_m - 1 + e^{-ik_m l}\right)f(k_m^2) + 2k_m\left(1 - e^{-ik_m l}\right)r(k_m^2)\Big\}, \tag{2.41}$$

where $P'(k)$ is the derivative of $P(k)$ with respect to $k$, and $k_m$ are the nonzero roots of $P(k)$ in the region $\mathbb{C}\setminus(D^+\cup D^-)$ of the complex $k$-plane.

**Remark 5.** *Evaluating expression (2.41) at $t=0$ and replacing $u(x,0)$ and $\hat{u}_0(k_m)$ by $q(x)$ and $\hat{q}(k)$, respectively, we obtain the analogue of Cohen's spectral representation (2.26), namely the spectral representation*

$$q(x)=\sum_{\substack{k_m\\ k_m\neq 0}} e^{ik_m x}\frac{ik_m}{lP'(k_m)}\left\{\left[(2i-lk_m)e^{-ik_m l}-2i\right]\hat{q}(k_m)-lk_m e^{-ik_m l}\hat{q}(-k_m)\right\}, \tag{2.42}$$

*which is the appropriate one for solving the IBVP (2.29) via classical spectral analysis.*

**Remark 6 (Weyl's asymptotic law).** *The roots $k_m$, $m\in\mathbb{Z}$, of $P(k)$ that appear in representation (2.41) can be regarded as the eigenvalues of the one-dimensional Laplace operator, $\partial_{xx}$. It is straightforward to see that $P(2n\pi/l)=0$; thus*

$$k_{2n}=\frac{2n\pi}{l}\quad \forall n\in\mathbb{Z}. \tag{2.43a}$$

*Furthermore, it is shown in [12] that*

$$k_{2n-1}=\frac{(2n-1)\pi}{l}+\mathcal{O}\left(\frac{1}{n}\right),\quad n\in\mathbb{Z},\quad |n|\to\infty. \tag{2.43b}$$

*An immediate consequence of equations (2.43) is the so-called Weyl's asymptotic law for the eigenvalues of the Laplace operator, that is,*

$$\lim_{|k_m|\to\infty}\frac{k_m^2\, l^2}{m^2\,\pi^2}=1. \tag{2.44}$$

### 2.1.4 ▪ Oblique Neumann Conditions on the Quarter Plane

The unified transform method extends in a straightforward manner to IBVPs in 2+1 dimensions. Below, we consider the two-dimensional heat equation on the quarter plane posed with *oblique Neumann* boundary conditions:

$$u_t=u_{xx}+u_{yy},\quad u=u(x,y,t),\quad 0<x<\infty,\ 0<y<\infty,\ t>0, \tag{2.45a}$$

$$u(x,y,0)=u_0(x,y),\quad 0<x<\infty,\quad 0<y<\infty, \tag{2.45b}$$

$$\cos\beta_1\, u_x(x,0,t)-\sin\beta_1\, u_y(x,0,t)=F_1(x,t),\quad 0<x<\infty,\ t>0, \tag{2.45c}$$

$$\cos\beta_2\, u_y(0,y,t)+\sin\beta_2\, u_x(0,y,t)=F_2(y,t),\quad 0<y<\infty,\ t>0, \tag{2.45d}$$

where $F_1(x,t)$ and $F_2(y,t)$ are known functions and the angles $\beta_1$ and $\beta_2$ are shown in Figure 2.5.

Conditions (2.45c) and (2.45d) resemble the oblique Robin conditions of problem (2.8) and, in particular, they are *nonseparable*.

We employ once again steps 1–3 of the unified transform method to solve the IBVP (2.45). The complete derivation for the IBVP (2.45) can be found in [17].

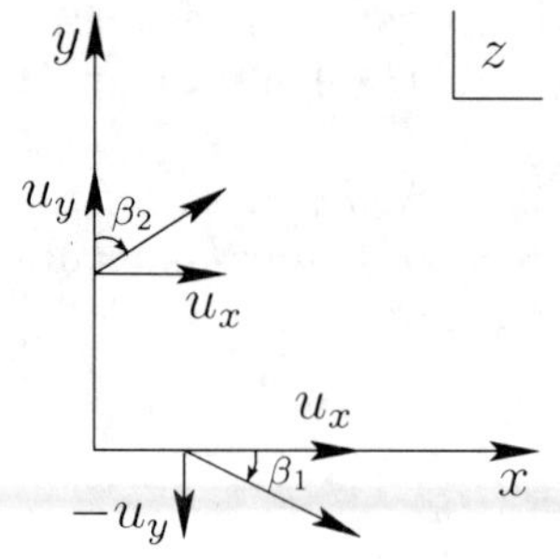


**Figure 2.5.** *The angles $\beta_1$ and $\beta_2$.*

Steps 1 and 2 are the same as for the one-dimensional case of Section 2.1.2, the only difference being that instead of Green's theorem we now apply Gauss's theorem on the divergence form. Supplementing conditions (2.45c) and (2.45d) with the expressions

$$s_1\, u_x(x,0,t) + c_1\, u_y(x,0,t) = Q_1(x,t), \qquad 0 < x < \infty,\ t > 0 \tag{2.46a}$$

$$s_2\, u_y(0,y,t) - c_2\, u_x(0,y,t) = Q_2(y,t), \qquad 0 < y < \infty,\ t > 0, \tag{2.46b}$$

where the functions $Q_1, Q_2$ are unknown, simplifies our computations a great deal. Eventually, we obtain the global relation

$$\begin{aligned} e^{\omega t}\,\hat{u}(k_1,k_2,t) = \hat{u}_0(k_1,k_2) &+ \left[\left(c_2 - s_2\frac{k_1}{k_2}\right) q_2(\omega,k_2,t) - \left(s_2 + c_2\frac{k_1}{k_2}\right) f_2(\omega,k_2,t)\right] \\ &- \left[\left(c_1 + s_1\frac{k_2}{k_1}\right) q_1(\omega,k_1,t) - \left(s_1 - c_1\frac{k_2}{k_1}\right) f_1(\omega,k_1,t)\right] \\ &- \left(\frac{k_1}{k_2} + \frac{k_2}{k_1}\right)\tilde{u}(\omega,t), \qquad k_1 \in \mathbb{C}^-,\ k_2 \in \mathbb{C}^-, \end{aligned} \tag{2.47}$$

where

$$\omega = \omega(k_1,k_2) = k_1^2 + k_2^2, \tag{2.48}$$

$$\cos\beta_j = c_j, \quad \sin\beta_j = s_j, \quad j = 1,2, \tag{2.49}$$

and

$$\hat{u}_0(k_1,k_2) = \int_0^\infty dx \int_0^\infty dy\, e^{-ik_1x - ik_2y}\, u_0(x,y), \tag{2.50a}$$

$$f_1(\omega,k_1,t) = \int_0^t ds \int_0^\infty dx\, e^{-ik_1x + \omega s} F_1(x,s), \tag{2.50b}$$

$$f_2(\omega,k_2,t) = \int_0^t ds \int_0^\infty dy\, e^{-ik_2y + \omega s} F_2(y,s), \tag{2.50c}$$

$$q_1(\omega,k_1,t) = \int_0^t ds \int_0^\infty dx\, e^{-ik_1x + \omega s} Q_1(x,s), \tag{2.50d}$$

$$q_2(\omega,k_2,t) = \int_0^t ds \int_0^\infty dy\, e^{-ik_2y + \omega s} Q_2(y,s), \tag{2.50e}$$

$$\tilde{u}(\omega,t) = \int_0^t ds\, e^{\omega s}\, u(0,0,s). \tag{2.50f}$$

Inverting the global relation and performing our usual contour deformations yield the integral representation

$$
\begin{aligned}
u(x,y,t) = {} & \frac{1}{(2\pi)^2}\int_{-\infty}^{\infty} dk_1 \int_{-\infty}^{\infty} dk_2\, e^{ik_1x+ik_2y-\omega t}\, \hat{u}_0(k_1,k_2) \\
& - \frac{1}{(2\pi)^2}\int_{\partial D^+} dk_1 \int_{-\infty}^{\infty} dk_2\, e^{ik_1x+ik_2y-\omega t}\left[\left(s_2 + c_2\frac{k_1}{k_2}\right)f_2(\omega,k_2,t) + \frac{k_1}{k_2}\tilde{u}(\omega,t)\right. \\
& \qquad\qquad \left. - \left(c_2 - s_2\frac{k_1}{k_2}\right)q_2(\omega,k_2,t)\right] \\
& + \frac{1}{(2\pi)^2}\int_{-\infty}^{\infty} dk_1 \int_{\partial D^+} dk_2\, e^{ik_1x+ik_2y-\omega t}\left[\left(s_1 - c_1\frac{k_2}{k_1}\right)f_1(\omega,k_1,t) - \frac{k_2}{k_1}\tilde{u}(\omega,t)\right. \\
& \qquad\qquad \left. - \left(c_1 + s_1\frac{k_2}{k_1}\right)q_1(\omega,k_1,t)\right].
\end{aligned}
\tag{2.51}
$$

**3. Elimination of the unknowns.** The integral representation (2.51) involves the unknown functions $q_1(\omega,k_1,t)$, $q_2(\omega,k_2,t)$, and $\tilde{u}(\omega,t)$, defined by equations (2.50d), (2.50e), and (2.50f), respectively. Since $\omega$ is invariant under the transformations $k_j \mapsto -k_j$, $j=1,2$, the global relation (2.47) can be supplemented with the following identities:

$$
\begin{aligned}
e^{\omega t}\hat{u}(-k_1,k_2,t) = {} & \hat{u}_0(-k_1,k_2) + \left[\left(c_2 + s_2\frac{k_1}{k_2}\right)q_2(\omega,k_2,t) - \left(s_2 - c_2\frac{k_1}{k_2}\right)f_2(\omega,k_2,t)\right] \\
& - \left[\left(c_1 - s_1\frac{k_2}{k_1}\right)q_1(\omega,-k_1,t) - \left(s_1 + c_1\frac{k_2}{k_1}\right)f_1(\omega,-k_1,t)\right] \\
& + \left(\frac{k_1}{k_2} + \frac{k_2}{k_1}\right)\tilde{u}(\omega,t), \quad k_1 \in \mathbb{C}^+,\ k_2 \in \mathbb{C}^-,
\end{aligned}
\tag{2.52a}
$$

$$
\begin{aligned}
e^{\omega t}\hat{u}(k_1,-k_2,t) = {} & \hat{u}_0(k_1,-k_2) + \left[\left(c_2 + s_2\frac{k_1}{k_2}\right)q_2(\omega,-k_2,t) - \left(s_2 - c_2\frac{k_1}{k_2}\right)f_2(\omega,-k_2,t)\right] \\
& - \left[\left(c_1 - s_1\frac{k_2}{k_1}\right)q_1(\omega,k_1,t) - \left(s_1 + c_1\frac{k_2}{k_1}\right)f_1(\omega,k_1,t)\right] \\
& + \left(\frac{k_1}{k_2} + \frac{k_2}{k_1}\right)\tilde{u}(\omega,t), \quad k_1 \in \mathbb{C}^-,\ k_2 \in \mathbb{C}^+,
\end{aligned}
\tag{2.52b}
$$

and

$$
\begin{aligned}
e^{\omega t}\hat{u}(-k_1,-k_2,t) = {} & \hat{u}_0(-k_1,-k_2) + \left[\left(c_2 - s_2\frac{k_1}{k_2}\right)q_2(\omega,-k_2,t) - \left(s_2 + c_2\frac{k_1}{k_2}\right)f_2(\omega,-k_2,t)\right] \\
& - \left[\left(c_1 + s_1\frac{k_2}{k_1}\right)q_1(\omega,-k_1,t) - \left(s_1 - c_1\frac{k_2}{k_1}\right)f_1(\omega,-k_1,t)\right] \\
& - \left(\frac{k_1}{k_2} + \frac{k_2}{k_1}\right)\tilde{u}(\omega,t), \quad k_1 \in \mathbb{C}^+,\ k_2 \in \mathbb{C}^+.
\end{aligned}
\tag{2.52c}
$$

Solving (2.52a) for $q_2(\omega,k_2,t)$, solving (2.52b) for $q_1(\omega,k_1,t)$, and inserting the resulting expressions into the integral representation (2.51), we obtain

$$\begin{aligned}
u(x,y,t) &= \frac{1}{(2\pi)^2}\int_{-\infty}^{\infty} dk_1 \int_{-\infty}^{\infty} dk_2\, e^{ik_1x+ik_2y-\omega t}\,\hat{u}_0(k_1,k_2)\\
&\quad -\frac{1}{(2\pi)^2}\int_{\partial D^+} dk_1 \int_{-\infty}^{\infty} dk_2\, e^{ik_1x+ik_2y-\omega t}\Bigg\{\left(s_2+c_2\frac{k_1}{k_2}\right)f_2(\omega,k_2,t)+\frac{k_1}{k_2}\tilde{u}(\omega,t)\\
&\qquad -\frac{c_2k_2-s_2k_1}{c_2k_2+s_2k_1}\Bigg[e^{\omega t}\hat{u}(-k_1,k_2,t)-\hat{u}_0(-k_1,k_2)\\
&\qquad\quad +\left(s_2-c_2\frac{k_1}{k_2}\right)f_2(\omega,k_2,t)+\left(c_1-s_1\frac{k_2}{k_1}\right)q_1(\omega,-k_1,t)\\
&\qquad\quad -\left(s_1+c_1\frac{k_2}{k_1}\right)f_1(\omega,-k_1,t)-\left(\frac{k_1}{k_2}+\frac{k_2}{k_1}\right)\tilde{u}(\omega,t)\Bigg]\Bigg\}\\
&\quad +\frac{1}{(2\pi)^2}\int_{-\infty}^{\infty} dk_1 \int_{\partial D^+} dk_2\, e^{ik_1x+ik_2y-\omega t}\Bigg\{\left(s_1-c_1\frac{k_2}{k_1}\right)f_1(\omega,k_1,t)-\frac{k_2}{k_1}\tilde{u}(\omega,t)\\
&\qquad +\frac{c_1k_1+s_1k_2}{c_1k_1-s_1k_2}\Bigg[e^{\omega t}\hat{u}(k_1,-k_2,t)-\hat{u}_0(k_1,-k_2)\\
&\qquad\quad +\left(s_2-c_2\frac{k_1}{k_2}\right)f_2(\omega,-k_2,t)-\left(c_2+s_2\frac{k_1}{k_2}\right)q_2(\omega,-k_2,t)\\
&\qquad\quad -\left(s_1+c_1\frac{k_2}{k_1}\right)f_1(\omega,k_1,t)-\left(\frac{k_1}{k_2}+\frac{k_2}{k_1}\right)\tilde{u}(\omega,t)\Bigg]\Bigg\}.
\end{aligned} \tag{2.53}$$

The contributions of the terms involving $\hat{u}(-k_1,k_2,t)$ and $\hat{u}(k_1,-k_2,t)$ vanish. Indeed, the relevant integrands are analytic and, furthermore, since $x+\xi\geq 0$ and $y+\eta\geq 0$, they also decay exponentially at infinity; hence, by Cauchy's theorem and Jordan's lemma the corresponding integrals vanish.

**The relation between $\beta_1$ and $\beta_2$.** The representation (2.53) still involves the unknown functions $q_1(\omega,-k_1,t)$ and $q_2(\omega,-k_2,t)$. It turns out that we can eliminate these functions by using equation (2.52c). However, this can be achieved only if the angles $\beta_1$ and $\beta_2$ are such that the combination of $q_1(\omega,-k_1,t)$ and $q_2(\omega,-k_2,t)$ appearing in equation (2.53) is proportional to the combination of these functions appearing in equation (2.52c) *for all* complex values of the spectral variables $k_1$ and $k_2$, i.e.,

$$\begin{aligned}
&\frac{c_2k_2-s_2k_1}{c_2k_2+s_2k_1}\left(c_1-s_1\frac{k_2}{k_1}\right)q_1(\omega,-k_1,t)-\frac{c_1k_1+s_1k_2}{c_1k_1-s_1k_2}\left(c_2+s_2\frac{k_1}{k_2}\right)q_2(\omega,-k_2,t)\\
&=\lambda\left[\left(c_1+s_1\frac{k_2}{k_1}\right)q_1(\omega,-k_1,t)-\left(c_2-s_2\frac{k_1}{k_2}\right)q_2(\omega,-k_2,t)\right],
\end{aligned} \tag{2.54}$$

for some constant $\lambda$ which does not depend on $k_1$ and $k_2$. By equating the coefficients of powers of $k_1$ and $k_2$, we find that equation (2.54) is satisfied *for all* $k_1$ and $k_2$ if the following three equations are satisfied:

$$(c_1c_2+s_1s_2)(\lambda-1)=0,\quad c_1s_2(\lambda+1)=0,\quad c_2s_1(\lambda+1)=0. \tag{2.55}$$

We have the following cases:

1. $\lambda=-1$. Then $c_1c_2+s_1s_2=\cos(\beta_1-\beta_2)=0$ or, equivalently,

$$\beta_1=\beta_2+(2n+1)\frac{\pi}{2},\quad n=-2,-1,0,1. \tag{2.56}$$

2. $\lambda=1$. Then $c_1s_2=c_2s_1=0$.

   a. If $c_1=c_2=0$, then

$$\beta_1=(2m+1)\frac{\pi}{2},\ m=0,1,\qquad \beta_2=(2n+1)\frac{\pi}{2},\ n=0,1. \tag{2.57a}$$

   b. If $s_2=s_1=0$, then

$$\beta_1=m\pi,\ m=0,1,\qquad \beta_2=n\pi,\ n=0,1. \tag{2.57b}$$

3. If $|\lambda|\neq 1$, then (2.55) cannot be satisfied for any choice of the angles $\beta_1,\beta_2$.

Equations (2.57) are actually the conditions that the boundary data vectors shown in Figure 2.5 are *orthogonal*, whereas condition (2.56) forces these vectors to be parallel.

Since both of the cases (2.57) reduce the boundary conditions of the IBVP (2.45) to separable ones, we choose to focus on the case (2.56). Note that (2.56) in combination with the oblique Neumann conditions (2.45c) and (2.45d) enforces the compatibility requirements

$$F_1(0,t)=F_2(0,t),\quad n=-1,1, \tag{2.58a}$$

and

$$F_1(0,t)=-F_2(0,t),\quad n=-2,0. \tag{2.58b}$$

After appropriate contour deformations and use of equation (2.52c), the integral representation (2.53) in the case (2.56) becomes

$$\begin{aligned}
u(x,y,t)&=\frac{1}{(2\pi)^2}\int_{-\infty}^{\infty}dk_1\int_{-\infty}^{\infty}dk_2\,e^{ik_1x+ik_2y-\omega t}\,\hat{u}_0(k_1,k_2)\\
&\quad-\frac{1}{(2\pi)^2}\int_{\partial D^+}dk_1\int_{-\infty}^{\infty}dk_2\,e^{ik_1x+ik_2y-\omega t}\Bigg\{\frac{2k_1}{c_2k_2+s_2k_1}f_2(\omega,k_2,t)\\
&\quad+\frac{c_2k_2-s_2k_1}{c_2k_2+s_2k_1}\left[\hat{u}_0(-k_1,k_2)+\left(s_1+c_1\frac{k_2}{k_1}\right)f_1(\omega,-k_1,t)\right]\Bigg\}\\
&\quad-\frac{1}{(2\pi)^2}\int_{-\infty}^{\infty}dk_1\int_{\partial D^+}dk_2\,e^{ik_1x+ik_2y-\omega t}\Bigg\{\frac{2k_2}{c_1k_1-s_1k_2}f_1(\omega,k_1,t)\\
&\quad+\frac{c_1k_1+s_1k_2}{c_1k_1-s_1k_2}\left[\hat{u}_0(k_1,-k_2)-\left(s_2-c_2\frac{k_1}{k_2}\right)f_2(\omega,-k_2,t)\right]\Bigg\}\\
&-\frac{1}{(2\pi)^2}\int_{\partial D^+}dk_1\int_{\partial D^+}dk_2\,e^{ik_1x+ik_2y-\omega t}\Big[\hat{u}_0(-k_1,-k_2)\\
&\quad+\left(s_1-c_1\frac{k_2}{k_1}\right)f_1(\omega,-k_1,t)-\left(s_2+c_2\frac{k_1}{k_2}\right)f_2(\omega,-k_2,t)\Big]\\
&+u^*(x,y,t),
\end{aligned} \tag{2.59}$$

where $u^*(x,y,t)$ is defined by

$$u^*(x,y,t) = -\frac{1}{(2\pi)^2}\int_{\partial D^+} dk_1 \int_{-\infty}^{\infty} dk_2\, e^{ik_1x+ik_2y-\omega t}\, \frac{c_2k_2 - s_2k_1}{c_2k_2+s_2k_1}\left(\frac{k_1}{k_2}+\frac{k_2}{k_1}\right)\tilde{u}(\omega,t)$$

$$-\frac{1}{(2\pi)^2}\int_{-\infty}^{\infty} dk_1 \int_{\partial D^+} dk_2\, e^{ik_1x+ik_2y-\omega t}\, \frac{c_1k_1 + s_1k_2}{c_1k_1-s_1k_2}\left(\frac{k_1}{k_2}+\frac{k_2}{k_1}\right)\tilde{u}(\omega,t) \quad (2.60)$$

and is the only unknown, since it involves the unknown function $\tilde{u}(\omega,t)$.

We have the following two cases:

1. $c_1s_1 \leq 0$. In this case, $c_2s_2 \geq 0$ and hence $(c_2k_2+s_2k_1)(c_1k_1-s_1k_2) \neq 0$ for all $k_1, k_2 \in \mathbb{C}^+$. Thus, both integrands on the right-hand side of (2.60) are analytic for $k_1, k_2 \in \mathbb{C}^+$. Hence, we can deform the contour of integration with respect to $k_2$ in the first double integral, as well as the contour of integration with respect to $k_1$ in the second double integral, to the contour $\partial D^+$. Then, due to condition (2.56),

$$\frac{c_2k_2 - s_2k_1}{c_2k_2+s_2k_1} + \frac{c_1k_1 + s_1k_2}{c_1k_1-s_1k_2} = \frac{2(k_1^2+k_2^2)(c_1c_2+s_1s_2)}{(c_1k_1-k_2s_1)(c_2k_2+k_1s_2)} \equiv 0,$$

and therefore $u^*(x,y,t)$ vanishes.

2. $c_1s_1 > 0$. In this case condition (2.56) implies $c_2s_2 < 0$; hence the integrands in (2.60) have poles at $k_2 = k_1 \cot\beta_1$ for $k_1, k_2 \in \mathbb{C}^+$. The contours of integration with respect to $k_1$ can be deformed to the boundary $\partial E^+$, shown in Figure 2.6. Thus, noting also that condition (2.56) is now equivalent to

$$\tan\beta_1 \tan\beta_2 = -1,$$

we find

$$u^*(x,y,t) = -\frac{1}{(2\pi)^2}\int_{\partial E^+} dk_1 \int_{-\infty}^{\infty} dk_2\, e^{ik_1x+ik_2y-\omega t}\, \frac{c_2k_2 - s_2k_1}{c_2k_2+s_2k_1}\left(\frac{k_1}{k_2}+\frac{k_2}{k_1}\right)\tilde{u}(\omega,t)$$

$$+\frac{1}{(2\pi)^2}\int_{\partial E^+} dk_1 \int_{\partial D^+} dk_2\, e^{ik_1x+ik_2y-\omega t}\, \frac{c_2k_2 - s_2k_1}{c_2k_2+s_2k_1}\left(\frac{k_1}{k_2}+\frac{k_2}{k_1}\right)\tilde{u}(\omega,t). \quad (2.61)$$

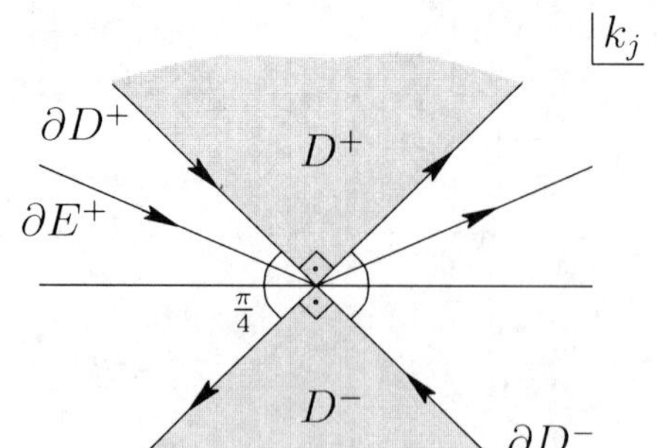


**Figure 2.6.** *The contours $\partial D^+$, $\partial D^-$, and $\partial E^+$.*

Cauchy's theorem in the region $\mathbb{C}^+ \setminus D^+$ of the $k_2$-plane for the first term of equation (2.61) yields

$$
\begin{aligned}
u^*(x,y,t) = &-\frac{1}{(2\pi)^2}\int_{\partial E^+} dk_1 \int_{\partial D^+} dk_2\, e^{ik_1x+ik_2y-\omega t}\,\frac{c_2k_2-s_2k_1}{c_2k_2+s_2k_1}\left(\frac{k_1}{k_2}+\frac{k_2}{k_1}\right)\tilde{u}(\omega,t)\\
&+\frac{1}{i\pi}\csc^2\beta_1 \int_{\partial E^+} dk_1\, e^{ik_1(x+y\cot\beta_1)-\omega^* t}\, k_1\tilde{u}(\omega^*,t)\\
&+\frac{1}{(2\pi)^2}\int_{\partial E^+} dk_1 \int_{\partial D^+} dk_2\, e^{ik_1x+ik_2y-\omega t}\,\frac{c_2k_2-s_2k_1}{c_2k_2+s_2k_1}\left(\frac{k_1}{k_2}+\frac{k_2}{k_1}\right)\tilde{u}(\omega,t),
\end{aligned}
$$

where the second term above corresponds to the residues from the poles at $k_2 = k_1\cot\beta_1$ and

$$
\omega^* = k_1^2\csc^2\beta_1. \tag{2.62}
$$

Hence, for $c_1s_1 > 0$ we have

$$
u^*(x,y,t) = \frac{1}{i\pi}\csc^2\beta_1 \int_{\partial E^+} dk_1\, e^{ik_1(x+y\cot\beta_1)-\omega^* t}\, k_1\tilde{u}(\omega^*,t). \tag{2.63}
$$

In order to determine $\tilde{u}(\omega^*,t)$, we need to employ appropriate version(s) of the global relations (2.52) evaluated at $k_2 = k_1\cot\beta_1$. Recalling that $q_1$ and $q_2$ are *unknown* and yield a nonzero contribution due to the exponential $e^{\omega^*(t-s)}$, we see that the only versions that may be used are (2.52a) and (2.52b), since in these identities the coefficients of $q_1$ and $q_2$ vanish at $k_2 = k_1\cot\beta_1$. However, equations (2.52a) and (2.52b) are valid for $\{k_1 \in \mathbb{C}^+,\, k_2 \in \mathbb{C}^-\}$ and $\{k_1 \in \mathbb{C}^-,\, k_2 \in \mathbb{C}^+\}$, respectively, and thus they cannot be employed since we have $k_2 = k_1\cot\beta_1$ and $\cot\beta_1 > 0$.

Consequently, for $c_1s_1 > 0$ the representation (2.59) contains the function $u^*(x,y,t)$ defined by equation (2.60). Equation (2.63) expresses $u^*(x,y,t)$ in terms of the unknown function $\tilde{u}(\omega^*,t)$, i.e., the function $\tilde{u}(\omega,t)$ evaluated at $k_2 = k_1\cot\beta_1$. On the other hand, for $c_1s_1 \le 0$ the function $u^*(x,y,t)$ vanishes and the representation (2.59) depends entirely on given functions.

The above are summarized in the following proposition.

**Proposition 2.3 (Solution).** *Let $c_j = \cos\beta_j$, $s_j = \sin\beta_j$, $j = 1,2$, where $\beta_1$, $\beta_2$ are given angles (see Figure 2.5) that satisfy condition* (2.56).

1. *If $c_1s_1 \le 0$, then the solution to the IBVP* (2.45) *is given by*

$$
\begin{aligned}
u(x,y,t) = &\frac{1}{(2\pi)^2}\int_{-\infty}^{\infty} dk_1 \int_{-\infty}^{\infty} dk_2\, e^{ik_1x+ik_2y-\omega t}\,\hat{u}_0(k_1,k_2)\\
&-\frac{1}{(2\pi)^2}\int_{\partial D^+} dk_1 \int_{-\infty}^{\infty} dk_2\, e^{ik_1x+ik_2y-\omega t}\left\{\frac{2k_1}{c_2k_2+s_2k_1}f_2(\omega,k_2,t)\right.\\
&\left.+\frac{c_2k_2-s_2k_1}{c_2k_2+s_2k_1}\left[\hat{u}_0(-k_1,k_2)+\left(s_1+c_1\frac{k_2}{k_1}\right)f_1(\omega,-k_1,t)\right]\right\}
\end{aligned} \tag{2.64}
$$

$$
\begin{aligned}
&-\frac{1}{(2\pi)^2}\int_{-\infty}^{\infty} dk_1 \int_{\partial D^+} dk_2\, e^{ik_1x+ik_2y-\omega t}\left\{\frac{2k_2}{c_1k_1-s_1k_2} f_1(\omega,k_1,t)\right.\\
&\quad\left.+\frac{c_1k_1+s_1k_2}{c_1k_1-s_1k_2}\left[\hat{u}_0(k_1,-k_2)-\left(s_2-c_2\frac{k_1}{k_2}\right)f_2(\omega,-k_2,t)\right]\right\}\\
&-\frac{1}{(2\pi)^2}\int_{\partial D^+} dk_1 \int_{\partial D^+} dk_2\, e^{ik_1x+ik_2y-\omega t}\left[\hat{u}_0(-k_1,-k_2)\right.\\
&\quad\left.+\left(s_1-c_1\frac{k_2}{k_1}\right)f_1(\omega,-k_1,t)-\left(s_2+c_2\frac{k_1}{k_2}\right)f_2(\omega,-k_2,t)\right],
\end{aligned}
$$

*where* $\hat{u}_0, f_1, f_2$ *are defined by equations* (2.50) *and* $\partial D^+$ *is shown in Figure* 2.6.

2. *If* $c_1 s_1 > 0$, *then we assume the additional boundary condition*

$$u(0,0,t) = h(t), \quad t > 0. \tag{2.65}$$

*In this case, the solution* (2.64) *involves the additional term* $u^*(x,y,t)$,

$$u^*(x,y,t) = \frac{1}{i\pi}\csc^2\beta_1 \int_{\partial D^+} dk_1\, e^{ik_1(x+y\cot\beta_1)-\omega^* t}\, k_1\tilde{u}(\omega^*,t), \tag{2.66}$$

*where the function* $\tilde{u}(\omega^*,t)$ *is defined by*

$$\tilde{u}(\omega^*,t) = \int_0^t ds\, e^{\omega^* s} h(s), \quad \omega^* = k_1^2\csc^2\beta_1. \tag{2.67}$$

## Appendix: Verification of the Solution Formulae

In the applied mathematics literature, a PDE is solved under the *assumption* of existence. In order to eliminate this assumption one should prove a posteriori that the expression obtained by this approach satisfies the PDE and the given initial and boundary conditions. However, this is not done in the applied literature. Actually, this verification is *not* straightforward because any representation obtained via the usual transform methods is *nonuniformly convergent* at the boundary. As noted in the introduction, a major advantage of the unified method is that it constructs a solution which is uniformly convergent at the boundary. Thus, it is straightforward, at least formally, to verify that the solution satisfies the given PDE and the given data.

Here, we illustrate this feature by considering the integral expression (2.24a) derived for the heat equation on the half-line with given initial data and the oblique Robin boundary condition (2.8c):

$$
\begin{aligned}
u(x,t) &= \frac{1}{2\pi}\int_{-\infty}^{\infty} e^{ikx-k^2t}\hat{u}_0(k)dk\\
&+\frac{1}{2\pi}\int_{\partial D^+} e^{ikx-k^2t}\,\frac{2ik\,\tilde{\delta}(k^2,t)-(k^2+i\alpha k-\beta)\,\hat{u}_0(-k)}{k^2-i\alpha k-\beta}\,dk,
\end{aligned}\tag{2.68}
$$

where $\hat{u}_0(k)$ is the Fourier transform of the initial data $u_0(x)$, the function $\tilde{\delta}(k^2,t)$ is defined by equation (2.20), and the contour $\partial D^+$ is depicted in Figure 2.4.

First, we note that for *any* $T > t$ we have

$$\int_{\partial D^+} e^{ikx-k^2 t} \frac{2ik\,\tilde{\delta}(k^2,t)}{k^2 - i\alpha k - \beta} dk = \int_{\partial D^+} e^{ikx-k^2 t} \frac{2ik\,\tilde{\delta}(k^2,T)}{k^2 - i\alpha k - \beta} dk. \tag{2.69}$$

The above equation is a direct consequence of Cauchy's theorem and Jordan's lemma applied in the region $D^+$. Consequently, the expression (2.68) can be written in the following *equivalent* form:

$$\begin{aligned} u(x,t) = &\frac{1}{2\pi}\int_{-\infty}^{\infty} e^{ikx-k^2 t}\hat{u}_0(k)dk \\ &+ \frac{1}{2\pi}\int_{\partial D^+} e^{ikx-k^2 t} \frac{2ik\,\tilde{\delta}(k^2,T) - (k^2 + i\alpha k - \beta)\,\hat{u}_0(-k)}{k^2 - i\alpha k - \beta} dk. \end{aligned} \tag{2.70}$$

**The Heat Equation.** The integral expression of $u(x,t)$ in the form (2.70) depends on $x$ and $t$ only through the exponential $e^{ikx-k^2 t}$, which is a particular solution of the heat equation; thus the expression (2.70) satisfies the heat equation.

**Initial Condition.** Evaluating the expression (2.68) at $t = 0$, we find

$$\begin{aligned} u(x,0) = &\frac{1}{2\pi}\int_{-\infty}^{\infty} e^{ikx}\hat{u}_0(k)dk \\ &+ \frac{1}{2\pi}\int_{\partial D^+} e^{ikx} \frac{2ik\,u_0(0) - (k^2 + i\alpha k - \beta)\,\hat{u}_0(-k)}{k^2 - i\alpha k - \beta} dk. \end{aligned} \tag{2.71}$$

The integrand of the integral along the contour $\partial D^+$ is bounded and analytic for all $k \in \mathbb{C}^+$; hence by Cauchy's theorem and Jordan's lemma this integral vanishes. Thus,

$$u(x,0) = \frac{1}{2\pi}\int_{-\infty}^{\infty} e^{ikx}\hat{u}_0(k)dk = u_0(x), \tag{2.72}$$

where the last equality above follows by the definition (2.11b) of the inverse Fourier transform.

**Boundary Condition.** We will compute the derivatives $u_t(0,t), u_x(0,t)$, as well as the function $u(0,t)$, by using the expression (2.70):

$$\begin{aligned} u_t(0,t) = &\frac{1}{2\pi}\int_{-\infty}^{\infty} e^{-k^2 t}\left(-k^2\right)\hat{u}_0(k)dk \\ &+ \frac{1}{2\pi}\int_{\partial D^+} e^{-k^2 t}\left(-k^2\right) \frac{2ik\,\tilde{\delta}(k^2,T) - (k^2 + i\alpha k - \beta)\,\hat{u}_0(-k)}{k^2 - i\alpha k - \beta} dk, \end{aligned} \tag{2.73a}$$

$$\begin{aligned} u_x(0,t) = &\frac{1}{2\pi}\int_{-\infty}^{\infty} e^{-k^2 t}\, ik\,\hat{u}_0(k)dk \\ &+ \frac{1}{2\pi}\int_{\partial D^+} e^{-k^2 t}\, ik\, \frac{2ik\,\tilde{\delta}(k^2,T) - (k^2 + i\alpha k - \beta)\,\hat{u}_0(-k)}{k^2 - i\alpha k - \beta} dk, \end{aligned} \tag{2.73b}$$

and

$$u(0,t)=\frac{1}{2\pi}\int_{-\infty}^{\infty} e^{-k^2 t}\,\hat{u}_0(k)dk$$
$$+\frac{1}{2\pi}\int_{\partial D^+} e^{-k^2 t}\,\frac{2ik\,\tilde{\delta}(k^2,T)-(k^2+i\alpha k-\beta)\,\hat{u}_0(-k)}{k^2-i\alpha k-\beta}\,dk. \tag{2.73c}$$

Taking the combination $u_t(0,t)+\alpha u_x(0,t)+\beta u(0,t)$, we find

$$u_t(0,t)+\alpha u_x(0,t)+\beta u(0,t)=\frac{1}{2\pi}\int_{-\infty}^{\infty} e^{-k^2 t}\left(-k^2+i\alpha k+\beta\right)\hat{u}_0(k)dk$$
$$+\frac{1}{2\pi}\int_{\partial D^+} e^{-k^2 t}\left[\left(k^2+i\alpha k-\beta\right)\hat{u}_0(-k)-2ik\,\tilde{\delta}(k^2,T)\right]dk. \tag{2.74}$$

The change of variables $k\mapsto -k$ in the first integral on the right-hand side of the above equation yields

$$u_t(0,t)+\alpha u_x(0,t)+\beta u(0,t)=\frac{1}{2\pi}\int_{-\infty}^{\infty} e^{-k^2 t}\left(-k^2-i\alpha k+\beta\right)\hat{u}_0(-k)dk$$
$$+\frac{1}{2\pi}\int_{\partial D^+} e^{-k^2 t}\left[\left(k^2+i\alpha k-\beta\right)\hat{u}_0(-k)-2ik\,\tilde{\delta}(k^2,T)\right]dk. \tag{2.75}$$

Furthermore, Cauchy's theorem and Jordan's lemma for $k\in\mathbb{C}^+\setminus D^+$ imply that we can deform the contour of the first integral from the real line to $\partial D^+$, i.e.,

$$u_t(0,t)+\alpha u_x(0,t)+\beta u(0,t)=\frac{1}{2\pi}\int_{\partial D^+} e^{-k^2 t}\left(-k^2-i\alpha k+\beta\right)\hat{u}_0(-k)dk$$
$$+\frac{1}{2\pi}\int_{\partial D^+} e^{-k^2 t}\left[\left(k^2+i\alpha k-\beta\right)\hat{u}_0(-k)-2ik\,\tilde{\delta}(k^2,T)\right]dk. \tag{2.76}$$

Hence, recalling the definition (2.20) of $\tilde{\delta}(k^2,T)$, we find

$$u_t(0,t)+\alpha u_x(0,t)+\beta u(0,t)=-\frac{1}{2\pi}\int_{\partial D^+} e^{-k^2 t}\,2ik\left[\int_0^T e^{k^2 s}\gamma(s)ds+u_0(0)\right]dk. \tag{2.77}$$

Letting $k^2=il$, equation (2.77) becomes

$$u_t(0,t)+\alpha u_x(0,t)+\beta u(0,t)=\frac{1}{2\pi}\int_{-\infty}^{\infty} e^{-ilt}\left[\int_0^T e^{ils}\gamma(s)ds+u_0(0)\right]dl$$
$$=\gamma(t), \tag{2.78}$$

where the last equality follows by using Cauchy's theorem in the lower half complex $l$-plane and by the definition of Dirac's $\delta$-function:

$$\delta(t)=\frac{1}{2\pi}\int_{-\infty}^{\infty} e^{il(s-t)}dl. \tag{2.79}$$

Hence, the oblique Robin boundary condition (2.8c) has been verified.

# Bibliography

[1] A. S. Fokas, A unified transform method for solving linear and certain nonlinear PDEs, Proc. R. Soc. A **453** (1997), 1411–1443.

[2] A. S. Fokas, On the integrability of linear and nonlinear PDEs, J. Math. Phys. **41** (2000), 4188–4237.

[3] A. S. Fokas, A new transform method for evolution partial differential equations, IMA J. Appl. Math. **67** (2002), 559–590.

[4] A. S. Fokas and P. A. Treharne, Initial-boundary value problems for linear PDEs with variable coefficients, Math. Proc. Cambridge Philos. Soc. **143** (2007), 221–242.

[5] B. Pelloni, The spectral representation of two-point boundary-value problems for third-order linear evolution partial differential equations, Proc. R. Soc. A **461** (2005), 2965–2984.

[6] B. Pelloni and D. A. Smith, Spectral theory of some mon-self-adjoint linear differential operators, Proc. R. Soc. A **469** (2013), 2154.

[7] D. A. Smith, Well-posed two-point initial-boundary value problems with arbitrary boundary conditions, Math. Proc. Cambridge Philos. Soc. **152** (2012), 473–496.

[8] A. S. Fokas and E. A. Spence, Synthesis, as opposed to separation, of variables, SIAM Rev. **54** (2012), 291–324.

[9] B. Deconinck, T. Trogdon, and V. Vasan, The method of Fokas for solving linear partial differential equations, SIAM Rev. **56** (2014), 159–186.

[10] D. S. Cohen, An integral transform associated with boundary conditions containing an eigenvalue parameter, SIAM J. Appl. Math. **14** (1966), 1164–1175.

[11] A. S. Fokas and B. Pelloni, Integral transforms, spectral representation and the d-bar problem, Proc. R. Soc. A **456** (2000), 805–833.

[12] D. Mugnolo and S. Nicaise, Well-Posedness and Spectral Properties of Heat and Wave Equations with Non-Local Conditions, arXiv:1112.0415 (2012).

[13] N. Flyer and A. S. Fokas, A hybrid analytical-numerical method for solving evolution partial differential equations: I. The half-line, Proc. R. Soc. A **464** (2008), 1823–1849.

[14] G. M. Dujardin, Asymptotics of linear initial-boundary value problems with periodic boundary data on the half-line and finite intervals, Proc. R. Soc. Lond. Ser. A **465** (2009), 3341–3360.

[15] A. S. Fokas and K. Kalimeris, The heat equation in the interior of an equilateral triangle, Stud. Appl. Math. **124** (2010), 283–305.

[16] D. Mantzavinos and A. S. Fokas, The unified method for the heat equation: I. Non-separable boundary conditions and non-local constraints in one dimension, Eur. J. Appl. Math. **24** (2013), 857–886.

[17] D. Mantzavinos and A. S. Fokas, The Unified Transform Method for the Heat Equation: II. Non-Separable Boundary Conditions in Two Dimensions (submitted).

[18] R. E. Langer, A problem in diffusion or in the flow of heat for a solid in contact with a fluid, Tohoku Math. J. **35** (1932), 360–375.

## 2.2 ▪ The Unified Transform for Linear IBVPs: A Spectral Interpretation

*D. A. Smith*

The research leading to these results has received funding from the European Union's Seventh Framework Programme (FP7-REGPOT-2009-1) under grant agreement 245749. A version of this section previously appeared in [21].

### 2.2.1 ▪ Introduction

Consider the Dirichlet problem for the heat equation on a finite interval,

$$\begin{aligned} (\partial_t - \partial_{xx})q(x,t) &= 0, && (x,t)\in(0,1\times(0,T), \\ q(x,0) &= q_0(x), && x\in[0,1], \\ q(0,t) = q(1,t) &= 0, && t\in[0,T], \end{aligned}$$

where $q_0$ is some known initial datum, compatible with the boundary conditions. In solving this problem by separation of variables, one begins with the assumption that $q(x,t) = \xi(x)\tau(t)$ but rapidly discovers that this ansatz is incompatible with general initial data and so replaces it with the more general assumption that $q$ can be expressed as a series of such functions $\xi$, with corresponding $\tau$. Of course, these $\xi$ are precisely the eigenfunctions of a certain self-adjoint differential operator,

$$S:\{f\in C^\infty[0,1]: f(0)=f(1)=0\}\to C^\infty[0,1], \qquad Sf = -\frac{\mathrm{d}^2 f}{\mathrm{d}x^2},$$

so it is no surprise that it is possible to write the solution in such a form. Crucially, it is the general results on completeness of eigenfunctions and convergence of eigenfunction series expansions for such operators that ensure the series representation is valid.

However, it is easy to construct a similar problem, for which the corresponding differential operator is non-self-adjoint. Consider the following.

**Problem 1.**

$$\begin{aligned} (\partial_t + \partial_{xxx})q(x,t) &= 0, && (x,t)\in(0,1)\times(0,T), && \text{(2.80a)} \\ q(x,0) &= q_0(x), && x\in[0,1], && \text{(2.80b)} \\ q(0,t) = q(1,t) = q_x(1,t) &= 0, && t\in[0,T], && \text{(2.80c)} \end{aligned}$$

for which the corresponding operator

$$S:\{f\in C^\infty[0,1]: f(0)=f(1)=f'(1)=0\}\to C^\infty[0,1], \qquad Sf = -i\frac{\mathrm{d}^3 f}{\mathrm{d}x^3}, \tag{2.81}$$

is formally self-adjoint but is non-self-adjoint because of its domain. In this case, one cannot rely on the classical theorems for the completeness of the eigenfunctions in the space of initial data.

Indeed, it turns out that while the (generalized) eigenfunctions do form a complete system [18], the corresponding expansions do not converge [13] except in very special cases [12], so an ansatz such as the classical separation of variables series representation cannot be valid. Nevertheless, the unified transform method can be used to obtain a solution to problem (2.80). In this work, we describe how the solution representation thence derived can be interpreted as an expansion in certain spectral objects, which are defined as a certain abstraction of eigenfunctions and the generalized eigenfunctions of Gelfand.

### Initial-Boundary Value Problems

In this work, we concentrate on two examples of non-self-adjoint IBVPs. We study Problem 1, described above, and also the following.
**Problem 2.**

$$(\partial_t + \partial_{xxx})q(x,t) = 0, \qquad (x,t) \in (0,\infty) \times (0,T), \tag{2.82a}$$
$$q(x,0) = q_0(x), \qquad x \in (0,\infty), \tag{2.82b}$$
$$q(0,t) = 0, \qquad t \in (0,T). \tag{2.82c}$$

The corresponding spatial differential operator is

$$S : \{f \in \mathscr{E}[0,\infty) : f(0) = 0\} \to \mathscr{S}[0,\infty), \qquad Sf = -i\frac{\mathrm{d}^3 f}{\mathrm{d}x^3}, \tag{2.83}$$

where for some arbitrarily small $\epsilon > 0$

$$E = \{f \in \mathscr{S}[0,\infty) : f(x) = O(e^{-\epsilon x}) \text{ as } x \to \infty\} \tag{2.84}$$

and $\mathscr{S}[0,\infty)$ is the Schwartz class of rapidly decaying smooth functions restricted to the half-line. Our reasons for working in $\mathscr{E}[0,\infty)$ rather than $\mathscr{S}[0,\infty)$ are discussed after the proof of Theorem 2.8.

Problem 1 is posed on the finite interval and is studied in detail in [7], whereas Problem 2 is posed on the half-line, the case covered by [17]. Here we study the two problems together, to draw attention to the similarities between the finite interval and half-line frameworks.

It should be noted that problems such as these are open to analysis via a Laplace transform in time. However, as described in [6, Appendix C], this method presents a number of difficulties, which do not occur with the unified transform method, particularly when one studies problems of still higher order or considers certain inhomogeneous boundary conditions. The surveys [2, 8] give good accounts of the differences between the unified transform method and classical methods.

The purpose of this work is to give a spectral interpretation of the success of the unified transform method. In order to motivate and inform the discussion, we shall refer to the familiar spectral understanding of the classical Fourier methods, which comprise an analysis of the spatial part of the PDE. The Laplace transform in the temporal part of the equation is not so clearly analogous to the unified method and is not discussed hereafter.

### Layout of Paper

In Section 2.2.2, we give the solution of Problems 1 and 2, as obtained through the unified transform method. We then reformulate these solution representations as a transform-inverse transform pair, whereupon the unified transform method may be seen as the method for deriving said transform pair, tailor-made for the particular IBVP of interest.

In Section 2.2.3, we consider the spectral meaning of the transform pair. Motivated by a discussion of Gelfand's generalized eigenfunctions as applied to the IBVP for the self-adjoint Dirichlet heat operator, we define a new class of spectral functionals, whose properties are more suited to non-self-adjoint operators such as (2.81) and (2.83). We discuss the characteristics of the new definition and show how the transform pair provides a new spectral diagonalization result for these operators.

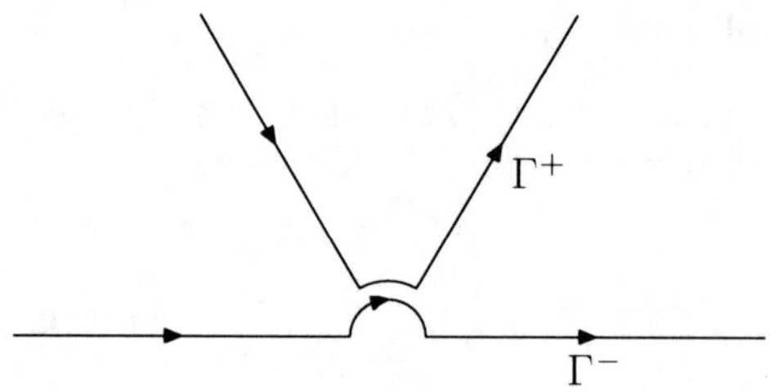


**Figure 2.7.** *Contours for Problem 2.*

### 2.2.2 ▪ Unified Transform Method

The unified transform method [4], as applied to half-line IBVPs for linear evolution equations, was first described in [3] and formalized in [9]. The method was extended to problems on the finite interval in [5], and detailed well-posedness results were established first for simple, uncoupled boundary conditions and then in full generality [15, 20].

In particular, [9] established rigorously that the solution of a linear IBVP may be obtained without any assumptions beyond well-posedness of the problem. This is in contrast to classical methods, where stronger assumptions on the form of the solution are imposed. Indeed, as discussed above, the d'Alembert method of separation of variables requires that the eigenfunctions of the spatial differential operator form a complete system in the space of admissible initial data and the expansion of the initial datum in these eigenfunctions converges. In applying the method of Fourier transforms, one implicitly assumes that the generalized eigenfunctions form a complete system in the space of initial data and that the corresponding expansion converges. This is certainly a weaker ansatz than completeness for ordinary eigenfunctions, and it is valid whenever the operator happens to be self-adjoint [11], but it is not guaranteed for non-self-adjoint operators; formal self-adjointness is insufficient.

When the unified transform method is applied to Problem 2, the solution is obtained in the form of a contour integral,

$$q(x,t)=\frac{1}{2\pi}\int_{\Gamma^+} e^{i\lambda x+i\lambda^3 t}\zeta(\lambda;q_0)\mathrm{d}\lambda+\frac{1}{2\pi}\int_{\Gamma^-} e^{i\lambda x+i\lambda^3 t}\zeta^-(\lambda;q_0)\mathrm{d}\lambda, \tag{2.85}$$

where $\Gamma^+$ is the boundary of the domain $\lambda\in\mathbb{C}^+ : \mathrm{Re}(-i\lambda^3)<0$ and $|\lambda|>1\}$ as shown in Figure 2.7, and $\Gamma^-$ is $\mathbb{R}$ perturbed away from 0 along a small (radius $<\epsilon$) semicircular arc in $\mathbb{C}^+$.

For $\lambda\in\mathbb{C}$,

$$\zeta^+(\lambda;q_0)=\alpha\hat{q}_0(\alpha\lambda)+\alpha^2\hat{q}_0(\alpha^2\lambda), \tag{2.86a}$$

$$\zeta^-(\lambda;q_0)=\hat{q}_0(\lambda), \tag{2.86b}$$

$\alpha$ is the cube root of unity $e^{2\pi i/3}$ and $\hat{q}_0$ is the Fourier transform

$$\hat{q}_0(\lambda)=\int_0^\infty e^{-i\lambda x}q_0(x)\mathrm{d}x, \qquad \lambda\in\mathbb{C}: \mathrm{Im}(\lambda)<\epsilon. \tag{2.87}$$

Applying the unified transform method to Problem 1, one obtains the following contour integral representation of the solution:

$$q(x,t)=\frac{1}{2\pi}\int_{\Gamma^+} e^{i\lambda x+i\lambda^3 t}\frac{\zeta^+(\lambda;q_0)}{\Delta(\lambda)}\mathrm{d}\lambda+\frac{1}{2\pi}\int_{\Gamma^-} e^{i\lambda(x-1)+i\lambda^3 t}\frac{\zeta^-(\lambda;q_0)}{\Delta(\lambda)}\mathrm{d}\lambda, \tag{2.88}$$

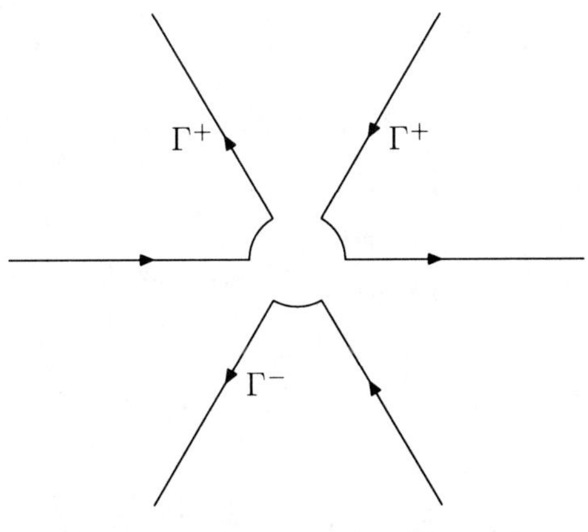


**Figure 2.8.** *Contours for Problem* 1.

where $\Gamma^{\pm}$ are the boundaries of the sectors $\{\lambda \in \mathbb{C}^{\pm} : \mathrm{Im}(\lambda^3) > 0 \text{ and } |\lambda| > 1\}$ as shown in Figure 2.8.

For $\lambda \in \mathbb{C}$,

$$\Delta(\lambda) = e^{-i\lambda} + \alpha e^{-i\alpha\lambda} + \alpha^2 e^{-i\alpha^2\lambda}, \tag{2.89a}$$

$$\zeta^{+}(\lambda; q_0) = \hat{q}_0(\lambda)(\alpha e^{-i\alpha\lambda} + \alpha^2 e^{-i\alpha^2\lambda}) - (\alpha\hat{q}_0(\alpha\lambda) + \alpha^2\hat{q}_0(\alpha^2\lambda))e^{-i\lambda}, \tag{2.89b}$$

$$\zeta^{-}(\lambda; q_0) = -\hat{q}_0(\lambda) - \alpha\hat{q}_0(\alpha\lambda) - \alpha^2\hat{q}_0(\alpha^2\lambda). \tag{2.89c}$$

$\alpha$ is the root of unity $e^{2\pi i/3}$ and $\hat{q}_0(\lambda)$ is the Fourier transform:

$$\hat{q}_0(\lambda) = \int_0^1 e^{-i\lambda x} q_0(x)\mathrm{d}x, \qquad \lambda \in \mathbb{C}. \tag{2.90}$$

The solution of each problem involves contour integrals of sums of Fourier transforms of the initial data. This is in contrast to the classical Fourier series expansion one might expect from studying the finite interval Dirichlet heat problem. Indeed, many finite interval IBVPs for higher-order PDEs have solutions which may be expressed as a series of (generalized) eigenfunctions in a similar manner. Having solved such a problem using the unified transform method, obtaining a solution representation similar to equation (2.88), one may use Jordan's lemma to perform a contour deformation, closing each component of $\Gamma^{\pm}$ onto the poles of the integrand. A residue calculation then yields a series representation of the solution, which corresponds to the classical eigenfunction expansion [1, 20]. But for some odd-order problems, including Problem 1, this contour deformation and series representation is impossible [16].

### Transform-Inverse Transform Pair

Some problems similar to those studied here can be solved using the method of Fourier transforms. The classical Fourier sine and Fourier cosine transforms are typically used to solve the heat equation with Dirichlet and Neumann boundary conditions, respectively. Indeed a different transform (and corresponding inverse transform) is required for each IBVP.

One may view the unified transform method as a way of determining an applicable transform pair for a given IBVP; we may rewrite the solution representations (2.85) and (2.88) in terms of Fourier-type integral transforms as

$$q(x,t) = f_x\left(e^{i\lambda^3 t} F_\lambda(q_0)\right), \tag{2.91}$$

where

$$f(x) \mapsto F(\lambda): \quad F_\lambda(f) = \begin{cases} \int_0^1 \phi^+(x,\lambda) f(x)\mathrm{d}x & \text{if } \lambda \in \Gamma^+, \\ \int_0^1 \phi^-(x,\lambda) f(x)\mathrm{d}x & \text{if } \lambda \in \Gamma^-, \end{cases} \tag{2.92a}$$

$$F(\lambda) \mapsto f(x): \quad f_x(F) = \left\{ \int_{\Gamma^+} + \int_{\Gamma^-} \right\} e^{i\lambda x} F(\lambda)\mathrm{d}\lambda, \qquad x \in [0,1], \text{ or } x \in [0,\infty), \tag{2.92b}$$

the contours $\Gamma^\pm$ are the appropriate contours for the problem, as defined above, and the kernels $\phi^\pm$ are given by

$$\phi^+(x,\lambda) = \frac{1}{2\pi\Delta(\lambda)}\left[e^{-i\lambda x}(\alpha e^{-i\alpha\lambda} + \alpha^2 e^{-i\alpha^2\lambda}) - (\alpha e^{-i\alpha\lambda x} + \alpha^2 e^{-i\alpha^2\lambda x})e^{-i\lambda}\right], \tag{2.92c}$$

$$\phi^-(x,\lambda) = \frac{-e^{-i\lambda}}{2\pi\Delta(\lambda)}\left[e^{-i\lambda x} + \alpha e^{-i\alpha\lambda x} + \alpha^2 e^{-i\alpha^2\lambda x}\right] \tag{2.92d}$$

for Problem 1 and

$$\phi^+(x,\lambda) = \frac{1}{2\pi}\left[\alpha e^{-i\alpha\lambda x} + \alpha^2 e^{-i\alpha^2\lambda x}\right], \tag{2.92e}$$

$$\phi^-(x,\lambda) = \frac{1}{2\pi} e^{-i\lambda x} \tag{2.92f}$$

for Problem 2.

Although the above is easy to see formally, it must be established that the maps $f(x) \mapsto F(\lambda)$ and $F(\lambda) \mapsto f(x)$ are truly a transform-inverse transform pair.

**Proposition 2.4.**

(i) *Suppose $F_\lambda$ and $f_x$ are defined by equations* (2.92a)–(2.92d). *Then for all $f \in C^\infty[0,1]$ with $f(0) = f(1) = f_x(1) = 0$ and for all $x \in [0,1]$, $f_x(F_\lambda(f)) = f(x)$.*

(ii) *Suppose $F_\lambda$ and $f_x$ are defined by equations* (2.92a), (2.92b), (2.92e), *and* (2.92f). *Then for all $f \in \mathscr{E}[0,\infty)$ with $f(0) = 0$ and for all $x \in [0,\infty)$, $f_x(F_\lambda(f)) = f(x)$.*

***Proof.***

(i) The definition of the transform pair implies

$$f_x(F_\lambda(f)) = \frac{1}{2\pi}\left\{\int_{\Gamma^+} + \int_{\Gamma_0}\right\} e^{i\lambda x}\frac{\zeta^+(\lambda)}{\Delta(\lambda)}\mathrm{d}\lambda + \frac{1}{2\pi}\int_{\Gamma^-} e^{i\lambda(x-1)}\frac{\zeta^-(\lambda)}{\Delta(\lambda)}\mathrm{d}\lambda, \tag{2.93}$$

where $\zeta^\pm$ and $\Delta$ are given by equations (2.89) and the contours $\Gamma^+$ and $\Gamma^-$ are shown in Figure 2.8.

The fastest-growing exponentials of the form $e^{\pm i\alpha^j\lambda}$ in the sectors exterior to $\Gamma^\pm$ are indicated in Figure 2.9. Each of these exponentials occurs in $\Delta$, and integration by parts shows that the fastest-growing terms in $\zeta^\pm$ are the exponentials shown in Figure 2.9 multiplied by $\lambda^{-2}$. Hence the ratio $\zeta^+(\lambda)/\Delta(\lambda)$ decays for large $\lambda$ within the sector

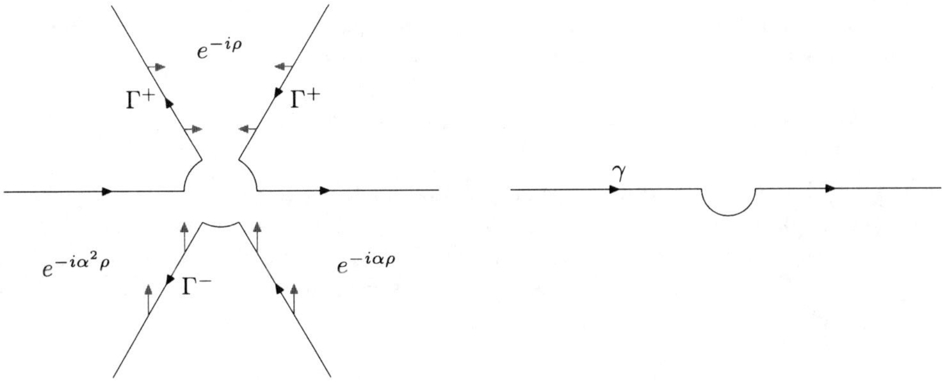


**Figure 2.9.** *Contour deformation for Problem* 1 *(finite interval).*

$\pi/3 \leq \arg\lambda \leq 2\pi/3$ and the ratio $\zeta^-(\lambda)/\Delta(\lambda)$ decays for large $\lambda$ within the sectors $-\pi \leq \arg\lambda \leq -2\pi/3$, $-\pi/3 \leq \arg\lambda \leq 0$. The relevant integrands are meromorphic functions with poles only at the zeros of $\Delta$. The distribution theory of zeros of exponential polynomials [14] implies that the only poles occur within the sets bounded by $\Gamma^\pm$.

The above observations and Jordan's lemma allow us to deform the relevant contours to the contour $\gamma$ shown in Figure 2.9; the finer arrows in Figure 2.9 indicate the deformation direction. Hence equation (2.93) simplifies to

$$f_x(F_\lambda(f)) = \frac{1}{2\pi}\int_\gamma \frac{e^{i\lambda x}}{\Delta(\lambda)}\left(\zeta^+(\lambda) - e^{-i\lambda}\zeta^-(\lambda)\right)\mathrm{d}\lambda. \tag{2.94}$$

Equations (2.89) imply

$$\left(\zeta^+(\lambda) - e^{-i\lambda}\zeta^-(\lambda)\right) = \hat{f}(\lambda)\Delta(\lambda), \tag{2.95}$$

where $\hat{f}$ is the Fourier transform of a piecewise smooth function supported on $[0,1]$. Hence the integrand on the right-hand side of equation (2.94) is an entire function, so we can deform the contour onto the real axis. The usual Fourier inversion theorem completes the proof.

(ii) The definition of the transform pair implies

$$f_x(F_\lambda(f)) = \frac{1}{2\pi}\int_{\Gamma^+} e^{i\lambda x}\zeta^+(\lambda)\mathrm{d}\lambda + \frac{1}{2\pi}\int_{\Gamma^-} e^{i\lambda x}\hat{f}(\lambda)\mathrm{d}\lambda, \tag{2.96}$$

where $\zeta$ is given by equation (2.86a) and the contours $\Gamma^+$ and $\Gamma^-$ are shown in Figure 2.7.

As $\lambda \to \infty$ from within the closed sector $\{\lambda : \frac{\pi}{3} \leq \arg(\lambda)\frac{2\pi}{3}\}$, the exponentials $e^{-i\alpha\lambda}$ and $e^{-i\alpha^2\lambda}$ are bounded. Moreover, integration by parts and the boundary condition $f(0) = 0$ yield $\hat{f}(\alpha\lambda)$, $\hat{f}(\alpha^2\lambda) = O(\lambda^{-2})$. These Fourier transforms are also holomorphic in the same sector; hence, by Jordan's lemma, the integral over $\Gamma^+$ vanishes. The integrand $e^{i\lambda x}\hat{f}(\lambda)$ is analytic for $\lambda$ with $\mathrm{Im}(\lambda) < \epsilon$ and hence admits a deformation of the contour $\Gamma^-$ onto $\mathbb{R}$. The validity of the usual Fourier transform on $\mathscr{S}[0,\infty)$ completes the proof. □

**Theorem 2.5.** *The transform pair $f(x) \mapsto F(\lambda)$ and $F(\lambda) \mapsto f(x)$ as defined above is the "correct" transform pair for the IBVP, in the sense that*

$$q(x,t) = f_x\left(e^{i\lambda^3 t} F_\lambda(q_0)\right). \tag{2.97}$$

Theorem 2.5 follows trivially from equations (2.85) and (2.88) by changing to the transform notation. In view of Proposition 2.4, this theorem is simply a statement that one may solve these problems using the unified transform method, as has been established in [3, 5]. A direct proof of this theorem is given in [7, 17] for Problems 1 and 2, respectively.

## 2.2.3 ▪ Spectral Interpretation

The concept of a generalized eigenfunction and the completeness of these spectral objects for self-adjoint operators is well known. Therefore it might be expected that, at least where the spatial differential operator is self-adjoint, the unified transform method might yield a solution expressed as an expansion in generalized eigenfunctions. To begin this section, we sketch a classical argument used to justify the importance of generalized eigenfunctions for the (self-adjoint) half-line Dirichlet heat operator. Below, we use a similar argument to show that the concept of generalized eigenfunctions is inadequate for our purposes and propose a definition for a new class of spectral objects, the augmented eigenfunctions. This is followed by a detailed discussion of the new definition, together with a list of known diagonalization results for spatial differential operators similar to (2.81) and (2.83). Finally, we provide proofs of the spectral results for the operators (2.81) and (2.83).

### Generalized Eigenfunctions

Suppose we seek traditional eigenfunctions of the spatial differential operator $S$ associated with the Dirichlet problem for the heat equation and given by

$$S : \{f \in \mathscr{S}[0,\infty) : f(0) = 0\} \to \mathscr{S}[0,\infty), \qquad Sf = -\frac{\mathrm{d}^2 f}{\mathrm{d}x^2}. \tag{2.98}$$

Then $-f''(x) = \lambda^2 f(x)$ implies $f(x) = Ae^{i\lambda x} + Be^{-i\lambda x}$ for some constants $A$, $B$ and the boundary condition yields $B = -A$, and hence

$$f(x) = A' \sin(\lambda x). \tag{2.99}$$

But, for $f \in S[0,\infty)$, we must have $A' = 0$ so there are no nonzero eigenfunctions of $S$.

Instead, following Gelfand and coauthors [10, 11], one must consider eigen*functionals* or *generalized eigenfunctions*, defined as functionals $F_\lambda \in (\mathscr{S}[0,\infty))'$ for which there exists some corresponding eigenvalue $\lambda^2 \in \mathbb{C}$ with

$$F_\lambda(Sf) = \lambda^2 F_\lambda(f) \qquad \forall\, f \in \mathscr{S}[0,\infty). \tag{2.100}$$

Searching for generalized eigenfunctions of the half-line Dirichlet heat operator, we find that the definition equation (2.100) holds when $\lambda$ is any real number and

$$F_\lambda(f) = \frac{1}{2\pi}\int_0^\infty \sin(\lambda x) f(x)\mathrm{d}x. \tag{2.101}$$

It is immediate that the generalized eigenfunctions we identify are complete in the space of initial data in the sense that if $F_\lambda(f) = 0$ for all $\lambda \in \mathbb{R}$, then $f = 0$. Hence it is reasonable to expect that one may express the solution of the Dirichlet problem for the heat equation as an expansion in these spectral objects. But that is the result of the classical Fourier transform approach. Indeed, the map $f(x) \mapsto F(\lambda)$ given by equation (2.101) is precisely the Fourier sine transform.

Moreover, it is a result of [11] that for *any* self-adjoint operator, the generalized eigenfunctions form a complete system, so it should be possible to represent the solution of any IBVP as an expansion in the generalized eigenfunctions of the associated spatial differential operator, provided that operator is self-adjoint. However, not all differential operators are self-adjoint. Indeed, within the class of well-posed IBVPs, those with self-adjoint operators are exceptional. In particular, both Problems 1 and 2 have non-self-adjoint operators.

Of course, this observation does not necessarily preclude the possibility of either of the specific non-self-adjoint operators (2.81) and (2.83) having a complete system of generalized eigenfunctions or that arbitrary expansions in the generalized eigenfunctions converge. However, it is argued below not only that the transforms $F_\lambda$ defined above are not generalized eigenfunctions but that no integral transform could be composed of generalized eigenfunctions of $S$.

It is said that the generalized eigenfunctions of a self-adjoint operator provide a spectral representation, in the sense that they diagonalize the operator. That is, a differential operator can be expressed, via its generalized eigenfunctions, as a multiplication operator. Hence it is reasonable to ask the following questions:

1. Let $f(x) \mapsto F(\lambda)$ given by equation (2.92) be the transform used to solve the IBVP. Is $(F_\lambda)_{\lambda \in \Gamma^+ \cup \Gamma^-}$ a system of generalized eigenfunctions of $S$? If not, then do these functionals have any properties similar to spectral functions?

2. Is the system $(F_\lambda)_{\lambda \in \Gamma^+ \cup \Gamma^-}$ complete in the space of initial data and does the expansion of an arbitrary initial datum in the system $(F_\lambda)_{\lambda \in \Gamma^+ \cup \Gamma^-}$ converge?

3. Can it be said that the system $(F_\lambda)_{\lambda \in \Gamma^+ \cup \Gamma^-}$ of functionals diagonalizes $S$ in any sense?

These questions are discussed below.

### Augmented Eigenfunctions: Motivation and Definition

Consider now the spatial differential operator (2.83) associated with Problem 2. Let us seek generalized eigenfunctions $E_\lambda$ of $S$ that represent integral transforms, that is, generalized eigenfunctions

$$E_\lambda S f = \lambda^3 E_\lambda f \tag{2.102}$$

of the form

$$E_\lambda f = \int_0^\infty \phi(x,\lambda) f(x) \mathrm{d}x, \tag{2.103}$$

for some integral kernel $\phi(x,\lambda)$.

Integrating by parts on the left-hand side of equation (2.102), we obtain

$$\begin{aligned}
\phi(0,\lambda) &= 0, \\
\phi'(0,\lambda) &= 0, \\
\phi'''(x,\lambda) &= (-i\lambda)^3 \phi(x,\lambda).
\end{aligned}$$

Hence

$$\phi(x,\lambda) = A(\lambda)\left[e^{-i\lambda x} + \alpha^2 e^{-i\alpha\lambda x} + \alpha e^{-i\alpha^2\lambda x}\right]$$

for some $A$. Note that by stressing the $\lambda$ dependence of $A$ we do not intend to suggest that $A$ is expected to be an analytic function of $\lambda$, merely that this coefficient may vary with $\lambda$. But if $A(\lambda)$ is nonzero, then the integral (2.103) defining $E_\lambda$ will diverge for general $f \in \mathscr{E}[0,\infty)$. Hence $S$ has no nonzero generalized eigenfunctions of this type.

The first part of question 1 has already been answered in the negative. In the following definition, we provide an answer to the more open-ended second part of question 1.

**Definition 2.6.** *Let $I$ be an open real interval and let $C$ be a space of functions defined on the closure of $I$ with sufficient smoothness and decay conditions. Let $\Phi \subseteq C$ and let $L : \Phi \to C$ be a linear operator. Let $\gamma$ be an oriented contour in $\mathbb{C}$ and let $E = \{E_\lambda : \lambda \in \gamma\}$ be a family of functionals $E_\lambda \in C'$. Suppose there exist corresponding* remainder *functionals $R_\lambda \in C'$ and* eigenvalues $z(\lambda)$ *such that*

$$E_\lambda(L\phi) = z(\lambda)E_\lambda(\phi) + R_\lambda(\phi) \qquad \forall\, \phi \in \Phi,\ \forall\, \lambda \in \gamma. \tag{2.104}$$

*If*

$$\int_\gamma e^{i\lambda x} R_\lambda(\phi)\mathrm{d}\lambda = 0 \qquad \forall\, \phi \in \Phi,\ \forall\, x \in I, \tag{2.105}$$

*then we say $E$ is a family of* type I augmented eigenfunctions *of $L$ up to integration along $\gamma$.*

*If*

$$\int_\gamma \frac{e^{i\lambda x}}{z(\lambda)} R_\lambda(\phi)\mathrm{d}\lambda = 0 \qquad \forall\, \phi \in \Phi,\ \forall\, x \in I, \tag{2.106}$$

*then we say $E$ is a family of* type II augmented eigenfunctions *of $L$ up to integration along $\gamma$.*

In claiming this definition completes an answer to question 1, we are not only claiming Theorem 2.8 (our $F_\lambda$ are augmented eigenfunctions of the relevant spatial differential operators) but also that augmented eigenfunctions can be called "spectral objects" in some reasonable sense; see below.

The answer to question 2 is positive. Indeed, evaluating the result of Theorem 2.5 at $t = 0$ immediately yields the desired completeness and convergence results. Question 3 is answered below with three senses of diagonalization in equations (2.111), (2.114), and (2.116).

### Augmented Eigenfunctions: Discourse on the Definition

In the earlier discussion of the generalized eigenfunctions (2.101) of the half-line Dirichlet heat operator (2.98), we noted that these spectral objects represent the Fourier sine transform. This is the transform tailor-made to solve the IBVP, because

(i) the transform is *defined* on the space of admissible initial data,

(ii) the transformed data (often called the spectral data) has a particularly *simple time evolution*, and

(iii) the transform is *invertible* on the relevant space so the solution at a later time can be reconstructed from the time-evolved spectral data.

At least for the purposes of solving IBVPs, it is these properties of generalized eigenfunctions that are essential.

Property (ii) follows from equation (2.100), usually taken to be the definition of a generalized eigenfunction. Indeed, taking spatial Fourier sine transforms of the heat equation,

$$q_t(x,t) = \left(\frac{\mathrm{d}^2 q}{\mathrm{d}x^2}\right)(x,t), \tag{2.107}$$

one obtains

$$\hat{q}_t(\lambda,t) = \widehat{\left(\frac{\mathrm{d}^2 q}{\mathrm{d}x^2}\right)}(\lambda,t). \tag{2.108}$$

Now equation (2.100) allows us to rewrite the right-hand side as $-\lambda^2\hat{q}(\lambda,t)$ which specifies a simple ordinary differential equation for the time evolution of the spectral data. The solution is

$$\hat{q}(\lambda,t) = e^{-\lambda^2 t}\hat{q}(0,\lambda), \tag{2.109}$$

which is property (ii). Hence we consider property (ii) as a consequence of the deeper property that

(ii′) the transform *maps* the spatial differential operator *to a multiplication operator* on the transformed (spectral) data.

In solving an IBVP, properties (i) and (ii′) would be of little use without a method for reconstructing the solution from the time-evolved spectral data, that is, property (iii). *The central idea of augmented eigenfunctions is that a suitable strengthening of property* (iii) *permits a weakening of property* (ii′) *to the point of tautology.* Modeled on these observations, we now discuss the definition of augmented eigenfunctions of each type.

Suppose $E = \{E_\lambda : \lambda \in \gamma\}$ is a family of type I augmented eigenfunctions of $L$ up to integration along $\gamma$. Then equation (2.104) implies

$$\int_\gamma e^{i\lambda x} E_\lambda(L\phi)\mathrm{d}\lambda = \int_\gamma e^{i\lambda x} z(\lambda) E_\lambda(\phi)\mathrm{d}\lambda + \int_\gamma e^{i\lambda x} R_\lambda(\phi)\mathrm{d}\lambda. \tag{2.110}$$

But equation (2.105) guarantees that the second integral on the right-hand side of equation (2.110) evaluates to zero. Hence

$$\int_\gamma e^{i\lambda x} E_\lambda(L\phi)\mathrm{d}\lambda = \int_\gamma e^{i\lambda x} z(\lambda) E_\lambda(\phi)\mathrm{d}\lambda, \tag{2.111}$$

and it appears formally that the type I augmented eigenfunctions describe an integral transform which diagonalizes $L$ and $\int_\gamma e^{i\lambda x}\cdot \mathrm{d}\lambda$ represents the corresponding inverse transform.

In the definition of type I augmented eigenfunctions, property (ii′) has been reduced merely to the definition (2.104) of the remainder functionals, whereas property (iii) has been strengthened; not only is $\int_\gamma e^{i\lambda x}\cdot \mathrm{d}\lambda$ an inverse transform but, according to equation (2.105), when it is applied to the remainder functionals (here considered as a function of $\lambda \in \mathbb{C}$, with $\phi$ a parameter) it must return 0.

There are some additional conditions required in order to make this rigorous, not least the convergence of the remaining two integrals in equation (2.110) and an argument that this is a genuine transform pair, but this is no different from the situation for generalized eigenfunctions. Indeed, the guarantee that generalized eigenfunctions can diagonalize an operator [11] only holds for a certain class of operators: self-adjoint operators. However, a theorem of [7] combined with results in [19, 20] shows that

(a) each operator $S$ of the form

$$S : \{f \in C^\infty[0,1] : n \text{ linearly independent boundary conditions}\} \to C^\infty[0,1], \tag{2.112a}$$

$$Sf = \left(\frac{-i\mathrm{d}}{\mathrm{d}x}\right)^n f, \tag{2.112b}$$

for which $n$ is even and an associated IBVP is well-posed, belongs to the class of operators which can be diagonalized by a complete family of their type I augmented eigenfunctions.

Now suppose instead that $E = \{E_\lambda : \lambda \in \gamma\}$ is a family of type II augmented eigenfunctions of $L$ up to integration along $\gamma$. Then, by equation (2.104),

$$\int_\gamma e^{i\lambda x} \frac{1}{z(\lambda)} E_\lambda(L\phi)\mathrm{d}\lambda = \int_\gamma e^{i\lambda x} E_\lambda(\phi)\mathrm{d}\lambda + \int_\gamma e^{i\lambda x} \frac{1}{z(\lambda)} R_\lambda(\phi)\mathrm{d}\lambda, \tag{2.113}$$

and equation (2.106) ensures the final integral vanishes,

$$\int_\gamma e^{i\lambda x} \frac{1}{z(\lambda)} E_\lambda(L\phi)\mathrm{d}\lambda = \int_\gamma e^{i\lambda x} E_\lambda(\phi)\mathrm{d}\lambda. \tag{2.114}$$

So formally the type II augmented eigenfunctions specify an integral transform and $\int_\gamma e^{i\lambda x} \cdot \mathrm{d}\lambda$ provides an inverse transform, as in the case of type I augmented eigenfunctions. But this time the transform pair does not diagonalize $S$ in the expected way. Instead, equation (2.114) *represents $S$ with diagonalized inverse.*

Type II augmented eigenfunctions have the same weakening (2.104) of property (ii′) as type I augmented eigenfunctions. However, the strengthening of property (iii), equation (2.106), now insists that the inverse transform of the *ratio* of the remainder functional to the eigenvalue evaluate to 0. It is clear how the form of diagonalization (2.114) arises from this.

Again, there is much in this formal discussion to be justified rigorously. In particular, in investigating the convergence of the relevant integrals it becomes clear why this division by the eigenvalue is necessary. Indeed, for the operators we study here, certain families of type II augmented eigenfunctions fail to be type I augmented eigenfunctions precisely because the convergence of the integral (2.105) fails. Nevertheless, the rigorous arguments may be constructed for several classes of operators, listed below.

**Theorem 2.7.** *The operators defined by* (2.81) *or* (2.83) *can be diagonalized in the sense of equation* (2.114) *by complete systems of their respective type* II *augmented eigenfunctions.*

The proof of this theorem appears below. This result is a special case of the more general results in [7, 17] summarized below:

(b) If $S$ is the operator (2.112) for any $n \geq 2$, the boundary conditions include $f(0) = 0 = f(1)$, and an associated IBVP is well-posed, then $S$ can be diagonalized in the sense of equation (2.114) by a complete system of its type II augmented eigenfunctions.

(c) If $S$ is the operator

$$S:\{f\in\mathscr{E}[0,\infty): N \text{ linearly independent boundary conditions}\}\to\mathscr{S}[0,\infty), \tag{2.115a}$$

$$Sf=\left(\frac{-i\mathrm{d}}{\mathrm{d}x}\right)^n f \tag{2.115b}$$

for any $n\geq 2$ and an associated IBVP is well-posed, then $S$ can be diagonalized in the sense of equation (2.114) by a complete system of its type II augmented eigenfunctions.

It is not always possible to find a complete system of either type I or type II augmented eigenfunctions. However, for a wide class of operators $L$ it is possible to find a complete system $\{E_\lambda:\lambda\in\gamma^{(\mathrm{I})}\cup\gamma^{(\mathrm{II})}\}$ of functionals where some, $\{E_\lambda:\lambda\in\gamma^{(\mathrm{I})}\}$, are type I augmented eigenfunctions and the rest, $\{E_\lambda:\lambda\in\gamma^{(\mathrm{II})}\}$, are type II augmented eigenfunctions. In such a situation, $L$ is diagonalized in the sense that

$$\int_{\gamma^{(\mathrm{I})}} e^{i\lambda x}E_\lambda L\phi\,\mathrm{d}\lambda=\int_{\gamma^{(\mathrm{I})}} e^{i\lambda x}z(\lambda)E_\lambda\phi\,\mathrm{d}\lambda \qquad \forall\,\phi\in\Phi,\ \forall\,x\in(0,1), \tag{2.116a}$$

$$\int_{\gamma^{(\mathrm{II})}} \frac{1}{z(\lambda)}e^{i\lambda x}E_\lambda L\phi\,\mathrm{d}\lambda=\int_{\gamma^{(\mathrm{II})}} e^{i\lambda x}E_\lambda\phi\,\mathrm{d}\lambda \qquad \forall\,\phi\in\Phi,\ \forall\,x\in(0,1). \tag{2.116b}$$

This includes the result for the general two-point differential operator with monomial symbol:

(d) If $S$ is the operator (2.112) for any $n\geq 2$ with a well-posed associated IBVP, then $S$ can be diagonalized in the sense of (2.116) by a complete system of type I and type II augmented eigenfunctions.

### Augmented Eigenfunctions: Application to Problems 1 and 2

**Theorem 2.8.**

(i) *Suppose $S$ is defined by equation* (2.81), $\Gamma^\pm$ *are the contours shown in Figure* 2.8, *and the functionals $F_\lambda$ are defined by equations* (2.92a), (2.92c), (2.92d). *Then* $\{F_\lambda^+:\lambda\in\Gamma^+\}\cup\{F_\lambda^-:\lambda\in\Gamma^-\}$ *are type* II *augmented eigenfunctions of $S$ up to integration along* $\Gamma^+\cup\Gamma^-$, *with eigenvalues* $\lambda^3$.

(ii) *Suppose $S$ is defined by equation* (2.83), $\Gamma^\pm$ *are the contours shown in Figure* 2.7, *and the functionals $F_\lambda$ are defined by equations* (2.92a), (2.92e), (2.92f). *Then* $\{F_\lambda^+:\lambda\in\Gamma^+\}\cup\{F_\lambda^-:\lambda\in\Gamma^-\}$ *are type* II *augmented eigenfunctions of $S$ up to integration along* $\Gamma^+\cup\Gamma^-$, *with eigenvalues* $\lambda^3$.

***Proof.***
(i) Integration by parts yields

$$F_\lambda^\pm(Sf)=\lambda^3F_\lambda^\pm(f)+R_\lambda^\pm(f) \qquad \forall\,\lambda\in\Gamma^\pm,\ \forall\,f\in\mathscr{D}(S), \tag{2.117}$$

where

$$R_\lambda^+(f)=-\frac{i}{2\pi}f''(0)-\frac{\lambda}{2\pi}f'(0), \tag{2.118}$$

$$R_\lambda^-(f)=-e^{-i\lambda}\frac{i}{2\pi}f''(1). \tag{2.119}$$

That is, equation (2.104) holds with eigenvalues $\lambda^3$ and remainder functionals entire in $\lambda$. Moreover, $R_\lambda^\pm/\lambda^3$ are meromorphic functions of $\lambda$, each with no pole in the sectors lying to the left of $\Gamma^\pm$. As $R_\lambda^+$ is decaying like $O(\lambda^{-2})$ as $\lambda \to \infty$, Jordan's lemma yields

$$\int_{\Gamma^+} e^{i\lambda x} \frac{R_\lambda^+(f)}{\lambda^3} \mathrm{d}\lambda = 0. \tag{2.120}$$

Similarly,

$$\int_{\Gamma^-} e^{i\lambda x} \frac{R_\lambda^-(f)}{\lambda^3} \mathrm{d}\lambda = \int_{\Gamma^-} e^{-i\lambda(1-x)} \left[ -\frac{i}{2\pi\lambda^3} f''(1) \right] \mathrm{d}\lambda = 0. \tag{2.121}$$

(ii) Integration by parts yields

$$F_\lambda^\pm(Sf) = \lambda^3 F_\lambda^\pm(f) + R_\lambda^\pm(f) \qquad \forall\, \lambda \in \Gamma^\pm,\ \forall\, f \in \mathcal{D}(S), \tag{2.122}$$

where

$$R_\lambda^+(f) = \frac{i}{2\pi} f''(0) - \frac{\lambda}{2\pi} f'(0), \tag{2.123}$$

$$R_\lambda^-(f) = -\frac{i}{2\pi} f''(0) + \frac{\lambda}{2\pi} f'(0). \tag{2.124}$$

That is, equation (2.104) holds with eigenvalues $\lambda^3$ and remainder functionals entire in $\lambda$. Moreover, $R_\lambda^\pm/\lambda^3$ are meromorphic functions, each with no pole to the left of $\Gamma^\pm$, and $O(\lambda^{-2})$ decay as $\lambda \to \infty$. Hence, by Jordan's lemma,

$$\int_{\Gamma^\pm} e^{i\lambda x} \frac{R_\lambda^\pm(f)}{\lambda^3} \mathrm{d}\lambda = 0. \quad \square \tag{2.125}$$

The reason we work in $\mathcal{E}[0,\infty)$ for Problem 2 now becomes clear. Suppose that we wish to work in the full Schwartz space $\mathcal{S}[0,\infty)$. Then our forward transform is not defined on all of $\Gamma^-$ (see equations (2.92a) and (2.92f) and Figure 2.7) so we must replace $\Gamma^-$ with some new contour, $\gamma^-$ say, which lies wholly in the closure of the lower half plane. But then equation (2.125) fails, as we pick up a contribution from the pole at zero.

***Proof of Theorem* 2.7.** We give the proof for the operator (2.81). The proof for the operator (2.83) is similar.

Theorem 2.8(i) gives the family of type II augmented eigenfunctions. Moreover, these augmented eigenfunctions describe the forward transform used in Theorem 2.5. Evaluating that result at $t = 0$, we obtain completeness of the system of functionals.

The integrals

$$\int_{\Gamma^\pm} e^{i\lambda x} F_\lambda^\pm f \, \mathrm{d}\lambda \tag{2.126}$$

converge as they are the integrals appearing in Theorem 2.5, for $t = 0$. Thus the integrals in equation (2.114) converge. $\square$

## Bibliography

[1] D. Chilton, An Alternative Approach to Two-Point Boundary Value Problems for Linear Evolution PDEs and Applications, PhD thesis, University of Reading, 2006.

[2] B. Deconinck, T. Trogdon, and V. Vasan, The method of Fokas for solving linear partial differential equations, SIAM Rev. **56** (2014), 159–186.

[3] A. S. Fokas, A unified transform method for solving linear and certain nonlinear PDEs, Proc. R. Soc. Lond. Ser. A **453** (1997), 1411–1443.

[4] A. S. Fokas, A Unified Approach to Boundary Value Problems, CBMS-NSF Regional Conf. Ser. in Appl. Math., AMS, Providence, RI (2008).

[5] A. S. Fokas and B. Pelloni, Two-point boundary value problems for linear evolution equations, Math. Proc. Cambridge Philos. Soc. **131** (2001), 521–543.

[6] A. S. Fokas and B. Pelloni, A transform method for linear evolution PDEs on a finite interval, IMA J. Appl. Math. **70** (2005), 564–587.

[7] A. S. Fokas and D. A. Smith, Evolution PDEs and Augmented Eigenfunctions. I. Finite Interval, arXiv:1303.2205v2[math. SP] (2014).

[8] A. S. Fokas and E. A. Spence, Synthesis, as opposed to separation, of variables, SIAM Rev. **54** (2012), 291–324.

[9] A. S. Fokas and L. Y. Sung, Initial-boundary value problems for linear dispersive evolution equations on the half-line (unpublished), 1999.

[10] I. M. Gelfand and G. E. Shilov, Generalized Functions Volume 3: Theory of Differential Equations, Academic Press, New York, 1967.

[11] I. M. Gelfand and N. Ya. Vilenkin, Generalized Functions Volume 4: Applications of Harmonic Analysis, Academic Press, New York, 1964.

[12] J. W. Hopkins, Some convergent developments associated with irregular boundary conditions, Trans. Amer. Math. Soc. **20** (1919), 245–259.

[13] D. Jackson, Expansion problems with irregular boundary conditions, Proc. Amer. Acad. Arts Sci. **51** (1915), 383–417.

[14] E. Langer, The zeros of exponential sums and integrals, Bull. Amer. Math. Soc. **37** (1931), 213–239.

[15] B. Pelloni, Well-posed boundary value problems for linear evolution equations on a finite interval, Math. Proc. Cambridge Philos. Soc. **136** (2004), 361–382.

[16] B. Pelloni, The spectral representation of two-point boundary-value problems for third-order linear evolution partial differential equations, Proc. R. Soc. Lond. Ser. A **461** (2005), 2965–2984.

[17] B. Pelloni and D. A. Smith, Evolution PDEs and augmented eigenfunctions. II. Half line, arXiv:1408:3657[math. AP] (2014).

[18] A. A. Shkalikov, The completeness of eigenfunctions and associated functions of an ordinary differential operator with irregular-separated boundary conditions, Funct. Anal. Appl. **10** (1976), 305–316.

[19] D. A. Smith, Spectral Theory of Ordinary and Partial Linear Differential Operators on Finite Intervals, PhD thesis, University of Reading, 2011.

[20] D. A. Smith, Well-posed two-point initial-boundary value problems with arbitrary boundary conditions, Math. Proc. Cambridge Philos. Soc. **152** (2012), 473–496.

[21] D. A. Smith, The Unified Transform Method for Linear Initial-Boundary Value Problems: A Spectral Interpretation, arXiv:1408.3659 (2014).

# Chapter 3

# Evolution Problems: Nonlinear

*G. Biondini, A. S. Fokas, and D. Shepelsky*

## 3.1 ▪ Comparison of Two Approaches to the Initial-Boundary Value Problem for the Nonlinear Schrödinger Equation on the Half-Line with Robin Boundary Conditions

### 3.1.1 ▪ Introduction

Linear evolution partial differential equations (PDEs) with *second*-order spatial derivatives formulated on the half-line with Dirichlet, or Neumann, or Robin boundary conditions (BCs) can be solved by employing an appropriate transform pair in the spatial variable. For example, the heat equation, with either Dirichlet or Neumann BCs at $x = 0$, can be solved by either the sine or the cosine $x$-transforms. It is well known that these transforms can be obtained from the Fourier transform on the line by using either an odd or an even extension. Hence, an equivalent approach to solving the above initial-boundary value problems (IBVPs) is to first use an odd or an even extension to the full line and then to employ the Fourier transform.

For a particular class of BCs called linearizable, IBVPs for integrable *nonlinear* PDEs formulated on the half-line can be solved with the same level of efficiency as the corresponding initial value problems (IVPs) on the line. In particular, the celebrated nonlinear Schrödinger (NLS) equation

$$iq_t + q_{xx} - 2\nu|q|^2 q = 0 \tag{3.1}$$

(where $\nu$ equals $+1$ or $-1$ identifies the defocusing and focusing cases, respectively, and subscripts $x$ and $t$ denote partial differentiation throughout) admits the following linearizable BCs [19]:

$$q(0,t) = 0, \tag{3.2a}$$

$$q_x(0,t) = 0, \tag{3.2b}$$

$$q_x(0,t) - \alpha q(0,t) = 0, \tag{3.2c}$$

with $\alpha$ an arbitrary real constant. The linearized version of the NLS equation, namely the equation

$$iq_t + q_{xx} = 0, \tag{3.3}$$

with any of the BCs (3.2), just like the heat equation, can be solved by an appropriate extension to the full line. Hence, it is natural to expect that such extensions are also applicable for the NLS equation. This is indeed the case: equation (3.1) with either the BC (3.2a) or (3.2b) was solved in [2] using an odd or an even extension, whereas equation (3.1) with the BC (3.2c) was solved in [3, 8, 16, 20] using a slightly more complicated extension.

On the other hand, *linear* evolution PDEs formulated on the half-line involving a *third*-order spatial derivative *cannot* be solved by an $x$-transform (or, equivalently, by an extension to the full line). However, such problems *can* be solved by the unified method introduced by one of the authors in [9] and further developed in [10, 11, 12]. The situation is similar for integrable nonlinear evolution PDEs: the Korteweg–deVries equation in the case of dominant surface tension, namely the equation

$$q_t + q_x + 6qq_x - q_{xxx} = 0, \tag{3.4}$$

admits the following class of linearizable BCs [15]:

$$q(0,t) = q_{xx}(0,t) = 0 \tag{3.5a}$$

and

$$q(0,t) = \alpha, \qquad q_{xx}(0,t) = \alpha + 3\alpha^2, \tag{3.5b}$$

where again $\alpha$ is an arbitrary real constant. Equation (3.4) with either (3.5a) or (3.5b) cannot be solved by an extension to the full line but can be solved by the unified method of [9].

In summary, for those linearizable IBVPs for which the associated linearized PDE can be solved by an $x$-transform (or, equivalently, by an extension to the full line), there exist two alternative approaches: (a) extension to the full line and (b) the unified method of [9]. On the other hand, for those problems for which the associated linearized PDEs cannot be solved by an $x$-transform, there exists only the approach of [9]. The main goal of this work is to review and compare the two approaches (a) and (b), namely the extension to the full line and the implementation of the unified method, for the particular case of the NLS equation with linearizable BCs.

Throughout this work the overbar denotes complex conjugation, subscripts 1 and 2 denote, respectively, the first and second column vectors of a $2 \times 2$ matrix, subscripts $i, j$ denote a corresponding matrix element, and $I$ is the $2 \times 2$ identity matrix. We assume that the initial condition $q(x,0)$ decays sufficiently fast as $x \to \infty$ so that the relevant integral equations are well defined. The precise functional class for the NLS equation was discussed in [11, 13].

### 3.1.2 ▪ The NLS Equation with Homogeneous Robin BCs

The following notation will be used:

$$\sigma_3 = \mathrm{diag}(1,-1), \qquad \sigma_{\mathrm{up}} = \mathrm{diag}(1,0), \qquad \sigma_{\mathrm{dn}} = \mathrm{diag}(0,1), \tag{3.6a}$$

$$\theta(x,t,k) = kx + 2k^2 t, \tag{3.6b}$$

$$Q_o(x) = \begin{pmatrix} 0 & q_o(x) \\ \nu \overline{q_o(x)} & 0 \end{pmatrix}. \tag{3.6c}$$

Since the solution of the IBVP via the extension to the full line makes use of the results for the IVP, it is convenient to briefly recall the results for the latter problem. Theorem 3.1 below expresses the results of [21] on the IVP using the language of Riemann–Hilbert problems (RHPs) [1, 7].

**Theorem 3.1 (IVP [21, 1, 7]).** *Let $q(x,t)$ be the solution of the IVP for the NLS equation* (3.1) *on the full line. Let $a_{\mathrm{ivp}}(k)$ and $b_{\mathrm{ivp}}(k)$ denote the spectral functions which are defined in terms of $q_o(x)$ as follows:*

$$\begin{pmatrix} b_{\mathrm{ivp}}(k) \\ a_{\mathrm{ivp}}(k) \end{pmatrix} = \lim_{x\to-\infty} \begin{pmatrix} e^{2ikx} & 0 \\ 0 & 1 \end{pmatrix} w(x,k), \qquad k\in\mathbb{R}, \tag{3.7}$$

*where $w(x,k)$ is the unique 2-component vector defined as the solution of*

$$(\partial_x + 2ik\sigma_{\mathrm{up}})w = Q_o(x)w, \quad x\in\mathbb{R}, \quad \operatorname{Im} k\geq 0, \qquad \lim_{x\to\infty} w(x,k) = \begin{pmatrix} 0 \\ 1 \end{pmatrix}. \tag{3.8}$$

*Further, suppose that $a_{\mathrm{ivp}}(k)$, which admits analytic continuation onto the upper-half plane, has $J$ simple zeros $k_1,\dots,k_J$, with corresponding norming constants $b_1,\dots,b_J$, defined by the relations*

$$w(x,k_j) = b_j\, e^{2ik_j x}\, \tilde{w}(x,k_j), \qquad x\in\mathbb{R}, \tag{3.9}$$

*where $\tilde{w}(x,k)$ is the unique 2-component vector defined as the solution of*

$$(\partial_x - 2ik\sigma_{\mathrm{dn}})\tilde{w} = Q_o(x)\tilde{w}, \quad x\in\mathbb{R}, \quad \operatorname{Im} k\geq 0, \qquad \lim_{x\to-\infty} \tilde{w}(x,k) = \begin{pmatrix} 1 \\ 0 \end{pmatrix}. \tag{3.10}$$

*If $\nu=1$, then $J=0$. Define the function $q(x,t)$ by*

$$q(x,t) = 2i \lim_{k\to\infty} [k\,M_{1,2}(x,t,k)], \qquad x\in\mathbb{R}, \quad t>0, \tag{3.11}$$

*where $M(x,t,k)$ is the unique $2\times2$ matrix, sectionally meromorphic in $\mathbb{C}\setminus\mathbb{R}$, defined by the following conditions for $x\in\mathbb{R}$ and $t>0$:*

(i) *$M(x,t,k) = I + O(1/k)$ as $k\to\infty$.*

(ii) *If $\nu=-1$, for all $j=1,\dots,J$ the second column of $M$ has a simple pole at $k=k_j$ and the first column of $M$ has a simple pole at $k=\bar{k}_j$, with residue conditions*

$$\operatorname*{Res}_{k=k_j}[M_2(x,t,k)] = \frac{b_j\, e^{2i\theta(x,t,k_j)}}{a'_{\mathrm{ivp}}(k_j)} M_1(x,t,k_j), \tag{3.12a}$$

$$\operatorname*{Res}_{k=\bar{k}_j}[M_1(x,t,k)] = -\frac{\bar{b}_j\, e^{2i\theta(x,t,\bar{k}_j)}}{\overline{a'_{\mathrm{ivp}}(k_j)}} M_2(x,t,\bar{k}_j), \tag{3.12b}$$

*where primes denote differentiation with respect to $k$.*

(iii) *$M$ satisfies the jump condition*

$$M_-(x,t,k) = M_+(x,t,k)\,G(x,t,k), \qquad k\in\mathbb{R}, \tag{3.13}$$

*with $M = M_+$ and $M = M_-$ in the upper half and lower half of the complex k-plane, respectively, and where the jump matrix $G(x,t,k)$ is given by*

$$G(x,t,k) = \begin{pmatrix} 1 & -\gamma(k)\,\mathrm{e}^{-2i\theta(x,t,k)} \\ \overline{\nu\gamma(k)}\,\mathrm{e}^{2i\theta(x,t,k)} & 1-\nu|\gamma(k)|^2 \end{pmatrix}, \qquad x\in\mathbb{R},\ t>0,\ k\in\mathbb{R}, \tag{3.14}$$

*with $\gamma(k) = b_{\text{ivp}}(k)/\overline{a_{\text{ivp}}(k)}$.*

*Then $q(x,t)$ solves the NLS equation, and $q(x,t) = q_o(x)$.*

We now turn our attention to the IBVP. The following theorem summarizes the results obtained in [13] via the unified method in the special case of linearizable BCs.

**Theorem 3.2 (IBVP via the unified method [13]).** *Let $q(x,t)$ be the solution of the IBVP for the NLS equation* (3.1) *for $0 < x < \infty$ with one of the linearizable BCs* (3.2) *at $x = 0$. Let $a(k)$ and $b(k)$ denote the spectral functions which are defined by*

$$\begin{pmatrix} b(k) \\ a(k) \end{pmatrix} = w(0,k), \qquad \operatorname{Im} k \geq 0, \tag{3.15}$$

*where the 2-component vector $w(x,k)$ is the unique solution of the problem*

$$(\partial_x + 2ik\sigma_{\text{up}})\,w = Q(x,0)\,w, \quad 0<x<\infty, \quad \operatorname{Im} k\geq 0, \qquad \lim_{x\to\infty} w(x,k) = \begin{pmatrix} 0 \\ 1 \end{pmatrix}. \tag{3.16}$$

*Furthermore, let $\Delta(k)$ be the function defined by*

$$\Delta(k) = a(k)\overline{a(-\bar{k})} + \nu f(k) b(k)\overline{b(-\bar{k})}, \qquad \operatorname{Im} k\geq 0, \tag{3.17}$$

*where*

(a) *for the Dirichlet BCs* (3.2a), $f(k) = -1$;

(b) *for the Neumann BCs* (3.2b), $f(k) = 1$;

(c) *for the Robin BCs* (3.2c),

$$f(k) = (2k - i\alpha)/(2k + i\alpha). \tag{3.18}$$

*Suppose that $a(k)$ has $J_1$ simple zeros, $k_1,\ldots,k_{J_1}$, in the first quadrant and $\Delta(k)$ has $J_2$ simple zeros, $\lambda_1,\ldots,\lambda_{J_2}$, in the second quadrant. Then*

$$q(x,t) = 2i \lim_{k\to\infty} [k M_{1,2}(x,t,k)], \qquad 0<x<\infty, \quad t>0, \tag{3.19}$$

*where $M(x,t,k)$ is the unique $2\times 2$ matrix, sectionally meromorphic in $\mathbb{C}\setminus(\mathbb{R}\cup i\mathbb{R})$, defined by the following conditions for all $0<x<\infty$, $t>0$:*

(i) *$M(x,t,k) = I + O(1/k)$ as $k\to\infty$.*

(ii) *For all $j = 1,\ldots,J_1$, the first and the second columns of $M$ have simple poles respectively at $k = k_j$ and at $k = \bar{k}_j$, with residue conditions*

$$\operatorname*{Res}_{k=k_j}[M_1(x,t,k)] = \frac{\mathrm{e}^{2i\theta(x,t,k_j)}}{a'(k_j)b(k_j)} M_2(x,t,k_j), \tag{3.20a}$$

$$\operatorname*{Res}_{k=\bar{k}_j}[M_2(x,t,k)] = \frac{\nu\,\mathrm{e}^{-2i\theta(x,t,\bar{k}_j)}}{\overline{a'(k_j)b(k_j)}} M_1(x,t,\bar{k}_j). \tag{3.20b}$$

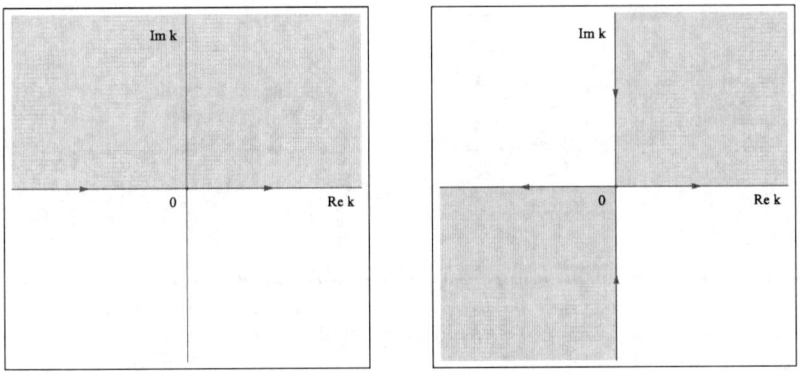


**Figure 3.1.** *The contours for the RHPs in Theorem 3.1 (left) and Theorem 3.2 (right). The domains of definition of $M_+(x,t,k)$ and $M_-(x,t,k)$ are indicated in gray and white, respectively.*

*Similarly, for all $j = 1,\dots,J_2$, the first and the second columns of $M$ have simple poles respectively at $k = \lambda_j$ and at $k = \bar\lambda_j$, with residue conditions*

$$\operatorname*{Res}_{k=\lambda_j}[M_1(x,t,k)] = -\nu f(\lambda_j)\frac{\overline{b(-\bar\lambda_j)}\,e^{2i\theta(x,t,\lambda_j)}}{a(\lambda_j)\Delta'(\lambda_j)}M_2(x,t,\lambda_j), \tag{3.21a}$$

$$\operatorname*{Res}_{k=\bar\lambda_j}[M_2(x,t,k)] = -\overline{f(\lambda_j)}\frac{b(-\bar\lambda_j)\,e^{-2i\theta(x,t,\bar\lambda_j)}}{\overline{a(\lambda_j)\Delta'(\lambda_j)}}M_1(x,t,\bar\lambda_j). \tag{3.21b}$$

(iii) *$M$ satisfies the jump condition*

$$M_-(x,t,k) = M_+(x,t,k)\,G(x,t,k), \qquad k \in \mathbb{R} \cup i\mathbb{R}, \tag{3.22}$$

*with $M = M_+$ and $M = M_-$ in the first and third quadrants and in the second and fourth quadrants of the complex $k$-plane, respectively (i.e., the contour $\mathbb{R} \cup i\mathbb{R}$ is oriented as in Figure 3.1), where the jump matrix $G(x,t,k)$ is given by $G_1,\dots,G_4$ for $\arg k = \pi/2$, $\pi$, $3\pi/2$, and $2\pi$, respectively, with*

$$G_1(x,t,k) = \begin{pmatrix} 1 & 0 \\ \Gamma(k)\,e^{2i\theta(x,t,k)} & 1 \end{pmatrix}, \quad G_3(x,t,k) = \begin{pmatrix} 1 & -\nu\overline{\Gamma(\bar k)}\,e^{-2i\theta(x,t,k)} \\ 0 & 1 \end{pmatrix}, \tag{3.23a}$$

$$G_4(x,t,k) = \begin{pmatrix} 1 & -\gamma(k)\,e^{-2i\theta(x,t,k)} \\ \nu\overline{\gamma(\bar k)}\,e^{2i\theta(x,t,k)} & 1-\nu|\gamma(k)|^2 \end{pmatrix}, \tag{3.23b}$$

$$G_2 = G_3G_4^{-1}G_1 = \begin{pmatrix} 1-\nu|\gamma(k)-\nu\overline{\Gamma(\bar k)}|^2 & (\gamma(k)-\nu\overline{\Gamma(\bar k)})\,e^{-2i\theta(x,t,k)} \\ -\nu(\overline{\gamma(\bar k)}-\nu\Gamma(k))\,e^{2i\theta(x,t,k)} & 1 \end{pmatrix}, \tag{3.23c}$$

*and with*

$$\gamma(k) = \frac{b(k)}{a(k)}, \quad k \in \mathbb{R}, \tag{3.24}$$

$$\Gamma(k) = -\nu f(k)\frac{\overline{b(-\bar{k})}}{a(k)\Delta(k)}$$
$$= -\nu\frac{\overline{b(-\bar{k})}}{a(k)}\frac{2k-i\alpha}{(2k+i\alpha)a(k)\overline{a(-\bar{k})}+\nu(2k-i\alpha)b(k)\overline{b(-\bar{k})}}, \quad k\in\mathbb{R}^-\cup i\mathbb{R}^+. \tag{3.25}$$

**Remark 7.** *The formulation of the above RHP appears more involved than that of the RHP of Theorem* 3.1*. Nonetheless, Theorem* 3.2 *yields the solution of the IBVP with the same efficiency as Theorem* 3.1 *yields the solution of the IVP.*

**Remark 8.** *Historically, the IBVP for the NLS equation on the half-line with the homogeneous Robin BCs* (3.2c) *was first studied in* [8]*. There, the solution of the IBVP was obtained using a simple, explicit extension to a k-dependent potential, namely* $q_{\rm ext}(x,k) = q(x)$ *for* $x \geq 0$ *and* $q_{\rm ext}(x,k) = f(k)q(-x)$ *for* $x < 0$ *(this extension was later used in* [5] *to study the soliton interactions). At the beginning of the* 1990s *it was also realized* [3, 19, 20] *that certain IBVPs with linearized BCs can be studied using an appropriate Bäcklund transformation, which allows the continuation of the given initial data to the full line, thus reducing the study of an IBVP to the study of the associated IVP without introducing an additional k dependence on the potential. Recently, the Bäcklund transformation approach for the NLS equation was rigorously revisited in* [6] *and placed within the modern RHP framework. In particular, this allowed the application of the nonlinear steepest descent method for oscillatory RHPs and the detailed study of several asymptotic and stability questions (e.g., see* [17]*). The following theorem summarizes the solution of the IBVP via Bäcklund transformation.*

**Theorem 3.3 (IBVP via extension [3, 20]).** *Let* $q(x,t)$ *be the solution of the IBVP for the NLS equation on the half-line* $x > 0$ *with one of the linearizable BCs* (3.2) *at* $x = 0$*. Let* $q_{\rm o,ext}(x) = q(x)$ *for all* $x \geq 0$ *and define* $q_{\rm o,ext}(x)$ *for all* $x < 0$ *as follows:*

(a) *for the Dirichlet BCs* (3.2a), $q_{\rm o,ext}(x) = -q_{\rm o}(-x)$ *(i.e.,* $q_{\rm o,ext}$ *is the odd extension of* $q_{\rm o}$*);*

(b) *for the Neumann BCs* (3.2b), $q_{\rm o,ext}(x) = q_{\rm o}(-x)$ *(i.e.,* $q_{\rm o,ext}$ *is the even extension of* $q_{\rm o}$*);*

(c) *for the Robin BCs* (3.2c), $q_{\rm o,ext}(x)$ *is defined by the unique solution of the system*

$$\frac{d}{dx}[q_{\rm o,ext}(x) - q_{\rm o}(-x)] + p(x)(q_{\rm o,ext}(x) + q_{\rm o}(-x)) = 0, \tag{3.26a}$$

$$\frac{dp}{dx} = \nu(|q_{\rm o,ext}(x)|^2 - |q_{\rm o}(-x)|^2), \tag{3.26b}$$

*with* $p(0) = \alpha$.

*Also, let* $a_{\rm ext}(k)$ *and* $b_{\rm ext}(k)$ *be defined as in equation* (3.7) *with* $q_{\rm o}(x)$ *replaced by* $q_{\rm o,ext}(x)$*. Then for all* $x \geq 0$ $q(x,t)$ *is given by equation* (3.11)*, where* $M(x,t,k)$ *is the unique solution of the RHP in Theorem* 3.1 *with* $a_{\rm ivp}(k)$ *and* $b_{\rm ivp}(k)$ *replaced by* $a_{\rm ext}(k)$ *and* $b_{\rm ext}(k)$.

**Remark 9.** *Equations* (3.26) *constitute a nonlinear system of ordinary differential equations (ODEs). This system was linearized directly in* [4]*, and it was linearized in* [6] *using the spectral problem associated with the NLS equation.*

**Remark 10.** *Theorems* 3.2 *and* 3.3 *express the solution of the same IBVP via two different RHPs. It is then natural to ask whether there is a relation between these two RHPs. We next show how to establish such a relation. For simplicity, we deal with the case in which there is* *no discrete spectrum (i.e., $J_1 = J_2 = 0$ in Theorem 3.2). We first note that in this case $\Gamma(k)$ is analytic and bounded for all $k \in \mathbb{C}^+$ and continuous up to the boundary. On the other hand, the exponentials in the jump matrices $G_1(x,t,k)$ and $G_3(x,t,k)$ in equation (3.23a) are bounded in the first and fourth quadrants, respectively. Thus, we can deform the RHP in Theorem 3.2, whose jump is on the cross $\mathbb{R} \cup i\mathbb{R}$, to an RHP with a jump on the real axis as described in the following lemma.*

**Lemma 3.4 (Deformation).** *Let $M(x,t,k)$ be the unique solution of the RHP in Theorem 3.2 with $J_1 = J_2 = 0$, and let*

$$\hat{M}(x,t,k) = \begin{cases} M(x,t,k), & k \in \mathbb{C}_{\mathrm{II}} \cup \mathbb{C}_{\mathrm{III}}, \\ M(x,t,k) G_1(x,t,k), & k \in \mathbb{C}_{\mathrm{I}}, \\ M(x,t,k) G_3^{-1}(x,t,k), & k \in \mathbb{C}_{\mathrm{IV}}. \end{cases} \tag{3.27}$$

*Then $\hat{M}(x,t,k)$ satisfies the jump condition*

$$\hat{M}_-(x,t,k) = \hat{M}_+(x,t,k) e^{-i\theta(x,t,k)\sigma_3} \hat{G}_o(k) e^{i\theta(x,t,k)\sigma_3}, \quad k \in \mathbb{R}, \tag{3.28}$$

*where*

$$\hat{G}_o(k) = \begin{pmatrix} 1 & -\gamma_o(k) \\ \nu\overline{\gamma_o(k)} & 1 - \nu|\gamma_o(k)|^2 \end{pmatrix} \tag{3.29}$$

*with*

$$\gamma_o(k) = \gamma(k) - \nu\overline{\Gamma(k)} = \frac{(2k - i\alpha)b(k)a(-k) + (2k + i\alpha)b(-k)a(k)}{(2k - i\alpha)\overline{a(k)}a(-k) + \nu(2k + i\alpha)\overline{b(k)}b(-k)}. \tag{3.30}$$

**Remark 11.** *The deformation of the RHP from the cross to the real axis was originally discussed in [14] (but note that $G_1$ was mistakenly written as $G_1^{-1}$ in [14]).*

**Remark 12.** *If the analytic continuation of the denominator in equation (3.30) has zeros in $\mathbb{C}^-$, then the formulation of the above RHP, normalized by the condition $\hat{M}(x,t,k) \to I$ as $k \to \infty$, is to be complemented by the residue conditions at these points.*

Comparing equation (3.30) with (3.14) suggests the interpretation $\gamma_o(k) = b_o(k)/\overline{a_o(k)}$, where $a_o(k)$ and $b_o(k)$ are the scattering coefficients associated with an appropriate IVP on the full line. We next show that such a correspondence does indeed exist.

**Lemma 3.5.** *Let $\{a(k), b(k)\}$ and $\{a_{\mathrm{ext}}(k), b_{\mathrm{ext}}(k)\}$ be, respectively, the scattering coefficients defined in Theorem 3.2 and in Theorem 3.3. For all $k \in \mathbb{R}$ the following relations hold:*

$$a_{\mathrm{ext}}(k) = \frac{2k - i\beta}{2k - i\alpha}\left(a(k)\overline{a(-k)} + \nu\frac{2k - i\alpha}{2k + i\alpha} b(k)\overline{b(-k)}\right), \tag{3.31a}$$

$$b_{\mathrm{ext}}(k) = \frac{2k + i\beta}{2k - i\alpha}\left(a(k)b(-k) + \frac{2k - i\alpha}{2k + i\alpha} b(k)a(-k)\right), \tag{3.31b}$$

*where*

$$\beta = \begin{cases} \alpha & \text{if } \alpha > 0 \wedge a(i\alpha/2) \neq 0 \text{ or } \alpha < 0 \wedge b(-i\alpha/2) = 0, \\ -\alpha & \text{if } \alpha < 0 \wedge b(-i\alpha/2) \neq 0 \text{ or } \alpha > 0 \wedge a(i\alpha/2) = 0. \end{cases} \tag{3.32}$$

**Remark 13.** *The relations* (3.31) *(which appeared first in* [20]*) were derived explicitly in* [6], *where equation* (3.32) *was also first obtained. In Section* 3.1.3 *we provide an alternative, more direct derivation. It is important to note that in* [3, 4, 6] *it was also shown that* $\beta = (-1)^S \alpha$, *where S is the number of discrete eigenvalues on the positive imaginary axis. By assumption, this number is zero in our case.*

We are now ready to establish the main result of this work, which is the explicit connection between the solution of the RHPs in Theorems 3.2 and 3.3.

**Theorem 3.6.** *Let* $\hat{M}(x,t,k)$ *be the solution of the deformed RHP in Lemma* 3.4, *and let* $M_{\text{ext}}(x,t,k)$ *be the solution of the RHP in Theorem* 3.1 *with* $q_o(x)$ *replaced by* $q_{o,\text{ext}}(x)$, *as defined as in Theorem* 3.3. *For all* $k \in \mathbb{C}$, $M_{\text{ext}}(x,t,k) = \hat{M}(x,t,k)$.

**Proof.** Following Theorem 3.3, the jump condition for the RHP defined by the extended potential has the form of equation (3.13) with $\gamma(k)$ replaced by $\gamma_{\text{ext}}(k) = b_{\text{ext}}(k)/\overline{a_{\text{ext}}(k)}$. Using equations (3.30) and (3.31), it is straightforward to verify that $\gamma_{\text{ext}}(k) = \gamma_o(k)$ for $k \in \mathbb{R}$. Thus, the jump condition for the RHPs in Theorem 3.3 and Lemma 3.4 are identical. Since both RHPs also satisfy the same normalization as $k \to \infty$ (namely, $M(x,t,k) = I + O(1/k)$ as $k \to \infty$), the uniqueness of the solution implies that the solutions of the two RHPs coincide for all $k \in \mathbb{C}$. □

**Remark 14.** *By construction, the scattering coefficients* $a_{\text{ext}}(k)$ *and* $b_{\text{ext}}(k)$ *enjoy the following symmetries:*

$$\overline{a_{\text{ext}}(-\bar{k})} = a_{\text{ext}}(k), \quad k \in \mathbb{C}^+, \qquad b_{\text{ext}}(-k) = b_{\text{ext}}(k)\frac{2k - i\beta}{2k + i\beta}, \quad k \in \mathbb{R}. \tag{3.33}$$

**Remark 15.** *A rigorous approach for general IBVPs was developed in* [13], *where the following were shown:* (i) *Given appropriate initial data and BCs (including the linearizable BCs* (3.2)*), it is possible to construct spectral functions which satisfy a certain global algebraic relation. This construction involves a nonlinear Volterra integral equation which admits a global solution.* (ii) *Given spectral functions satisfying the above global algebraic relation, the function* $q(x,t)$ *defined in terms of the RHP in Theorem* 3.2 *exists globally, solves the NLS equation, and satisfies the given initial conditions and BCs. The validity of this construction as* $t \to \infty$ *depends on the technical assumption that the boundary values* $q(0,t)$ *and* $q_x(0,t)$ *decay as* $t \to \infty$. *For linearizable BCs, this condition is equivalent to requiring that no stationary solitons (i.e., purely imaginary eigenvalues) be present.*

**Remark 16.** *In the case* $\alpha < 0$, *if* $b(-i\alpha/2) \neq 0$, *then* $a_{\text{ext}}(-i\alpha/2) = -|b(-i\alpha/2)|^2 < 0$. *On the other hand,* $a_{\text{ext}}(iy) \to 1$ *as* $y \to +\infty$. *Noticing that* $a_{\text{ext}}(iy) \in \mathbb{R}$ *for all* $y > 0$, *it follows that* $a_{\text{ext}}(k)$ *must have at least one zero for* $k \in i\mathbb{R}^+$. *Consequently, an asymptotic analysis similar to that in* [6] *implies that stationary solitons generated by these zeros dominate the large-t asymptotics in the direction along the t-axis, which violates the assumption of decaying behavior of* $q(0,t)$ *and* $q_x(0,t)$.

**Remark 17.** *The analysis of the deformed RHP of Lemma* 3.4 *with the jump conditions* (3.28) *on the real axis allows one to verify directly that* $q(x,t)$ *satisfies the given BCs* [18]. *(This is achieved by employing the symmetries of the scattering coefficients of the deformed problem.) This allows one to verify directly that the solution obtained with the formalism of* [13] *satisfies the given BCs, avoiding the above-mentioned a priori restriction that the boundary values decay as* $t \to \infty$.

**Remark 18.** *The explicit, k-dependent extension of the potential used in* [8] *introduces a few technical complications on the machinery of the inverse scattering transform (IST) used to solve the IVP, because for* $\alpha \neq 0$ *the extended potential has simple poles at* $k = \pm i\alpha/2$. *Nonetheless, one can still carry out the IST. The scattering eigenfunctions and the analytic scattering coefficients also acquire simple poles at* $k = \pm i\alpha/2$. *Moreover, by construction the scattering coefficients of the extended problem possess the symmetry* (3.33), *with the ratio* $(k+i\beta)/(k-i\beta)$ *replaced by the function* $f(k)$ *defined in equation* (3.18). *On the other hand, for* $\alpha \neq 0$ *this extension does not correctly deal with solutions for which the discrete spectrum contains an odd number of purely imaginary eigenvalues in either half plane, since the scattering coefficients obtained from the extended potential do not possess the correct symmetries* [4].

### 3.1.3 ▪ Relation between the Half-Line and the Full-Line Scattering Problem

Here we concisely derive the relations (3.31) between the scattering coefficients of the half-line problem and those of the full-line problem.

Recall that the NLS equation (3.1) is the compatibility condition of the matrix Lax pair [21]

$$\phi_x = (-ik\sigma_3 + Q(x,t))\phi, \tag{3.34a}$$
$$\phi_t = (-2ik^2\sigma_3 + H(x,t,k))\phi, \tag{3.34b}$$

with

$$Q(x,t) = \begin{pmatrix} 0 & q(x,t) \\ \nu\overline{q(x,t)} & 0 \end{pmatrix}, \qquad H(x,t,k) = 2kQ(x,t) - iQ_x(x,t)\sigma_3 - i\nu|q(x,t)|^2\sigma_3. \tag{3.35}$$

The Jost solutions are the solutions of (3.34) with specified asymptotic behavior as $x \to \pm\infty$ or at $x = 0$; namely, for all $k \in \mathbb{R}$,

$$\phi_-(x,t,k) = e^{-i\theta(x,t,k)\sigma_3} + o(1), \qquad x \to -\infty, \tag{3.36a}$$
$$\phi_o(0,0,k) = I, \tag{3.36b}$$
$$\phi_+(x,t,k) = e^{-i\theta(x,t,k)\sigma_3} + o(1), \qquad x \to +\infty. \tag{3.36c}$$

(Of course $\phi_-(x,t,k)$ is only defined when the potential is given over the full line.) The relations between these fundamental matrix solutions of the scattering problem define the scattering matrices $A(k)$ and $A_{\text{ivp}}(k)$ for the half-line problem and the full-line problem, respectively. Namely, for all $k \in \mathbb{R}$,

$$\phi_+(x,t,k) = \phi_0(x,t,k)A(k), \tag{3.37a}$$
$$\phi_+(x,t,k) = \phi_-(x,t,k)A_{\text{ivp}}(k), \tag{3.37b}$$

with the above definitions chosen consistently with [13]. The symmetries of the scattering problem imply

$$A(k) = \begin{pmatrix} \overline{a(k)} & b(k) \\ \nu\overline{b(k)} & a(k) \end{pmatrix}, \qquad A_{\text{ivp}}(k) = \begin{pmatrix} \overline{a_{\text{ivp}}(k)} & b_{\text{ivp}}(k) \\ \nu\overline{b_{\text{ivp}}(k)} & a_{\text{ivp}}(k) \end{pmatrix}, \tag{3.38}$$

with $a(k)$, $b(k)$, and $a_{\text{ivp}}(k)$ analytic in the upper-half plane [1, 13].

Now suppose $q(x,t)$ and $\tilde{q}(x,t)$ both solve the NLS equation (3.1), and the corresponding eigenfunctions are related by the dressing transformation

$$\tilde{\phi}(x,t,k) = B(x,t,k)\phi(x,t,k). \tag{3.39}$$

Then the Darboux matrix $B(x,t,k)$ solves the ODE

$$B_x + ik[\sigma_3, B] + BQ - \tilde{Q}B = 0, \qquad (3.40)$$

where $[A,B] = AB - BA$ is the matrix commutator. Looking for $B$ in the form

$$B(x,t,k) = 2ikI + O(1)$$

as $k \to \infty$, one obtains

$$B(x,t,k) = 2ikI + \sigma_3(\tilde{Q}(x,t) - Q(x,t)) + p(x,t)\sigma_3, \qquad (3.41a)$$

where

$$p(x,t) = p_o + \nu \int_0^x \big(q(x,t)\overline{q(x,t)} - \tilde{q}(x,t)\overline{\tilde{q}(x,t)}\big)\,\mathrm{d}y, \qquad (3.41b)$$

and $p_o$ is an arbitrary constant. The corresponding auto-Bäcklund transformation between the potentials $q(x,t)$ and $\tilde{q}(x,t)$ is

$$\tilde{q}_x - q_x + (q + \tilde{q})\,p = 0. \qquad (3.42)$$

Equation (3.42) together with an "initial condition" at some fixed value of $x$ (say, $x = 0$) determines completely $\tilde{q}(x,t)$ in terms of $q(x,t)$. Then, if $q(x,t)$ is given for $x \geq 0$, by imposing the mirror symmetry

$$\tilde{q}(x,t) = q(-x,t), \qquad (3.43)$$

the Bäcklund transformation (3.42) allows one to define an extended potential for all $x \in \mathbb{R}$. Choosing $p_o = \alpha$, one obtains equations (3.26). Moreover, equation (3.42) evaluated at $x = 0$ yields the Robin BC (3.2c).

Equation (3.39) implies that Jost solutions of the original and transformed scattering problems are related by

$$\tilde{\phi}_\pm(x,t,k) = B(x,t,k)\,\phi_\pm(x,t,k)\,B_\infty^{-1}(k), \qquad k \in \mathbb{R}, \qquad (3.44)$$

where $B_\infty(k) = \lim_{x\to\pm\infty} B(x,t,k)$. (It is straightforward to show that the two limits coincide and are independent of $t$.) Moreover, the mirror symmetry implies

$$\tilde{\phi}_\pm(x,t,k) = \sigma_3\phi_\mp(-x,t,-k)\,\sigma_3, \qquad k \in \mathbb{R}. \qquad (3.45)$$

Combining equations (3.44) and (3.45) one obtains a relation between the Jost solutions of the scattering problem on the full line with the extended potential:

$$\phi_\mp(-x,t,-k) = \sigma_3 B(x,t,k)\,\phi_\pm(x,t,k)\,B_\infty^{-1}(k)\sigma_3, \qquad k \in \mathbb{R}. \qquad (3.46)$$

Evaluating equation (3.46) at $x = 0$ and using equations (3.37), one also obtains a relation between the scattering matrix of the full-line problem and that of the half-line problem:

$$A_{\mathrm{ivp}}(k) = (2ik\sigma_3 - \beta)A^{-1}(-k)(2ik\sigma_3 - \alpha)^{-1}A(k), \qquad k \in \mathbb{R}, \qquad (3.47)$$

where $\beta = \lim_{x\to\infty} p(x,t)$. The entries of equation (3.47) yield the relations (3.31) between the corresponding scattering coefficients. Equation (3.32) can be obtained by evaluating the Bäcklund transformation (3.39) at $k = \pm i\alpha/2$ (namely, the values of $k$ for which $\det B(x,t,k) = 0$) and looking at the asymptotic behavior as $x \to \infty$.

# Bibliography

[1] M. J. Ablowitz, B. Prinari, and A. D. Trubatch, Discrete and Continuous Nonlinear Schrödinger Systems, London Math. Soc. Lecture Note Ser. **302**, Cambridge University Press (2003).

[2] M. J. Ablowitz and H. Segur, The inverse scattering transform: Semi-infinite interval, J. Math. Phys. **16** (1975), 1054–1056.

[3] R. F. Bikbaev and V. O. Tarasov, Initial-boundary value problem for the nonlinear Schrödinger equation, J. Phys. A **24** (1991), 2507–2516.

[4] G. Biondini and A. Bui, On the nonlinear Schrödinger equation on the half line with linearizable boundary conditions, Stud. Appl. Math. **129** (2012), 249–271.

[5] G. Biondini and G. Hwang, Solitons, boundary value problems and a nonlinear method of images, J. Phys. A **42** (2009), 205207.

[6] P. Deift and J. Park, Long-time asymptotics for solutions of the NLS equation with a delta potential and even initial data, Int. Math. Res. Not. IMRN **2011** (2011), 5505–5624.

[7] L. D. Faddeev and L. A. Takhtajan, Hamiltonian Methods in the Theory of Solitons, Springer, Berlin (1987).

[8] A. S. Fokas, An initial-boundary value problem for the nonlinear Schrödinger equation, Phys. D **35** (1989), pp 167–185.

[9] A. S. Fokas, A unified transform method for solving linear and certain nonlinear PDEs, Proc. Roy. Soc. Lond. Ser. A **453** (1997), 1411–1443.

[10] A. S. Fokas, On the integrability of certain linear and nonlinear partial differential equations, J. Math. Phys. **41** (2000), 4188–4237.

[11] A. S. Fokas, Integrable nonlinear evolution equations on the half line, Commun. Math. Phys. **230** (2002), 1–39.

[12] A. S. Fokas, The generalized Dirichlet-to-Neumann map for certain nonlinear evolution PDEs, Comm. Pure Appl. Math. **58** (2005), 639–670.

[13] A. S. Fokas, A. R. Its, and L-Y Sung, The nonlinear Schrödinger equation on the half-line, Nonlinearity **18** (2005), 1771–1822.

[14] A. S. Fokas and S. Kamvissis, Zero-dispersion limit for integrable equations on the half line with linearizable data, Abstr. Appl. Anal. **2004** (2004), 361–370.

[15] B. Gürel, M. Gürses, and I. T. Habibullin, Boundary value problems compatible with symmetries, Phys. Lett. A **190** (1994), 231–237.

[16] I. T. Habibullin, Integrable initial boundary value problem, Theoret. and Math. Phys. **86** (1991), 28–36.

[17] J. Holmer and M. Zvorski, Breathing pattern in nonlinear relaxation, Nonlinearity, **22** (2009), 1259–1301.

[18] A. Its and D. Shepelsky, Initial boundary value problem for the focusing NLS equation with Robin boundary condition: Half-line approach, Proc. Roy. Soc. A **469** (2013), 20120199.

[19] E. K. Skylanin, Boundary conditions for integrable equations, Funct. Anal. Appl. **21** (1987), 86–87.

[20] V. O. Tarasov, The integrable initial-boundary value problem on a semiline: Nonlinear Schrödinger equation and sine-Gordon equation, Inverse Problems **7** (1991), 435–449.

[21] V. E. Zakharov and A. B. Shabat, Exact theory of two-dimensional self-focusing and one-dimensional self-modulation of waves in nonlinear media, Sov. Phys. JETP **34** (1972), 62–69.

## Chapter 4

# Elliptic Problems: Linear

*A. C. L. Ashton*

The author is grateful for the support of Pembroke College, University of Cambridge.

## 4.1 ▪ Elliptic Boundary Value Problems via the Fokas Method

### 4.1.1 ▪ Introduction

The unified method for boundary value problems was introduced by A. S. Fokas in the late nineties [7, 6] and has since been applied to a variety of problems. It can be interpreted as the Fourier analogue of the classical boundary integral techniques. The analysis takes place in the "spectral space" rather than on the physical space on which the boundary value problem is posed. The unified approach also gives a concrete realization of the fundamental principle of Ehrenpreis [5], which states that solutions to constant coefficient PDEs on convex domains can be written as a superposition of exponential solutions.

The implementation of the unified method has been largely formal in nature. One would *assume* that the global relation determines the unknown boundary data. This assumption could then be checked a posteriori. A rigorous study of this method was initiated in [1, 2]. Here we give an overview of these results and demonstrate how they can be used to formulate effective numerical implementations of the unified method for elliptic boundary value problems.

### 4.1.2 ▪ The Global Relation for Arbitrary Convex Domains

Let $\Omega \subset \mathbf{R}^n$ be an arbitrary convex domain with boundary $\Gamma = \partial\Omega$. In this discussion we will confine our attention to the case in which $\Gamma$ is smooth, but we emphasize that the arguments presented can easily be adapted for Lipschitz domains.

We will concern ourselves with the class of elliptic boundary value problems

$$-\Delta u + \beta^2 u = 0 \quad \text{in } \Omega, \tag{4.1a}$$

$$u = f \quad \text{on } \Gamma. \tag{4.1b}$$

We will assume $f$ and its tangential derivative belong to $L^2(\Gamma)$, although lower regularity data can be assumed when $\Gamma$ is smooth. Our analysis includes the cases of Laplace ($\beta = 0$), Helmholtz ($\beta^2 < 0$), and modified Helmholtz ($\beta^2 > 0$). Associated with these boundary value problems is the symbol of the differential operator $P(D) = D^2 + \beta^2$, where $D = -i\partial$, and it is given by

$$P(\lambda) = \lambda^2 + \beta^2, \tag{4.2}$$

where $\lambda = (\lambda_1, \ldots, \lambda_n) \in \mathbf{C}^n$. Denote the zero set of this polynomial by $Z_P$, so that

$$Z_P = \{\lambda \in \mathbf{C}^n : P(\lambda) = 0\}. \tag{4.3}$$

We will often denote an arbitrary element of $Z_P$ by $\zeta$, so that $P(\zeta) = 0$.

A standard approach used to solve the boundary value problem (4.1) is to invoke so-called *boundary integral methods*. Rather than solve the problem for the function $u$ directly, one first constructs the *unknown* boundary values, in this case the Neumann data $\partial u/\partial \nu|_\Gamma$, where $\nu$ denotes the outward normal on $\Gamma$. Given the boundary value problem (4.1), the map which sends the known Dirichlet data $f$ to the unknown Neumann data $\partial u/\partial \nu|_\Gamma$ is called the Dirichlet–Neumann map. The Dirichlet–Neumann map is realized using the so-called *Steklov–Poincaré operator* $S$:

$$\left.\frac{\partial u}{\partial \nu}\right|_\Gamma = Sf. \tag{4.4}$$

This map is well understood. For instance, it is known that if $\Gamma$ is Lipschitz, then $S : H^{s+1/2}(\Gamma) \to H^{s-1/2}(\Gamma)$ for each $-\frac{1}{2} \leq s \leq \frac{1}{2}$ (see [9, Chapter 4] for a discussion of this and related results).

A concrete realization of the Steklov–Poincaré operator is found as follows. If $E_P(x,y)$ denotes the fundamental solution for the operator $P(D)$, i.e., $P(D)E_P(x,y) = \delta_y(x)$, then a careful application of Green's second identity gives rise to

$$u(x) = \int_\Gamma \left[ u(y)\frac{\partial E_P}{\partial \nu}(x,y) - E_P(x,y)\frac{\partial u}{\partial \nu}(y) \right] d\Gamma(y), \quad x \in \Omega, \tag{4.5}$$

for any function $u = u(x)$ such that $P(D)u = 0$ in $\Omega$. Assuming $u$ is a solution to the boundary value problem (4.1) and taking nontangential limits $x \to \Gamma$ one finds

$$f = \mathrm{DL}_{\Gamma,P}(f) + \tfrac{1}{2}f - \mathrm{SL}_{\Gamma,P}(g), \tag{4.6}$$

where $\mathrm{SL}_{\Gamma,P}$ and $\mathrm{DL}_{\Gamma,P}$ denote the operators associated with the single- and double-layer potentials for $(\Gamma, P)$ and $g \equiv \partial u/\partial \nu|_\Gamma$. Denoting these operators by $V$ and $K$, respectively, we find

$$Vg = -\left(\tfrac{1}{2}I - K\right)f. \tag{4.7}$$

The properties of the single-layer potential operator $V$ are well known [9]: in particular it is known that $V^{-1}$ is a bounded linear operator (between appropriate function spaces). Consequently, we have the representation of the Steklov–Poincaré operator

$$S = -V^{-1}\left(\tfrac{1}{2}I - K\right). \tag{4.8}$$

Using the well-established properties of the single- and double-layer potentials allows one to obtain $S : H^{s+1/2}(\Gamma) \to H^{s-1/2}(\Gamma)$ for Lipschitz domains for each $-\frac{1}{2} \leq s \leq \frac{1}{2}$.

Let us now discuss an entirely different approach. We again use Green's second identity, but now we use the exponential function $\exp(-\mathrm{i}\zeta\cdot x)$ with $\zeta\in Z_P$ in place of the fundamental solution $E_P$. Owing to the fact $P(D)\exp(-\mathrm{i}\zeta\cdot x)=0$ for $\zeta\in Z_P$ we find

$$0=\int_\Gamma e^{-\mathrm{i}\zeta\cdot x}\left[\mathrm{i}(\zeta\cdot\nu)f(x)+\frac{\partial u}{\partial\nu}(x)\right]\mathrm{d}\Gamma(x),\quad \zeta\in Z_P. \tag{4.9}$$

This is the global relation—if the solution to the boundary value problem (4.1) exists and is sufficiently regular, then this equation must hold for all $\zeta\in Z_P$. The question of importance is whether the converse holds: if we find a function $g$ on $\Gamma$ such that

$$0=\int_\Gamma e^{-\mathrm{i}\zeta\cdot x}\left[\mathrm{i}(\zeta\cdot\nu)f(x)+g(x)\right]\mathrm{d}\Gamma(x) \tag{4.10}$$

for each $\zeta\in Z_P$, is it necessarily true that $g=Sf$? Can the global relation be used in place of the classical boundary integral equations to effectively study the Dirichlet–Neumann map? The answer to both these questions is yes.

The proof that the global relation for linear elliptic PDEs completely describes the Dirichlet–Neumann map appeared in [1]. Here we sketch the proof—there are essentially three steps.

(a) Let $g$ and $f$ satisfy the global relation and consider the function

$$U(\lambda)=-\frac{1}{P(\lambda)}\int_\Gamma e^{-\mathrm{i}\lambda\cdot x}\left[\mathrm{i}(\lambda\cdot\nu)f(x)+g(x)\right]\mathrm{d}\Gamma(x),\quad \lambda\in\mathbf{C}^n. \tag{4.11}$$

It is clear that $U$ is holomorphic away from $Z_P$ and, owing to the global relation, it can be shown that $U$ extends to a holomorphic function on $Z_P$, i.e., it is entire. One can go further and show that $U$ satisfies an estimate of the form

$$|U(\lambda)|\lesssim(1+|\lambda|)\,e^{H_\Omega(\operatorname{Im}\lambda)}, \tag{4.12}$$

where $H_\Omega(x)=\sup_{y\in\Omega}(x\cdot y)$ is the supporting function of the domain $\Omega$.

(b) We now have an entire function $U=U(\lambda)$ which obeys the estimate (4.12) for every $\lambda\in\mathbf{C}^n$. Appealing to the famous Paley–Wiener–Schwartz theorem from the theory of Fourier analysis in several variables we conclude that there exists a distribution $u$ supported in the closure of $\Omega$ such that $U=\hat{u}$, where the hat denotes the (distributional) Fourier transform. So there is a compactly supported distribution $u$ such that

$$\hat{u}(\lambda)=-\frac{1}{P(\lambda)}\int_\Gamma e^{-\mathrm{i}\lambda\cdot x}\left[\mathrm{i}(\lambda\cdot\nu)f(x)+g(x)\right]\mathrm{d}\Gamma(x),\quad \lambda\in\mathbf{C}^n. \tag{4.13}$$

(c) Invoking the inverse Fourier transform for compactly supported distributions, we find

$$u(x)=\int_\Gamma\left[u(y)\frac{\partial E_P}{\partial\nu}(x,y)-E_P(x,y)g(y)\right]\mathrm{d}\Gamma(y),\quad x\in\Omega, \tag{4.14}$$

and zero for $x$ outside the closure of $\Omega$. Clearly $u$ satisfies $P(D)u=0$ in $\Omega$. By taking nontangential limits onto $\Gamma$ from outside and inside $\Omega$ we find

$$u_-=\mathrm{DL}_{\Gamma,P}(f)+\tfrac{1}{2}f-\mathrm{SL}_{\Gamma,P}(g),$$
$$u_+=\mathrm{DL}_{\Gamma,P}(f)-\tfrac{1}{2}f-\mathrm{SL}_{\Gamma,P}(g),$$

where $u^{\pm}$ denotes the limits taken from above and below $\Gamma$. Since $u$ is only supported in the closure of $\Omega$ we have $u^+ = 0$, and hence $u^- = f$. So $u$ is precisely the (unique) solution to the boundary value problem (4.1) and $g$ must correspond to the unknown Neumann data, i.e., $g = Sf$.

This establishes that the global relation completely classifies the Dirichlet–Neumann map for the boundary value problem (4.1). The question as to whether the global relation can be used *effectively* to study this map is a rather more involved question. This will be the subject of the next section. For simplicity, we address the problem only in the case of polygonal domains. However, the basic methodology can be extended to the more general case.

### 4.1.3 ▪ The Global Relation for Convex Polygons

We work on a polygon $\Omega \subset \mathbf{R}^2 \simeq \mathbf{C}$ with vertices $\{z_i\}_{i=1}^n$ and sides $\Gamma_i = (z_i, z_{i+1})$ with $z_{n+1} = z_1$. We write $\alpha_i = \arg(z_{i+1} - z_i)$ for the angle the side $\Gamma_i$ makes with the positive real axis and $\Delta_{ij} = \alpha_i - \alpha_j$. Also set $2\sigma_i = |\Gamma_i|$ for the length of the side $\Gamma_i$.

We are given real-valued boundary data $f_i \in H^1(\Gamma_i)$ for $i = 1, \dots, n$, meaning that $f_i$ and its first tangential derivative are square integrable along the edge $\Gamma_i$. In addition we impose continuity of the boundary data at the vertices—without this extra regularity at the vertices one cannot rule out the presence of nonintegrable singularities. We are interested in the Dirichlet–Neumann map associated with the classical Dirichlet problem

$$-\Delta u + 4\beta^2 u = 0 \quad \text{in } \Omega, \tag{4.15a}$$

$$u = f_i \quad \text{on } \Gamma_i \text{ for } i = 1, \dots, n. \tag{4.15b}$$

That is to say, we want to reconstruct the unknown Neumann boundary values from the known Dirichlet boundary data. The factor of 4 has been included for notational convenience. Set

$$v_\beta(z, \bar z; \lambda) = e^{-\mathrm{i}\lambda z + \mathrm{i}\beta^2 \bar z/\lambda}, \quad \lambda \in \mathbf{C} \setminus \{0\}, \tag{4.16}$$

where $z = x + \mathrm{i}y$, $(x, y)$ being the usual Cartesian coordinates on $\mathbf{R}^2$. Then $v_\beta$ is of the form $\exp(-\mathrm{i}\zeta \cdot x)$ with $\zeta \in Z_P$, and it satisfies (4.15a) for all $\lambda \in \mathbf{C} \setminus \{0\}$. Indeed

$$v_\beta = \exp\left[-\mathrm{i}\zeta_1(\lambda)x - \mathrm{i}\zeta_2(\lambda)y\right], \tag{4.17}$$

where

$$\zeta_1(\lambda) = \lambda - \frac{\beta^2}{\lambda}, \quad \zeta_2(\lambda) = \mathrm{i}\left(\lambda + \frac{\beta^2}{\lambda}\right). \tag{4.18}$$

Clearly $\zeta_1^2 + \zeta_2^2 + 4\beta^2 = 0$. As in the previous section we set $g \equiv \partial u/\partial \nu|_\Gamma$. Suppressing the $(\bar z, z)$ dependence the global relation becomes

$$\int_\Gamma v_\beta(\lambda) g \, \mathrm{d}\Gamma = F(\lambda), \tag{4.19}$$

where the function $F$ is known and determined completely by the Dirichlet data on each side $\Gamma_i$. Explicitly

$$F(\lambda) \equiv -\mathrm{i} \int_\Gamma (\zeta \cdot \nu) v_\beta(\lambda) f \, \mathrm{d}\Gamma. \tag{4.20}$$

It will be convenient to have a local description of the edges $\Gamma_i$. Let us introduce the local parametrizations $\psi_i : [-\sigma_i, \sigma_i] \to \Gamma_i$ with

$$\psi_i(\tau) = \frac{1}{2\sigma_i}\Big[(\sigma_i+\tau)z_{i+1} + (\sigma_i-\tau)z_i\Big] \equiv m_i + \tau e^{\mathrm{i}\alpha_i}, \tag{4.21}$$

where $m_i = \frac{1}{2}(z_i + z_{i+1})$ is the midpoint of the side $\Gamma_i$. For a function $f : \Gamma_i \to \mathbb{C}$ we write its pullback by $\psi_i$ by $\psi_i^*(f)(\tau) = f(\psi(\tau))$. Let us set $\varphi_i(\tau) = \psi_i^*(\partial u/\partial \nu)(\tau)$ for the pullback of the unknown Neumann boundary values on $\Gamma_i$. The global relation (4.19) now becomes

$$\sum_{j=1}^{n} e^{-\mathrm{i}\lambda m_j + \mathrm{i}\beta^2 \bar{m}_j/\lambda} \hat{\varphi}_j\left(\lambda e^{\mathrm{i}\alpha_j} - \frac{\beta^2}{\lambda e^{\mathrm{i}\alpha_j}}\right) = F(\lambda), \quad \lambda \in \mathbb{C}\setminus\{0\}, \tag{4.22}$$

where we have introduced the Fourier transform

$$\mathscr{F} : \varphi_i \mapsto \hat{\varphi}_i(\lambda) = \int_{-\sigma_i}^{\sigma_i} e^{-\mathrm{i}\lambda\tau}\varphi_i(\tau)\,d\tau. \tag{4.23}$$

It will be convenient to have a more symmetric form of the global relation. To this end we multiply (4.22) by $e^{\mathrm{i}\lambda m_i - \mathrm{i}\beta^2 \bar{m}_i/\lambda}$ for some $1 \le i \le n$ and replace $\lambda$ with $\lambda e^{-\mathrm{i}\alpha_i}$. We obtain the $n$ equations

$$\sum_{j=1}^{n} e^{\mathrm{i}e^{-\mathrm{i}\alpha_i}(m_i - m_j)\lambda - \mathrm{i}e^{\mathrm{i}\alpha_i}(\bar{m}_i - \bar{m}_j)\beta^2/\lambda}\hat{\varphi}_i\left(\lambda e^{-\mathrm{i}\Delta_{ij}} - \frac{\beta^2}{\lambda e^{-\mathrm{i}\Delta_{ij}}}\right) = e^{\mathrm{i}\lambda m_i - \mathrm{i}\beta^2\bar{m}_i/\lambda}F(e^{-\mathrm{i}\alpha_i}\lambda) \tag{4.24}$$

for $i = 1, \ldots, n$, each of which must hold for all $\lambda \in \mathbb{C}\setminus\{0\}$. Introduce the vectors $\Phi(\lambda) = (\hat{\varphi}_1(\lambda), \ldots, \hat{\varphi}_n(\lambda))^t$ and $\Psi(\lambda)$ defined by

$$\Psi_i(\lambda) = e^{\mathrm{i}\lambda m_i - \mathrm{i}\beta^2\bar{m}_i/\lambda}F(e^{-\mathrm{i}\alpha_i}\lambda), \quad i = 1, \ldots, n, \tag{4.25}$$

and the operator $T(\beta)$ defined by

$$[T(\beta)\Phi]_i(\lambda) = \sum_{j=1}^{n} e^{\mathrm{i}e^{-\mathrm{i}\alpha_i}(m_i - m_j)\lambda - \mathrm{i}e^{\mathrm{i}\alpha_i}(\bar{m}_i - \bar{m}_j)\beta^2/\lambda}\Phi_i\left(\lambda e^{-\mathrm{i}\Delta_{ij}} - \frac{\beta^2}{\lambda e^{-\mathrm{i}\Delta_{ij}}}\right), \tag{4.26}$$

for $i = 1, \ldots, n$. Then the global relation is simply

$$[T(\beta)\Phi](\lambda) = \Psi(\lambda), \quad \lambda \in \mathbb{C}\setminus\{0\}. \tag{1.27}$$

Our aim will be to analyze this linear equation for $\Phi$ and hence $g$. This can be viewed as an abstract operator equation $T(\beta)\Phi = \Psi$, and to proceed we must analyze the functional analytic properties of the operator $T(\beta)$.

### 4.1.4 ▪ Analysis of the Operator $T(\beta)$

In this section we prove certain functional analytic properties of the operator $T(\beta)$ which will be relevant to solving the global relation (4.27). These results are taken from [2, 3, 4]. Before we do this let us briefly recap some basic results from operator theory.

Given a bounded linear operator between Hilbert spaces $A : H_1 \to H_2$ we consider the linear problem $A\mathbf{x} = \mathbf{y}$, where $\mathbf{x} \in H_1$ is unknown and $\mathbf{y} \in H_2$ is known. The three questions one needs to consider are the following:

(a) Does a solution exist?

(b) Is the solution unique?

(c) Does the solution depend continuously on the known data?

These three questions can be answered by demonstrating that (a) $\mathbf{y} \in R(A)$, where $R(A)$ denotes the range of $A$; (b) $A$ is injective; (c) $R(A)$ is closed. Injectivity is usually the simplest to prove, i.e., if $A\mathbf{x} = 0$, then $\mathbf{x} = 0$, so we will assume this is the case for the sake of this discussion. The remaining problems are usually more difficult.

If one establishes that $R(A)$ is closed, then much of the work is done. Indeed, from Banach's closed range theorem we know that $R(A)$ is closed if and only if

$$R(A) = N(A^*)^{\perp} \equiv \{\mathbf{y} \in H_2 : \langle \mathbf{y}, \mathbf{z} \rangle_{H_2} = 0 \text{ for every } \mathbf{z} \in N(A^*)\}, \tag{4.28}$$

where $A^*$ denotes the adjoint of $A$ and $N(A^*)$ denotes its null space. It is often easy to check whether $\mathbf{y} \in N(A^*)^{\perp}$, so existence can be established. We also then have continuity: since $R(A) \subseteq H_2$ is closed, it follows that

$$H_3 = \left(R(A), \langle \cdot, \cdot \rangle_{H_2}\right) \tag{4.29}$$

is a Hilbert space in its own right and $A : H_1 \to H_3$ is a continuous bijection. By Banach's bounded inverse theorem $A^{-1} : H_3 \to H_1$ is also bounded, i.e., continuous. So the solution depends continuously on the known data, answering (c).

It is possible to deal with questions (a) and (c) in one fell swoop if one can prove that the operator $A$ is *bounded below*, i.e., $||A\mathbf{x}||_{H_2} \gtrsim ||\mathbf{x}||_{H_1}$ for all $\mathbf{x} \in H_1$. Indeed, if $A$ is bounded below, then it is clearly injective. What is more, if $\{\mathbf{y}_n = A\mathbf{x}_n\}_{n\geq 1}$ is a sequence in $R(A)$ such that $\mathbf{y}_n \to \mathbf{y} \in H_2$, then we have

$$||\mathbf{x}_n - \mathbf{x}_m||_{H_1} \lesssim ||A\mathbf{x}_n - A\mathbf{x}_m||_{H_2} = ||\mathbf{y}_n - \mathbf{y}_m||_{H_2}. \tag{4.30}$$

And since $\{\mathbf{y}_n\}_{n\geq 1}$ converges it follows that the sequence $\{\mathbf{x}_n\}_{n\geq 1}$ is Cauchy, so it converges by the completeness of $H_1$. We deduce that $\mathbf{y} = A\mathbf{x}$ for some $\mathbf{x} \in H_1$, and hence $R(A)$ is closed.

We will now address these problems for the operator $T(\beta)$. We will show that on specified function spaces, the operator $T(\beta)$ is bounded below for all but countably many $\beta$ which are related to the Dirichlet eigenvalues of the Laplacian. This is contained in Theorem 4.1, which answers questions (b) and (c) for the operator $T(\beta)$. Our first task is to establish the relevant function spaces to work on.

The components of our unknown vector $\Phi$ consist of the Fourier transform of real-valued functions supported on an interval of the form $[-\sigma, \sigma]$. It is natural to work with the Paley–Wiener spaces

$$PW^{\sigma}_{\text{sym}} = \mathscr{F} L^2_{\mathbf{R}}[-\sigma, \sigma] \tag{4.31}$$

consisting of the Fourier transform of real, square-integrable functions supported on $[-\sigma, \sigma]$. As a consequence of the classical Paley–Wiener theorem, this space can be identified with

$$\left\{ f : \mathbf{C} \to \mathbf{C} \text{ entire}, \int_{-\infty}^{\infty} |f(x)|^2 \, dx < \infty, \ |f(\lambda)| \lesssim e^{\sigma|\lambda|}, \ f(\lambda) = f^{\star}(\lambda) \right\}, \tag{4.32}$$

where $f^{\star}(\lambda) \equiv \overline{f(-\bar{\lambda})}$. We set

$$X = PW^{\sigma_1}_{\text{sym}} \times \cdots \times PW^{\sigma_n}_{\text{sym}} \tag{4.33}$$

and note that $X \simeq L^2(\Gamma)$. The norm on $X$ is defined by

$$||\Phi||_X^2 = \sum_{i=1}^{n} ||\Phi_i||_{2,-}^2, \tag{4.34}$$

where here and throughout $||\cdot||_{2,-}$ denotes the usual $L^2$-norm on the negative real axis. Due to the symmetry $\Phi^* = \Phi$ for $\Phi \in X$, this norm is equivalent to that in which the integrals run over the whole real axis. Finally set $I = (-\infty, -1)$ and let $Y = L^2(I)^{\times n}$.

The main result in this section is contained in the following theorem.

**Theorem 4.1.** *For $\beta \in \mathbb{C}$ the operator $T(\beta) : X \to Y$ is upper semi-Fredholm. In particular, if $4\beta^2$ is not a Dirichlet eigenvalue of the Laplacian, then $T(\beta)$ is bounded below, i.e., $||T(\beta)\Phi||_Y \gtrsim_\beta ||\Phi||_X$ for all $\Phi \in X$.*

Recall that a linear operator $A : X \to Y$ between Banach spaces is called upper semi-Fredholm if the range of $A$, $R(A)$, is closed in $Y$ and the null space, $N(A)$, is finite dimensional.

We will need several lemmas to prove our main theorem. The following pseudo-compactness lemma [2, Lemma 2] will be of central importance.

**Lemma 4.2.** *Any norm-bounded sequence in $X$ contains a subsequence that converges pointwise and locally uniformly to an element of $X$ that obeys the same norm bound.*

**Lemma 4.3.** *For $\beta \in \mathbb{C}$ the operator $T(\beta) : X \to Y$ is continuous.*

***Proof.*** We only sketch the argument, for it is similar to that found in [2, Lemma 3]. For $\Phi \in X$ it is clear that $[T(\beta)\Phi](\lambda)$ is analytic on $I$, so we need only look at the large $|\lambda|$ behavior. We may write

$$\begin{aligned} &e^{\mathrm{i}e^{-\mathrm{i}\alpha_i}(m_i - m_j)\lambda - \mathrm{i}e^{\mathrm{i}\alpha_i}(\bar{m}_i - \bar{m}_j)\beta^2/\lambda}\Phi_i\left(\lambda e^{-\mathrm{i}\Delta_{ij}} - \frac{\beta^2}{\lambda e^{-\mathrm{i}\Delta_{ij}}}\right) \\ &\qquad = \int_{-\sigma_j}^{\sigma_j} e^{\mathrm{i}e^{-\mathrm{i}\alpha_i}(m_i - m_j - \tau e^{\mathrm{i}\alpha_j})\lambda - \mathrm{i}e^{\mathrm{i}\alpha_i}(\bar{m}_i - \bar{m}_j - \tau e^{-\mathrm{i}\alpha_j})\beta^2/\lambda}\hat{\Phi}_j(-\tau)\,\mathrm{d}\tau. \end{aligned}$$

We may effectively treat $\beta$ as zero since the terms involving $\beta$ do not affect the large $|\lambda|$ behavior. By convexity one sees that for $\tau \in [-\sigma_j, \sigma_j]$ we have

$$0 \le \arg\left(e^{-\mathrm{i}\alpha_i}(m_j + \tau e^{\mathrm{i}\alpha_j} - m_i)\right) \le \pi, \tag{4.35}$$

and the inequality is strict if $j \ne i, i \pm 1$. Applying Cauchy–Schwarz and using $\hat{\Phi}_j \in L^2[-\sigma_j, \sigma_j]$, we see that for $j \ne i, i \pm 1$ this integral is exponentially small in $|\lambda|$ as $\lambda \to -\infty$. It remains to estimate

$$\left| \sum_{j=i,i\pm 1} e^{\mathrm{i}e^{-\mathrm{i}\alpha_i}(m_i - m_j)\lambda - \mathrm{i}e^{\mathrm{i}\alpha_i}(\bar{m}_i - \bar{m}_j)\beta^2/\lambda}\Phi_i\left(\lambda e^{-\mathrm{i}\Delta_{ij}} - \frac{\beta^2}{\lambda e^{-\mathrm{i}\Delta_{ij}}}\right)\right|. \tag{4.36}$$

The $i = j$ term is simply $\Phi(\lambda + \beta^2/\lambda)$, so the relevant norm in this case is

$$\int_I \left|\Phi_i\left(\lambda - \frac{\beta^2}{\lambda}\right)\right|^2 \mathrm{d}\lambda. \tag{4.37}$$

Again it is straightforward to demonstrate that this is dominated by $||\Phi_i||_2^2$. For the $j = \pm i$ terms, we again effectively treat $\beta$ as zero. In this case the relevant terms are, after a simple change of variables, of the form

$$e^{i\sigma_j \lambda} \int_0^{2\sigma_j} e^{-|\lambda| w_j \tau} h_j(\tau)\,d\tau \tag{4.38}$$

for appropriate $h_j, w_j$ with $\operatorname{Re} w_j > 0$. These correspond to the Laplace transform (along a ray) of a compactly supported, square-integrable function. It is well known that the Laplace transform defined a bounded linear map from $L^2(0,\infty)$ to itself, from which it is straightforward to deduce that the $L^2(I)$-norm of each of these integrals is bounded by $||\Phi_j||_{2,-}^2$. We conclude that

$$\int_I |[T(\beta)\Phi]_i(\lambda)|^2\,d\lambda \lesssim \sum_{j=1}^n ||\Phi_j||_{2,-}^2 \tag{4.39}$$

for each $i = 1, \ldots, n$, and hence $||T(\beta)\Phi||_Y \lesssim ||\Phi||_X$. □

It will be convenient to split $T(\beta)$ as

$$T(\beta) = T(0) + \left[T(\beta) - T(0)\right] \equiv T(0) + K(\beta). \tag{4.40}$$

**Lemma 4.4.** *$T(0) : X \to Y$ is injective and has closed range.*

The proof of this result follows largely from the ideas used in [2, Theorem 3]. There it was shown that

$$\sum_{i=1}^n \int_{-\infty}^0 |[T(0)\Phi]_i(\lambda)|^2\,d\lambda \gtrsim ||\Phi||_X^2, \tag{4.41}$$

and we will use this in our proof.

***Proof.*** It is well known that an operator between Banach spaces is bounded below if and only if it is injective and has closed range. We proceed to show that $T(0)$ is bounded below, i.e., $||T(0)\Phi||_Y \gtrsim ||\Phi||_X$ for all $\Phi \in X$.

Suppose $T(0)$ is not bounded below. Then there exists a sequence of unit vectors $\{\Phi_m\}_{m\geq 1}$ in $X$ such that $T(0)\Phi_m \to 0$ in $Y$. By Lemma 4.2 (and going to a subsequence if necessary) this sequence converges locally uniformly to some $\Phi \in X$ with $||\Phi||_X \leq 1$. Since $T(0)\Phi_m \to 0$ in $Y$, it follows that $T(0)\Phi_m(\lambda) \to 0$ almost everywhere on $I$. Consequently, $T(0)\Phi(\lambda) = 0$ almost everywhere on $I$ and on all of C by analytic continuation. Using (4.41) we deduce that $\Phi = 0$, i.e., $\Phi_m(\lambda) \to 0$ locally uniformly. Hence

$$\lim_{m\to\infty} \sum_{i=1}^n \int_{-1}^0 |[T(0)\Phi_m]_i(\lambda)|^2\,d\lambda = 0. \tag{4.42}$$

Using (4.41) and the fact that $T(0)\Phi_m \to 0$ in $Y$ we find

$$\lim_{m\to\infty} ||\Phi_m||_X^2 \lesssim \lim_{m\to\infty} \left[ ||T(0)\Phi_m||_Y^2 + \sum_{i=1}^n \int_{-1}^0 |[T(0)\Phi_m]_i(\lambda)|^2\,d\lambda \right] = 0, \tag{4.43}$$

contradicting $||\Phi_m||_X = 1$. Hence $T(0) : X \to Y$ is bounded below. □

**Lemma 4.5.** *$K(\beta): X \to Y$ is compact.*

***Proof.*** Let $\{\Phi_m\}_{m\geq 1}$ be a bounded sequence in $X$. We will show that $\{K(\beta)\Phi_m\}_{m\geq 1}$ contains a convergent subsequence. Going to a subsequence if necessary, we know that $\Phi_m(\lambda) \to \Phi(\lambda)$ locally uniformly for some $\Phi \in X$ with $||\Phi||_X \leq ||\Phi_m||_X$. Set $\Psi_m = \Phi_m - \Phi$. We claim $K(\beta)\Psi_m \to 0$ in $Y$. The $i$th component of $[K(\beta)\Psi_m](\lambda)$ is

$$\sum_{j=1}^{n} \int_{-\sigma_j}^{\sigma_j} e^{ie^{-i\alpha_i}(m_i - m_j - e^{i\alpha_j}\tau)\lambda} \left[ e^{-ie^{i\alpha_i}(\bar{m}_i - \bar{m}_j - e^{-i\alpha_j}\tau)\beta^2/\lambda} - 1 \right] (\hat{\Psi}_m)_j(-\tau)\, \mathrm{d}\tau. \tag{4.44}$$

We have the elementary *uniform* (in $\lambda$) estimate

$$\left| e^{-ie^{i\alpha_i}(\bar{m}_i - \bar{m}_j - e^{-i\alpha_j}\tau)\beta^2/\lambda} - 1 \right| \lesssim_\beta \frac{1}{|\lambda|} \tag{4.45}$$

for $\lambda \in I$. Using Cauchy–Schwarz we deduce

$$|[K(\beta)\Psi_m]_i(\lambda)|^2 \lesssim_\beta \frac{1}{|\lambda|^2} \tag{4.46}$$

for $\lambda \in I$. Since the right-hand side is integrable on $I$ and $[K(\beta)\Psi_m](\lambda)$ tends to zero locally uniformly, the dominated convergence theorem implies $||K(\beta)\Psi_m||_Y \to 0$, so $\{K(\beta)\Phi_m\}_{m\geq 1}$ has a convergent subsequence. Hence $K(\beta): X \to Y$ is compact. □

***Proof of Theorem* 4.1.** By Lemma 4.4 we know that $T(0): X \to Y$ has closed range and $N(T(0)) = \{0\}$, so it is clearly upper semi-Fredholm. From Lemma 4.5 it follows that

$$T(\beta) = T(0) + K(\beta) \tag{4.47}$$

is a compact perturbation of an upper semi-Fredholm operator. Recalling standard results from perturbation theory of linear operators [8, Theorem 5.26] we deduce that $T(\beta): X \to Y$ is also upper semi-Fredholm. If $4\beta^2$ is not an eigenvalue, then $T(\beta)$ is injective with closed range and hence bounded below. □

With these results it is now possible to treat $T(\beta)$ as a Fredholm operator between $X$ and $Z$ for some Banach space $Z \neq Y$. Since the range of $T(\beta)$, $R(T(\beta))$, is closed in $Y$, the normed space

$$Z_\beta = (R(T(\beta)), ||\cdot||_Y) \tag{4.48}$$

is also a Banach space and clearly $T(\beta): X \to Z_\beta$ is a surjection. So in this case $T(\beta)$ becomes a Fredholm operator with index equal to its nullity. In particular, when $4\beta^2$ is not a Dirichlet eigenvalue of the Laplacian the operator $T(\beta): X \to Z_\beta$ is a Fredholm operator of index zero, which is particularly well understood.

### 4.1.5 ▪ A New Numerical Approach

We know that if $4\beta^2$ is not a Dirichlet eigenvalue of the Laplacian on the convex polygon $\Omega$, then the solution to the boundary value problem (4.15) can be reduced to the problem $T(\beta)\Phi = \Psi$. This equation will have a unique solution that depends continuously on the boundary data if $\Psi \in R(T(\beta))$.

It turns out that for boundary data with $f_i \in H^1(\Gamma_i)$ for $i = 1, \ldots, n$ which are continuous at the vertices, one can show that $\Psi$, as defined in (4.25), belongs to $N(T(\beta)^*)^\perp$.

Since $R(T(\beta))$ is closed, this immediately tells us $\Psi \in R(T(\beta))$, and so the solution to the problem $T(\beta)\Phi = \Psi$ for $\Phi \in X$ is guaranteed. Here we address the very practical problem of obtaining (approximations) to this $\Phi \in X$, given the vector $\Psi$.

It will be advantageous to first express our problem in terms of a weak formulation. The weak of the problem $T(\beta)\Phi = \Psi$ is given by

$$\langle T(\beta)\Phi, \Theta' \rangle_Z = \langle \Psi, \Theta' \rangle_Z \quad \forall \Theta' \in Z_\beta. \tag{4.49}$$

If we define the bilinear form $a : X \times X \to \mathbb{C}$ by

$$a_\beta(\Phi, \Theta) = \langle T(\beta)\Phi, T(\beta)\Theta \rangle_Z \tag{4.50}$$

and the linear form $\ell(\Theta) = \langle \Psi, T(\beta)\Theta \rangle_Z$, then the weak formulation can be written as

$$a_\beta(\Phi, \Theta) = \ell(\Theta) \quad \forall \Theta \in X. \tag{4.51}$$

Owing to the results of Theorem 4.1 the bilinear form $a$ is bounded and coercive, i.e.,

$$\text{(a) } |a_\beta(\Phi, \Theta)| \lesssim_\beta ||\Phi||_X ||\Theta||_X, \quad \text{(b) } |a_\beta(\Phi, \Phi)| \gtrsim_\beta ||\Phi||_X^2 \tag{4.52}$$

for all $\Phi, \Theta \in X$. Boundedness in (a) follows from the definition of the bilinear form $a$ and the Cauchy–Schwarz inequality, whereas (b) follows from the fact that $T(\beta)$ is bounded below

$$|a_\beta(\Phi, \Phi)| = ||T(\beta)\Phi||_Z^2 \gtrsim_\beta ||\Phi||_X^2 \tag{4.53}$$

by Theorem 4.1. An application of the Lax–Milgram lemma gives rise to the following.

**Theorem 4.6.** *The weak problem* (4.51) *admits a unique solution.*

To obtain a numerical approximation to this solution we can implement a standard Galerkin approach. Let $X_n$, $n = 1, 2, \ldots$, be finite-dimensional approximations to $X$ with bases $\{\mathbf{e}_i\}_{i=1}^n$. We now seek a solution $\Phi^n = \sum_{i=1}^n c_i \mathbf{e}_i$ to the weak problem (4.51) for $\Theta$ restricted to $X_n$. This gives rise to the linear algebra problem

$$A_{ij} c_j = l_i \quad 1 \le i, j \le n, \tag{4.54}$$

where $A_{ij} = a_\beta(\mathbf{e}_i, \mathbf{e}_j)$ and $l_i = \ell(\mathbf{e}_i)$. The matrix $A$ is nondegenerate, owing to the fact that $a$ is coercive on $X$ and hence also on each $X_n$. So we can solve this linear algebra problem for the unknown coefficients $c_i$. The resulting solution $\Phi^n$ is guaranteed to approximate the true solution $\Phi$ to the full problem. This follows from Céa's lemma. The proof of this lemma is simple enough to reproduce here. Let $\Theta^n$ denote an arbitrary element of $X_n$. Using the coercivity and boundedness of $a$ we have

$$\begin{aligned}
||\Phi - \Phi^n||_X^2 &\lesssim a(\Phi - \Phi^n, \Phi - \Phi^n) \\
&= a(\Phi - \Phi^n, \Phi - \Theta^n) + a(\Phi - \Phi^n, \Theta^n - \Phi^n) \\
&= a(\Phi - \Phi^n, \Phi - \Theta^n) \\
&\lesssim ||\Phi - \Phi^n||_X ||\Phi - \Theta^n||_X,
\end{aligned}$$

where in the second line we used the fact $a(\Phi - \Phi^n, \Theta^n - \Phi^n) = 0$ for all $\Theta^n, \Phi^n \in X_n$. This follows from the observation

$$a(\Phi - \Phi^n, \Theta^n) = a(\Phi, \Theta^n) - a(\Phi^n, \Theta^n) = \ell(\Theta^n) - \ell(\Theta^n) = 0 \tag{4.55}$$

for every $\Theta^n \in X_n$. So we arrive at the final estimate

$$||\Phi - \Phi^n||_X \lesssim ||\Phi - \Theta^n||_X \quad \text{for each } \Theta^n \in X_n. \tag{4.56}$$

This tells us the error between $\Phi^n$ and the true solution $\Phi$ is bounded by the best approximation for $\Phi$ in the subspace $X_n$. Consequently this error can be made arbitrarily small by using sufficiently large $n$.

The spectral convergence of this Galerkin scheme has been demonstrated in [3].

# Bibliography

[1] A. C. L. Ashton, On the rigorous foundations of the Fokas for linear elliptic PDEs, Proc. Roy. Soc. A **468** (2012), 1325–1331.

[2] A. C. L. Ashton, The spectral Dirichlet–Neumann map for Laplace's equation on a convex polygon, SIAM J. Math. Anal. **45** (2013), 3575–3591.

[3] A. C. L. Ashton and K. Crooks, Numerical study of the spectral Dirichlet-Neumann map (submitted).

[4] A. C. L. Ashton & K. Crooks, Eigenvalues of the Laplacian on convex polygons via the unified method (submitted).

[5] L. Ehrenpreis. Fourier Analysis in Several Complex Variables, Wiley-Interscience, New York, 1970.

[6] A. S. Fokas, Two–dimensional linear partial differential equations in a convex polygon, Proc. Roy. Soc. A **457** (2006), 371–393.

[7] A. S. Fokas, A Unified Approach to Boundary Value Problems, SIAM (2008).

[8] T. Kato, Perturbation Theory for Linear Operators, Springer-Verlag, New York, 1995.

[9] W. C. H. McLean, Strongly Elliptic Systems and Boundary Integral Equations, Cambridge University Press, Cambridge, UK, 2000.

# Chapter 5
# Elliptic Problems: Nonlinear

*B. Pelloni*

## 5.1 ▪ The Elliptic sine-Gordon Equation: A Nonlinear Elliptic Integrable PDE

### 5.1.1 ▪ Introduction

The elliptic sine-Gordon equation is the nonlinear PDE

$$q_{xx} + q_{yy} = \sin q, \quad q = q(x,y), \; x,y \in \mathbb{R}. \tag{5.1}$$

This equation appears as a mathematical model of several physical phenomena. For example, it was derived in the context of the theory of Josephson effects, superconductors, and spin waves in ferromagnets; see [35].

From a strictly mathematical point of view, the elliptic sine-Gordon equation is the simplest and most fundamental example of an nonlinear elliptic *integrable* PDE, in the sense of admitting a Lax pair formulation. Therefore the tools developed in connection with the so-called inverse scattering transform can be used for its analysis, at least in principle. The fundamental problems with using inverse scattering approach is that the problems posed for this PDE are always *boundary value problems*, and hence one has to turn instead to the unified transform approach, introduced by Fokas in the late 1990s.

The inverse scattering analysis for this integrable equation had been considered in [7] for a problem posed on $\mathbb{R}^2$ with prescribed periodic behavior at infinity, and in [35] for the problem on a half plane $y \geq 0$, with all boundary values prescribed, but constrained by a certain nonlinear relation. Due to the limitations of the inverse scattering approach when dealing with boundary value problems, in these works the solution is not constructed effectively. Special exact solutions for the problem posed in the whole of $\mathbb{R}^2$ were found in the 1980s by various authors (see the references in [35]).

Recent progress was made using the unified transform [43, 45, 29, 30]. In this series of works, the equation was considered on a half plane, quarter plane, and semistrip. Particular cases of boundary conditions that yield an explicit representation for the solution were identified and analyzed. In this paper, we summarize this recent progress to underline the features specific to this nonlinear elliptic case, and we give a new classification of boundary conditions on the semistrip that satisfy a necessary condition for yielding that a boundary value problem can be effectively linearized. This classification is based on formulating the equation in terms of an alternative Lax pair.

## 5.1.2 ▪ Lax Pair/Divergence Formulation

The elliptic sine-Gordon equation is the compatibility condition of the following Lax pair for the $2 \times 2$ matrix-valued function $\Psi(x,y,\lambda)$, $\lambda \in \mathbb{C}$:

$$\Psi_x + \frac{\Omega(\lambda)}{2}[\sigma_3, \Psi] = Q(x,y,\lambda)\Psi, \tag{5.2}$$

$$\Psi_y + \frac{\omega(\lambda)}{2}[\sigma_3, \Psi] = iQ(x,y,-\lambda)\Psi, \tag{5.3}$$

where

$$\Omega(\lambda) = \frac{1}{2i}\left(\lambda - \frac{1}{\lambda}\right), \qquad \omega(\lambda) = \frac{1}{2}\left(\lambda + \frac{1}{\lambda}\right), \tag{5.4}$$

$$Q(x,y,\lambda) = \frac{i}{4}\begin{pmatrix} \frac{1}{\lambda}(1-\cos q) & q_x - iq_y + \frac{i\sin q}{\lambda} \\ [q_x - iq_y - \frac{i\sin q}{\lambda} & -\frac{1}{\lambda}(1-\cos q) \end{pmatrix}, \qquad q = q(x,y). \tag{5.5}$$

Equations (5.2) and (5.3) can be written as the single equation

$$d\left(e^{(\Omega(\lambda)x+\omega(\lambda)y)\frac{\widehat{\sigma_3}}{2}}\right)\Psi(x,y,\lambda) = W(x,y,\lambda), \tag{5.6}$$

where the differential form $W$ is given by

$$W(x,y,\lambda) = e^{(\Omega(\lambda)x+\omega(\lambda)y)\frac{\widehat{\sigma_3}}{2}}\left(Q(x,y,\lambda)\Psi(x,y,\lambda)dx + iQ(x,y,-\lambda)\Psi(x,y,\lambda)dy\right), \tag{5.7}$$

and $\widehat{\sigma_3}$ acts on a $2 \times 2$ matrix $A$ by

$$\widehat{\sigma_3}A = [\sigma_3, A], \text{ hence } e^{\widehat{\sigma_3}x}A = e^{\sigma_3 x}Ae^{-\sigma_3 x} = \begin{pmatrix} A_{11} & e^{2x}A_{12} \\ e^{-2x}A_{21} & A_{22} \end{pmatrix}.$$

Equation (5.6) implies that if $Q(x,y)$ is a solution of equation (5.1) in a given simply connected domain $D$, then the following equation characterizes a function $\Psi_j(x,y,\lambda)$ which solves both equations (5.2) and (5.3):

$$\Psi_j(x,y,\lambda) = I + \int_{(x_j,y_j)}^{(x,y)} e^{-(\Omega(\lambda)(x-\xi)+\omega(\lambda)(y-\eta))\widehat{\sigma_3}} W_j(\xi,\eta,\lambda), \quad (x,y),\,(x_j,y_j) \in D, \tag{5.8}$$

where $(x_j, y_j)$ is a fixed point in $D$ and $W_j$ is the differential form defined by (5.7) with $\Psi$ replaced by $\Psi_j$ and $(x,y)$ replaced by $(\xi,\eta)$. It is shown in [20] that if $D$ is the interior of a convex polygon, then the proper choice for $(x_j, y_j)$ is the collection of all the vertices of the polygon.

The closure of the form $W$ implies that its integral along the boundary of this (or any convex) polygon must be zero. The resulting identity is called the *global relation*.

### 5.1.3 ▪ The Half Plane $y > 0$

In this section we consider equation (5.1) posed in the upper-half plane, i.e., in the domain $\{-\infty < x < \infty, 0 < y < \infty\}$. On this domain, a problem is well-posed (i.e., it admits a unique solution) if one condition is prescribed for $y = 0$, and appropriate smoothness and decay at infinity are assumed.

The first advantage of the unified transform approach is the fact that it algorithmically, and rather naturally, leads to the analysis the global relation. This relation allows us to characterize any pair of functions $g_0(x)$ $g_1(x)$ that constitute the boundary values at $y = 0$ of a function $q(x,y)$ solving equation (5.1) in the half plane. Indeed, this can be the case only if certain transforms of these functions satisfy the global relation.

**Definition 5.1 (Spectral functions).** *Given a pair of functions $g_0(x)$, $g_1(x)$ such that $g_0 - 2\pi m \in \mathbf{H}^1(\mathbb{R})$, $m \in \mathbb{Z}$, and $g_1(x) \in \mathbf{H}^1(\mathbb{R})$, we define the associated* spectral functions $a(\lambda)$ *and $b(\lambda)$ by*

$$a(\lambda) = 1 - \frac{1}{4}\int_{-\infty}^{\infty} \left\{ \frac{i}{\lambda}(1-\cos g_0(\xi))m_1(\xi,\lambda) + \left[ -\frac{1}{\lambda}\sin g_0(\xi) + i\dot{g}_0(\xi) + g_1(\xi) \right] m_2(\xi,\lambda) \right\} d\xi,$$
$$\operatorname{Im}\lambda \geq 0, \tag{5.9}$$

$$b(\lambda) = -\frac{1}{4}\int_{-\infty}^{\infty} e^{-2\omega(\lambda)\xi} \left\{ \frac{i}{\lambda}(\cos g_0(\xi) - 1)m_2(\xi,\lambda) + \left[ \frac{1}{\lambda}\sin g_0(\xi) + i\dot{g}_0(\xi) + g_1(\xi) \right] m_1(\xi,\lambda) \right\} d\xi,$$
$$\lambda \in \mathbb{R}, \tag{5.10}$$

*where $(m_1(x,\lambda),\ m_2(x,\lambda))$ denotes the solution of the following system of ODEs:*

$$\begin{cases} (m_1)_x = \frac{i}{\lambda}[1-\cos g_0(x)]m_1 - [\frac{1}{\lambda}\sin g_0(x) - i\dot{g}_0(x) - g_1(x)]m_2, \\ (m_2)_x + 2\omega(\lambda)m_2 = [\frac{1}{\lambda}\sin g_0(x) + i\dot{g}_0(x) + g_1(x)]m_1 - \frac{i}{\lambda}[1-\cos g_0(x)]m_2, \\ \lim_{x\to\infty}(m_1,m_2) = (1,0), \qquad x\in\mathbb{R},\ \lambda\in\mathbb{C}^+. \end{cases} \tag{5.11}$$

**Definition 5.2 (Admissibility and global relation).** *A pair of functions $g_0(x)$, $g_1(x)$ as in Definition* 5.1 *is* admissible *if the associated spectral functions $a(\lambda)$, $b(\lambda)$ satisfy the* global relation, *i.e., the following constraints:*

$$a(\lambda) = 1,\ 0 \leq arg(\lambda) \leq \pi, \qquad b(\lambda) = 0,\ \lambda \geq 0. \tag{5.12}$$

The main existence theorem, assuming admissibility, is the following.

**Theorem 5.3.** *Let the functions $g_0(x)$, $g_1(x)$ be such that $g_0 - 2\pi m \in \mathbf{H}^1(\mathbb{R})$, $m \in \mathbb{Z}$, and $g_1(x) \in \mathbf{H}^1(\mathbb{R})$.*

*Define $a(\lambda)$ and $b(\lambda)$ as in Definition* 5.2, *and assume that the pair $(g_0(x),\ g_1(x))$ is admissible. Define the following Riemann–Hilbert (RH) problem in terms of $b(\lambda)$:*

$$\Psi^-(x,y,\lambda) = \Psi^+(x,y,\lambda)J(x,y,\lambda), \qquad \lambda\in\mathbb{R}, \quad \Psi = I + O\left(\frac{1}{\lambda}\right),\ \lambda\to\infty, \tag{5.13}$$

*where*

$$J = \begin{pmatrix} 1 & b(-\lambda)e^{-\theta(x,y,\lambda)} \\ -b(\lambda)e^{\theta(x,y,\lambda)} & 1 \end{pmatrix}, \qquad \theta(x,y,\lambda) = 2(\omega(\lambda)x + \Omega(\lambda)y). \tag{5.14}$$

*If the $\mathbf{H}^1$ norm of the data $g_0(x)$, $g_1(x)$ is sufficiently small, the above RH problem admits a unique solution $\Psi(x,y,\lambda)$.*

*Let the function* $q(x,y)$, $x \in \mathbb{R}$, $0 < y < \infty$, *be defined in terms of this unique solution by*

$$iq_x + q_y = -\lim_{\lambda\to\infty}(i\lambda\Psi)_{12}, \quad \cos q(x,y) = 1 - \lim_{\lambda\to\infty} 4i\lambda\left(\frac{\partial\Psi}{\partial x}\right)_{22} - 2\lim_{\lambda\to\infty}(\lambda\Psi)_{12}^2. \tag{5.15}$$

*Then* $q(x,y)$ *solves the elliptic sine-Gordon equation* (5.1) *in the half plane* $y > 0$, *and furthermore*

$$q(x,0) = g_0(x), \qquad q_y(x,0) = g_1(x), \quad x \in \mathbb{R}. \tag{5.16}$$

The global relation, in this case equation (5.12), characterizes any pair of functions $(g_0, g_1)$ that are admissible as boundary values of solutions of (5.1) in the half plane (see also the article of A. C. L. Ashton in this book).

Conversely, using the global relation it is possible to characterize the Dirichlet-to-Neumann map, i.e., to express $g_1$ in terms of $g_0$. Hence the Dirichlet problem for the half plane is always linearizable.

More precisely, we characterize uniquely the unknown function $\chi(x)$ in terms of the given function $g_0(x)$, where we use the notation

$$\chi(x,y) = iq_x(x,y) + q_y(x,y), \qquad \chi(x) = i\dot{g}_0(x) + q_y(x,0). \tag{5.17}$$

**Theorem 5.4.** *Let* $q(x,y)$ *satisfy the elliptic sine-Gordon equation* (5.1), *for* $-\infty < x < \infty$, $0 < y < \infty$, *with prescribed Dirichlet boundary condition (within* $2\pi$ *multiples)*

$$q(x,0) = g_0(x), \quad g_0(x) - 2\pi m \in \mathbf{H}^1(\mathbb{R})\,(\text{some } m \in \mathbb{Z}). \tag{5.18}$$

*The Neumann boundary value is characterized by*

$$\begin{aligned} q_y(x,0)\cos\frac{g_0(x)}{2} &= -i\dot{g}_0(x)\cos\frac{g_0(x)}{2}, \qquad x \in \mathbb{R}, \\ &+ \frac{1}{\pi}\int_{\partial\mathbb{R}} \Omega(l)\left\{\int_x^\infty \left[\sin g_0(\xi)m_1\left(\xi,-\frac{1}{l}\right) + i(\cos g_0(\xi)-1)m_2\left(\xi,-\frac{1}{l}\right)\right]\mathrm{e}^{-2\omega(l)(\xi-x)}d\xi \right. \\ &\left. + \int_{-\infty}^x \left[\sin g_0(\xi)n_1\left(\xi,-\frac{1}{l}\right) + i(\cos g_0(\xi)-1)n_2\left(\xi,-\frac{1}{l}\right)\right]\mathrm{e}^{-2\omega(l)(\xi-x)}d\xi\right\} d\lambda, \end{aligned} \tag{5.19}$$

*where the vectors* $(m_1(x,\lambda), m_2(x,\lambda))$ *and* $(n_1(x,\lambda), n_2(x,\lambda))$ *satisfy the ODEs*

$$\begin{cases} (m_1)_x = \frac{i}{\lambda}(1-\cos g_0(x))m_1 - [\frac{1}{\lambda}\sin g_0(x) - \chi(x)]m_2, \\ (m_2)_x + 2\omega(\lambda)m_2 = [\frac{1}{\lambda}\sin g_0(x) + \chi(x)]m_1 - \frac{i}{\lambda}(1-\cos g_0(x))m_2, \end{cases} \quad \lim_{x\to\infty}(m_1,m_2) = (1,0),$$

$$x \in \mathbb{R}, \ \lambda \in \mathbb{C}^+,$$

$$\begin{cases} (n_1)_x = \frac{i}{\lambda}(1-\cos g_0(x))n_1 - [\frac{1}{\lambda}\sin g_0(x) - \chi(x)]n_2, \\ (n_2)_x + 2\omega(\lambda)n_2 = [\frac{1}{\lambda}\sin g_0(x) + \chi(x)]n_1 - \frac{i}{\lambda}(1-\cos g_0(x))n_2, \end{cases} \quad \lim_{x\to-\infty}(n_1,n_2) = (1,0),$$

$$x \in \mathbb{R}, \ \lambda \in \mathbb{C}^-.$$

The proof of this theorem relies on the validity of the global relation and on the analyticity properties of the spectral functions in the $\lambda$ complex plane.

### 5.1.4 ▪ The Quarter Plane $x > 0$, $y > 0$

For the boundary value problem posed in the quarter plane, one condition must be prescribed for $x = 0$ and one for $y = 0$.

The spectral formulation depends on four spectral functions, $a_0(\lambda)$, $b_0(\lambda)$, $a_1(\lambda)$, $b_1(\lambda)$, two corresponding to the two boundary values on each side. Namely, let $d_1(y)$, $u_1(y)$,

$d_2(x)$, $u_2(x)$ be smooth functions defined on $[0,\infty)$ that decay at infinity and are compatible at the corner $(x,y)=(0,0)$.

Define the column vectors $(a_0(\lambda), b_0(\lambda))^\tau$, $(a_1(\lambda), b_1(\lambda))^\tau$ by

$$\begin{pmatrix} a_0(\lambda) \\ b_0(\lambda) \end{pmatrix} = \psi_3(0,\lambda),\ \mathrm{Im}(\lambda)\geq 0, \qquad \begin{pmatrix} a_1(\lambda) \\ b_1(\lambda) \end{pmatrix} = \psi_1(0,\lambda),\ \mathrm{Re}(\lambda)\geq 0, \tag{5.20}$$

where $\psi_1$, $\psi_3$ are the second column vectors of solutions (5.8), i.e., are the unique solutions of the following ODEs:

$$\psi_3(x,\lambda)_x + \Omega(\lambda)\begin{pmatrix} 0 & 0 \\ 0 & 1 \end{pmatrix}\psi_3(x,\lambda) = Q_0(x,0,\lambda)\psi_3(x,\lambda),$$

$$\psi_1(y,\lambda)_y + \omega(\lambda)\begin{pmatrix} 0 & 0 \\ 0 & 1 \end{pmatrix}\psi_1(y,\lambda) = iQ_0(0,y,-\lambda)\psi_1(y,\lambda),$$

$$\lim_{x\to\infty}\psi_3(x,\lambda)\to(1,0)^\tau, \qquad \lim_{y\to\infty}\psi_1(y,\lambda)\to(1,0)^\tau,$$

where

$$Q_0(x,0,\lambda) = \frac{id_2'(x)+u_2(x)}{2}\sigma_1 - \frac{i}{4\lambda}(\sin d_2(x))\sigma_2 + \frac{i}{4\lambda}(1-\cos d_2(x))\sigma_3, \tag{5.21}$$

$$Q_0(0,y,\lambda) = \frac{iu_1(y)+d_1'(y)}{2}\sigma_1 - \frac{i}{4\lambda}(\sin d_1(y))\sigma_2 + \frac{i}{4\lambda}(1-\cos d_1(y))\sigma_3, \tag{5.22}$$

where $\sigma_i$, $i=1,2,3$, are the usual Pauli matrices.

**Definition 5.5.** *The set of functions* $\{d_1(y), u_1(y), d_2(x), u_2(x\}$ *is called* admissible *if the associate spectral functions* $a_0(\lambda), b_0(\lambda), a_1(\lambda), b_1(\lambda)$ *satisfy the* global relation*:*

$$a_0(\lambda)=a_1(\lambda), \qquad b_0(\lambda)=b_1(\lambda), \quad \text{for } \lambda:\ \mathrm{Re}(\lambda)\geq 0,\ \mathrm{Im}(\lambda)\geq 0.$$

Under the assumption of admissibility, we can derive a main existence theorem analogous to Theorem 5.3. We express some of the quantities in this theorem in the variables $(z,\bar z)=(x+iy, x-iy)$, a natural choice in the case of the quarter plane.

**Theorem 5.6.** *Consider the Dirichlet boundary value problem for the elliptic sine-Gordon equation in the quarter plane*

$$q_{xx}+q_{yy}=\sin q, \quad q=q(x,y), \quad x\geq 0,\ y\geq 0, \tag{5.23}$$

$$q(0,y)=d_1(y), \qquad q(x,0)=d_2(x), \qquad d_1(0)=d_2(0), \tag{5.24}$$

*where* $d_1$, $d_2$ *are in the Schwartz class* $\mathcal{S}[0,\infty)$.

*Suppose that there exist two functions* $u_1(y)$, $u_2(x)$ *such that the set* $\{d_1, d_2\, u_1, u_2\}$ *is admissible in the sense of Definition* 5.5. *Assume also that* $a(\lambda)\neq 0$, $\lambda\in(1)$.

*Consider the complex variable* $z=x+iy$, $x\geq 0$, $y\geq 0$, *and define* $\Psi(z,\bar z,\lambda)$ *as the solution of the following RH problem:*

$$\Psi_-(z,\bar z,\lambda)=\Psi_+(z,\bar z,\lambda)J(z,\bar z,\lambda), \quad \lambda\in\Gamma, \tag{5.25}$$

*where* $\Gamma=\mathbb{R}\cup i\mathbb{R}$ *and*

$$J(z,\bar z,\lambda)=J^\alpha(z,\bar z,\lambda), \text{ if } arg(\lambda)=\alpha, \quad \alpha=0, \frac{\pi}{2}, \pi, \frac{3\pi}{2}, \tag{5.26}$$

*with*

$$J^0=\begin{pmatrix} 1 & -\frac{\bar b_0}{a_0}e^{i\theta(z)} \\ 0 & 1 \end{pmatrix}, \quad J^{\pi/2}=\begin{pmatrix} 1 & -\frac{\bar b_1}{a_0}e^{i\theta(z)} \\ 0 & 1 \end{pmatrix}, \quad J^\pi=\begin{pmatrix} 1 & 0 \\ \frac{b_0}{\bar a_1}e^{-i\theta(z)} & 1 \end{pmatrix},$$

*and*

$$J^{3\pi/2} = J_\pi^{-1} J_0 J_{\pi/2}, \tag{5.27}$$

*where the functions* $a_0(\lambda)$, $a_1(\lambda)$, $b_0(\lambda)$, $b_1(\lambda)$ *are given by* (5.20), $\bar{b} = \overline{b(\bar{\lambda})}$, *and*

$$\theta(z) = \frac{1}{2}\left[\lambda z - \frac{\bar{z}}{\lambda}\right]. \tag{5.28}$$

*Then the function* $q(x,y)$ *defined by*

$$\cos q(z,\bar{z}) = 1 + \frac{2}{\pi}\frac{\partial}{\partial \bar{z}} \int_{\Gamma_-} (\Psi_+)_{12} \frac{b(\lambda)}{a(-\lambda)} e^{-\frac{i}{2}(\lambda z - \frac{\bar{z}}{\lambda})} d\lambda, \tag{5.29}$$

*where* $\Gamma_- = \Gamma \cap \mathbb{C}^-$, $b = b_0$ *on* $\mathbb{R}^-$, $b = b_1$ *on* $i\mathbb{R}^-$, $a(-\lambda) = a_0(-\lambda) = a_1(-\lambda)$ *for* $\lambda \in (3)$, *satisfies* $q_{xx} + q_{yy} = \sin q$, *as well as the boundary conditions*

$$q(0,y) = d_1(y) \;\; q_x(0,y) = u_1(y) \quad q(x,0) = d_2(x), \; ; \; q_y(x,0) = u_2(x).$$

## Linearizable Boundary Conditions

The analysis of the conditions that allow the problem to be effectively linearized, similar to the one we present in the next section for the case of a semistrip domain, yields that the only boundary conditions that could be linearizable are constant. Indeed, consider that the $2 \times 2$ matrix-valued function $\Phi_1(y,\lambda)$ obtained from (5.8) at the $(0,\infty)$ vertex satisfies the ODE

$$\begin{aligned} &(\Phi_1)_y + \frac{\omega(\lambda)}{2}[\sigma_3, \Phi_2] = iQ(0,y,-\lambda)\Phi_1, \qquad 0 < y < \infty, \\ &\lim_{y\to\infty} \Phi_2(y,\lambda) = I, \end{aligned} \tag{5.30}$$

where $Q(x,y,\lambda)$ is defined in (5.5), and no specific boundary condition has been imposed. Then the spectral functions $a_1(\lambda)$, $b_1(\lambda)$ are defined as the first column vector of $\Phi_1(0,\lambda)$.

Letting

$$\Phi_1(y,\lambda) = M(y,\lambda) e^{\frac{\omega(\lambda)}{2}\sigma_3(y-L)} \tag{5.31}$$

it follows that $M$ satisfies the ODE

$$\begin{aligned} &(M)_y = VM, \\ &\lim_{y\to\infty} M(y,\lambda) = I, \qquad 0 < y < \infty, \quad \mathrm{Re}(\lambda) \geq 0, \end{aligned} \tag{5.32}$$

where $V = iQ_0(0,y,-\lambda) - \sigma_3 \frac{\omega(\lambda)}{2}$, namely

$$V(y,\lambda) = \frac{1}{4}\begin{pmatrix} -(\lambda + \frac{\cos q(0,y)}{\lambda}) & -q_x(0,y) + iq_y(0,y) + \frac{i \sin q(0,y)}{\lambda} \\ -q_x(0,y) + iq_y(0,y) - \frac{i \sin q(0y)}{\lambda} & \lambda + \frac{\cos q(0,y)}{\lambda} \end{pmatrix}. \tag{5.33}$$

We now consider the following question: find a matrix $R = R(\lambda)$ depending only on the spectral parameter $\lambda$, such that

$$V(y, \frac{1}{\lambda}) = R(\lambda) V(y,\lambda) R(\lambda)^{-1}. \tag{5.34}$$

A straightforward computation shows that the only choice of boundary conditions that allows such a matrix to exist *independent of* $x, y$ is a Dirichlet condition of the form $q(0,y) = d$, $d \neq 0$.

The same construction for the matrix-valued functions $\Phi_3(x,\lambda)$ implies that the only choice of boundary conditions allowing an invariance of a form analogous to (5.34) is $q(x,0) = 0$.

But the requirement of the decay of the solution as $x, y \to \infty$ then forces these conditions to be zero, and then the problem admits the unique zero solution.

Therefore, there appear to be no linearizable boundary conditions for the quarter plane problem. Indeed, this is the case even if we consider the alternative and substantially different Lax pair considered in the following section.

### 5.1.5 ▪ Boundary Value Problems in a Semistrip

In this section we will analyze boundary value problems posed in the semistrip

$$\mathcal{S} = \{0 < x < \infty, \qquad 0 < y < L\},$$

where $L$ is a positive finite constant. The sides $\{y = L,\ 0 < x < \infty\}$, $\{x = 0,\ 0 < y < L\}$, and $\{y = 0,\ 0 < x < \infty\}$ will be referred to as sides (1), (2), and (3), respectively; see Figure 5.1.

The formal analysis of this problem is given in [30]. Suppose that (5.1) is supplemented with appropriate boundary conditions on the boundary of the semistrip $\mathcal{S}$, so that there exists a unique solution $q(x,y)$. Then this solution can be expressed in terms of the solution of a $2 \times 2$ matrix RH problem with jumps on the union of the real and imaginary axes of the $\lambda$ complex plane. The "jump matrices" are expressed in terms of the *spectral functions* $\{a_j(\lambda), b_j(\lambda)\}$, $j = 1, 2, 3$, that are uniquely characterized via the solution of certain linear Volterra integral equations, in terms of the Dirichlet and Neumann boundary values. Namely, $\{a_1, b_1\}$, $\{a_2, b_2\}$, and $\{a_3, b_3\}$ are uniquely determined in terms of $\{q(x,L), q_y(x,L)\}$, $\{q(0,y), q_x(0,y)\}$, and $\{q(x,0), q_y(x,0)\}$, respectively. However, for a well-posed problem only a subset of these boundary values are prescribed as boundary conditions. Thus, in order to compute the spectral functions in terms of the given boundary conditions, one must first determine the unknown boundary values; i.e., one must characterize the Dirichlet-to-Neumann map. The solution of this problem, which makes crucial use of the global relation, yields in general a nonlinear map.

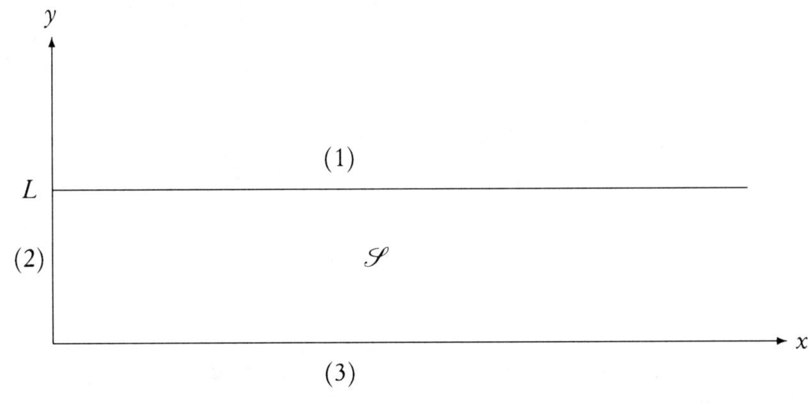


**Figure 5.1.** *The semistrip S.*

In [30] we showed that there exists a choice of boundary conditions, namely $q(x,0) = q(x,L) = 0$, $q(0,y) = d$, for which the latter nonlinear map linearizes and can be given explicitly (up to some unknown residue conditions). Here, we characterize additional boundary conditions with this property. Namely, we show that the conditions that have a chance of being linearizable (based on the fact that for these it is possible to conjugate the relevant spectral matrices) are the following:

$$\text{(a)}\quad q(x,0)=q(x,L)=0,\qquad q_x(0,y)=\alpha\cos\frac{q(0,y)}{2}+\beta\sin\frac{q(0,y)}{2}; \tag{5.35}$$

$$\text{(b)}\quad q_y(x,0)=\gamma\sin\frac{q(x,0)}{2},\qquad q_y(x,L)=\delta\sin\frac{q(x,L)}{2},\qquad q(0,y)=d; \tag{5.36}$$

$$\text{(c)}\quad q_y(x,0)=\gamma\sin\frac{q(x,0)}{2},\qquad q_y(x,L)=\delta\sin\frac{q(x,L)}{2},$$
$$q(0,y)=\alpha\cos\frac{q(0,y)}{2}+\beta\sin\frac{q(0,y)}{2}, \tag{5.37}$$

where $\alpha,\beta,\gamma,\delta$, and $d$ are given real nonzero constants. The choice $\gamma=\delta=0$, $d\neq 0$ yields the boundary conditions of [30]. In what follows, we first summarize the general results of [30], in particular setting down the same notation as in that paper, and then derive the conditions (a)–(c) and briefly discuss the case (a).

### Spectral Functions and Global Relation

We introduce the notation

$$S_1(\lambda)=\Psi_1(0,L,\lambda),\quad S_2(\lambda)=\Psi_2(0,0,\lambda),\quad S_3(\lambda)=\Psi_1(0,0,\lambda), \tag{5.38}$$

where the eigenfunctions $\Psi_i$, $i=1,2,3$, correspond to the three vertices of the semistrip.

Then

$$S_i(\lambda)=\begin{pmatrix} a_i(\lambda) & b_i(-\lambda)\\ b_i(\lambda) & a_i(-\lambda)\end{pmatrix},\qquad i=1,2,3.$$

The spectral functions $\{a_1(\lambda),\ b_1(\lambda)\}$, $\{a_2(\lambda),\ b_2(\lambda)\}$, and $\{a_3(\lambda),\ b_3(\lambda)\}$ are defined in terms of $\{q(x,L),\ q_y(x,L)\}$, $\{q(0,y),\ q_x(0,y)\}$, and $\{q(x,0),\ q_y(x,0)\}$, respectively, through the following equations:

$$S_1(\lambda)=\Phi_1(0,\lambda),\Phi_1(x,\lambda)=I-\int_x^\infty e^{\Omega(\lambda)(\xi-x)\frac{\widehat{\sigma_3}}{2}}Q(\xi,L,\lambda)\Phi_1(\xi,\lambda)d\xi,$$
$$\lambda\in(\mathbb{C}^+,\mathbb{C}^-),\quad 0<x<\infty, \tag{5.39}$$

$$S_2(\lambda)=\Phi_2(0,\lambda),\Phi_2(y,\lambda)=I-i\int_y^L e^{\omega(\lambda)(\eta-y)\frac{\widehat{\sigma_3}}{2}}Q(0,\eta,-\lambda)\Phi_2(\eta,\lambda)d\eta,$$
$$\lambda\in\mathbb{C},\quad 0<y<L, \tag{5.40}$$

$$S_3(\lambda)=\Phi_3(0,\lambda),\Phi_3(x,\lambda)=I-\int_x^\infty e^{\Omega(\lambda)(\xi-x)\frac{\widehat{\sigma_3}}{2}}Q(\xi,0,\lambda)\Phi_3(\xi,\lambda)d\xi,$$
$$\lambda\in(\mathbb{C}^+,\mathbb{C}^-),\quad 0<x<\infty. \tag{5.41}$$

The global relation relating the spectral functions $a_i(\lambda)$, $b_i(\lambda)$, $i=1,2,3$, is given in this case by

$$e^{\frac{\omega(\lambda)}{2}L\widehat{\sigma_3}}S_1(\lambda)=S_2(\lambda)^{-1}S_3(\lambda). \tag{5.42}$$

The first column vector of this equation yields the following scalar global relations:

$$a_1(\lambda) = a_2(-\lambda)a_3(\lambda) - b_2(-\lambda)b_3(\lambda), \qquad \lambda \in \mathbb{C}^+, \tag{5.43}$$

$$b_1(\lambda)e^{-\omega(\lambda)L} = a_2(\lambda)b_3(\lambda) - a_3(\lambda)b_2(\lambda), \qquad \lambda \in \mathbb{C}^+. \tag{5.44}$$

## Main Representation Theorem

**Theorem 5.7.** *Suppose a subset of the boundary values* $\{q(x,L), q_y(x,L)\}$, $\{q(x,0), q_y(x,0)\}$, $0 < x < \infty$, *and* $\{q(y,0), q_x(y,0)\}$, $0 < y < L$, *satisfying sufficient smoothness assumptions, are prescribed as boundary conditions. Suppose that these prescribed boundary conditions are admissible such that the global relations* (5.43) *and* (5.44) *hold.*

*Define the spectral functions* $\{a_j, b_j\}$, $j = 1,..,3$, *through equations* (5.39)–(5.41).

*Define* $\Psi(x,y,\lambda)$ *as the solution of the following* $2 \times 2$ *matrix Riemann-Hilbert problem:*

- *The function* $\Psi(x,y,\lambda)$ *is a sectionally meromorphic function of* $\lambda$ *away from* $\mathbb{R} \cup i\mathbb{R}$.
- $\Psi = I + O\left(\frac{1}{\lambda}\right)$ *as* $\lambda \to \infty$, *and*

$$\Psi_+(x,y,\lambda) = \Psi_-(x,y,\lambda)J(x,y,\lambda), \qquad \lambda \in \mathbb{R} \cup i\mathbb{R},$$

  *where* $\Psi = \Psi_+$ *for* $\lambda$ *in the first or third quadrant, and* $\Psi = \Psi_-$ *for* $\lambda$ *in the second or fourth quadrant of the complex* $\lambda$ *plane, and* $J$ *is defined in terms of* $\{a_j, b_j\}$ *by equations* (5.48) *below.*

- *The possible poles of the second column of* $\Psi$ *occur at* $\lambda = \zeta_j$, $j = 1,\ldots,N_2$, *in the first quadrant and at* $\lambda = \lambda_j$, $j = 1,\ldots,N_1$, *in the second quadrant of the complex* $\lambda$ *plane.*

  *The possible poles of the first column of* $\Psi$ *occur at* $\lambda = -\lambda_j$ $(j = 1,\ldots,N_1)$ *and* $\lambda = -\zeta_j$ $(j = 1,\ldots,N_2)$.

*Then* $\Psi$ *exists and is unique, provided that the* $\mathbf{H}^1$*-norm of the spectral functions is sufficiently small.*

*Define* $q(x,y)$ *in terms of* $\Psi(x,y,\lambda)$ *by*

$$q_x - iq_y = 2 \lim_{\lambda\to\infty} (\lambda\Psi)_{21}, \tag{5.45}$$

$$\cos q(x,y) = 1 + 4i(\lim_{\lambda\to\infty} (\lambda\Psi_x)_{11}) - 2(\lim_{\lambda\to\infty} (\lambda\Psi)_{21})^2. \tag{5.46}$$

*Then* $q(x,y)$ *solves* (5.1). *Furthermore,* $q(x,y)$ *evaluated at the boundary yields the functions used for the computation of the spectral functions.*

The jump matrix $J$ is defined by

$$J(x,y,\lambda) = J^\alpha(x,y,\lambda), \text{ if } arg(\lambda) = \alpha, \quad \alpha = 0, \frac{\pi}{2}, \pi, \frac{3\pi}{2}, \tag{5.47}$$

where

$$J^0 = \begin{pmatrix} \frac{a_2(\lambda)}{a_1(-\lambda)a_3(\lambda)} & \frac{b_3(-\lambda)}{a_3(\lambda)}e^{-\theta(x,y,\lambda)} \\ -\frac{e^{-\omega(\lambda)L}b_1(\lambda)}{a_1(-\lambda)}e^{\theta(x,y,\lambda)} & 1 \end{pmatrix},$$

$$J^{\pi/2} = \begin{pmatrix} 1 & \frac{b_2(-\lambda)}{a_1(\lambda)a_3(\lambda)}e^{-\theta(x,y,\lambda)} \\ 0 & 1 \end{pmatrix}, \qquad J^{3\pi/2} = \begin{pmatrix} 1 & 0 \\ \frac{b_2(\lambda)}{a_1(-\lambda)a_3(-\lambda)}e^{\theta(x,y,\lambda)} & 1 \end{pmatrix},$$

and

$$J^{\pi} = J^{3\pi/2}(J^{0})^{-1}J^{\pi/2}, \tag{5.48}$$

where

$$\theta(x,y,\lambda) = \Omega(\lambda)x + \omega(\lambda)y. \tag{5.49}$$

### Linearizable Boundary Conditions

By a construction analogous to the one in section 5.1.4, it is shown in [30] that the only choice of boundary conditions that allow such a matrix to exist *independently of* $x, y$ is a Dirichlet condition of the form $q(x,0) = q(x,L) = 0$, $q(0,y) = d$, $d \neq 0$.

The result, which we summarize below, is based on the fact that in this case, the spectral functions satisfy some additional symmetry relations, in particular

$$a_2\left(\frac{1}{\lambda}\right) = \frac{a_2(\lambda) - F(\lambda)b_2(\lambda) + F(\lambda)e^{-\omega(\lambda)L}b_2(-\lambda) - F(\lambda)^2 e^{-\omega(\lambda)L}a_2(-\lambda)}{1 - F(\lambda)^2},$$

$$b_2\left(\frac{1}{\lambda}\right) = \frac{b_2(\lambda) - F(\lambda)a_2(\lambda) + F(\lambda)e^{-\omega(\lambda)L}a_2(-\lambda) - F(\lambda)^2 e^{-\omega(\lambda)L}b_2(-\lambda)}{1 - F(\lambda)^2},$$

$$\lambda \in \mathbb{C}\setminus\{0\}, \tag{5.50}$$

where $F(\lambda)$ is the known function

$$F(\lambda) = i\frac{1-\lambda^2}{1+\lambda^2}\tan\frac{d}{2}. \tag{5.51}$$

Manipulating these symmetry relations, and using the global relation and the condition of unit determinant, one can prove the following [30].

**Theorem 5.8.** *The RH problem defined in Theorem* 5.7 *and characterized by the jump matrices* $\{J^{\pi/2}, J^{3\pi/2}, J^{0}, J^{\pi}\}$ *defined in equation* (5.48) *can be mapped to an equivalent RH problem with the following jump matrices:*

$$\tilde{J}^{\pi/2}(x,y,\lambda) = \begin{pmatrix} 1 & (1+e^{\omega(\lambda)L})G(\lambda)e^{-\theta(x,y,\lambda)} \\ 0 & 1 \end{pmatrix},$$

$$\tilde{J}^{3\pi/2}(x,y,\lambda) = \begin{pmatrix} 1 & 0 \\ -(1+e^{-\omega(\lambda)L})G(\lambda)e^{\theta(x,y,\lambda)} & 1 \end{pmatrix},$$

$$\tilde{J}^{0}(x,y,\lambda) = \begin{pmatrix} 1 - G^2(\lambda)e^{-\omega(\lambda)L} & G(\lambda)e^{-\theta(x,y,\lambda)} \\ -G(\lambda)e^{-\omega(\lambda)L}e^{\theta(x,y,\lambda)} & 1 \end{pmatrix}, \tag{5.52}$$

*where the known function* $G(\lambda)$ *is given by* $G(\lambda) = F(\lambda)\tanh\left(\frac{L\omega}{2}\right)$.

The explicit equivalence between the two problems depends on the *unknown* function $h(\lambda) = a_3(\lambda)^2 - b_3(\lambda)^2$. More significantly, the solution $\tilde{M}$ of the new RH problem is a sectionally meromorphic function with simple poles at each point in the set $P_G$ of poles of $G(\lambda)$. At these points the residue conditions are given by

$$\operatorname*{Res}_{\lambda^*}[\tilde{M}(x,y,\lambda)]_2 = -\left(\operatorname*{Res}_{\lambda^*} G\right)\left(1+\frac{1}{h(\lambda^*)}\right)e^{-\theta(x,y,\lambda^*)}[\tilde{M}(x,y,\lambda^*)]_1, \qquad \lambda^* \in P_G \cap \mathbb{C}^+, \tag{5.53}$$

$$\operatorname*{Res}_{\lambda^*}[\tilde{M}(x,y,\lambda)]_1 = \left(\operatorname*{Res}_{\lambda^*} G\right)\left(1+\frac{1}{h(-\lambda^*)}\right)e^{\theta(x,y,\lambda^*)}[\tilde{M}(x,y,\lambda^*)]_2, \qquad \lambda^* \in P_G \cap \mathbb{C}^-. \tag{5.54}$$

Hence they are *not explicitly known*.

Despite this, the theorem above is a very significant result giving an almost explicit representation for this particular boundary value problem.

To test for the existence of other potentially linearizable boundary condition, we now consider an alternative Lax pair. This is inspired by the results in [15] for the evolution sine-Gordon equation.

### An Alternative Lax Pair

The elliptic sine-Gordon equation can be alternatively written as the compatibility condition of the following Lax pair:

$$\mathcal{M}_x = \mathcal{U}\mathcal{M}, \quad \mathcal{M}_y = \mathcal{V}\mathcal{M},$$

$$\mathcal{U}(x,y,\lambda) = \frac{1}{4}\begin{pmatrix} -q_y & \lambda e^{\frac{iq}{2}} - \frac{1}{\lambda}e^{-\frac{iq}{2}} \\ -\lambda e^{-\frac{iq}{2}} + \frac{1}{\lambda}e^{\frac{iq}{2}} & q_y \end{pmatrix}, \tag{5.55}$$

$$\mathcal{V}(x,y,\lambda) = \frac{1}{4}\begin{pmatrix} q_x & i\left[\lambda e^{\frac{iq}{2}} + \frac{1}{\lambda}e^{-\frac{iq}{2}}\right] \\ i\left[-\lambda e^{-\frac{iq}{2}} - \frac{1}{\lambda}e^{\frac{iq}{2}}\right] & -q_x \end{pmatrix}. \tag{5.56}$$

Indeed,

$$\mathcal{V}_x - \mathcal{U}_y + [\mathcal{V},\mathcal{U}] = 0 \Longleftrightarrow q_{xx} + q_{yy} = \sin q.$$

It can be checked that $\det(\mathcal{U})(-\frac{1}{\lambda}) = \det(\mathcal{U})(\lambda)$ and $\det(\mathcal{V})(\frac{1}{\lambda}) = \det(\mathcal{V})(\lambda)$. Hence, as in Section 5.1.4, we consider the question of finding conjugating matrices $N_i$ ($i = 1,2,3$ referring to each side of the semistrip; see Figure 5.1) such that

$$\mathcal{U}\left(x,0,-\frac{1}{\lambda}\right)N_3 = N_3\mathcal{U}(x,0,\lambda),$$
$$\mathcal{V}\left(0,y,\frac{1}{\lambda}\right)N_2 = N_2\mathcal{V}(0,y,\lambda),$$
$$\mathcal{U}\left(x,L,-\frac{1}{\lambda}\right)N_0 = N_0\mathcal{U}(x,L,\lambda).$$

Taking into account the fact that the two functions $q$, $q_y$ must decay as $x \to \infty$, it turns out that such matrices $N_i$ can be found only if the boundary conditions satisfy conditions of the form involved in (5.35)–(5.37). This is the content of the following two lemmas.

**Lemma 5.9.** *Let the given boundary conditions at $y = 0$ and $y = L$ be*

$$q_y(x,0) = \gamma \sin\frac{q(x,0)}{2}, \quad q_y(x,L) = \delta \sin\frac{q(x,L)}{2}, \qquad \gamma,\delta \in \mathbb{C}. \tag{5.57}$$

*Let*

$$N_3(\lambda)=\begin{pmatrix} \frac{i\gamma}{\lambda+\frac{1}{\lambda}} & 1 \\ -1 & \frac{i\gamma}{\lambda+\frac{1}{\lambda}} \end{pmatrix}, \qquad N_1(\lambda)=\begin{pmatrix} \frac{i\delta}{\lambda+\frac{1}{\lambda}} & 1 \\ -1 & \frac{i\delta}{\lambda+\frac{1}{\lambda}} \end{pmatrix}, \tag{5.58}$$

*Then*

$$\mathcal{U}(x,L,-\frac{1}{\lambda})N_1(\lambda)=N_1(\lambda)\mathcal{U}(x,L\lambda), \qquad \mathcal{U}(x,0,-\frac{1}{\lambda})N_3(\lambda)=N_3(\lambda)\mathcal{U}(x,0\lambda).$$

**Lemma 5.10.** *Let the given boundary condition at $x=0$ be of the form*

$$q_x(0,y)=\alpha\cos\frac{q(0,y)}{2}+\beta\sin\frac{q(0,y)}{2}, \qquad \alpha,\ \beta\in\mathbb{C}. \tag{5.59}$$

*Let*

$$N_2(\lambda)=\begin{pmatrix} -\frac{i\alpha}{\lambda+\frac{1}{\lambda}}-\frac{\beta}{\lambda-\frac{1}{\lambda}} & 1 \\ -1 & -\frac{\beta}{\lambda-\frac{1}{\lambda}}+\frac{i\alpha}{\lambda+\frac{1}{\lambda}} \end{pmatrix}. \tag{5.60}$$

*Then*

$$\mathcal{V}\left(0,y,\frac{1}{\lambda}\right)N_2(\lambda)=N_2(\lambda)\mathcal{V}(0,y,\lambda). \tag{5.61}$$

It follows that the combination of the above boundary conditions and of the Dirichlet conditions $q(x,0)=q(x,L)$, $q(,y)=d$ satisfy the necessary condition to be linearizable.

The condition that the solution decays implies that all given boundary condition must also decay as $x\to\infty$.

Moreover, note that the homogeneous combinations

$$q_y(x,0)=\gamma\sin\frac{q(x,0)}{2}, \quad q_y(x,L)=\delta\sin\frac{q(x,L)}{2}, \quad q(0,y)=0$$

and

$$q(x,0)=q(x,L)=0, \qquad q_x(0,y)=\beta\sin\frac{q(x,L)}{2}$$

admit only the trivial solution $q(x,y)=0$.

In summary, the list of potentially linearizable boundary conditions is given by the conditions (5.35)–(5.37).

### The Boundary Conditions (5.35)

We now concentrate on the particular boundary conditions (5.35), namely

$$q(x,0)=q(x,L)=0, \qquad q_x(0,y)=\alpha\cos\frac{q(0,y)}{2}+\beta\sin\frac{q(0,y)}{2}.$$

In this case, equations (5.39) and (5.41) simplify as follows:

$$\begin{pmatrix} (\Phi_1(x,\lambda))_{11} \\ (\Phi_1(x,\lambda))_{21} \end{pmatrix}=\begin{pmatrix} 1 \\ 0 \end{pmatrix}-\frac{1}{4}\int_x^\infty \begin{pmatrix} q_y(\xi,L)(\Phi_1(\xi,\lambda))_{11} \\ e^{\Omega(\lambda)(x-\xi)}q_y(\xi,L)(\Phi_1(\xi,\lambda))_{21} \end{pmatrix}d\xi,$$
$$0<x<\infty, \qquad \mathrm{Im}(\lambda)\geq 0, \tag{5.62}$$

$$\begin{pmatrix} (\Phi_3(x,\lambda))_{11} \\ (\Phi_3(x,\lambda))_{21} \end{pmatrix}=\begin{pmatrix} 1 \\ 0 \end{pmatrix}-\frac{1}{4}\int_x^\infty \begin{pmatrix} q_y(\xi,0)(\Phi_3(\xi,\lambda))_{11} \\ e^{\Omega(\lambda)(x-\xi)}q_y(\xi,0)(\Phi_3(\xi,\lambda))_{21} \end{pmatrix}d\xi,$$
$$0<x<\infty, \qquad \mathrm{Im}(\lambda)\geq 0. \tag{5.63}$$

In equations (5.62) and (5.63), the only dependence on $\lambda$ is through $\Omega(\lambda)$. Thus, since $\Omega(-\frac{1}{\lambda}) = \Omega(\lambda)$, it follows that the vector functions $((\Phi_1(x,\lambda))_{11}, (\Phi_1(x,\lambda))_{21})$ and $((\Phi_3(x,\lambda))_{11}, (\Phi_3(x,\lambda))_{21})$ satisfy the same symmetry properties. Hence, computing these vectors at $x = 0$, we obtain

$$a_j\left(-\frac{1}{\lambda}\right) = a_j(\lambda), \quad b_j\left(-\frac{1}{\lambda}\right) = b_j(\lambda), \qquad j = 1,3, \quad \mathrm{Im}(\lambda) \geq 0. \tag{5.64}$$

We now consider the boundary values on the side (2).

Let $M(y,\lambda)$ be defined by in analogy to (5.31) and let the matrix $V$ be given by (5.33). The function $M$ satisfies the ODE

$$M_y = VM, \tag{5.65}$$

$$M(L,\lambda) = I, \qquad 0 < y < L, \quad \mathrm{Re}(\lambda) \geq 0. \tag{5.66}$$

We denote by $\mathcal{M}$ the solution of the following ODE:

$$\mathcal{M}_y = \mathcal{V}(y,\lambda)\mathcal{M}, \tag{5.67}$$

$$\mathcal{M}(L,\lambda) = I, \qquad 0 < y < L, \quad \mathrm{Re}(\lambda) \geq 0, \tag{5.68}$$

with

$$\mathcal{V}(y,\lambda) = \frac{1}{4}\begin{pmatrix} q_x(0,y) & i\left[\lambda e^{\frac{iq(0,y)}{2}} + \frac{1}{\lambda}e^{-\frac{iq}{2}}\right] \\ i\left[-\lambda e^{-\frac{iq(0,y)}{2}} - \frac{1}{\lambda}e^{\frac{iq(0,y)}{2}}\right] & -q_x(0,y) \end{pmatrix}. \tag{5.69}$$

Note that (5.61) implies that $\mathcal{M}$ satisfies

$$\mathcal{M}\left(y,\frac{1}{\lambda}\right)N_2(\lambda) = N_2(\lambda)\mathcal{M}(y,\lambda).$$

Our aim is to relate $M$ and $\mathcal{M}$. This is the content of the next lemma.

**Lemma 5.11.** *Define the matrix $H$ by*

$$H(y) = \begin{pmatrix} e^{\frac{iq(0,y)}{4}} & e^{\frac{iq(0,y)}{4}} \\ ie^{-\frac{iq(0,y)}{4}} & -ie^{\frac{-iq(0,y)}{4}} \end{pmatrix}. \tag{5.70}$$

*Then $H(y)$ satisfies*

$$H_y(y) = \mathcal{V}(y,\lambda)H(y) - H(y)V(y,\lambda).$$

The proof is a straightforward computation when one takes into account the prescribed boundary conditions. This result on the function $H$ allows us to use the invariance relation (5.61) to derive the transformation properties of the spectral functions $a_2(\lambda)$, $b_2(\lambda)$.

**Proposition 5.12.** *Let the matrix-valued function $F$ be defined by*

$$F(y,\lambda) = H(y)^{-1}N_2(\lambda)H(y) = \begin{pmatrix} i\cos\frac{q(0,y)}{2} - \frac{\beta}{\lambda-\frac{1}{\lambda}} & -\sin\frac{q(0,y)}{2} - \frac{i\alpha}{\lambda+\frac{1}{\lambda}} \\ \sin\frac{q(0,y)}{2} - \frac{i\alpha}{\lambda+\frac{1}{\lambda}} & -i\cos\frac{q(0,y)}{2} - \frac{\beta}{\lambda-\frac{1}{\lambda}} \end{pmatrix}, \tag{5.71}$$

*where $N_2$ is given by* (5.60) *and $H$ by* (5.70). *Then the solution $M$ of* (5.66) *satisfies*

$$M\left(y,\frac{1}{\lambda}\right) = F(y,\lambda)M(y,\lambda)F(L,\lambda)^{-1}. \tag{5.72}$$

The proof is analogous to the one given in [15], after remarking that the relation (5.72) holds identically for $y = L$.

Equation (5.72) and the ansatz $\Phi_2(y,\lambda) = M(y,\lambda)e^{\frac{\omega(\lambda)}{2}\sigma_3(y-l)}$ imply

$$\Phi_2\left(y,\frac{1}{\lambda}\right) = F(y,\lambda)\Phi_2(y,\lambda)\left[e^{-\frac{\omega(\lambda)}{2}\widehat{\sigma_3}(y-L)}F(L,\lambda)^{(-1)}\right]. \tag{5.73}$$

Equations (5.40) and (5.73) evaluated at $y = 0$ yield

$$\begin{pmatrix} a_2(\frac{1}{\lambda}) & b_2(-\frac{1}{\lambda}) \\ b_2(\frac{1}{\lambda}) & a_2(-\frac{1}{\lambda}) \end{pmatrix} = F(0,\lambda)\begin{pmatrix} a_2(\lambda) & b_2(-\lambda) \\ b_2(\lambda) & a_2(-\lambda) \end{pmatrix}\left[e^{\frac{\omega(\lambda)}{2}\widehat{\sigma_3}L}F(L,\lambda)^{-1}\right]. \tag{5.74}$$

Assuming that $q(0,0) = q(0,L) = 0$ and setting

$$\Lambda_+ = \lambda + \frac{1}{\lambda}, \quad \Lambda_- = \lambda - \frac{1}{\lambda}, \quad \hat{a}_2 = a_2(-\lambda), \quad \hat{b}_2 = b_2(-\lambda),$$

we find

$$F(0,\lambda) = \begin{pmatrix} i - \frac{\beta}{\Lambda_-} & -\frac{i\alpha}{\Lambda_+} \\ -\frac{i\alpha}{\Lambda_+} & -i - \frac{\beta}{\Lambda_-} \end{pmatrix}, \qquad e^{\frac{\omega(\lambda)}{2}\widehat{\sigma_3}L}F(L,\lambda)^{-1} = \frac{1}{\Delta}\begin{pmatrix} -i - \frac{\beta}{\Lambda_-} & \frac{i\alpha}{\Lambda_+}e^{\omega(\lambda)L} \\ \frac{i\alpha}{\Lambda_+}e^{-\omega(\lambda)L} & i - \frac{\beta}{\Lambda_-} \end{pmatrix}, \tag{5.75}$$

where $\Delta$ is the determinant of $F(L,\lambda)$,

$$\Delta = 1 - \left(\frac{\alpha^2}{\Lambda_+^2} + \frac{\beta^2}{\Lambda_-^2}\right),$$

and therefore

$$\left(a_2(\tfrac{1}{\lambda})b_2(\tfrac{1}{\lambda})\right) = \frac{1}{\Delta}\begin{pmatrix} \left(1 + \frac{\beta^2}{\Lambda_-^2}\right)a_2 + \frac{i\alpha}{\Lambda_+}\left(i - \frac{\beta}{\Lambda_-}\right)e^{-\omega(\lambda)L}\hat{b}_2 + \frac{\alpha^2}{(\Lambda_+)^2}e^{-\omega(\lambda)L}\hat{a}_2 + \frac{i\alpha}{\Lambda_+}\left(i + \frac{\beta}{\Lambda_-}\right)b_2 \\ \frac{i\alpha}{\Lambda_+}\left(i + \frac{\beta}{\Lambda_-}\right)a_2 + \frac{\alpha^2}{\Lambda_+^2}e^{-\omega(\lambda)L}\hat{b}_2 - \frac{i\alpha}{\Lambda_+}\left(i + \frac{\beta}{\Lambda_-}\right)e^{-\omega(\lambda)L}\hat{a}_2 + \left(i + \frac{\beta}{\Lambda_-}\right)^2 b_2 \end{pmatrix}. \tag{5.76}$$

Setting

$$F_1 = \frac{\alpha}{\Lambda_+}, \qquad F_2 = \frac{\beta}{\Lambda_-},$$

this becomes the following system:

$$a_2\left(\frac{1}{\lambda}\right) = \frac{1}{1-(F_1^2+F_2^2)}\left((1+F_2^2)a_2 - (F_1 + iF_1F_2)e^{-\omega(\lambda)L}\hat{b}_2 + F_1^2 e^{-\omega(\lambda)L}\hat{a}_2 + (iF_1F_2 - F_1)b_2\right), \tag{5.77}$$

$$b_2\left(\frac{1}{\lambda}\right) = \frac{1}{1-(F_1^2+F_2^2)}\left((iF_1F_2 - F_1)a_2 + F_1^2 e^{-\omega(\lambda)L}\hat{b}_2 + (F_1 - iF_1F_2)e^{-\omega(\lambda)L}\hat{a}_2 + (i+F_2)^2 b_2\right). \tag{5.78}$$

This is an additional symmetry condition, similar to the one valid in the case considered in [30]. Coupled with the global relation, these conditions are therefore a starting point for the effective linearization of the associated boundary value problem.

**Remark 19 (The case $\beta = 0$).** The case $\beta = 0$, which still gives rise to a nonconstant case of boundary condition along the $x = 0$ side, yields the symmetry condition

$$\left(a_2(\tfrac{1}{\lambda})b_2(\tfrac{1}{\lambda})\right) = \frac{1}{\Delta}\begin{pmatrix} a_2 - \frac{\alpha}{\Lambda_+}e^{-\omega(\lambda)L}\hat{b}_2 + \left(\frac{\alpha}{\Lambda_+}\right)^2 e^{-\omega(\lambda)L}\hat{a}_2 - \frac{\alpha}{\Lambda_+}b_2 \\ -\frac{\alpha}{\Lambda_+}a_2 + \left(\frac{\alpha}{\Lambda_+}\right)^2 e^{-\omega(\lambda)L}\hat{b}_2 + \frac{\alpha}{\Lambda_+}e^{-\omega(\lambda)L}\hat{a}_2 - b_2 \end{pmatrix}, \quad \Delta = 1 - \left(\frac{\alpha}{\Lambda_+}\right)^2. \tag{5.79}$$

Setting $F = \alpha/\Lambda_+$, this relation can be written as

$$\left(\, a_2(\tfrac{1}{\lambda})b_2(\tfrac{1}{\lambda})\, \right) = \frac{1}{1-F^2}\begin{pmatrix} a_2 - Fe^{-\omega(\lambda)L}\hat{b}_2 + F^2e^{-\omega(\lambda)L}\hat{a}_2 - Fb_2 \\ -Fa_2 + F^2e^{-\omega(\lambda)L}\hat{b}_2 + Fe^{-\omega(\lambda)L}\hat{a}_2 - b_2 \end{pmatrix}. \tag{5.80}$$

Apart from signs the same relation is found for the particular case of constant linearizable conditions considered in [30]. Therefore we expect that the analysis of that paper can be extended to yield an effective linearization of the boundary value problem corresponding to these conditions, namely $q(x,0) = q(x,L) = 0$, $q_x(0,y) = \alpha\cos\frac{q(0,y)}{2}$, $0 < y < L$, in terms of the parameters $\alpha$ and $L$.

# Bibliography


[1] M. J Ablowitz and A. S. Fokas, Introduction and Applications of Complex Variables, 2nd ed., Cambridge University Press, Cambridge, UK, 2003.

[2] Y. Antipov and A. S. Fokas, The modified Helmholtz equation in a semi-strip, Math. Proc. Cambridge Philos. Soc. **138** (2005), 339–365.

[3] D. ben-Avraham and A. S. Fokas, The modified Helmholtz equation in a triangular domain and an application to diffusion-limited coalescence, Phys. Rev. E **64** (2001), 016114-6.

[4] J. L. Bona, and A. S. Fokas, Initial-boundary-value problems for linear and integrable nonlinear dispersive partial differential equations, Nonlinearity **21** (2008), T195–T203.

[5] A. Boutet de Monvel, A. S. Fokas and D. Shepelsky, The modified KdV equation on the half-line, J. Inst. Math. Jussieu **3** (2004), 139–164.

[6] A. Boutet de Monvel, A. S. Fokas and D. Shepelsky, The analysis of the global relation for the nonlinear Schrödinger equation on the half-line, Lett. Math. Phys. **65** (2003), 199–212.

[7] A. B. Borisov and V. V. Kiseliev, Inverse problems for an elliptic sine-Gordon equation with an asymptotic behaviour of the cnoidal type, Inverse Problems **5** (1989), 959–982.

[8] D. Crowdy and A. S. Fokas, Explicit integral solutions for the plane elastostatic semi-strip, Proc. R. Soc. Lond. Ser. A **460** (2004), 1285–1309.

[9] G. Dassios and A. S. Fokas, The basic elliptic equations in an equilateral triangle, Proc. R. Soc. Lond. Ser. A **461** (2005), 2721–2748.

[10] P. Deift, Orthogonal Polynomials and Random Matrices: A Riemann-Hilbert approach, American Mathematical Society, Providence, RI (2000).

[11] G. M. Dujardin, Asymptotics of linear initial boundary value problems with periodic boundary data on the half-line and finite intervals, Proc. R. Soc. Lond. Ser. A **465** (2009), 3341–3360.

[12] N. Flyer and A. S. Fokas, A hybrid analytical numerical method for solving evolution partial differential equations. I. The half-line, Proc. R. Soc. **464** (2008), 1823–1849.

[13] A. S. Fokas, A unified transform method for solving linear and certain nonlinear PDE's, Proc. Roy. Soc. Ser. A **453** (1997), 1411–1443.

[14] A. S. Fokas, A Unified Approach to Boundary Value Problems, CBMS-NSF Regional Conf. Ser. in Appl. Math., SIAM (2008).

[15] A. S. Fokas, Linearizable initial boundary value problems for the sine-Gordon equation on the half line, Nonlinearity **17** (2004), 1521–1534.

[16] A. S. Fokas, Two dimensional linear PDE's in a convex polygon, Proc. Roy. Soc. Lond. Ser. A **457** (2001), 371–393.

[17] A. S. Fokas, Nonlinear evolution equations on the half line, Comm. Math. Phys. **230** (2002), 1–39.

[18] A. S. Fokas, The generalized Dirichlet-to-Neumann map for certain nonlinear evolution PDEs, Comm. Pure Appl. Math. **58** (2005), 639–670.

[19] A. S. Fokas, A New Transform Method for Evolution PDEs, IMA J. Appl. Math. **67** (2002), 559–590.

[20] A. S. Fokas, Two Dimensional Linear PDEs in a Convex Polygon, Proc. Roy. Soc. A **457** (2001).

[21] A. S. Fokas and K. Kalimeris, Eigenvalues of the Laplace operator in the interior of an equilateral triangle (submitted).

[22] A. S. Fokas and A. R. Its, An initial-boundary value problem for the sine-Gordon equation in laboratory coordinates, Theoret. and Math. Phys. **92** (1993), 387–403.

[23] A. S. Fokas and A. R. Its, The linearization of the initial-boundary value problem of the nonlinear Schrödinger equation, SIAM J. Math. Anal. **27** (1996), 738–764.

[24] A. S. Fokas and A. R. Its, The nonlinear Schrödinger equation on the interval, J. Phys. A. **37** (2004), 6091–6114.

[25] A. S. Fokas, A. R. Its, and L. Y. Sung, The nonlinear Schrödinger equation on the half-line, Nonlinearity **18** (2005), 1771–1822.

[26] A. S. Fokas and J. Lennels, Boundary Value Problems for the Stationary Axisymmetric Einstein Equations: A Rotating Disk (preprint).

[27] A. S. Fokas and C. Menyuk, Integrability and self-similarity in transient stimulated Raman scattering, J. Nonlinear Sci. **9** (1999), 1–31.

[28] A. S. Fokas and B. Pelloni, A transform method for linear evolution PDEs on a finite interval, IMA J. Appl. Math. **70** (2005), 1–24.

[29] A. S. Fokas and B. Pelloni, The Dirichlet-to-Neumann map for the elliptic sine-Gordon equation, Nonlinearity **25** (2012), 1011–1031.

[30] A. S. Fokas, J. Lenells, and B. Pelloni, Explicit solutions for the elliptic sine-Gordon equation in a semistrip, J. Nonlinear Sci. **23** (2013), 241–282.

[31] A. S. Fokas and E. A. Spence, Novel Analytical and Numerical Methods for Elliptic Boundary Value Problems, in Highly Oscillatory Problems, Cambridge University Press, Cambridge, UK (2009).

[32] A. S. Fokas and X. Zhou, On the solvability of Painleve II and IV, Comm. Math. Phys. **144** (1992), 601–622.

[33] A. S. Fokas, and M. Zyskin, The fundamental differential form and boundary value problems, Quart. J. Mech. Appl. Math. **55** (2001), 457–479.

[34] F. D.Gakhov, Boundary Value Problems, Pergamon Press, Elmsford, NY (1966).

[35] E. S. Gutshabash and V. D. Lipovskii, Boundary value problem for the two-dimensional elliptic sine-Gordon equation and its applications to the theory of the stationary Josephson effect, J. Math. Sci. **68** (1994), 197–201.

[36] P. D. Lax, Integrals of nonlinear equations of evolution and solitary waves, Comm. Pure Appl. Math. **21** (1968), 467–490.

[37] J. Lenells and A. S. Fokas, An integrable generalization of the nonlinear Schrödinger equation on the half-line and solitons, Inverse Problems **25** (2009), 115006.

[38] J. Lenells and A. S. Fokas, Explicit soliton asymptotics for the Korteweg-de-Vries equation on the half-line (submitted).

[39] V. D. Lipovskii and S. S. Nikulichev, On nonlinear Debye screening, Vestn. LGU Ser. Fiz. Khim. **4** (1988), 61–65.

[40] B. Pelloni, The spectral representation of two-point boundary value problems for linear evolution equations, Proc. R. Soc. A **461** (2004), 2965–2984.

[41] B. Pelloni, Well posed boundary value problems for linear evolution equations in finite intervals, Math. Proc. Cambridge Philos. Soc. **136** (2004), 361–382.

[42] B. Pelloni, The asymptotic behaviour of the solution of boundary value problems for the sine-Gordon equation on a finite interval, J. Nonlinear Math. Phys. **12** (2005), 518–529.

[43] B. Pelloni, Spectral analysis of the elliptic sine-Gordon equation in the quarter plane, Theoret. and Math. Phys. **160** (2009), 1031–1041.

[44] B. Pelloni and D.A. Pinotsis, The Klein-Gordon equation on the half line: A Riemann-Hilbert approach, J. Nonlinear Math. Phys. **15** (2008), 334–344.

[45] B. Pelloni and D.A. Pinotsis, The elliptic sine-Gordon equation in a half plane, Nonlinearity **23** (2010), 77–88.

[46] S. A. Smitheman, E. A. Spence, and A. S. Fokas, A spectral collocation method for the Laplace and modified Helmholtz equations in a convex polygon, IMA J. Numer. Anal. **30** (2010), 1184–1205.

[47] E. A. Spence and A. S. Fokas, A new transform method I: Domain-dependent fundamental solutions and integral representations, Proc. Roy. Soc. A **466** (2010), 2259–2281.

[48] E. A. Spence and A. S. Fokas, A new transform method II: The global relation and boundary value problems in polar coordinates, Proc. Roy. Soc. A **466** (2010), 2283–2307.

[49] P. A. Treharne and A. S. Fokas, The Generalized Dirichlet to Neumann map for the KdV Equation on the Half-Line, J. Nonlinear Sci. **18** (2008), 191–217.

[50] V. E. Zakharov and A. B. Shabat, A scheme for integrating the nonlinear equations of numerical physics by the method of the inverse scattering problem I, Funct. Anal. Appl. **8** (1974), 226–235.

[51] V. E. Zakharov and A. B. Shabat, A scheme for integrating the nonlinear equations of numerical physics by the method of the inverse scattering problem II, Funct. Anal. Appl. **13** (1979), 166–174.

[52] X. Zhou, The Riemann–Hilbert problem and inverse scattering, SIAM J. Math. Anal. **20** (1989), 966–986.

# Part II

# Between Theory and Applications

# Chapter 6

# Overview of Variational Formulations for Linear Elliptic PDEs

*E. A. Spence*

For useful discussions and comments, the author thanks Anthony Ashton (University of Cambridge), Simon Chandler-Wilde (University of Reading), Thanasis Fokas (University of Cambridge), Ivan Graham (University of Bath), Ralf Hiptmair (ETH Zürich), Paul Martin (Colorado School of Mines),[1] Peter Monk (University of Delaware), and especially Andrea Moiola (University of Reading).

The author was supported by EPSRC grant EP/1025995/1.

## 6.1 ▪ "When All Else Fails, Integrate by Parts": New and Old Variational Formulations for Linear Elliptic PDEs

### 6.1.1 ▪ Introduction

Although the motto "when all else fails, integrate by parts" is applicable in a wide range of situations, the author first heard it in the context of the finite element method (FEM). The motto is relevant here since this method is based on the weak form of the PDE that one is trying to solve, and the weak form is obtained by multiplying the PDE by a test function and integrating by parts.

The weak form of a PDE is an example of a *variational problem*: given a Hilbert space $\mathcal{H}$, a sesquilinear form $a(\cdot,\cdot) : \mathcal{H} \times \mathcal{H} \to \mathbb{C}$, and a continuous antilinear functional $F : \mathcal{H} \to \mathbb{C}$,

$$\text{find } u \in \mathcal{H} \text{ such that } a(u,v) = F(v) \text{ for all } v \in \mathcal{H}. \tag{6.1}$$

We highlight immediately that the term "variational problem" is also used to describe the problem of minimizing a functional (as in the calculus of variations). We see the link between these two notions of variational problem in Section 6.1.5 below (in Lemma 6.26), but we emphasize that this article is concerned only with problems of the form (6.1).

[1]Regarding the title of this paper, Paul Martin told the author that Fritz Ursell was famous for asking the question of graduate students and seminar speakers "Have you tried using Green's theorem?"

There are several different ways of converting a boundary value problem (BVP) for a linear PDE into a variational problem of the form (6.1), and many of them are based on integration by parts; this brings us to the first goal of the paper.

> Goal 1: To give an overview of variational formulations for second-order linear elliptic PDEs based on multiplying by a test function and integrating by parts (or, equivalently, based on Green's identities).

We restrict our attention to the particular second-order linear elliptic PDE

$$\Delta u + \lambda u = -f, \tag{6.2}$$

where $\lambda \in \mathbb{R}$, although many of the ideas that we describe also apply to general second-order linear elliptic PDEs, higher-order elliptic PDEs, and other linear PDEs. When $\lambda = 0$, (6.2) is Poisson's equation (and when $f = 0$ it is Laplace's equation), when $\lambda > 0$ it is the Helmholtz equation, and when $\lambda < 0$ it is often called the modified Helmholtz equation. We concentrate on the Helmholtz equation, since of these three equations it currently commands the most attention from the research community.

The variational formulations that we discuss are

- the standard variational formulation, i.e., the weak form of the PDE (which is the basis of the FEM),
- *Discontinuous Galerkin (DG) methods* and *Trefftz-discontinuous Galerkin (TDG) methods*, including the *ultra-weak variational formulation (UWVF)*,
- a variational formulation based on a quadratic functional introduced by Després in [34],
- *boundary integral equations* (BIEs) (which are the basis of the boundary element method), and
- the *null-field method.*

This paper appears in a collection of articles about the so-called *unified transform method* or *Fokas transform method* introduced by Fokas in 1997 [49] and developed further by Fokas and collaborators since then (see the monograph [48] and the review papers [53] and [25]). This method was introduced in the context of certain nonlinear PDEs called *integrable* PDEs (with the defining property that they possess a so-called *Lax pair* formulation), but the method is also applicable to linear PDEs; this brings us to the second goal of the paper.

> Goal 2: To show how the Fokas transform method applied to second-order linear elliptic PDEs can be placed into the framework established in Goal 1.

The heart of this paper is the "map" shown in Figure 6.1. To understand the map, note that $\mathscr{L} := \Delta + \lambda$ and recall that Green's first identity (G1) is

$$\int_D \overline{v}\mathscr{L}u = \int_{\partial D} \overline{v}\frac{\partial u}{\partial n} - \int_D \left(\nabla u \cdot \overline{\nabla v} - \lambda u \overline{v}\right), \tag{6.3}$$

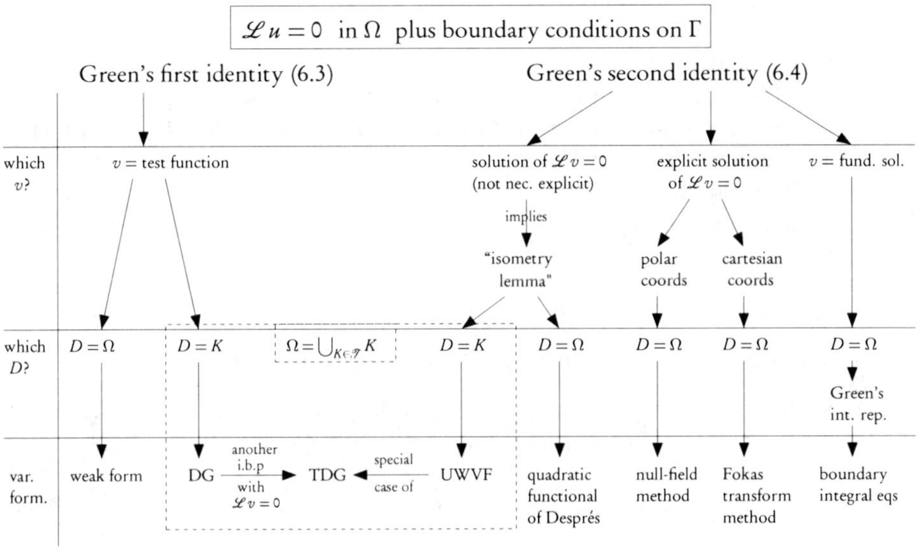


**Figure 6.1.** *"Map" of the variational formulations discussed in this paper.*

and Green's second identity (G2) is

$$\int_D \left(\overline{v}\mathscr{L}u - u\overline{\mathscr{L}v}\right) = \int_{\partial D}\left[\overline{v}\frac{\partial u}{\partial n} - u\overline{\frac{\partial v}{\partial n}}\right]. \tag{6.4}$$

G1 therefore corresponds to multiplying the differential operator by (the complex conjugate of) a test function $v$ and integrating by parts once (i.e., moving one derivative from $u$ onto $v$), and G2 corresponds to multiplying by a test function and integrating by parts twice (i.e., moving two derivatives onto $v$).

To make the integration by parts completely explicit, recall that the divergence theorem

$$\int_D \nabla\cdot\mathbf{F} = \int_{\partial D}\mathbf{F}\cdot\mathbf{n} \tag{6.5}$$

applied with $\mathbf{F} = \phi\mathbf{G}$ gives the integration by parts formula

$$\int_D \phi\nabla\cdot\mathbf{G} = \int_{\partial D}\phi\mathbf{G}\cdot\mathbf{n} - \int_D \mathbf{G}\cdot\nabla\phi, \tag{6.6}$$

where $\mathbf{n}$ is the outward-pointing unit normal vector to $D$. Letting $\phi = \overline{v}$ and $\mathbf{G} = \nabla u$ in (6.6), we obtain G1 (6.3) (with the $\lambda u\overline{v}$ term on each side removed).

The methods in the dashed box in Figure 6.1 require the notion of a triangulation $\mathscr{T}$ of $\Omega$ (where $\Omega$ is the domain in which the PDE is posed), i.e., $\overline{\Omega}$ is divided into a finite number of subsets (not necessarily triangles) satisfying certain conditions; see Definition 6.1 below.

Finally, observe that Figure 6.1 concerns the homogeneous PDE $\mathscr{L}u = 0$. This is because, whereas methods based on G1 can be used to solve BVPs involving the inhomogeneous PDE $\mathscr{L}u = -f$, methods based on G2 are limited (either by definition or in practice) to BVPs involving the homogeneous PDE $\mathscr{L}u = 0$.

**Outline of the Paper.** In Section 6.1.2 we define some notation and recap some standard results (mainly concerning function spaces). In Section 6.1.3 we define some BVPs for the operator $\mathscr{L} := \Delta + \lambda$.

In Section 6.1.4 we derive Green's first and second identities and then discuss whether the BVPs introduced in Section 6.1.3 are self-adjoint; we do this because BVPs for the Helmholtz equation in exterior domains are not self-adjoint (because of the Sommerfeld radiation condition) and this has important implications for the variational formulations discussed in Section 6.1.10.

In Section 6.1.5 we recap standard results about variational problems. When one encounters a variational problem of the form (6.1) the following two questions naturally arise:

Q1. Does the variational problem (6.1) have a solution and, if so, is the solution unique?

Q2. Can we obtain a bound on the solution $u$ in terms of $F$?

Furthermore, the variational problem (6.1) is the basis of the Galerkin method. Indeed, given a finite-dimensional subspace of $\mathcal{H}$, $\mathcal{H}_N$, the Galerkin method is

$$\text{find } u_N \in \mathcal{H}_N \text{ such that } a(u_N, v_N) = F(v_N) \text{ for all } v_N \in \mathcal{H}_N. \tag{6.7}$$

This leads us to a third question:

Q3. Do the Galerkin equations (6.7) have a unique solution (after imposing that $N \geq N_0$, for some $N_0$, if necessary)? If so, is the solution *quasi-optimal*, i.e., does there exist a $C_{qo} > 0$ such that

$$\|u - u_N\|_{\mathcal{H}} \leq C_{qo} \min_{v_N \in \mathcal{H}_N} \|u - v_N\|_{\mathcal{H}} \tag{6.8}$$

(again, after imposing that $N \geq N_0$ if necessary)?[2]

The results outlined in Section 6.1.5 focus on the key properties of *continuity*, *coercivity*, and *coercivity up to a compact perturbation*, since using these concepts we can give conditions under which the answers to the questions Q1, Q2, and Q3 are all "yes."

In Sections 6.1.6–6.1.10 we describe the variational formulations shown in Figure 6.1. Our goals in these sections are to

1. derive the variational formulations and
2. discuss, and sometimes prove, whether the formulations are continuous, coercive, or coercive up to a compact perturbation (and thus establish which of the results in Section 6.1.5 are applicable).

We go through the variational formulations more or less in the order that they appear in Figure 6.1 (reading left to right). The only exception is that we discuss Green's integral representation and BIEs before the null-field method and the Fokas transform method (this is because we use some integral equation results in the section on the null-field method). Note that we do *not* discuss the relative merits of the different variational formulations (since this would make the paper substantially longer).

In Section 6.1.11 we conclude by discussing the two goals of the paper in a wider context.

[2]Strictly speaking, quasi optimality is a property of a *sequence* of Galerkin solutions, $(u_N)_{N\in\mathbb{Z}}$, corresponding to a sequence of finite-dimensional subspaces $(\mathcal{H}_N)_{N\in\mathbb{Z}}$ (since for a single $u_N$ one can always find a $C_{qo}$ such that (6.8) holds, and thus we are really interested in obtaining a $C_{qo}$ such that (6.8) holds for all sufficiently large $N$). Nevertheless, when (6.8) holds we will be slightly cavalier and describe both the Galerkin solution $u_N$ and the Galerkin method itself as quasi-optimal.

**A Note on Style and Rigor.** When writing this paper, I tried to strike a balance between keeping everything mathematically honest and not getting bogged down in technicalities. The outcome is that the results are all stated rigorously, but I hope that even a reader who is not fully comfortable with Sobolev spaces (such as, perhaps, a graduate student) will still be able to understand the ideas behind the variational formulations.

For the reader trying to see the "forest from the trees" regarding the function spaces needed in the variational formulations, since all the variational formulations discussed in the paper arise from Green's identities, the definitions of the function spaces are all consequences of the conditions on $u$ and $v$ needed for G1 (6.3) to hold, and these are given in Lemma 6.6 (see also Remark 23).

## 6.1.2 ▪ Notation and Background Results

**Differential Operators.** We defined above $\mathscr{L} := \Delta + \lambda$ for $\lambda \in \mathbb{R}$. It will be useful later to have specific notation for this operator when we restrict $\lambda$ to be either $> 0$ or $< 0$. Therefore, for $k, \mu > 0$, we define

$$\mathscr{L}_k := \Delta + k^2 \quad \text{and} \quad \mathscr{L}_\mu := \Delta - \mu^2$$

(so $\mathscr{L}_k$ is the operator in the Helmholtz equation and $\mathscr{L}_\mu$ is the operator in the modified Helmholtz equation).

**Notation for Domains.** We use the word "domain" to mean a connected open set, and we let $D$ denote a generic bounded Lipschitz domain (see, e.g., [73, Definition 3.28] for the definition of Lipschitz).[3]

In this paper we consider BVPs in both bounded and unbounded Lipschitz domains (and we call these *interior* and *exterior* BVPs, respectively). We use $\Omega$ to denote the domain in which the PDE is posed in interior BVPs and $\Omega_+$ to denote the domain in exterior BVPs.

**Notation for Interior BVPs.** Let $\Omega \subset \mathbb{R}^d$ ($d = 2$ or $3$) be a bounded Lipschitz domain. Let $\Gamma := \partial\Omega$, and let $\mathbf{n}$ be the outward-pointing unit normal vector to $\Omega$. As usual,

$$L^2(\Omega) := \left\{ v : \Omega \to \mathbb{C} : v \text{ is Lebesgue-measurable and } \int_\Omega |v|^2 < \infty \right\}.$$

For $v : \Omega \to \mathbb{C}$, let $\partial^\alpha v$ denote the weak derivative of $v$ with multi-index $\alpha$. Let

$$H^m(\Omega) := \left\{ v \in L^2(\Omega) : \partial^\alpha v \text{ exists and is in } L^2(\Omega) \text{ for all } |\alpha| \le m \right\}$$

and

$$H^1(\Omega, \Delta) := \left\{ v \in H^1(\Omega) : \Delta v \text{ exists in a weak sense and is in } L^2(\Omega) \right\}.$$

For the definition of the Sobolev spaces $H^s(\Gamma)$ for $|s| \le 1$ see, e.g., [73, Page 96], [18, Section A.3].

For $u \in C^\infty(\overline{\Omega}) := \{ v|_{\overline{\Omega}} : v \in C^\infty(\mathbb{R}^d) \}$ we define the trace of $u$, $\gamma u$, by $\gamma u = u|_\Gamma$. Recall that this operator extends to a bounded linear operator from $H^1(\Omega)$ to $H^{1/2}(\Gamma)$, i.e., there exists a $C > 0$ such that

$$\|\gamma v\|_{H^{1/2}(\Gamma)} \le C \|v\|_{H^1(\Omega)} \quad \text{for all } v \in H^1(\Omega); \tag{6.9}$$

[3] Note that some authors, including McLean [73], allow both connected *and* disconnected sets in the definition of "domain."

see, e.g., [73, Theorem 3.37]. We also have the multiplicative trace inequality, i.e., there exists a $C_1 > 0$ such that

$$\|\gamma v\|^2_{L^2(\Gamma)} \le C_1 \|v\|_{L^2(\Omega)} \|v\|_{H^1(\Omega)} \quad \text{for all } v \in H^1(\Omega); \tag{6.10}$$

see [62, Theorem 1.5.1.10, last formula on Page 41], [15, Theorem 1.6.6].

Let $\partial_n u$ denote the normal-derivative trace on $\Gamma$. Recall that if $u \in H^2(\Omega)$, then $\partial_n u := \mathbf{n}\cdot\gamma(\nabla u)$ and for $u \in H^1(\Omega,\Delta)$, $\partial_n u$ is defined as an element of $H^{-1/2}(\Gamma)$ so that Green's first identity holds; see Lemma 6.6 below. We sometimes call $\gamma u$ the *Dirichlet trace* of $u$ and $\partial_n u$ the *Neumann trace* of $u$.

**Notation for Exterior BVPs.** Let $\Omega_-$ be a bounded Lipschitz open set such that the open complement $\Omega_+ := \mathbb{R}^d \setminus \overline{\Omega_-}$ is connected (this condition rules out $\Omega_-$ being, e.g., an annulus). Let $\Gamma := \partial\Omega_-$, and let $\mathbf{n}$ denote the outward-pointing unit normal vector to $\Omega_-$ (thus $\mathbf{n}$ points *into* $\Omega_+$). Given $R > \sup_{\mathbf{x}\in\Omega_-} |\mathbf{x}|$, let $B_R := \{\mathbf{x} : |\mathbf{x}| < R\}$, $\Omega_R := \Omega_+ \cap B_R$, and $\Gamma_R := \partial B_R = \{\mathbf{x} : |\mathbf{x}| = R\}$.

Let

$$H^1_{\text{loc}}(\Omega_+) := \left\{v : \chi v \in H^1(\Omega_+) \text{ for every } \chi \in C^\infty_{\text{comp}}(\overline{\Omega_+})\right\},$$

where $C^\infty_{\text{comp}}(\overline{\Omega_+}) := \{v|_{\overline{\Omega_+}} : v \in C^\infty_{\text{comp}}(\mathbb{R}^d)\}$. Let

$$H^1_{\text{loc}}(\Omega_+,\Delta) := \left\{v : v \in H^1_{\text{loc}}(\Omega_+) \text{ and } \Delta v|_G \in L^2(G) \text{ for every bounded measurable set } G \subset \Omega_+\right\}.$$

Let $\gamma_\pm$ denote the exterior and interior traces from $\Omega_\pm$ to $\Gamma$, which satisfy $\gamma_+ : H^1_{\text{loc}}(\Omega_+) \to H^{1/2}(\Gamma)$ and $\gamma_- : H^1(\Omega_-) \to H^{1/2}(\Gamma)$. Let $\partial_n^\pm$ denote the exterior and interior normal-derivative traces, which satisfy $\partial_n^+ : H^1_{\text{loc}}(\Omega_+,\Delta) \to H^{-1/2}(\Gamma)$ and $\partial_n^- : H^1(\Omega_-,\Delta) \to H^{-1/2}(\Gamma)$.

**Inequalities.** We have the inequality

$$2ab \le \frac{a^2}{\varepsilon} + \varepsilon b^2 \quad \text{for } a,b,\varepsilon > 0 \tag{6.11}$$

and its consequence

$$(a+b) \le \sqrt{2}\sqrt{a^2+b^2}. \tag{6.12}$$

We recall the Cauchy–Schwarz inequality

$$\left|\int_D u\,\overline{v}\right| \le \|u\|_{L^2(D)} \|v\|_{L^2(D)} \quad \text{for all } u,v \in L^2(D) \tag{6.13}$$

(we use this inequality with $D$ equal either $\Omega$ or $\Gamma$).

**Miscellaneous.** We write $a \lesssim b$ if $a \le Cb$ for some $C > 0$ that is independent of all parameters of interest (these will usually be $k$ and $\mu$, but also sometimes $h$). We write $a \gtrsim b$ if $b \lesssim a$, and $a \sim b$ if both $a \gtrsim b$ and $a \lesssim b$.

We display the surface and volume measures in integrals only when it might be ambiguous as to which variable the integration is respect to (this turns out to be only in Sections 6.1.9 and 6.1.10).

As one can see from Figure 6.1, some of the variational formulations require the notion of a *triangulation* of $\Omega$ (also called a *mesh* or *partition*).

**Definition 6.1 (Triangulation).** *Following* [20, Page 61], *we say that a finite collection of sets $\mathcal{T}$ is a triangulation of $\Omega$ if the following properties hold:*

1. $\overline{\Omega} = \bigcup_{K\in\mathcal{T}} K$.
2. *Each $K \in \mathcal{T}$ is closed and its interior, $\mathring{K}$, is nonempty and connected.*
3. *If $K_1, K_2 \in \mathcal{T}$ and $K_1 \neq K_2$, then $\mathring{K}_1 \cap \mathring{K}_2 = \emptyset$.*
4. *Each $K \in \mathcal{T}$ is Lipschitz.*

One then defines $h_K := \operatorname{diam}(K) = \max_{\mathbf{x},\mathbf{y}\in\overline{K}} |\mathbf{x}-\mathbf{y}|$ and $h := \max_{K\in\mathcal{T}} h_K$. One usually thinks of a family of triangulations $\mathcal{T}_h$, with $0 < h \leq h_0$ for some $h_0$. Let $\rho_K$ be the diameter of the largest ball contained in $K$ (so $\rho_K \leq h_K$). The family $\mathcal{T}_h$ is *regular* (or *nondegenerate*) if $h_K \lesssim \rho_K$ for all $K \in \mathcal{T}_h$ and for all $0 < h \leq h_0$.

### 6.1.3 ▪ BVPs for the Operator $\mathcal{L} := \Delta + \lambda$

**Definition 6.2 (Interior Dirichlet problem).** *Let $\Omega$ be a bounded Lipschitz domain. Given $g_D \in H^{1/2}(\Gamma)$ and $f \in L^2(\Omega)$, we say that $u \in H^1(\Omega)$ satisfies the* interior Dirichlet problem *(IDP) if*

$$\mathcal{L}u = -f \quad \text{in } \Omega \quad \text{and} \quad \gamma u = g_D \quad \text{on } \Gamma. \tag{6.14}$$

Some remarks on the IDP follow:

- The most general situation is where $f \in (H^1(\Omega))^*$ (where $(H^1(\Omega))^*$ is the antidual space to $H^1(\Omega)$) instead of $f \in L^2(\Omega)$.
- The PDE in (6.14) is understood as holding in a distributional sense, i.e.,
$$\int_\Omega u\,\mathcal{L}\phi = -\int_\Omega f\phi \quad \text{for all } \phi \in C^\infty_{\text{comp}}(\Omega).$$
- The uniqueness of the IDP depends on $\lambda$. Indeed, if $\lambda \leq 0$, then the solution is unique (this can be proved using Green's first identity). On the other hand, if $\lambda = \lambda_j$, where $\lambda_j$ is the $j$th Dirichlet eigenvalue of the negative Laplacian in $\Omega$, i.e., there exists a $u_j \in H^1(\Omega)\setminus\{0\}$ such that
$$-\Delta u_j = \lambda_j u_j \quad \text{in } \Omega \quad \text{and} \quad \gamma u_j = 0 \quad \text{on } \Gamma, \tag{6.15}$$
then the solution of the IDP is not unique.

The only exterior problem that we consider is the exterior Dirichlet problem (EDP) for the Helmholtz equation.

**Definition 6.3 (Exterior Dirichlet problem for Helmholtz).** *Let $\Omega_-$ be a bounded Lipschitz open set such that the open complement $\Omega_+ := \mathbb{R}^d \setminus \overline{\Omega_-}$ is connected. Given $g_D \in H^{1/2}(\Gamma)$ and $f \in L^2(\Omega_+)$ with compact support, we say that $u \in H^1_{\text{loc}}(\Omega_+)$ satisfies the EDP for the Helmholtz equation if*

$$\mathcal{L}_k u := \Delta u + k^2 u = -f \quad \text{in } \Omega_+, \tag{6.16}$$

*$\gamma_+ u = g_D$ on $\Gamma$, and $u$ satisfies the Sommerfeld radiation condition*

$$\frac{\partial u}{\partial r}(\mathbf{x}) - ik u(\mathbf{x}) = o\left(\frac{1}{r^{(d-1)/2}}\right) \tag{6.17}$$

*as $r := |\mathbf{x}| \to \infty$, uniformly in $\mathbf{x}/r$.*

**Remark 20 (How should (6.16) and (6.17) be understood?).** *As in the case of the IDP, the PDE* (6.16) *is understood in a distributional sense. To impose the pointwise condition* (6.17), *we need $u$ to be in $C^1(\mathbb{R}^d \setminus \Omega_R)$ for some $R > 0$. This is ensured, however, by interior regularity of the Helmholtz operator, which implies that $u$ is $C^\infty$ outside the support of $f$ and away from $\Gamma$ (see Remark* 21 *below).*

*An alternative way of defining the EDP is to fix $R$ such that $\operatorname{supp} f \subset B_R$ and impose the PDE* (6.16) *in $\Omega_R$ in its weak form. The sesquilinear form of that variational formulation involves an operator on $\Gamma_R$ that (roughly speaking) encodes the information that $u$ satisfies the Sommerfeld radiation condition outside $\Omega_R$. One can then show that the solution of this variational problem has an extension from $\Omega_R$ to $\Omega_+$ that satisfies the EDP defined by Definition* 6.3*; for more details see, e.g., the discussion in* [94, Section 4.3] *(this discussion is for the exterior impedance problem; however, the case of the EDP is almost identical).*

A further remark on the EDP follows:

- The solution of the EDP for the Helmholtz equation is unique for all $k$; see [22, Theorem 3.13], [18, Corollary 2.9].

The majority of practical applications of the Helmholtz equation involve posing the PDE in exterior domains. Although there are many standard ways of dealing numerically with both unbounded domains and the radiation condition, most investigations of numerical methods for solving the Helmholtz equation begin by considering the *interior impedance problem* (IIP). This problem has the advantage that it is posed on a bounded domain, but (unlike the IDP) the solution is unique for all $k$ (as it is for the EDP); see Theorem 6.5 below.

**Definition 6.4 (Interior impedance problem for Helmholtz).** *Let $\Omega$ be a bounded Lipschitz domain. Given $g \in L^2(\Gamma)$, $f \in L^2(\Omega)$, and $\eta \in \mathbb{R} \setminus \{0\}$, we say that $u \in H^1(\Omega)$ satisfies the IIP if*

$$\mathcal{L}_k u := \Delta u + k^2 u = -f \quad \text{in } \Omega \tag{6.18}$$

*and*

$$\partial_n u - \mathrm{i}\eta\gamma u = g \quad \text{on } \Gamma. \tag{6.19}$$

Some remarks on the IIP follow:

- We could consider the more general situation where $f \in (H^1(\Omega))^*$ and $g \in H^{-1/2}(\Gamma)$.

- The PDE (6.18) is understood in a distributional sense, and the boundary condition (6.19) is understood as saying that $\partial_n u$ (as an element of $H^{-1/2}(\Gamma)$) equals the $L^2$-function $g + \mathrm{i}\eta\gamma u$.

- One often chooses $\eta = k$; this is because the impedance boundary condition is often used as an approximation to the radiation condition. For example, an approximation of the EDP is the following: given $g_D \in H^{1/2}(\Gamma)$ and $f \in L^2(\Omega_+)$ with compact support, choose $R$ such that $\operatorname{supp} f \subset B_R$ and find $u \in H^1(\Omega_R)$ such that $\mathcal{L}_k u = -f$ in $\Omega_R$, $\gamma_+ u = g_D$ on $\Gamma$, and

$$\partial_n u - \mathrm{i}ku = 0 \quad \text{on } \Gamma_R.$$

**Theorem 6.5 (Uniqueness of the Helmholtz IIP).** *The solution of the Helmholtz IIP is unique.*

***Proof.*** This proof uses some results from later sections, namely Green's first identity (Lemma 6.6) and Green's integral representation (Theorem 6.41).

We need to show that if $\mathcal{L}_k u = 0$ in $\Omega$ and $\partial_n u - i\eta\gamma u = 0$ on $\Gamma$, then $u = 0$ in $\Omega$. Since $u \in H^1(\Omega, \Delta)$, Green's first identity ((6.24) below) with $v = u$ and $D = \Omega$ implies that

$$\int_\Gamma \overline{\gamma u}\, \partial_n u - \int_\Omega \left(|\nabla u|^2 - k^2|u|^2\right) = 0.$$

Using the boundary condition, we obtain

$$i\eta \int_\Gamma |\gamma u|^2 - \int_\Omega \left(|\nabla u|^2 - k^2|u|^2\right) = 0. \tag{6.20}$$

If $\eta \in \mathbb{R} \setminus \{0\}$, the imaginary part of (6.20) implies that $\gamma u = 0$, and then the boundary condition implies that $\partial_n u = 0$. Green's integral representation (Theorem 6.41) implies that if $\gamma u = 0$ and $\partial_n u = 0$, then $u = 0$ in $\Omega$. □

This theorem can also be understood as saying that, under the impedance boundary condition $\partial_n u - i\eta\gamma u = 0$ with $\eta \in \mathbb{R} \setminus \{0\}$, the eigenvalues of the negative Laplacian are complex.

**Remark 21 (Interior regularity for solutions of $\mathcal{L}u = -f$).** *The PDE $\mathcal{L}u = -f$ states that the Laplacian of $u$ equals a linear combination of $u$ and $f$, and so if $f \in C^m$, say, we might expect $u$ to be in $C^{m+2}$. The rigorous version of this observation is the following: if $f \in H^m(\Omega)$ and $\mathcal{L}u = -f$, then $u \in H^{m+2}(\widetilde{\Omega})$ for any $\widetilde{\Omega}$ that is compactly contained in $\Omega$ (i.e., there exists a compact set $\Omega'$ such that $\widetilde{\Omega} \subset \Omega' \subset \Omega$); see, e.g.,* [41, Section 6.3.1, Theorem 2]. *The Sobolev embedding theorem can then be used to prove that* (i) *if $f \in L^2(\Omega)$, then $u \in C(\Omega)$* [41, Section 5.6.3, Theorem 6], *and* (ii) *if $f \in C^\infty(\Omega)$, then $u \in C^\infty(\Omega)$* [41, Section 6.3.1, Theorem 3].

### 6.1.4 ▪ Green's Identities and Self-Adjointness

#### Green's Identities

**Informal Discussion.** We multiply $\mathcal{L}u$ by the complex conjugate of a test function $v$ (where by "test function" we mean a function with, as yet, unspecified properties) and move either one or two derivatives onto $v$ to obtain

$$\overline{v}\mathcal{L}u = \nabla \cdot \left[\overline{v}\nabla u\right] - \nabla u \cdot \overline{\nabla v} + \lambda u \overline{v} \tag{6.21}$$

and

$$\overline{v}\mathcal{L}u - u\overline{\mathcal{L}v} = \nabla \cdot \left[\overline{v}\nabla u - u\overline{\nabla v}\right], \tag{6.22}$$

respectively. (Note that one can obtain (6.22) from two copies of (6.21), with the roles of $u$ and $v$ swapped in the second copy.) We then integrate (6.21) and (6.22) over a domain $D$ (assuming $u$ is sufficiently smooth on $D$) and use the divergence theorem (6.5).

In the rest of the paper, we will be a bit cavalier and refer to both (6.21) and its integrated form as "G1", and similarly we refer to both (6.22) and integrated form as "G2".

**Statement of Green's First and Second Identities (G1 and G2, respectively).**
**Lemma 6.6 (Green's first identity (G1)).** *Let $D$ be a bounded Lipschitz domain. If $u \in H^1(D, \Delta)$ and $v \in H^1(D)$, then*

$$\int_D v\mathcal{L}u = \int_{\partial D} \gamma v\, \partial_n u - \int_D (\nabla u \cdot \nabla v - \lambda u v) \tag{6.23}$$

*and*

$$\int_D \overline{v}\mathscr{L}u = \int_{\partial D} \overline{\gamma v}\, \partial_n u - \int_D \left(\nabla u \cdot \overline{\nabla v} - \lambda u \overline{v}\right). \tag{6.24}$$

**Lemma 6.7 (Green's second identity (G2)).** *Let $D$ be a bounded Lipschitz domain. If $u$ and $v$ are both in $H^1(D,\Delta)$, then*

$$\int_D \left(v\mathscr{L}u - u\mathscr{L}v\right) = \int_{\partial D} \left[v\partial_n u - u\partial_n v\right] \tag{6.25}$$

*and*

$$\int_D \left(\overline{v}\mathscr{L}u - u\overline{\mathscr{L}v}\right) = \int_{\partial D} \left[\overline{\gamma v}\, \partial_n u - u\overline{\partial_n v}\right]. \tag{6.26}$$

**Remark 22 (How should one understand the integrals over $\partial D$?).** *By the results recapped in Section 6.1.2, when $u \in H^1(D,\Delta)$ and $v \in H^1(D)$, $\partial_n u \in H^{-1/2}(\partial D)$ and $\gamma v \in H^{1/2}(\partial D)$. The integral $\int_{\partial D} \overline{\gamma v}\, \partial_n u$ in (6.24) therefore does not make sense as a usual (Lebesgue) integral.*

*Recall that there exists a continuous sesquilinear form (see Definition 6.14 below for the definition of a sesquilinear form) $\langle\cdot,\cdot\rangle_{\partial D} : H^{-s}(\partial D)\times H^s(\partial D)\to\mathbb{C}$ for $|s|\leq 1$ such that*

$$\langle \phi,\psi\rangle_{\partial D} = \int_{\partial D} \phi\overline{\psi} \tag{6.27}$$

*when $\phi,\psi \in L^2(\partial D)$ (i.e., (6.27) defines $\langle\cdot,\cdot\rangle_{\partial D}$ for $s=0$). The integral $\int_{\partial D}\overline{\gamma v}\,\partial_n u$ in (6.24) should be understood as $\langle \partial_n u,\gamma v\rangle_{\partial D}$. (Therefore, if $\partial_n u \in L^2(\partial D)$, then $\langle \partial_n u,\gamma v\rangle_{\partial D} = \int_{\partial D}\overline{\gamma v}\,\partial_n u$ in the usual sense.) The notation $\langle\cdot,\cdot\rangle_{\partial D}$ is used for this sesquilinear form since $H^{-s}(\partial D)$ can be understood as a realization of the dual space of $H^s(\partial D)$, with $\langle\cdot,\cdot\rangle_{\partial D}$ then equivalent to the duality pairing $\langle\cdot,\cdot\rangle_{(H^s(\partial D))^*\times H^s(\partial D)}$; see* [73, Pages 76 and 98], [18, Page 279] *for more details.*

***Proof of Lemmas* 6.6 and 6.7.** It is sufficient to prove (6.23), since (6.24) follows from (6.23) by replacing $v$ by $\overline{v}$, (6.25) follows from (6.23), and (6.26) follows from (6.25).

The divergence theorem (6.5) holds when $\mathbf{F}\in C^1(\overline{D})$ [73, Theorem 3.34], and then (6.23) holds for $u \in H^2(D)$ and $v \in H^1(D)$ by (i) the density of $C^k(\overline{D})$ in $H^k(D)$ for $k\in\mathbb{N}$ [73, Page 77], (ii) the boundedness of the trace operator from $H^1(D)$ to $H^{1/2}(\partial D)$, (iii) the fact that $\partial_n u = \mathbf{n}\cdot\gamma(\nabla u)$ for $u\in H^2(D)$, and (iv) the Cauchy–Schwarz inequality (6.13). The proof of how to lessen the condition $u\in H^2(D)$ to $u\in H^1(D,\Delta)$ is given in, e.g., [18, Pages 280 and 281], [73, Lemma 4.3]; the key point is that the normal derivative $\partial_n u$ is *defined* for $u \in H^1(D,\Delta)$ so that (6.23) holds (with $\int_{\partial D}\partial_n u\,\overline{\gamma v}$ understood as $\langle \partial_n u,\gamma v\rangle_{\partial D}$ as discussed in Remark 22). □

**Remark 23 (If you remember only one thing from this section...).** *The key point to take from this section regarding function spaces is that if $u\in H^1(D)$ satisfies $\mathscr{L}u=-f$ (in a distributional sense) with $f\in L^2(D)$, then*

$$-\int_D f\overline{v} = \int_{\partial D} \overline{\gamma v}\, \partial_n u - \int_D \left(\nabla u \cdot \overline{\nabla v} - \lambda u \overline{v}\right) \tag{6.28}$$

*for all $v\in H^1(D)$ (with the integral over $\partial D$ understood as a duality pairing as discussed in Remark 22).*

### Self-Adjointness

**Definition 6.8 (Formal adjoint of a differential operator).** *If $\mathscr{L}$ is a general linear differential operator of order $p$, then its* formal adjoint, *denoted by $\mathscr{L}^*$, is defined so that the identity*

$$\overline{v}\mathscr{L}u - u\overline{\mathscr{L}^*v} = \nabla\cdot \mathbf{J}(u,v) \tag{6.29}$$

*holds with $\mathbf{J}(u,v)$ a sesquilinear form involving derivatives of $u$ and $v$ of order $p-1$ or less* [96, Section 5.7], [69, Section 4.3.2], [81, Section 5].

The identity (6.29) (which can be seen as a generalization of G2 (6.22)) is often called *Lagrange's identity* or *Lagrange's formula.*

If $\mathscr{L}^* = \mathscr{L}$, then $\mathscr{L}$ is *formally self-adjoint*. The identity G2 (6.22) therefore shows that the operator $\mathscr{L} := \Delta + \lambda$ is formally self-adjoint when $\lambda \in \mathbb{R}$. Note that formal self-adjointness is a condition on the differential operator itself (i.e., *without* boundary conditions), whereas self-adjointness is a condition on the BVP, i.e., on the differential operator *with* boundary conditions.

We now introduce the notion of adjoint boundary conditions. To keep things simple, we give the definition only for the second-order operator $\mathscr{L} := \Delta + \lambda$. This assumption ensures that the adjoint boundary conditions consist of one condition on each part of the boundary, but the same idea generalizes to higher-order equations (for which the adjoint boundary conditions can involve several conditions on each part of the boundary); see, e.g., [81, Section 1.6], [97, Chapter 3, Section 3]. Note that, up to now, $D$ has always been a bounded Lipschitz domain, but in this subsection (and this subsection only) we allow $D$ to be unbounded.

**Definition 6.9 (Adjoint boundary conditions for BVPs involving $\mathscr{L} := \Delta + \lambda$).** *Let $D$ be a Lipschitz domain (either bounded or unbounded) with outward-pointing normal vector $\nu$. Let $\mathscr{L} := \Delta + \lambda$ and let $u$ satisfy the BVP*

$$\mathscr{L}u = f \text{ in } D \quad \text{and} \quad \mathscr{B}u = 0 \text{ on } \partial D \tag{6.30}$$

*for some linear operator $\mathscr{B}$ involving the trace operators $\gamma$ and $\partial_\nu$. The* adjoint boundary-condition *operator $\mathscr{B}^*$ is such that*

$$\int_D \overline{v}\mathscr{L}u - u\overline{\mathscr{L}^*v} = 0 \tag{6.31}$$

*when $\mathscr{B}u = 0$ and $\mathscr{B}^*v = 0$.*

When $D$ is bounded the Lagrange identity (6.29) and the divergence theorem (6.5) imply that (6.31) is equivalent to

$$\int_{\partial D} \mathbf{J}(u,v)\cdot \nu = 0,$$

but when $D$ is unbounded things are more complicated (see Example 6.12 below).

**Definition 6.10 (Self-adjoint).** *A BVP is* self-adjoint *if the differential operator is formally self-adjoint (i.e., $\mathscr{L} = \mathscr{L}^*$) and the boundary-condition operator $\mathscr{B}$ equals the adjoint boundary-condition operator $\mathscr{B}^*$.*

**Example 6.11 (Self-adjointness of the Helmholtz IDP).** In this case $D = \Omega$, $\nu = \mathbf{n}$, $\mathscr{L}u = (\Delta + k^2)u$, and $\mathscr{B}u = \gamma u$. G2 (6.22) implies that $\mathscr{L}^* = \mathscr{L}$ and $\mathbf{J}(u,v) =$

$\left[\overline{v}\nabla u - u\overline{\nabla v}\right]$, and thus

$$\int_\Omega \overline{v}\mathscr{L}u - u\overline{\mathscr{L}^*v} = \int_{\partial\Omega} \mathbf{J}(u,v)\cdot\mathbf{n} = \int_{\partial\Omega}\left[\overline{\gamma v}\,\partial_n u - \gamma u\,\overline{\partial_n v}\right].$$

When $\mathscr{B}u = 0$, $\int_{\partial\Omega}\mathbf{J}(u,v)\cdot\mathbf{n} = \int_\Gamma \overline{\gamma v}\,\partial_n u$. The condition on $v$ that causes this last integral to vanish is $\gamma v = 0$, and thus $\mathscr{B}^*v = \gamma v$. Since $\mathscr{L} = \mathscr{L}^*$ and $\mathscr{B} = \mathscr{B}^*$, the Helmholtz IDP is self-adjoint. ■

**Example 6.12 (Non-self-adjointness of the Helmholtz EDP).** In this case $D = \Omega_+$, $\mathscr{L}u = (\Delta + k^2)u$, and the boundary-condition operator $\mathscr{B}$ is understood as the trace operator on $\Gamma$ and the Sommerfeld radiation condition (6.17) at infinity. G2 (6.22) again implies that $\mathscr{L}^* = \mathscr{L}$ and $\mathbf{J}(u,v) = \left[\overline{v}\nabla u - u\overline{\nabla v}\right]$.

To find $\mathscr{B}^*$ from the condition (6.31), we observe that

$$\int_{\Omega_+} \overline{v}\mathscr{L}u - u\overline{\mathscr{L}^*v} = \lim_{R\to\infty}\int_{\Omega_R} \overline{v}\mathscr{L}u - u\overline{\mathscr{L}^*v} = \lim_{R\to\infty}\int_{\partial\Omega_R} \mathbf{J}(u,v)\cdot\boldsymbol{\nu},$$

where $\boldsymbol{\nu}$ is the outward-pointing unit normal vector to $\Omega_R$ (so $\boldsymbol{\nu} = -\mathbf{n}$ on $\Gamma$ and $\boldsymbol{\nu} = \widehat{\mathbf{x}}$ on $\Gamma_R$). Therefore, from (6.29) and the divergence theorem (6.5),

$$\int_{\Omega_+} \overline{v}\mathscr{L}u - u\overline{\mathscr{L}^*v} = -\int_\Gamma \left[\overline{\gamma_+ v}\,\partial_n^+ u - \gamma_+ u\,\overline{\partial_n^+ v}\right] + \lim_{R\to\infty}\int_{\Gamma_R}\left[\overline{v}\frac{\partial u}{\partial r} - u\overline{\frac{\partial v}{\partial r}}\right]. \tag{6.32}$$

Just like in the case of the IDP, $\mathscr{B}^* = \gamma$ on $\Gamma$.

To determine $\mathscr{B}^*$ at infinity, we need to consider the integral over $\Gamma_R$ in (6.32). More specifically, $\mathscr{B}^*v$ will consist of the conditions on $v$ that ensure that

$$\lim_{R\to\infty}\int_{\Gamma_R}\left[\overline{v}\frac{\partial u}{\partial r} - u\overline{\frac{\partial v}{\partial r}}\right] = 0 \tag{6.33}$$

when $u$ satisfies the Sommerfeld radiation condition (6.17). The radiation condition (6.17) implies that, when $\mathbf{x}\in\Gamma_R$,

$$\frac{\partial u}{\partial r}(\mathbf{x}) - \mathrm{i}ku(\mathbf{x}) = o\left(\frac{1}{R^{(d-1)/2}}\right), \quad u(\mathbf{x}) = \mathcal{O}\left(\frac{1}{R^{(d-1)/2}}\right), \quad \text{and} \quad \frac{\partial u}{\partial r}(\mathbf{x}) = \mathcal{O}\left(\frac{1}{R^{(d-1)/2}}\right) \tag{6.34}$$

as $R\to\infty$ (see [22, Theorem 3.6] for how the second and third conditions in (6.34) follow from the first). Since $\partial u/\partial r - \mathrm{i}ku$ is smaller than $\partial u/\partial r$ as $r\to\infty$, it makes sense to introduce $\partial u/\partial r - \mathrm{i}ku$ in the integral in (6.33) by adding and subtracting $-\mathrm{i}ku\overline{v}$,

$$\int_{\Gamma_R}\left[\overline{v}\frac{\partial u}{\partial r} - u\overline{\frac{\partial v}{\partial r}}\right] = \int_{\Gamma_R}\left[\overline{v}\left(\frac{\partial u}{\partial r} - \mathrm{i}ku\right) - u\overline{\left(\frac{\partial v}{\partial r} + \mathrm{i}kv\right)}\right]. \tag{6.35}$$

Recalling that $\int_{\Gamma_R} = \mathcal{O}(R^{d-1})$, we see that the right-hand side of (6.35) tends to zero as $R\to\infty$ if $v$ satisfies the *adjoint radiation condition*

$$\frac{\partial v}{\partial r}(\mathbf{x}) + \mathrm{i}kv(\mathbf{x}) = o\left(\frac{1}{R^{(d-1)/2}}\right) \quad \text{and} \quad v(\mathbf{x}) = \mathcal{O}\left(\frac{1}{R^{(d-1)/2}}\right) \tag{6.36}$$

as $R\to\infty$. (In a way similar to how $u = \mathcal{O}(R^{-(d-1)/2})$ follows from $\partial u/\partial r - \mathrm{i}ku =$

$o(R^{-(d-1)/2})$ when $\mathscr{L}u = 0$ [22, Theorem 3.6], the second condition in (6.36) follows from the first when $\mathscr{L}^*v = 0$.)

Therefore, $\mathscr{B}^*v = \gamma v$ on $\Gamma$ and $\mathscr{B}^*$ equals the adjoint radiation condition (6.36) at infinity. ∎

Note that we sometimes call (6.17) the *outgoing* radiation condition and (6.36) the *incoming* radiation condition (with this terminology assuming that we create a solution of the wave equation from a solution of the Helmholtz equation by multiplying by $e^{-i\omega t}$).

In Section 6.1.10 we need the following lemma, which can be proved using the calculations at the end of Example 6.12.

**Lemma 6.13.** *If $u$ and $v$ both satisfy the Sommerfeld radiation condition* (6.17), *then*

$$\lim_{R\to\infty}\int_{\Gamma_R}\left[v\frac{\partial u}{\partial r}-u\frac{\partial v}{\partial r}\right]=0 \quad \textit{and} \quad \lim_{R\to\infty}\int_{\Gamma_R}\left[\overline{v}\frac{\partial u}{\partial r}-u\overline{\frac{\partial v}{\partial r}}\right]=2ik\lim_{R\to\infty}\int_{\Gamma_R}u\overline{v}\neq 0.$$

## 6.1.5 ▪ Recap of Variational Problems

In this section we recap some of the standard theory of variational problems, focusing on how to obtain the following three key things (which correspond to the three questions Q1–Q3 in Section 6.1.1):

K1. Existence and uniqueness of a solution to the variational problem (6.1).

K2. A bound on the solution $u$ in terms of $F$.

K3. Existence and uniqueness of a solution to the Galerkin equations (6.7) (when $N \geq N_0$, for some $N_0$, if necessary) and quasi optimality of the Galerkin solution, i.e., the bound

$$\|u-u_N\|_{\mathscr{H}} \leq C_{qo}\min_{v_N\in\mathscr{H}_N}\|u-v_N\|_{\mathscr{H}} \tag{6.37}$$

for some $C_{qo} > 0$ (again, when $N \geq N_0$ if necessary).

This theory is contained in many texts; we follow [86] and use mostly the same notation.

The variational problem (6.1) concerns a sesquilinear form mapping $\mathscr{H}\times\mathscr{H}\to\mathbb{C}$. We now consider the more general situation of a sesquilinear form mapping $\mathscr{H}_1\times\mathscr{H}_2\to\mathbb{C}$, where $\mathscr{H}_1$ and $\mathscr{H}_2$ are two (not necessarily equal) Hilbert spaces.

**Definition 6.14 (Sesquilinear form).** *Let $\mathscr{H}_1$ and $\mathscr{H}_2$ be Hilbert spaces over $\mathbb{C}$. The mapping $a(\cdot,\cdot):\mathscr{H}_1\times\mathscr{H}_2\to\mathbb{C}$ is called a* sesquilinear form *if it is linear in its first argument and antilinear in its second argument. That is, if $u_1,u_2\in\mathscr{H}_1$, $v_1,v_2\in\mathscr{H}_2$, and $\lambda_1,\lambda_2\in\mathbb{C}$, then*

$$a(\lambda_1u_1+\lambda_2u_2,v_1)=\lambda_1a(u_1,v_1)+\lambda_2a(u_2,v_1)$$

*and*

$$a(u_1,\lambda_1v_1+\lambda_2v_2)=\overline{\lambda_1}a(u_1,v_1)+\overline{\lambda_2}a(u_1,v_2).$$

The appropriate variational problem for a sesquilinear form $a:\mathscr{H}_1\times\mathscr{H}_2\to\mathbb{C}$ is then, given a continuous antilinear functional $F:\mathscr{H}_2\to\mathbb{C}$,

$$\boxed{\text{find } u\in\mathscr{H}_1 \text{ such that } a(u,v)=F(v) \quad \text{for all } v\in\mathscr{H}_2.} \tag{6.38}$$

The Hilbert space $\mathscr{H}_1$ is often called the *trial space* and $\mathscr{H}_2$ is then called the *test space*.

**Definition 6.15 (Bilinear form).** *Let $\mathcal{H}_1$ and $\mathcal{H}_2$ be Hilbert spaces over $\mathbb{R}$. The mapping $a(\cdot,\cdot) : \mathcal{H}_1 \times \mathcal{H}_2 \to \mathbb{R}$ is called a* bilinear form *if it is linear in both arguments. That is, if $u_1, u_2 \in \mathcal{H}_1$, $v_1, v_2 \in \mathcal{H}_2$, and $\lambda_1, \lambda_2 \in \mathbb{R}$, then*

$$a(\lambda_1 u_1 + \lambda_2 u_2, v_1) = \lambda_1 a(u_1, v_1) + \lambda_2 a(u_2, v_1)$$

*and*

$$a(u_1, \lambda_1 v_1 + \lambda_2 v_2) = \lambda_1 a(u_1, v_1) + \lambda_2 a(u_1, v_2).$$

In the rest of this section we focus only on sesquilinear forms (since the majority of variational problems in the rest of the paper involve these), but all the results below have analogues involving bilinear forms.

**Definition 6.16 (Continuity of sesquilinear forms).** *A sesquilinear form $a(\cdot,\cdot) : \mathcal{H}_1 \times \mathcal{H}_2 \to \mathbb{C}$ is* continuous *(or* bounded*) if there exists a $C_c < \infty$ such that*

$$|a(u,v)| \leq C_c \|u\|_{\mathcal{H}_1} \|v\|_{\mathcal{H}_2} \tag{6.39}$$

*for all $u \in \mathcal{H}_1$ and $v \in \mathcal{H}_2$. We define the smallest $C_c$ for which* (6.39) *holds to be the norm of $a(\cdot,\cdot)$ denoted by $\|a\|$.*

The next lemma shows that we can identify a sesquilinear form mapping $\mathcal{H}_1 \times \mathcal{H}_2 \to \mathbb{C}$ with a linear operator from $\mathcal{H}_1 \to \mathcal{H}_2^*$, where $\mathcal{H}_2^*$ is the antidual space of $\mathcal{H}_2$ (i.e., the set of all continuous, antilinear functionals on $\mathcal{H}_2$; see [86, Section 2.1.2 and Lemma 2.1.38]). Given $F \in \mathcal{H}_2^*$ we introduce the notation that

$$F(v) = \langle F, v \rangle_{\mathcal{H}_2^* \times \mathcal{H}_2},$$

where $\langle \cdot,\cdot \rangle_{\mathcal{H}_2^* \times \mathcal{H}_2}$ is called a *duality pairing*. (Note that $\langle \cdot,\cdot \rangle_{\mathcal{H}_2^* \times \mathcal{H}_2}$ is then a sesquilinear form on $\mathcal{H}_2^* \times \mathcal{H}_2$.) We use $L(\mathcal{H}_1, \mathcal{H}_2^*)$ to denote the set of all continuous (i.e., bounded) linear operators from $\mathcal{H}_1$ to $\mathcal{H}_2^*$.

**Lemma 6.17 (Sesquilinear forms $\longleftrightarrow$ linear operators).** *For every continuous sesquilinear form $a : \mathcal{H}_1 \times \mathcal{H}_2 \to \mathbb{C}$ there exists a unique $A \in L(\mathcal{H}_1, \mathcal{H}_2^*)$ such that*

$$a(u,v) = \langle Au, v \rangle_{\mathcal{H}_2^* \times \mathcal{H}_2}$$

*for all $u \in \mathcal{H}_1$, $v \in \mathcal{H}_2$, where $\langle \cdot,\cdot \rangle_{\mathcal{H}_2^* \times \mathcal{H}_2}$ denotes the duality pairing between $\mathcal{H}_2^*$ and $\mathcal{H}_2$. Furthermore,*

$$\|A\|_{\mathcal{H}_1 \to \mathcal{H}_2^*} \leq \|a\|.$$

***Proof.*** See [86, Lemma 2.1.38]. □

In what follows we make no distinction between a sesquilinear form $a : \mathcal{H}_1 \times \mathcal{H}_2 \to \mathbb{C}$ and the associated operator $A : \mathcal{H}_1 \to \mathcal{H}_2^*$ (i.e., we state results in terms of whichever object is most natural).

Given $A \in L(\mathcal{H}_1, \mathcal{H}_2^*)$ we say that $A$ is invertible if it is both injective and surjective (so there is an algebraic inverse $A^{-1}$). Observe that if $A^{-1} \in L(\mathcal{H}_2^*, \mathcal{H}_1)$, then K1 and K2 both hold, and vice versa.

Our goal in this section is to find conditions on $a(\cdot,\cdot)$ that ensure the properties K1–K3 above hold (to ensure K3 holds we might also need conditions on the finite-dimensional subspaces of the Galerkin method). We discuss three properties $a(\cdot,\cdot)$ can have that ensure K1–K3 hold (in various degrees). These three properties are that

(a) $a : \mathcal{H}_1 \times \mathcal{H}_2 \to \mathbb{C}$ satisfies the *inf-sup condition*,

(b) $a : \mathcal{H} \times \mathcal{H} \to \mathbb{C}$ is *coercive*, and

(c) $a : \mathcal{H} \times \mathcal{H} \to \mathbb{C}$ is *coercive up to a compact perturbation.*

Property (a) is the weakest, (b) is the strongest, and (c) is in the middle.

## The inf-sup Condition

**Definition 6.18 (The inf-sup condition).** *The sesquilinear form $a : \mathcal{H}_1 \times \mathcal{H}_2 \to \mathbb{C}$ satisfies the* inf-sup condition *if there exists a $\gamma > 0$ such that*

$$\inf_{u\in\mathcal{H}_1\setminus\{0\}} \sup_{v\in\mathcal{H}_2\setminus\{0\}} \frac{|a(u,v)|}{\|u\|_{\mathcal{H}_1}\|v\|_{\mathcal{H}_2}} \geq \gamma \tag{6.40a}$$

*and*

$$\sup_{u\in\mathcal{H}_1\setminus\{0\}} |a(u,v)| > 0 \quad \textit{for all } v \in \mathcal{H}_2 \setminus \{0\}. \tag{6.40b}$$

The following theorem is fundamental.

**Theorem 6.19 (K1 and K2 under the inf-sup condition).** *The following are equivalent:*

(a) *The sesquilinear form $a(\cdot,\cdot)$ satisfies the inf-sup condition* (6.40).

(b) *$A$ is invertible and $A^{-1} \in L(\mathcal{H}_2^*, \mathcal{H}_1)$ with $\|A^{-1}\|_{\mathcal{H}_2^*\to\mathcal{H}_1} \leq 1/\gamma$; i.e., for each $F \in \mathcal{H}_2^*$, the variational problem* (6.38) *has a unique solution which satisfies*

$$\|u\|_{\mathcal{H}_1} \leq \frac{1}{\gamma}\|F\|_{\mathcal{H}_2^*}. \tag{6.41}$$

***Proof.*** See [86, Theorem 2.1.44 and Remark 2.1.46]. □

**Remark 24.** *In functional analysis texts one often encounters the result that $B \in L(H_1, H_2)$ is invertible (with bounded inverse) if* (i) *$B$ is bounded below, i.e.,*

$$\|Bu\|_{H_2} \geq \gamma\|u\|_{H_1} \quad \textit{for all } u \in H_1, \tag{6.42}$$

*and* (ii) *the range of $B$ is dense in $H_2$; see, e.g.,* [2, Lemma 2.8]. *The first part of the inf-sup condition* (6.40a) *is equivalent to $A \in L(\mathcal{H}_1, \mathcal{H}_2^*)$ being bounded below (i.e.,* (6.42) *holds with $B$ replaced by $A$, $H_1$ replaced by $\mathcal{H}_1$, and $H_2$ replaced by $\mathcal{H}_2^*$). The second part* (6.40b) *ensures that the range of $A$ is dense in $\mathcal{H}_2^*$ (see* [86, Part (iii) of proof of Theorem 2.1.44]*).*

Before we state the next result (which gives conditions under which K3 holds), we need to define the discrete inf-sup condition. We first introduce some notation. For $i = 1,2$ let $(\mathcal{H}_N^i)_{N\in\mathbb{Z}^+}$ be a sequence of finite-dimensional, nested subspaces of $\mathcal{H}_i$, whose union is dense in $\mathcal{H}_i$; i.e., for all $N \geq 1$,

$$\mathcal{H}_N^i \subset \mathcal{H}_{N+1}^i, \quad \dim \mathcal{H}_N^i < \infty, \quad \overline{\bigcup_{N\in\mathbb{Z}^+} \mathcal{H}_N^i} = \mathcal{H}_i. \tag{6.43}$$

We assume that $\dim \mathcal{H}_N^i = N$ for $i = 1,2$.

The Galerkin equations for the variational problem (6.38) are then

$$\boxed{\text{find } u_N \in \mathcal{H}_N^1 \text{ such that } a(u_N, v_N) = F(v_N) \quad \text{for all } v_N \in \mathcal{H}_N^2.} \tag{6.44}$$

Note that

- Galerkin methods with different trial and test spaces, i.e., those of the form (6.44), are often called *Petrov–Galerkin* methods (with methods that use the same trial and test spaces, i.e., those of the form (6.7), then called *Bubnov–Galerkin* methods), and
- from the variational problem (6.38) and the Galerkin equations (6.44), one obtains the *Galerkin orthogonality* condition that

$$a(u-u_N, v_N) = 0 \quad \text{for all } v_N \in \mathcal{H}_N^2. \tag{6.45}$$

**Definition 6.20 (The discrete inf-sup condition).** *The sesquilinear form $a(\cdot,\cdot)$ and finite-dimensional subspaces $(\mathcal{H}_N^i)_{N\in\mathbb{Z}^+}$, $i=1,2$, satisfy the* discrete inf-sup condition *if, for each $N$, there exists a $\gamma_N > 0$ such that*

$$\inf_{u\in\mathcal{H}_N^1\setminus\{0\}} \sup_{v\in\mathcal{H}_N^2\setminus\{0\}} \frac{|a(u,v)|}{\|u\|_{\mathcal{H}_1}\|v\|_{\mathcal{H}_2}} \geq \gamma_N \tag{6.46a}$$

*and*

$$\sup_{u\in\mathcal{H}_N^1\setminus\{0\}} |a(u,v)| > 0 \quad \text{for all } v \in \mathcal{H}_N^2 \setminus \{0\}. \tag{6.46b}$$

Note that

- since $\mathcal{H}_N^1$ and $\mathcal{H}_N^2$ are finite dimensional, the infima and suprema in Definition 6.20 can be replaced by minima and maxima, respectively, and
- (although it is not important for the rest of this paper) the second condition (6.46b) can be obtained from the first (6.46a) (recalling that $N = \dim \mathcal{H}_N^1 = \dim \mathcal{H}_N^2$); see, e.g., [60, Page 16].

**Theorem 6.21 (K3 under discrete inf-sup condition).** *Assume that $a(\cdot,\cdot)$ is continuous and satisfies the inf-sup condition* (6.40). *Assume that there exists an $N_0$ such that $(\mathcal{H}_N^i)_{N\in\mathbb{Z}^+}$, $i=1,2$, satisfy the discrete inf-sup condition* (6.46) *when $N \geq N_0$. Then, for each $F \in \mathcal{H}_2^*$, the variational problem* (6.38) *has a unique solution, the Galerkin equations* (6.44) *have a unique solution $u_N \in \mathcal{H}_N^1$ when $N \geq N_0$, and*

$$\|u-u_N\|_{\mathcal{H}_1} \leq \left(\frac{\|a\|}{\gamma_N}\right) \min_{v_N\in\mathcal{H}_N^1} \|u-v_N\|_{\mathcal{H}_1} \tag{6.47}$$

*when $N \geq N_0$.*

***Proof.*** This result is usually proved with the factor $(1+\|a\|/\gamma_N)$ in front of the best approximation error in the quasi-optimality bound; see, e.g., [86, Theorem 4.2.1]. It was observed in [102], however, that a result of Kato on projection operators [68, Lemma 4] means that $(1+\|a\|/\gamma_N)$ can be replaced by $(\|a\|/\gamma_N)$ (see [26] for a good introduction to these results). Note that (6.47) implies that if $\gamma_N$ is independent of $N$, then $u_N \to u$ as $N \to \infty$. □

Having established that K1, K2, and K3 hold if the sesquilinear form satisfies the inf-sup condition (and the finite-dimensional spaces satisfy the discrete inf-sup condition), we now consider sesquilinear forms that correspond to *compact perturbations* of invertible operators and use Fredholm theory (see, e.g., [86, Section 2.1.3] for the definition of a compact operator).

Note for the next theorem that $Au = 0$ (as an element of $\mathcal{H}_2^*$) if and only if $a(u,v) = 0$ for all $v \in \mathcal{H}_2$.

**Theorem 6.22 (K1 and K2 for invertible + compact).** *Let $A \in L(\mathcal{H}_1, \mathcal{H}_2^*)$ be such that $A = B + T$, where $B \in L(\mathcal{H}_1, \mathcal{H}_2^*)$ is such that $B^{-1} \in L(\mathcal{H}_2^*, \mathcal{H}_1)$ (i.e., $b(\cdot,\cdot)$ satisfies the inf-sup condition) and $T \in L(\mathcal{H}_1, \mathcal{H}_2^*)$ is compact. If $A$ is injective, i.e.,*

$$Au = 0 \quad \text{implies that} \quad u = 0,$$

*then $A$ is invertible and $A^{-1} \in L(\mathcal{H}_2^*, \mathcal{H}_1)$ (i.e., $a(\cdot,\cdot)$ satisfies the inf-sup condition (6.40)).*

***Proof.*** This is proved using Fredholm theory in [86, Theorem 4.2.7] when $\mathcal{H}_1 = \mathcal{H}_2$, but the proof when $\mathcal{H}_1 \neq \mathcal{H}_2$ follows in the same way. (Note that this argument does not give us any information about the norm of $A^{-1}$, other than that it is finite.) □

Combining Theorems 6.22 and 6.21, we obtain the following corollary.

**Corollary 6.23 (K3 for invertible + compact under the discrete inf-sup condition).** *Let $A$ satisfy the assumptions of Theorem 6.22 and assume that there exists an $N_0$ such that $a(\cdot,\cdot)$ and $(\mathcal{H}_N^i)_{N\in\mathbb{Z}^+}$, $i = 1,2$, satisfy the discrete inf-sup condition (6.46) when $N \geq N_0$. Then, when $N \geq N_0$, the Galerkin equations (6.44) have a unique solution $u_N \in \mathcal{H}_N^1$ and the error estimate (6.47) holds.*

#### Coercivity

We now assume that $\mathcal{H}_1 = \mathcal{H}_2 = \mathcal{H}$.

**Definition 6.24.** *The sesquilinear form $a : \mathcal{H} \times \mathcal{H} \to \mathbb{C}$ is coercive if there exists an $\alpha > 0$ such that*

$$|a(v,v)| \geq \alpha \|v\|_{\mathcal{H}}^2 \quad \text{for all } v \in \mathcal{H}. \tag{6.48}$$

The property (6.48) is sometimes called "$\mathcal{H}$-ellipticity" (as in, e.g., [86, Page 39], [98, Section 3.2], and [66, Definition 5.2.2]), with "coercivity" then used to mean *either* that $a(\cdot,\cdot)$ is a compact perturbation of a coercive operator (as in, e.g., [98, Section 3.6] and [66, Section 5.2]) *or* that $a(\cdot,\cdot)$ satisfies a Gårding inequality (as in [86, Definition 2.1.54]).

**Remark 25 (The numerical range and alternative definitions of coercivity).** *Define the* numerical range *of $A$ by*

$$W(A) := \left\{ \frac{a(v,v)}{\|v\|_{\mathcal{H}}^2} : v \in \mathcal{H} \setminus \{0\} \right\}. \tag{6.49}$$

*This set is convex (see, e.g., [63, Theorem 1.1-2]) and so if $a(\cdot,\cdot)$ is coercive, then there exists a $\theta \in [0, 2\pi)$ such that*

$$\operatorname{Re}\left(e^{i\theta} a(v,v)\right) \geq \alpha \|v\|_{\mathcal{H}}^2 \quad \text{for all } v \in \mathcal{H}. \tag{6.50}$$

*Although the convexity of the numerical range is well known, the implication (6.50) does not usually appear in the literature on variational problems. (Note that (6.50) is sometimes used as a definition of coercivity; see [86, Equation (2.43)].)*

**Theorem 6.25 (The Lax–Milgram theorem).** *If $a : \mathcal{H} \times \mathcal{H} \to \mathbb{C}$ is a continuous and coercive sesquilinear form, then $A^{-1} \in L(\mathcal{H}^*, \mathcal{H})$ with $\|A^{-1}\|_{\mathcal{H}^* \to \mathcal{H}} \leq 1/\alpha$; i.e., for each*

*$F \in \mathcal{H}^*$, the variational problem* (6.1) *has a unique solution which satisfies*

$$\|u\|_{\mathcal{H}} \leq \frac{1}{\alpha} \|F\|_{\mathcal{H}^*}.$$

***Proof.*** One can *either* show that $a(\cdot,\cdot)$ satisfies the inf-sup condition (6.40) with $\gamma = \alpha$ and then obtain the result from Theorem 6.19 (see, e.g., [86, Proof of Lemma 2.1.51]) *or* prove the result directly (see, e.g., [41, Section 6.2.1], [73, Lemma 2.32], [66, Theorem 5.2.3], [98, Theorem 3.4]). □

Observe that the Lax–Milgram theorem therefore gives K1 and K2 for continuous and coercive sesquilinear forms.

The following lemma shows the link with the other definition of a "variational problem" mentioned in Section 6.1.1 (the problem of minimizing a functional).

**Lemma 6.26 (The link between symmetric coercive bilinear forms and minimizing functionals).** *Assume that $a : \mathcal{H} \times \mathcal{H} \to \mathbb{R}$ (note the change from a sesquilinear form to a bilinear form) is continuous, coercive, and symmetric (i.e., $a(u,v) = a(v,u)$ for all $u, v \in \mathcal{H}$). Then, given $F \in \mathcal{H}'$, the unique solution of the variational problem* (6.1) *is also a solution of the problem of minimizing the quadratic functional $J : \mathcal{H} \to \mathbb{R}$ defined by*

$$J(v) := \frac{1}{2}a(v,v) - F(v). \tag{6.51}$$

*Conversely, if $u \in \mathcal{H}$ minimizes* (6.51), *then $u$ solves the variational problem* (6.1).

***Proof.*** See [86, Proposition 2.1.53]. □

The following result gives K3 when $a(\cdot,\cdot)$ is continuous and coercive.

**Theorem 6.27 (Céa's lemma).** *Let $a : \mathcal{H} \times \mathcal{H} \to \mathbb{C}$ be continuous and coercive. If $\mathcal{H}_N$ is any finite-dimensional subspace of $\mathcal{H}$, then the Galerkin equations* (6.7) *have a unique solution and the quasi-optimality bound* (6.8) *holds with $C_{qo} = C_c/\alpha$.*

***Proof.*** This is proved in [86, Proposition 4.1.25] for particular $a(\cdot,\cdot)$ and $\mathcal{H}$, although the proof also applies to the general case; see also [98, Theorem 8.1], [20, Theorem 13.1]. □

### Coercivity up to a Compact Perturbation

We saw in Section 6.1.5 that Fredholm theory gives us K1 and K2 for compact perturbations of invertible operators (assuming that the resulting operator is injective) and that we can obtain K3 if the finite-dimensional subspaces satisfy the discrete inf-sup condition (usually under the assumption that $N$ is sufficiently large).

We see below that for compact perturbations of *coercive* (as opposed to just invertible) operators we can obtain K3 (if $N$ is large enough) *without* the finite-dimensional subspaces satisfying the discrete inf-sup condition.

**Definition 6.28 (Coercivity up to a compact perturbation).** *$A \in L(\mathcal{H}, \mathcal{H}^*)$ is a compact perturbation of a coercive operator if there exists a compact operator $T \in L(\mathcal{H}, \mathcal{H}^*)$ and $\beta > 0$ such that*

$$\left|\langle (A - T)v, v\rangle_{\mathcal{H}^* \times \mathcal{H}}\right| \geq \beta \|v\|^2_{\mathcal{H}} \quad \textit{for all } v \in \mathcal{H}$$

*(i.e., $A = B + T$, where $B$ is coercive).*

**Theorem 6.29 (K1, K2, and K3 for coercive + compact).** *Let $\mathcal{H}$ be a Hilbert space over $\mathbb{C}$ and let $(\mathcal{H}_N)_{N\in\mathbb{Z}^+}$ be a dense sequence of finite-dimensional nested subspaces (i.e., (6.43) holds). Let $A \in L(\mathcal{H},\mathcal{H}^*)$ be a compact perturbation of a coercive operator and assume that $A$ is injective. Then the following hold:*

(i) *$A$ is invertible and $A^{-1} \in L(\mathcal{H}^*,\mathcal{H})$.*

(ii) *There exists an $N_0 > 0$ such that, for all $N \geq N_0$ and all $F \in \mathcal{H}^*$, the Galerkin equations (6.7) have a unique solution and there exists a $C_{qo}$ (independent of $N$) such that the quasi-optimality bound (6.8) holds.*

***Proof.*** Part (i) follows from Theorem 6.22, since every coercive operator satisfies the inf-sup condition. See [86, Theorem 4.2.9] for the proof of (ii). This proof verifies that the discrete inf-sup condition holds (with $\gamma_N$ independent of $N$ for $N$ sufficiently large) by a contradiction argument and then uses Corollary 6.23. Note that the proof does not give an explicit value for $\gamma_N$, and so we do not get an explicit value for $C_{qo}$. □

A common situation in which $A$ is a compact perturbation of a coercive operator is when $a(\cdot,\cdot)$ satisfies a *Gårding inequality*.

**Definition 6.30 (Gårding inequality).** *Let $\mathcal{H}$ and $\mathcal{V}$ be Hilbert spaces such that $\mathcal{H} \subset \mathcal{V}$ and the inclusion map is continuous. We say that $a(\cdot,\cdot)$ satisfies a Gårding inequality if there exist $\alpha > 0$ and $C_{\mathcal{V}} \in \mathbb{R}$ such that*

$$\mathrm{Re}\, a(v,v) \geq \alpha \|v\|_{\mathcal{H}}^2 - C_{\mathcal{V}} \|v\|_{\mathcal{V}}^2 \quad \text{for all } v \in \mathcal{H}. \tag{6.52}$$

(Note that if $C_{\mathcal{V}} \leq 0$, then $a(\cdot,\cdot)$ is coercive, and so the definition is only really useful when $C_{\mathcal{V}} > 0$.)

The standard example is $\mathcal{H} = H^1(\Omega)$ (or a closed subspace of it such as $H_0^1(\Omega)$) and $\mathcal{V} = L^2(\Omega)$, and we encounter this situation in Section 6.1.6.

**Theorem 6.31 (Gårding inequality $\Longrightarrow$ coercive + compact).** *Let $\mathcal{H}$ and $\mathcal{V}$ be as in Definition 6.30, and assume further that the inclusion map is compact. If $a(\cdot,\cdot)$ satisfies a Gårding inequality, then $A : \mathcal{H} \to \mathcal{H}^*$ is a compact perturbation of a coercive operator.*

In the literature, the Gårding inequality (6.52) is usually stated only for the case where $\mathcal{H} = H^1(\Omega)$ or $H_0^1(\Omega)$ and $\mathcal{V} = L^2(\Omega)$ (exceptions are [73, Page 44] and [82, Theorem 5.4.5][4]). We have been unable to find an explicit proof of Theorem 6.31 in the general case, and so we give one here.

***Proof of Theorem* 6.31.** Let $\imath$ denote the inclusion map from $\mathcal{H}$ to $\mathcal{V}$. Define $T_1 : \mathcal{V} \to \mathcal{H}^*$ by

$$\langle T_1 u, v\rangle_{\mathcal{H}^*\times\mathcal{H}} = (u, \imath v)_{\mathcal{V}},$$

where $(\cdot,\cdot)_{\mathcal{V}}$ denotes the inner product on $\mathcal{V}$. Then, by the Cauchy–Schwarz inequality and the boundedness of $\imath$,

$$\left|\langle T_1 u, v\rangle_{\mathcal{H}^*\times\mathcal{H}}\right| \leq \|u\|_{\mathcal{V}} \|\imath\|_{\mathcal{H}\to\mathcal{V}} \|v\|_{\mathcal{H}};$$

this implies that $\|T_1 u\|_{\mathcal{H}^*} \leq \|u\|_{\mathcal{V}} \|\imath\|_{\mathcal{H}\to\mathcal{V}}$ and thus $T_1 : \mathcal{V} \to \mathcal{H}^*$ is continuous.

Next define $T_2 = T_1 \circ \imath : \mathcal{H} \to \mathcal{H}^*$. Observe that $T_2$ is compact since it is the composition of a continuous operator with a compact operator (see [86, Lemma 2.1.29]).

[4]In both these references $\mathcal{V}$ is used to denote the smaller space, and $\mathcal{H}$ is used to denote the larger one. Here we use the opposite notation, so that $A$ remains an operator from $\mathcal{H}$ to $\mathcal{H}^*$ (as it has been up to now).

Define $B : \mathcal{H} \to \mathcal{H}^*$ by

$$\langle Bu, v\rangle_{\mathcal{H}^*\times\mathcal{H}} := a(u,v) + C_{\mathcal{V}}(\imath u, \imath v)_{\mathcal{V}}.$$

Then, by (6.52), $\mathrm{Re}\langle Bu, v\rangle_{\mathcal{H}^*\times\mathcal{H}} \geq \alpha \|v\|^2_{\mathcal{H}}$, so $B$ is coercive. Furthermore, $B = A + C_{\mathcal{V}} T_2$, since

$$\langle T_2 u, v\rangle_{\mathcal{H}^*\times\mathcal{H}} = \langle T_1 \imath u, v\rangle_{\mathcal{H}^*\times\mathcal{H}} = (\imath u, \imath v)_{\mathcal{V}},$$

and thus $A$ is a compact perturbation of a coercive operator since $T_2$ is compact. □

If $a(\cdot,\cdot)$ is injective and satisfies a Gårding inequality (with $\mathcal{H}$ compactly contained in $\mathcal{V}$), then Theorems 6.31 and 6.29 give quasi optimality of the Galerkin method (i.e., K3), but *without* explicit expressions for $C_{qo}$ or $N_0$. The advantage of $a(\cdot,\cdot)$ satisfying a Gårding inequality (over just knowing that $A$ is a compact perturbation of a coercive operator) is that we can obtain information about $C_{qo}$ and $N_0$. We can do this by *either* verifying that the discrete inf-sup condition holds *or* proving quasi optimality directly. It turns out that the second method gives stronger results, and so we present only this method here (but see Remark 28 for a comparison of the results of the two methods).

**Theorem 6.32 (K3 for Gårding inequality).** *Let $\mathcal{H}$ and $\mathcal{V}$ be as in Definition* 6.30 *and assume further that the inclusion map is compact. Let $a : \mathcal{H}\times\mathcal{H} \to \mathbb{C}$ be continuous and injective and satisfy a Gårding inequality (so that by Theorems* 6.29 *and* 6.31 *the variational problem* (6.1) *has a unique solution).*

*Given $f \in \mathcal{V}$, define $S^* f \in \mathcal{H}$ as the solution of the variational problem*

$$a(v, S^* f) = (v, f)_{\mathcal{V}} \quad \textit{for all } v \in \mathcal{H} \tag{6.53}$$

*(observe that $(v,f)_{\mathcal{V}}$ is a linear functional on $\mathcal{H}$ since $\mathcal{H}$ is continuously embedded in $\mathcal{V}$). Let $(\mathcal{H}_N)_{N\in\mathbb{Z}^+}$ be a dense sequence of finite-dimensional, nested subspaces, and let*

$$\eta(\mathcal{H}_N) := \sup_{f\in\mathcal{V}} \min_{v_N\in\mathcal{H}_N} \frac{\|S^* f - v_N\|_{\mathcal{H}}}{\|f\|_{\mathcal{V}}}. \tag{6.54}$$

*If*

$$\eta(\mathcal{H}_N) \leq \frac{1}{C_c}\sqrt{\frac{\alpha}{2C_{\mathcal{V}}}}, \tag{6.55}$$

*then the Galerkin equations* (6.7) *have a unique solution which satisfies*

$$\|u - u_N\|_{\mathcal{H}} \leq \frac{2C_c}{\alpha} \min_{v_N\in\mathcal{H}_N} \|u - v_N\|_{\mathcal{H}} \tag{6.56}$$

*(i.e., quasi optimality with $C_{qo} = 2C_c/\alpha$).*

Note that the operator $S^* : \mathcal{V} \to \mathcal{H}$ in Theorem 6.32 is related to the adjoint of $A$ (hence the reason for the $^*$ notation). In what follows we refer to the variational problem (6.53) as the *adjoint variational problem*, although it is really the adjoint problem with a certain class of right-hand sides (see, e.g., [86, Exercise 2.1.41] for more on adjoint operators).

***Proof of Theorem* 6.32.** Since this argument usually appears in the literature only for the case when $\mathcal{H} = H^1(\Omega)$ and $\mathcal{V} = L^2(\Omega)$ (see, e.g., [15, Section 5.7], [40, Lemma 4.1]), we give the details for the general situation here.

We first assume that a solution $u_N$ exists. The Gårding inequality (6.52) applied to $u-u_N$ implies that

$$\alpha\,\|u-u_N\|_{\mathscr{H}}^2 - C_{\mathscr{V}}\,\|u-u_N\|_{\mathscr{V}}^2 \le \operatorname{Re} a(u-u_N, u-u_N).$$

By Galerkin orthogonality (6.45), the right-hand side of this inequality can be replaced by $\operatorname{Re} a(u-u_N, u-v_N)$ for any $v_N \in \mathscr{H}_N$. Using this fact along with continuity of $a(\cdot,\cdot)$, we find that

$$\alpha\,\|u-u_N\|_{\mathscr{H}}^2 - C_{\mathscr{V}}\,\|u-u_N\|_{\mathscr{V}}^2 \le C_c\,\|u-u_N\|_{\mathscr{H}}\,\|u-v_N\|_{\mathscr{H}} \qquad \text{for all } v_N \in \mathscr{H}_N.$$

Therefore, the quasi optimality (6.56) follows if we can show that

$$\sqrt{C_{\mathscr{V}}}\,\|u-u_N\|_{\mathscr{V}} \le \sqrt{\frac{\alpha}{2}}\,\|u-u_N\|_{\mathscr{H}}. \tag{6.57}$$

Now, by the definition of $S^*$ (6.53), Galerkin orthogonality (6.45), and continuity,

$$\begin{aligned}\|u-u_N\|_{\mathscr{V}}^2 &= a\big(u-u_N, S^*(u-u_N)\big) = a\big(u-u_N, S^*(u-u_N)-v_N\big)\\ &\le C_c\,\|u-u_N\|_{\mathscr{H}}\,\|S^*(u-u_N)-v_N\|_{\mathscr{H}}\end{aligned} \tag{6.58}$$

for any $v_N \in \mathscr{H}_N$. The definition of $\eta(\mathscr{H}_N)$ (6.54) implies that there exists a $w_N \in \mathscr{H}_N$ such that

$$\|S^*(u-u_N)-w_N\|_{\mathscr{H}} \le \eta(\mathscr{H}_N)\,\|u-u_N\|_{\mathscr{V}}$$

and using this fact in (6.58) we obtain that

$$\|u-u_N\|_{\mathscr{V}} \le C_c\,\eta(\mathscr{H}_N)\,\|u-u_N\|_{\mathscr{H}}. \tag{6.59}$$

Therefore, the condition (6.55) implies that (6.57), and thus also (6.56), holds.

We have so far assumed that $u_N$ exists. Recall that an $N\times N$ matrix $A$ is invertible if and only if $A$ has full rank, which is the case if and only if the only solution of $A\mathbf{x}=0$ is $\mathbf{x}=0$. Therefore, to show that $u_N$ exists, we need only show that $u_N$ is unique. Seeking a contradiction, suppose that there exists a $\widetilde{u}_N \in \mathscr{H}_N$ such that

$$a(\widetilde{u}_N, v_N) = 0 \qquad \text{for all } v_N \in \mathscr{H}_N.$$

Let $\widetilde{u}$ be such that

$$a(\widetilde{u}, v) = 0 \qquad \text{for all } v \in \mathscr{H}; \tag{6.60}$$

thus $\widetilde{u}_N$ is the Galerkin approximation to $\widetilde{u}$. Repeating the argument in the first part of the proof we see that if (6.55) holds, then the quasi optimality (6.56) holds (with $u$ replaced by $\widetilde{u}$ and $u_N$ replaced by $\widetilde{u}_N$). By assumption, the only solution to the variational problem (6.60) is $\widetilde{u}=0$, and then (6.56) implies that $\widetilde{u}_N=0$. We have therefore shown that the solution $u_N$ exists under the condition (6.55) and the proof is complete. □

**Remark 26 (The history of the argument in Theorem 6.32).** *The argument that obtains* (6.59) *from* (6.58) *was first introduced in the coercive case by Nitsche in* [83] *and Aubin in* [8, Theorem 3.1] *and is thus often referred to as the "Aubin-Nitsche lemma" or the "Aubin-Nitsche duality argument" (see, e.g.,* [20, Theorem 19.1]*). Schatz* [88] *was then the first to* *use this argument in conjunction with a Gårding inequality to prove quasi optimality of the corresponding Galerkin method.*

**Remark 27 (Can one obtain information about $N_0$ via the condition on $\eta(\mathcal{H}_N)$?).** *The usefulness of Theorem* 6.32 *depends on how easy it is to estimate $\eta(\mathcal{H}_N)$ and thus determine a bound on the threshold $N_0$ for quasi optimality to hold. The quantity $\eta(\mathcal{H}_N)$ measures how well the finite-dimensional subspaces $\mathcal{H}_N$ approximate $S^*f$; therefore, to estimate $\eta(\mathcal{H}_N)$ we need information about* (i) *the subspaces and* (ii) $S^*f$.

*Regarding* (ii), *for the standard variational formulations of BVPs involving the PDE $\mathcal{L}u = -f$ (discussed in Section* 6.1.6*), $\mathcal{H} = H^1(\Omega)$ or $H_0^1(\Omega)$ and $\mathcal{V} = L^2(\Omega)$. The solution of the adjoint variational problem* (6.53) *can then be shown to be the solution of the adjoint BVP with $f \in L^2(\Omega)$ and zero boundary conditions, and thus bounds on $S^*f$ in terms of $f$ can be obtained using PDE techniques. (We give the resulting bound on $\eta(\mathcal{H}_N)$ in the case of the Helmholtz IIP in Remark* 31 *below.) This argument relies on the fact that the antilinear functionals in the standard variational formulations are already of the form $(f,v)_{\mathcal{V}}$ for some $f \in \mathcal{V}$. This is not always the case; see, e.g., the variational formulation of the Helmholtz IDP in Section* 6.1.10. *For this variational formulation it is not clear what BVP (if any) the solution of the adjoint variational problem* (6.53) *corresponds to, and thus it is not clear how to obtain a bound on $\eta(\mathcal{H}_N)$ in this case.*

**Remark 28.** *As mentioned before Theorem* 6.32, *one can also use the Gårding inequality to verify that the discrete inf-sup condition* (6.46) *holds and obtain bounds on $C_{qo}$ and $N_0$ this way. Using an argument that also involves the adjoint variational problem* (6.53), *one finds that $a(\cdot,\cdot)$ satisfies the first discrete inf-sup condition* (6.46a) *if*

$$\eta(\mathcal{H}_N) \leq \frac{\alpha}{2C_c C_{\mathcal{V}}} \tag{6.61}$$

*(as with the argument in Theorem* 6.32, *the* 2 *in* (6.61) *can be replaced by any number* $> 1$*). Recall that (as noted under Definition* 6.20*) one can obtain* (6.46b) *from* (6.46a), *and thus under* (6.61) *both parts of the inf-sup condition hold.*

*The condition* (6.61) *is more restrictive than* (6.55). *Indeed, when $a(\cdot,\cdot)$ satisfies a Gårding inequality, the worst-case scenario is when $C_{\mathcal{V}}$ is large and $\alpha$ is small, and then $1/C_{\mathcal{V}}$ is smaller than $1/\sqrt{C_{\mathcal{V}}}$ and $\alpha/2$ is smaller than $\sqrt{\alpha/2}$.*

#### Advantages of Coercivity over the inf-sup Condition and Coercivity up to a Compact Perturbation

We noted above that, of the inf-sup condition, coercivity, and coercivity up to a compact perturbation, coercivity is the strongest property. Comparing the results obtained in the sections above entitled "The inf-sup Condition" and "Coercivity up to a Compact Perturbation," we see that coercivity has the following two advantages over the other two conditions:

1. The constant in the bound on $u$ in terms of $F$ is known explicitly (if $C_c$ and $\alpha$ are given explicitly).

2. We obtain quasi optimality for *any* finite-dimensional subspace (i.e., without any constraint on the subspace dimension) and with an explicit expression for $C_{qo}$.

There is a third advantage of coercivity over the other two conditions, and this concerns the solution of the Galerkin equations:

3. The matrix of the Galerkin method inherits analogous continuity and coercivity properties from $a(\cdot,\cdot)$. These allow one to prove results about the number of iterations required to solve the Galerkin equations using iterative methods. For symmetric, coercive sesquilinear forms one can use the conjugate gradient method, and then the well-known bounds on the number of iterations can be found in, e.g., [61, Chapter 3]. For nonsymmetric, coercive sesquilinear forms one must use more general iterative methods such as the generalized minimum residual method (GMRES), and in this case a result of Elman [38], [37, Theorem 3.3] then gives bounds on the number of iterations (see, e.g., [57, Section 1.3] and [95, Section 1.3] for alternative statements of this result and applications of it to Helmholtz problems).

## 6.1.6 ▪ Standard Variational Formulation (i.e., the Weak Form)

**Summary (linking to Figure 6.1):** G1 with $v$ a test function and $D=\Omega$.

We consider the IDP (Definition 6.2) in each of the three cases $\lambda=0$, $\lambda<0$, and $\lambda>0$ and the IIP for the Helmholtz equation (Definition 6.4).

### The Interior Dirichlet Problem

Since $\lambda$ is real, we can take $f$ and $g_D$ to be real-valued. Indeed, if $f$ and $g_D$ are complex, then the solution $u=u_1+\mathrm{i}u_2$, where the real-valued functions $u_1$ and $u_2$ are the solutions of the BVPs

$$\mathscr{L}u_1=-\operatorname{Re} f \ \text{ on } \Omega \quad \text{and} \quad \gamma u_1=\operatorname{Re} g \ \text{ on } \Gamma$$

and

$$\mathscr{L}u_2=-\operatorname{Im} f \ \text{ on } \Omega \quad \text{and} \quad \gamma u_2=\operatorname{Im} g \ \text{ on } \Gamma.$$

Having real-valued $f$ and $g_D$ means that we can restrict our attention to spaces of real-valued functions.

Multiplying the PDE $\mathscr{L}u=-f$ by a (real) test function $v$ and integrating by parts (i.e., using G1 (6.23) with $D=\Omega$), we obtain that

$$-\int_\Omega f\,v=\int_{\partial\Omega}\gamma v\,\partial_n u-\int_\Omega\left(\nabla u\cdot\nabla v-\lambda u v\right) \tag{6.62}$$

for all $v\in H^1(\Omega)$ (recall Remark 23).

We first consider the case $g_D=0$; see Remark 29 below for the case $g_D\neq 0$. Since $\gamma u=0$, it is natural to work in the Hilbert space $H_0^1(\Omega):=\{v\in H^1(\Omega):\gamma v=0\}$. If $v$ in (6.62) is in $H_0^1(\Omega)$, then the integral over $\Gamma$ vanishes, and (6.62) becomes the assertion that $a_D(u,v)=F(v)$, where

$$a_D(u,v):=\int_\Omega \nabla u\cdot\nabla v-\lambda u v \quad \text{and} \quad F(v):=\int_\Omega f v \tag{6.63}$$

(with the subscript $D$ standing for "Dirichlet"). Note that $a_D(\cdot,\cdot)$ is a bilinear form on $H_0^1(\Omega)\times H_0^1(\Omega)$, and $F(\cdot)$ is a linear functional on $H_0^1(\Omega)$.

The standard variational formulation of the IDP is therefore, given $f\in L^2(\Omega)$,

$$\boxed{\text{find } u\in H_0^1(\Omega) \text{ such that } a_D(u,v)=F(v) \text{ for all } v\in H_0^1(\Omega).} \tag{6.64}$$

We now turn our attention to the continuity and coercivity properties of $a_D(\cdot,\cdot)$, and we consider the three cases $\lambda=0$, $\lambda<0$, and $\lambda>0$ separately.

**Poisson's equation ($\lambda = 0$).** We impose the usual $H^1$-norm on the space $H_0^1(\Omega)$:

$$||v||^2_{H^1(\Omega)} := ||\nabla v||^2_{L^2(\Omega)} + ||v||^2_{L^2(\Omega)}. \tag{6.65}$$

By the Cauchy–Schwarz inequality (6.13), $a_D(\cdot,\cdot)$ is continuous with $C_c = 1$ in this norm.

Moving to the question of whether or not $a_D(\cdot,\cdot)$ is coercive, we recall the Poincaré inequality, namely that there exists a $C > 0$ such that

$$||v||_{L^2(\Omega)} \leq C\,||\nabla v||_{L^2(\Omega)} \qquad \text{for all } v \in H_0^1(\Omega) \tag{6.66}$$

[41, Section 5.6.1, Theorem 3], [15, Section 5.3]. This inequality implies that $a_D(\cdot,\cdot)$ is coercive with $\alpha = 1/(1+C^2)$, and existence and uniqueness of a solution to the variational problem (6.64) and continuous dependence of the solution on the data (i.e., properties K1 and K2 of Section 6.1.5) then follow from the Lax–Milgram theorem (Theorem 6.25). Existence and uniqueness of a quasi-optimal solution of the Galerkin equations (i.e., property K3) for any finite-dimensional subspace then follows from Céa's lemma (Theorem 6.27).

Since $a_D(\cdot,\cdot)$ is also symmetric, Lemma 6.26 implies that the unique solution of the variational problem (6.64) minimizes the functional

$$J(v) = \frac{1}{2}a_D(v,v) - F(v) = \frac{1}{2}\int_\Omega |\nabla v|^2 - f v;$$

this fact is called Dirichlet's principle [41, Sections 2.2.5, 8.1.2].

**The Modified Helmholtz Equation ($\lambda < 0$).** We now let $\lambda = -\mu^2$ for some $\mu > 0$ (so that $\mathscr{L} = \mathscr{L}_\mu$). It is then convenient to introduce the weighted norm

$$||v||^2_{1,\mu,\Omega} := ||\nabla v||^2_{L^2(\Omega)} + \mu^2\,||v||^2_{L^2(\Omega)}, \tag{6.67}$$

which is equivalent to the usual $H^1$-norm (6.65) when $\mu > 0$.

We weight the norm in this way for two reasons. The first is that if $v$ is a solution of $\mathscr{L}_\mu v = 0$, then we expect that $|\nabla v| \sim \mu|v|$ (this can be verified when $v$ is a separable solution of $\mathscr{L}_\mu v = 0$ in cartesian or polar coordinates) and then the two terms on the right-hand side of (6.67) are comparable. The second reason is that the constant $C_c$ for $a_D(\cdot,\cdot)$ is independent of $\mu$ in the norm $||\cdot||_{1,\mu,\Omega}$ (indeed, $C_c = 2$ in this case), but it is dependent on $\mu$ in the usual $H^1$-norm ($C_c = 2\mu^2$ in this case). To prove these facts about $C_c$, we use the Cauchy–Schwarz inequality (6.13) and the inequality (6.12).

Turning to coercivity we see that $a_D(v,v) = ||v||^2_{1,\mu,\Omega}$, and thus $a_D(\cdot,\cdot)$ is coercive with $\alpha = 1$ in the norm $||\cdot||_{1,\mu,\Omega}$; properties K1 and K2 then follow from the Lax–Milgram theorem and property K3 follows from Céa's lemma.

**The Helmholtz Equation ($\lambda > 0$).** We now let $\lambda = k^2$ for some $k > 0$ (so that $\mathscr{L} = \mathscr{L}_k$). For exactly the same reasons as in the case of the modified Helmholtz equation, we introduce the weighted norm

$$||v||^2_{1,k,\Omega} := ||\nabla v||^2_{L^2(\Omega)} + k^2\,||v||^2_{L^2(\Omega)}, \tag{6.68}$$

and in this norm $a_D(\cdot,\cdot)$ is continuous with $C_c = 2$.

The question of whether or not $a_D(\cdot,\cdot)$ is coercive is covered by the following two lemmas.

**Lemma 6.33 ($a_D(\cdot,\cdot)$ is coercive when $k$ is sufficiently small).** *Let $C$ be the constant in the Poincaré inequality (6.66). If*

$$k \leq \frac{1}{2C},$$

*then $a_D(\cdot,\cdot)$ is coercive (in the norm $\|\cdot\|_{1,k,\Omega}$) with $\alpha = 1/2$.*

***Proof.*** The definition of $a_D(\cdot,\cdot)$ and the Poincaré inequality (6.66) imply that, for any $v \in H_0^1(\Omega)$,

$$\begin{aligned} a_D(v,v) &= \|\nabla v\|^2_{L^2(\Omega)} + k^2\|v\|^2_{L^2(\Omega)} - 2k^2\|v\|^2_{L^2(\Omega)} \\ &\geq \|\nabla v\|^2_{L^2(\Omega)} + k^2\|v\|^2_{L^2(\Omega)} - 2k^2C^2\|\nabla v\|^2_{L^2(\Omega)}, \end{aligned} \tag{6.69}$$

and the result follows. □

**Lemma 6.34 ($a_D(\cdot,\cdot)$ is not coercive when $k$ is sufficiently large).** *Let $\lambda_1$ be the first Dirichlet eigenvalue of the negative Laplacian in $\Omega$. If $k^2 \geq \lambda_1$, then there exists a $v \in H_0^1(\Omega)$ with $a_D(v,v) = 0$.*

***Proof.*** If $\lambda_j$ is an eigenvalue of the negative Laplacian with eigenfunction $u_j \in H_0^1(\Omega)$, i.e., (6.15) holds, then

$$\left\|\nabla u_j\right\|^2_{L^2(\Omega)} - \lambda_j \left\|u_j\right\|^2_{L^2(\Omega)} = 0 \tag{6.70}$$

from G1 (6.62). The definition of $a_D(\cdot,\cdot)$ then implies that

$$a_D(u_j,u_j) = \left\|\nabla u_j\right\|^2_{L^2(\Omega)} - \lambda_j\left\|u_j\right\|^2_{L^2(\Omega)} + (\lambda_j - k^2)\left\|u_j\right\|^2_{L^2(\Omega)} = (\lambda_j - k^2)\left\|u_j\right\|^2_{L^2(\Omega)}. \tag{6.71}$$

Therefore if $k^2 = \lambda_1$, then $a_D(u_1,u_1) = 0$.

We now need to show that if $k^2 > \lambda_1$, then there exists a $v \in H_0^1(\Omega)$ with $a_D(v,v) = 0$. The expression (6.71) implies that if $\lambda_1 < k^2 < \lambda_j$, then

$$a_D(u_1,u_1) < 0 < a_D(u_j,u_j). \tag{6.72}$$

Since the numerical range of an operator is convex (see Remark 25), (6.72) implies there exists a $v \in H_0^1(\Omega)$ such that $a_D(v,v) = 0$. □

The equation (6.69) shows that $a_D(\cdot,\cdot)$ satisfies a Gårding inequality (see Definition 6.30) with $\mathcal{H} = H_0^1(\Omega)$, $\mathcal{V} = L^2(\Omega)$, $\alpha = 1$, and $C_{\mathcal{V}} = 2k^2$. Since $H_0^1(\Omega)$ is compactly contained in $L^2(\Omega)$ (i.e., the inclusion map from $H_0^1(\Omega)$ to $L^2(\Omega)$ is continuous and compact; see, e.g., [41, Section 5.7], [73, Theorem 3.27]), Theorem 6.31 shows that the operator associated with $a_D(\cdot,\cdot)$ is a compact perturbation of a coercive operator. Theorem 6.29 then implies that if $k^2 \neq \lambda_j$ for every $j = 1,2,\ldots$, the variational problem (6.64) has a unique solution which depends continuously on $f$ (i.e., properties K1 and K2 hold).

**Remark 29 (The case $g_D \neq 0$).** *We now show how to reduce the case $g_D \neq 0$ to the case $g_D = 0$. Given $g_D \in H^{1/2}(\Gamma)$, let $u_1 \in H^1(\Omega)$ be such that $\gamma u_1 = g_D$; the existence of such a $u_1$ is guaranteed by* [73, Theorem 3.37]. *With $u$ the solution of the IDP, we define $u_0 \in H_0^1(\Omega)$ by $u_0 := u - u_1$. Then $u_0$ satisfies the BVP*

$$\mathcal{L}u_0 = -f - \mathcal{L}u_1 \text{ in } \Omega \quad \text{and} \quad \gamma u_0 = 0 \text{ on } \Gamma.$$

*Multiplying the PDE by a test function $v \in H_0^1(\Omega)$ and integrating by parts, we find that the variational problem for $u_0$ is*

$$\boxed{\text{find } u_0 \in H_0^1(\Omega) \text{ such that } a_D(u_0, v) = F(v) - a_D(u_1, v) \text{ for all } v \in H_0^1(\Omega),}$$

*which is just the variational problem (6.64) with a different linear functional.*

### The IIP for the Helmholtz Equation

Since the impedance boundary condition (6.19) contains the complex number i, the solution of the IIP is (in general) complex-valued, and we therefore need to consider complex-valued function spaces.

Multiplying the PDE (6.18) by a test function and integrating by parts (i.e., using G1 (6.24) with $D = \Omega$) we obtain

$$-\int_\Omega f\,\overline{v} = \int_{\partial\Omega} \overline{\gamma v}\,\partial_n u - \int_\Omega \left(\nabla u \cdot \overline{\nabla v} - k^2 u \overline{v}\right) \tag{6.73}$$

for all $v \in H^1(\Omega)$. Using the impedance boundary condition (6.19) in the integral over $\Gamma$ we find that (6.73) is equivalent to $a_I(u, v) = F(v)$, where

$$a_I(u,v) := \int_\Omega \left(\nabla u \cdot \overline{\nabla v} - k^2 u \overline{v}\right) - i\eta \int_\Gamma \gamma u\, \overline{\gamma v} \quad \text{and} \quad F(v) := \int_\Omega f\,\overline{v} + \int_\Gamma g\,\overline{v} \tag{6.74}$$

(with the subscript $I$ standing for "impedance"). Note that $a_I(\cdot,\cdot)$ is a sesquilinear form on $H^1(\Omega) \times H^1(\Omega)$, and $F(\cdot)$ is an antilinear functional on $H^1(\Omega)$.

The variational formulation of the IIP is therefore given $f \in L^2(\Omega)$, $g \in L^2(\Gamma)$, and $\eta \in \mathbb{R} \setminus \{0\}$,

$$\boxed{\text{find } u \in H^1(\Omega) \text{ such that } a_I(u,v) = F(v) \text{ for all } v \in H^1(\Omega).} \tag{6.75}$$

As in the case of the Dirichlet problem, we use the weighted norm $\|\cdot\|_{1,k,\Omega}$ defined by (6.68). The Cauchy–Schwarz inequality (6.13) and the multiplicative trace inequality (6.10) imply that $a_I(\cdot,\cdot)$ is continuous with

$$C_c \sim \left(1 + \frac{|\eta|}{k}\right). \tag{6.76}$$

We now turn to coercivity and prove analogues of Lemmas 6.33 and 6.34. For the first, we need an appropriate analogue of the Poincaré inequality (6.66) for functions in $H^1(\Omega)$ and not just $H_0^1(\Omega)$; this is

$$\|v\|^2_{L^2(\Omega)} \le C\left(\|\nabla v\|^2_{L^2(\Omega)} + \|\gamma v\|^2_{L^2(\Gamma)}\right) \quad \text{for all } v \in H^1(\Omega). \tag{6.77}$$

The inequality (6.77) follows from the inequality

$$\|v\|_{L^2(\Omega)} \lesssim \left(\|\nabla v\|_{L^2(\Omega)} + \|\gamma v\|_{L^1(\Gamma)}\right) \quad \text{for all } v \in H^1(\Omega) \tag{6.78}$$

by using the Cauchy–Schwarz inequality (to show that $\|\gamma v\|_{L^1(\Gamma)} \lesssim \|\gamma v\|_{L^2(\Gamma)}$) and the inequality (6.12). The inequality (6.78) can be proved using the inequality

$$\|v - (v)_\Omega\|_{L^2(\Omega)} \lesssim \|\nabla v\|_{L^2(\Omega)} \quad \text{for all } v \in H^1(\Omega), \tag{6.79}$$

where $(v)_\Omega$ denotes the average of $v$ over $\Omega$ (see, e.g., [41, Section 5.8.1, Theorem 1] for the proof of (6.79) and [15, Section 5.3] for the proof of (6.78) from (6.79)).

**Lemma 6.35 ($a_I(\cdot,\cdot)$ is coercive when $k$ is sufficiently small).** *Let $C$ be the constant in the inequality* (6.77). *If*

$$2Ck^2 + C^2\frac{k^4}{|\eta|^2} \leq \frac{1}{2}, \tag{6.80}$$

*then $a_I(\cdot,\cdot)$ is coercive (in the norm $\|\cdot\|_{1,k,\Omega}$) with $\alpha = 1/2$. (Note that if $|\eta| = k$, then* (6.80) *reduces to the requirement that $k$ be sufficiently small.)*

*Proof* **(Sketch).** We have

$$\operatorname{Re} a_I(v,v) = \|\nabla v\|^2_{L^2(\Omega)} - k^2\|v\|^2_{L^2(\Omega)} \quad \text{and} \quad \operatorname{Im} a_I(v,v) = -\eta\|\gamma v\|^2_{L^2(\Gamma)}. \tag{6.81}$$

The result follows after forming $|a_I(v,v)|^2$ using the expressions (6.81) and then using the inequalities (6.77), (6.11), and (6.12) (in that order). □

For the IDP we knew in advance that $a_D(\cdot,\cdot)$ could not be coercive for all $k > 0$ since the variational problem (6.64) does not have a unique solution when $k^2 = \lambda_j$. In contrast, the solution of the IIP is unique for every $k > 0$ by Theorem 6.5 (assuming $\eta \in \mathbb{R}\setminus\{0\}$) and thus is it not immediately obvious whether $a_I(\cdot,\cdot)$ is coercive for every $k > 0$ or not.

**Lemma 6.36 ($a_I(\cdot,\cdot)$ is not coercive when $k$ is sufficiently large).** *Let $\lambda_1$ be the first Dirichlet eigenvalue of the negative Laplacian in $\Omega$. If $k^2 \geq \lambda_1$, then there exists a $v \in H^1(\Omega)$ with $a_I(v,v) = 0$.*

***Proof.*** If $v \in H^1_0(\Omega)$, then

$$a_D(v,v) = a_I(v,v)$$

(the integral over $\Gamma$ in $a_I(v,v)$ vanishes since $\gamma v = 0$). The result then follows from Lemma 6.34, since $H^1_0(\Omega) \subset H^1(\Omega)$. □

The definition of $a_I(\cdot,\cdot)$ implies that

$$\operatorname{Re} a_I(v,v) = \|v\|^2_{1,k,\Omega} - 2k^2\|v\|^2_{L^2(\Omega)},$$

and thus $a_I(\cdot,\cdot)$ satisfies a Gårding inequality with $\mathcal{H} = H^1(\Omega)$, $\mathcal{V} = L^2(\Omega)$, $\alpha = 1$, and $C_{\mathcal{V}} = 2k^2$. Since $H^1(\Omega)$ is compactly contained in $L^2(\Omega)$ [41, Section 5.7], [73, Theorem 3.27] and the solution of the IIP is unique for every $k > 0$ (by Theorem 6.5), Theorems 6.31 and 6.29 give properties K1, K2, and K3 (with K3 holding once the subspace dimension is large enough).

**Remark 30 (Finite-dimensional subspaces).** *The standard choices of finite-dimensional subspaces of $H^1(\Omega)$ (or $H^1_0(\Omega)$) are subspaces consisting of piecewise polynomials, i.e., polynomials supported on each element of a triangulation of $\Omega$. We note, however, that the* partition of unity finite element method *(PUFEM) for the Helmholtz equation introduced in* [74], [75], [10] *uses piecewise polynomials multiplied by local solutions of the homogeneous PDE (i.e., solutions of $\mathcal{L}_k v = 0$ on each element). This is an example of a "wave-based method"; see Remark* 35.

**Remark 31 (Illustration of Theorem 6.32—Quasi optimality of the Galerkin method for the IIP)).** *We now show how Theorem* 6.32 *can be used to find bounds on both the constant of quasi optimality $C_{qo}$ and the threshold $N_0$ after which quasi optimality holds.*

*Consider the case* $\eta = k$. *Then, with the explicit values of* $\alpha$ *and* $C_\gamma$ *given above, the condition* (6.55) *becomes*

$$\eta(\mathcal{H}_N) \leq \frac{1}{2kC_c} \tag{6.82}$$

*(and recall from* (6.76) *that* $C_c$ *is independent of* $k$*).*

*Let* $\mathcal{T}_h$ *be a family of regular triangulations of* $\Omega$ *(see Definition* 6.1*) and assume that each element of* $\mathcal{T}_h$ *is a simplex, i.e., a triangle in two dimensions and a tetrahedron in three dimensions. Fix* $p \in \mathbb{N}$ *and let* $\mathcal{H}_N := \{v \in C(\overline{\Omega}) : v|_K$ *is a polynomial of degree* $\leq p$ *for each* $K \in \mathcal{T}_h\}$ *(note that the subspace dimension,* $N$*, is then proportional to* $h^{-d}$*). Then, by* [89, Theorem 4.1],

$$\min_{v_N \in \mathcal{H}_N} \|w - v_N\|_{1,k,\Omega} \lesssim h\, \|w\|_{H^2(\Omega)} + hk\, \|w\|_{H^1(\Omega)}, \tag{6.83}$$

*where the omitted constant is independent of* $h$ *and* $k$*.*

*Given* $f \in L^2(\Omega)$*, let* $w = S^* f$ *(where* $S^* f$ *satisfies the variational problem* (6.53)*). Using the definition of* $a_I(\cdot,\cdot)$*, one can then show that* $w$ *satisfies the BVP*

$$\Delta w + k^2 w = -f \quad \text{in } \Omega \quad \text{and} \quad \partial_n w + ikw = 0 \quad \text{on } \Gamma$$

*(recalling that we're taking* $\eta = k$*). The question of bounding* $\eta(\mathcal{H}_N)$ *is then reduced to bounding the* $H^2$*- and* $H^1$*-norms of* $w$ *in terms of the* $L^2$*-norm of* $f$*, with the* $k$*-dependence of the constants known explicitly. It can be shown that, under some geometric assumptions, given* $k_0 > 0$*,*

$$\|w\|_{H^2(\Omega)} \lesssim k\, \|f\|_{L^2(\Omega)} \quad \text{and} \quad \|w\|_{H^1(\Omega)} \lesssim \|f\|_{L^2(\Omega)} \tag{6.84}$$

*for all* $k \geq k_0$ *(where the omitted constants are independent of* $k$*)* [74, Proposition 8.1.4], [23, Theorem 1] *(see also* [57, Remark 2.5, Theorem 2.9, Lemma 2.12]*). Using the bounds* (6.84) *along with* (6.83) *in the definition of* $\eta(\mathcal{H}_N)$ (6.54), *we find that*

$$\eta(\mathcal{H}_N) \lesssim hk.$$

*This last bound, along with* (6.82)*, implies that there exists a* $C > 0$ *(independent of* $h$ *and* $k$*) such that if*

$$hk^2 \leq C,$$

*then the quasi optimality bound* (6.56) *holds.*

### 6.1.7 ▪ Discontinuous Galerkin (DG) Formulation and Trefftz-Discontinuous Galerkin (TDG) Formulation

**Summary:** (effectively) G1 with $v$ a test function and $D = K$, where $K$ is an element of a triangulation of $\Omega$.

The term "discontinuous Galerkin method" encompasses many different methods, and so we first give a brief (and broad) overview of these. The overall idea is to use finite-dimensional spaces consisting of functions that are not continuous (and thus might not be in $H^1(\Omega)$). Since the solutions of the BVPs introduced in Section 6.1.3 are continuous (by Remark 21), their Galerkin approximations should ideally be continuous (or "close to continuous"). There are then three standard ways of imposing this continuity:

1. using *numerical fluxes*,
2. using *interior penalty terms*, or
3. using *Lagrange multipliers*.

(Note that there is some overlap between 1 and 2, since many interior penalty methods can be obtained via special choices of the numerical fluxes through the framework of DG methods established in [5].)

The plan of this section is the following. We sketch the DG approach based on 1 above (i.e., numerical fluxes) for the Helmholtz IIP. We do not state explicitly the variational problem because defining the relevant $\mathcal{H}$, $a(\cdot,\cdot)$, and $F(\cdot)$ would involve introducing a fairly substantial amount of new notation, with relatively little insight in return. (For the reader interested in the precise details, the discussion below contains appropriate references.) We then briefly discuss approaches based on 2 and 3 in Remarks 32 and 33, respectively. We also briefly mention a new DG-type method, the so-called *discontinuous Petrov–Galerkin* (DPG) method, in Remark 34.

In this section we discuss TDG methods for the Helmholtz IIP. The idea behind these methods is that solutions of $\mathcal{L}_k u = 0$ in $\Omega$ can be approximated well by piecewise solutions of the homogeneous Helmholtz equation, i.e., functions $v$ such that $\mathcal{L}_k v = 0$ on each $K$ of a triangulation of $\Omega$. (Note that requiring such a $v$ to be continuous on $\Omega$ is very restrictive, and thus the discontinuous setting is essential.) TDG methods arise from considering DG methods with these particular subspaces.

Finally, we note that in this section (and this section only) we do not use the notation $\gamma u$ for the trace of $u$, and write only $u$ instead. Similarly we write the normal derivative of $u$ as $\nabla u \cdot \mathbf{n}$ instead of $\partial_n u$. We do this to keep the notation consistent with that in the references we give.

### The DG Formulation

The starting point of the DG formulation is the IIP written as a first-order system. With $\boldsymbol{\sigma}$ defined by

$$ik\boldsymbol{\sigma} := \nabla u \quad \text{in } \Omega, \tag{6.85}$$

the IIP becomes

$$iku - \nabla\cdot\boldsymbol{\sigma} = \frac{1}{ik} f \quad \text{in } \Omega, \tag{6.86}$$

$$i\boldsymbol{\sigma}\cdot\mathbf{n} - i\eta\gamma u = g \quad \text{on } \Gamma. \tag{6.87}$$

Multiplying (6.85) by the complex conjugate of the vector test function $\tau$, multiplying (6.86) by the complex conjugate of the scalar test function $v$, integrating both equations over $K \in \mathcal{T}$ (where $\mathcal{T}$ is a triangulation of $\Omega$; see Definition 6.1), and integrating by parts (i.e., using (6.6)), we obtain

$$ik\int_K \boldsymbol{\sigma}\cdot\overline{\tau} + \int_K u\overline{\nabla\cdot\tau} - \int_{\partial K} u\,\overline{\tau}\cdot\mathbf{n} = 0 \tag{6.88a}$$

and

$$ik\int_K u\overline{v} + \int_K \boldsymbol{\sigma}\cdot\overline{\nabla v} - \int_{\partial K}\boldsymbol{\sigma}\cdot\mathbf{n}\overline{v} - \frac{1}{ik}\int_K f\overline{v} = 0. \tag{6.88b}$$

If $u \in H^1(K)$ and $\boldsymbol{\sigma} \in L^2(K)$, then the equation (6.88a) holds for all $\tau \in H(\mathrm{div};K) := \{\tau : \tau \in L^2(K), \nabla\cdot\tau \in L^2(K)\}$ and the equation (6.88b) holds for all $v \in H^1(K)$.

We now introduce finite-dimensional spaces $V_p(K)$ and $\mathbf{V}_p(K)$ such that $V_p(K) \subset H^1(K)$ and $\mathbf{V}_p(K) \subset H(\mathrm{div};K)$, and we replace $u, v$ by $u_p, v_p \in V_p(K)$ and replace $\boldsymbol{\sigma}, \tau$ by $\boldsymbol{\sigma}_p, \tau_p \in \mathbf{V}_p(K)$. Finally, we approximate the traces of $u$ and $\boldsymbol{\sigma}$ on $K$ by the *numerical fluxes* $\widehat{u}_p$ and $\widehat{\boldsymbol{\sigma}}_p$ (which will be specified later in terms of $u_p$ and $\nabla u_p$). Doing all this, we find that (6.88) becomes

$$ik\int_K \boldsymbol{\sigma}_p\cdot\overline{\boldsymbol{\tau}_p}+\int_K u_p\overline{\nabla\cdot\boldsymbol{\tau}_p}-\int_{\partial K}\widehat{u}_p\overline{\boldsymbol{\tau}_p}\cdot\mathbf{n}=0 \tag{6.89a}$$

and

$$ik\int_K u_p\overline{v}_p+\int_K \boldsymbol{\sigma}_p\cdot\overline{\nabla v_p}-\int_{\partial K}\widehat{\boldsymbol{\sigma}}_p\cdot\mathbf{n}\overline{v}_p-\frac{1}{ik}\int_K f\overline{v}=0. \tag{6.89b}$$

We now integrate by parts back in (6.89a) to obtain

$$ik\int_K \boldsymbol{\sigma}_p\cdot\overline{\boldsymbol{\tau}_p}-\int_K \nabla u_p\cdot\overline{\boldsymbol{\tau}_p}-\int_{\partial K}(\widehat{u}_p-u_p)\overline{\boldsymbol{\tau}_p}\cdot\mathbf{n}=0. \tag{6.90}$$

Observe that if $\widehat{u}_p=u_p$ and $ik\boldsymbol{\sigma}_p=\nabla u_p$, then (6.90) states that $0=0$.

We now recombine (6.89b) and (6.90) and seek to eliminate the variable $\boldsymbol{\sigma}$ and the test function $\boldsymbol{\tau}$. We do this by choosing $\boldsymbol{\tau}_p=\nabla v_p$ on each element, and to do this we need to assume that $\nabla_h V_p(K)\subset \mathbf{V}_p(K)$, where $\nabla_h$ denotes the elementwise gradient. Choosing $\boldsymbol{\tau}_p=\nabla v_p$ in (6.90) and substituting the resulting expression for $\int_K\boldsymbol{\sigma}_p\cdot\nabla v_p$ into (6.89b), we find that

$$\int_K \nabla u_p\cdot\overline{\nabla v_p}-k^2u_p\overline{v_p}-\int_{\partial K}(u_p-\widehat{u}_p)\overline{\nabla v_p}\cdot\mathbf{n}-ik\int_{\partial K}\widehat{\boldsymbol{\sigma}}_p\cdot\mathbf{n}\overline{v_p}=\int_K f\overline{v}. \tag{6.91}$$

Note that (a) we have succeeded in eliminating $\boldsymbol{\sigma}_p$ and $\boldsymbol{\tau}_p$, but $\widehat{\boldsymbol{\sigma}}_p$ remains, and (b) if $\widehat{u}_p=u_p$ and $ik\widehat{\boldsymbol{\sigma}}_p=\nabla u_p$, then (6.91) is just G1 (6.24) with $D=K$. (One can therefore understand (6.91) as arising from applying G1 on $K$ to $v_p$ and $u_p$ but, first, making $\nabla u_p$ on $\partial K$ a new variable $\widehat{\boldsymbol{\sigma}}_p/ik$ and, second, introducing a term on $\partial K$ with a new variable $\widehat{u}_p$ such that the term is zero when $\widehat{u}_p=u_p$.)

The equation (6.91) is the basis of the DG variational formulation. To obtain this variational formulation from (6.91) one must

(i) sum (6.91) over $K\in\mathcal{T}$,

(ii) decide on the definitions of $\widehat{\boldsymbol{\sigma}}_p$ and $\widehat{u}_p$ in terms of $u_p$ and $\nabla_h u_p$ on adjacent elements (these definitions will be different when $\partial K$ contains part of $\Gamma$ than when $\partial K\subset\Omega$), and

(iii) use the boundary condition (6.87) on $\Gamma$.

We omit the details; these can be found (for the Helmholtz IIP) in, e.g., [40, Section 6.3].

**Continuity and Coercivity Properties.** For certain choices of the numerical fluxes, the sesquilinear form of the DG formulation is continuous (in an appropriate norm) and satisfies a Gårding inequality; see, e.g., [5, Section 4.1–4.2], [84, Proposition 3.1], and [11, Theorem 3.3].

**Remark 32 (Interior penalty methods).** *Interior penalty methods impose continuity of the solution via terms that penalize the jump of the solution (and possibly its derivatives) across element edges. There are several different interior penalty methods for the Helmholtz equation,* *but we mention here the methods introduced in* [45], [46], *and* [47]. *The methods introduced in the first two papers do not fit in the framework of* [5] *(i.e., they do not arise from an appropriate choice of numerical fluxes), but those introduced in* [47] *do. The novelty of these methods is that, although the sesquilinear forms are not coercive, some of the consequences of coercivity hold; namely, the Galerkin equations have a unique solution without any constraint on the dimension of the (piecewise polynomial) approximation space, and error estimates can be obtained that are explicit in $k$, $h$, and $p$ (where $p$ is the polynomial degree)* [45, Remarks 4.3 and 5.1], [46, Remark 3.2].

*Note that one can also add similar penalty terms to the sesquilinear form of the standard variational formulation of the Helmholtz IIP ($a_I(\cdot,\cdot)$ defined by* (6.74)*), and some of the resulting methods share the features just described; see* [101, Corollary 3.5 and Theorem 4.4].

**Remark 33 (Methods using Lagrange multipliers).** *Probably the most well-known method for solving the Helmholtz equation that imposes continuity via Lagrange multipliers is the* discontinuous enrichment method *(DEM) (and related DG method) of Farhat and collaborators. The DEM, introduced in* [43], *consists of imposing the standard variational formulation on each element $K$ and then imposing interelement continuity weakly via Lagrange multipliers. The main idea in* [43] *is then to "enrich" the piecewise-polynomial approximation space with plane waves* $\exp(ik\mathbf{x}\cdot\mathbf{d})$, *where* $\mathbf{d}=(\cos\theta,\sin\theta)^T$. *In* [44] *(and subsequent papers) the polynomial part of the space was dropped, and then since there was no longer any "enrichment," the method was called a DG method (and abbreviated to DGM); see* [44, Section 1.2] *or* [64, Section 1.2.2] *for more details. (Note that with the choice of plane wave subspaces, the DG method of* [44] *falls under the category of TDG methods described below.)*

**Remark 34 (The DPG method).** *The DPG method applied to the Helmholtz IIP uses equations* (6.88) *to formulate a variational problem with unknowns $u$, $\boldsymbol{\sigma}$, $\widehat{u}$, and $\widehat{\boldsymbol{\sigma}}$ (where $\widehat{u}$ and $\widehat{\boldsymbol{\sigma}}$ are numerical fluxes); see* [27, Section 3.1], [28, Section 2.4]. *This variational problem has different trial and test spaces (i.e., it is of the form* (6.38) *with $\mathcal{H}_1\neq\mathcal{H}_2$, hence the "Petrov" in DPG). The key point of the DPG method is that, given finite-dimensional subspaces $\mathcal{H}_N^1$ of $\mathcal{H}_1$, the method defines so-called optimal test spaces $\mathcal{H}_N^2$ of $\mathcal{H}_2$; these optimal test spaces are such that, when the sesquilinear form satisfies the inf-sup condition* (6.40), *the discrete inf-sup condition* (6.46) *is* automatically *satisfied; see, e.g.,* [59, Proposition 5]. *These optimal test spaces admit functions with no continuity constraints across element interfaces, hence the "discontinuous" in "DPG"* [59, Definition 16], [29, Section 3]. *In practice, the optimal test spaces are not computed exactly (since this would require solving BVPs on each $K\in\mathcal{T}$) but instead are approximated* [59, Section 3]. *The DPG method can also be thought of as a least squares method in a nonstandard inner product; see, e.g.,* [59, Theorem 13], [29, Section 2].

### The TDG Formulation

The idea behind the TDG formulation is to assume that $V_p(K)$ satisfies the *Trefftz property*

$$\mathcal{L}_k v_p = 0 \text{ in } K \text{ for all } v_p\in V_p(K). \tag{6.92}$$

This property allows us to perform another integration by parts in (6.91); i.e., we use

$$\int_K \nabla u_p\cdot\overline{\nabla v_p} = -\int_K u_p\overline{\Delta v_p} + \int_{\partial K} u_p\overline{\nabla v_p}\cdot\mathbf{n}$$

to obtain

$$-\int_K \overline{(\Delta v_p + k^2 v_p)}u_p + \int_{\partial K}\widehat{u}_p\overline{\nabla v_p}\cdot\mathbf{n} - ik\int_{\partial K}\widehat{\boldsymbol{\sigma}}_p\cdot\mathbf{n}\overline{v_p} = \int_K f\overline{v}. \tag{6.93}$$

The fact that $V_p(K)$ satisfies the Trefftz property means that (6.93) becomes

$$\int_{\partial K} \widehat{u}_p \overline{\nabla v_p} \cdot \mathbf{n} - ik \int_{\partial K} \widehat{\boldsymbol{\sigma}}_p \cdot \mathbf{n} \overline{v_p} = \int_K f \overline{v}. \tag{6.94}$$

Just as (6.91) is the basis of the DG variational formulation for the IIP, (6.94) is the basis of the TDG variational formulation of the IIP. To obtain this variational formulation from (6.94) we carry out steps (i)–(iii) on page 122; see [40, Section 6.3] or [64, Section 4.2] for the details.

Although we have formulated the variational problem with $f \neq 0$, the TDG formulation is usually only considered for the case $f = 0$. This is because, whereas solutions of $\mathscr{L}_k u = 0$ are well approximated by functions in $V_p(K)$ [64, Chapter 3] (which was the original motivation for considering these spaces), solutions of the inhomogeneous Helmholtz equation $\mathscr{L}_k u = -f$ are (in general) poorly approximated by functions in $V_p(K)$ (see [64, Remark 3.5.11]).

**Continuity and Coercivity Properties.** The sesquilinear form of the TDG formulation is continuous and coercive in a norm consisting of jumps across element boundaries; see [65, Section 3.1], [64, Section 4.3], [16, Lemma 3.4]. The error estimates given by Céa's lemma (Theorem 6.27) can then be used to obtain estimates in the $L^2$-norm through duality arguments (i.e., arguments involving solutions of certain adjoint problems). The key result that allows one to do this in a Trefftz framework is [77, Theorem 3.1], with this result then used in [16, Section 4], [64, Section 4.3.1].

**Remark 35 (Wave-based methods).** *TDG methods for the Helmholtz equation come under the broad heading of "wave-based methods." Roughly speaking, there are two main classes of such methods. The first class consists of methods that use basis functions that are locally solutions of the PDEs (with these bases used either in a DG setting, as discussed above, or in a least squares setting); these methods are then called Trefftz methods. The second class consists of methods that use "modulated bases," i.e., the basis functions are local solutions of the PDEs multiplied by nonoscillatory functions (such as low-degree polynomials); the standard example of a method in this class is the PUFEM mentioned in Remark* 30. *We refer the reader to the theses* [64] *and* [70] *for good overviews of wave-based methods.*

## 6.1.8 ▪ The UWVF and the Quadratic Functional of Després

**Summary:** G2 with $v$ a solution of $\mathscr{L} v = 0$ can be used to prove the so-called *isometry lemma*. This lemma is then applied with $D = K$ (where $K$ is an element of a triangulation of $\Omega$) to obtain the UWVF or with $D = \Omega$ to obtain the quadratic functional.

### The Isometry Lemma

Both the UWVF and the quadratic functional introduced by Després are based on the following lemma, which is a consequence of G2 (6.26) when $\mathscr{L} v = 0$.

**Lemma 6.37 (Isometry lemma).** *Let $D$ be a bounded Lipschitz domain. If $u, v \in H^1(D, \Delta)$ with $\mathscr{L} u = 0$, $\mathscr{L} v = 0$, $\partial_n u \in L^2(\partial D)$, and $\partial_n v \in L^2(\partial D)$, then, for $\eta \in \mathbb{R}$,*

$$\int_{\partial D} (\partial_n u + i\eta\gamma u)\overline{(\partial_n v + i\eta\gamma v)} = \int_{\partial D} (\partial_n u - i\eta\gamma u)\overline{(\partial_n v - i\eta\gamma v)}. \tag{6.95}$$

***Proof.*** Expanding the brackets on both sides, we see that (6.95) reduces to two copies of G2 (6.26). □

This result is called the isometry lemma because when $u = v$, (6.95) becomes

$$||\partial_n u + i\eta\gamma u||^2_{L^2(\partial D)} = ||\partial_n u - i\eta\gamma u||^2_{L^2(\partial D)}; \tag{6.96}$$

i.e., the map from one impedance trace to the other is an isometry.

Although the isometry lemma is valid when $\mathscr{L} = \Delta + \lambda$ with $\lambda \in \mathbb{R}$, in what follows we only consider the Helmholtz equation (i.e., $\mathscr{L} = \mathscr{L}_k$). This is because, although the two formulations discussed in this section can be used to solve BVPs for the Laplace and modified Helmholtz equations, the formulations are most naturally applied to certain Helmholtz BVPs.

### The Ultra-weak Variational Formulation

We restrict our attention to the Helmholtz IIP. Furthermore, since the isometry lemma concerns solutions of the homogeneous Helmholtz equation, we take $f = 0$.

First, some notation. Let $\mathscr{T}$ be a triangulation of $\Omega$, with elements $\{K_j\}_{j=1}^N$. Let $\nu^{K_j}$ be the outward-pointing unit normal vector to $K_j$, let $\gamma_j$ be the trace on $\partial K_j$, and let $\partial_\nu^j$ be the normal derivative trace on $\partial K_j$. Let $\Sigma_{jl} := \overline{K_j} \cap \overline{K_l}$ with normal $\nu^{K_j}$ (so $\Sigma_{jl} = \Sigma_{lj}$ but the normal on $\Sigma_{lj}$ equals $\nu^{K_l} = -\nu^{K_j}$). Let $\Gamma_j := \partial K_j \cap \Gamma$. These definitions imply that

$$\int_{\partial K_j} = \sum_{l=1, l\neq j}^{N} \int_{\Sigma_{jl}} + \int_{\Gamma_j}. \tag{6.97}$$

The unknowns in the variational problem will be

$$\mathscr{X}_j := \partial_\nu^j u + i\eta\gamma_j u, \quad j = 1, \ldots, N.$$

Observe that if we know all the $\mathscr{X}_j$'s, then we know $\partial_n u + i\eta\gamma u$ on $\Gamma$, and combining this with the impedance boundary condition (6.19) we then know both $\partial_n u$ and $\gamma u$ on $\Gamma$. We can then find $u$ in $\Omega$ using Green's integral representation (Theorem 6.41 below).

We now assume that

(i) $\partial_\nu^j u \in L^2(\partial K_j)$ for all $j$, and

(ii) when $\Sigma_{jl} \neq \emptyset$, $\partial_\nu^j u = -\partial_\nu^l u$.

Regarding (i), $\partial_\nu^j u$ is always in $L^2(\partial K_j)$ for elements $K_j$ away from $\Gamma$ by interior regularity (see Remark 21). The reason we assume (i) is that we then have $\mathscr{X}_j \in L^2(\partial K_j)$ for all $j$. Regarding (ii), this is certainly the case if $u \in H^2(\Omega)$, since then $\partial_\nu^j u = \nu^{K_j} \cdot \gamma(\nabla u)$ and when $\Sigma_{jl} \neq \emptyset$, $\nu^{K_j} = -\nu^{K_l}$ (and the traces of $\nabla u$ from either side of $\Sigma_{jl}$ are equal).

Given $v^{K_j} : K_j \to \mathbb{C}$ with $\mathscr{L}_k v^{K_j} = 0$ and $\gamma_j v^{K_j}, \partial_\nu^j v^{K_j} \in L^2(\partial K_j)$ for all $j$, we define

$$\mathscr{Y}_j := \left(\partial_\nu^j v^{K_j} + i\eta\gamma_j v^{K_j}\right) \in L^2(\partial K_j). \tag{6.98}$$

We then define $F_j(\mathscr{Y}_j)$ to be the other impedance trace, i.e.,

$$F_j(\mathscr{Y}_j) := \left(-\partial_\nu^j v^{K_j} + i\eta\gamma_j v^{K_j}\right) \in L^2(\partial K_j).$$

Therefore $F_j : L^2(\partial K_j) \to L^2(\partial K_j)$ and is an isometry by Lemma 6.37. Applying the result of the isometry lemma (6.95) with $D = K_j$, $u = u$, and $v = v^{K_j}$, and rewriting the resulting equation using the notation introduced above and the expression (6.97), we obtain

$$\int_{\partial K_j} \mathscr{X}_j \overline{\mathscr{Y}_j} = \sum_{l=1, l\neq j}^{N} \int_{\Sigma_{jl}} (-\partial_\nu^j u + i\eta\gamma_j u)\overline{F_j(\mathscr{Y}_j)} + \int_{\Gamma_j} (-\partial_\nu^j u + i\eta\gamma_j u)\overline{F_j(\mathscr{Y}_j)}. \quad (6.99)$$

When $\Sigma_{jl} \neq \emptyset$, $\gamma_j u = \gamma_l u$, and then this fact, along with the assumption (ii) above, implies that

$$(-\partial_\nu^j u + i\eta\gamma_j u)\Big|_{\Sigma_{jl}} = (\partial_\nu^l u + i\eta\gamma_l u)\Big|_{\Sigma_{lj}} = \mathscr{X}_l.$$

The boundary condition (6.19) implies that

$$(-\partial_\nu^j u + i\eta\gamma_j u)\Big|_{\Gamma_j} = -g,$$

and using these last two equations in (6.99) we obtain

$$\int_{\partial K_j} \mathscr{X}_j \overline{\mathscr{Y}_j} = \sum_{l=1, l\neq j}^{N} \int_{\Sigma_{lj}} \mathscr{X}_l \overline{F_j(\mathscr{Y}_j)} - \int_{\Gamma_j} g \overline{F_j(\mathscr{Y}_j)}. \quad (6.100)$$

Define the Hilbert space $\mathscr{H}$ by

$$\mathscr{H} := \prod_{j=1}^{N} L^2(\partial K_j)$$

and let $\mathscr{X} := (\mathscr{X}_1, \dots, \mathscr{X}_N)$; assumption (i) then implies that $\mathscr{X} \in \mathscr{H}$. Define an inner product on $\mathscr{H}$ by

$$\langle \mathscr{X}, \mathscr{Y} \rangle := \sum_{j=1}^{N} \int_{\partial K_j} \mathscr{X}_j \overline{\mathscr{Y}_j}.$$

Then, summing up (6.100) over $j$, we find that

$$a(\mathscr{X}, \mathscr{Y}) = G(\mathscr{Y}) \quad \text{for all } \mathscr{Y} \text{ of the form (6.98)} \quad (6.101)$$

(i.e., impedance traces of piecewise solutions of $\mathscr{L}_k v = 0$), where

$$a(\mathscr{X}, \mathscr{Y}) := \langle \mathscr{X}, \mathscr{Y} \rangle - \sum_{j=1}^{N} \sum_{l=1, l\neq j}^{N} \int_{\Sigma_{lj}} \mathscr{X}_l \overline{F_j(\mathscr{Y}_j)}$$

and

$$G(\mathscr{Y}) := -\sum_{j=1}^{N} \int_{\Gamma_j} g \overline{F_j(\mathscr{Y}_j)}.$$

We then consider the variational problem

$$\boxed{\text{find } \mathscr{X} \in \mathscr{H} \text{ such that } \quad a(\mathscr{X}, \mathscr{Y}) = G(\mathscr{Y}) \quad \text{for all } \mathscr{Y} \in \mathscr{H}.} \quad (6.102)$$

Given $\mathscr{Y} \in \mathscr{H}$, there exist $\{v^{K_j}\}_{j=1}^N$ such that $\mathscr{L}_k v^{K_j} = 0$ in $K_j$ and $(\partial_\nu^j v^{K_j} + i\eta\gamma_j v^{K_j}) = \mathscr{Y}_j$ for $j = 1, \dots, N$ (using existence of a solution to the IIP on each $K_j$). Therefore, by (6.101), the $\mathcal{X}$ corresponding to the solution of the IIP in $\Omega$ satisfies the variational problem (6.102). The continuity and coercivity properties of $a(\cdot,\cdot)$ discussed below imply that the variational problem (6.102) has a unique solution, and thus the only solution of (6.102) is $\mathcal{X}$ given by $\mathcal{X}_j := \partial_\nu^j u + i\eta\gamma_j u$, $j = 1,\dots,N$, where $u$ is the solution of the IIP.

**Discretizing the Variational Problem (6.102).** Seeking to discretize (6.102) by choosing a finite-dimensional subspace $\mathcal{H}_N \subset \mathcal{H}$, we see that we need to be able to find easily $\{F_j(\mathcal{Y}_j)\}_{j=1}^N$ for $\mathcal{Y} \in \mathcal{H}_N$. Given a $\mathcal{Y} \in \mathcal{H}_N$, finding $\{F_j(\mathcal{Y}_j)\}_{j=1}^N$ means solving an impedance BVP on each $K_j$. These BVPs are less oscillatory than the original Helmholtz BVP (since the $K_j$ are smaller than $\Omega$) and are therefore easier to solve. However, we can make the BVPs on $K_j$ even easier to solve by letting $\mathcal{H}_N$ consist of impedance traces of explicit solutions of $\mathcal{L}_k v = 0$ on each $K_j$, since finding $\{F_j(\mathcal{Y}_j)\}_{j=1}^N$ is then straightforward. More precisely, for $K \in \mathcal{T}$ let $M_P^K$ be a finite-dimensional subspace of $P$ explicit solutions of $\mathcal{L}_k v = 0$ on $K$ (and we allow the number $P$ to vary from element to element). Let $\mathcal{H}_N := \prod_{j=1}^N \mathcal{H}_{N,j}$, where

$$\mathcal{H}_{N,j} := \left\{ \mathcal{Y}_j : \mathcal{Y}_j = \partial_\nu^j v + i\eta\gamma_j v \text{ for } v \in M_{P_j}^{K_j} \right\}.$$

Given $\mathcal{Y}_j \in \mathcal{H}_{N,j}$, we obtain $F_j(\mathcal{Y}_j)$ by finding the $v \in M_{P_j}^{K_j}$ that gave rise to that particular $\mathcal{Y}_j$ and then setting $F_j(\mathcal{Y}_j) = -\partial_\nu^j v + i\eta\gamma_j v$.

Common choices of the functions in $M_P^K$ are separable solutions of the Helmholtz equation, e.g., plane waves $\exp(ik\mathbf{x}\cdot\mathbf{d})$, where $\mathbf{d} = (\cos\theta, \sin\theta)^T$ (the separable solutions in Cartesian coordinates), spherical waves (the separable solutions in polar coordinates), or a mixture of the two; see, e.g., the review [71, Section 3] for more details.

**Continuity and Coercivity Properties.** With $\mathcal{H}_N$ constructed as above, the UWVF can be recast as a particular TDG method; this fact was noticed in [56, Section 2.6], [16, Section 2], and [58, Section 3]. The sesquilinear form is then continuous and coercive in a norm consisting of jumps across element boundaries; see the discussion in the section entitled "The TDG Formulation." (Note that a slightly weaker result was proved in the original analysis of the UWVF; see [17, Lemma 3.3, Equation 3.30].)

**Remark 36 (Connection between UWVF and least squares).** *When written in operator form, the UWVF is equivalent to a factorization of the normal equations for the least squares method consisting of minimizing the jumps in impedance traces across element boundaries; see* [71, Page 135]. *(This helps explain why the UWVF is better conditioned than the least squares method.)*

### The Quadratic Functional of Després

This variational formulation arises most naturally for problems with an impedance boundary condition (although it can be modified to work for other boundary conditions). It was introduced for the exterior impedance problem for the Helmholtz equation in [34] (see also [35] and [12]) and for the analogous problem for the time-harmonic Maxwell equations in [21].

Here we give the formulation for the Helmholtz IIP, and then we talk briefly about the exterior impedance problem at the end. Since the formulation is based on the isometry lemma (Lemma 6.37), it is applicable only when $f = 0$ and $\eta$ is a real constant.

The formulation is based on the isometry lemma with $u = v$, i.e., (6.96), and this result boils down to the fact that

$$\operatorname{Im} \int_{\partial D} \gamma u \, \overline{\partial_n u} = 0. \tag{6.103}$$

The equation (6.103) can be proved using only G1 (6.24) (i.e., one doesn't need G2). However, we keep this formulation under G2 in the map in Figure 6.1 because of the conceptual link with the UWVF.

Given $g \in L^2(\Gamma)$ and $\eta \in \mathbb{R} \setminus \{0\}$, define the quadratic functional $I(\cdot)$ by

$$I(v) := \frac{1}{2} \|\partial_n v - i\eta\gamma v\|^2_{L^2(\Gamma)} + \frac{1}{2} \|\partial_n v + i\eta\gamma v\|^2_{L^2(\Gamma)} - 2\operatorname{Re}\big(\partial_n v - i\eta\gamma v, g\big)_{L^2(\Gamma)} \tag{6.104}$$

and the space $\mathcal{H}$ by

$$\mathcal{H} := \left\{ v \in H^1(\Omega, \Delta) : \mathcal{L}_k v = 0,\ \partial_n v \in L^2(\Gamma) \right\} \tag{6.105}$$

(note that this is the space needed for the isometry lemma to hold).

**Theorem 6.38.** *Given $g \in L^2(\Gamma)$ and $\eta \in \mathbb{R} \setminus \{0\}$, let $u$ be the solution of Helmholtz IIP with $f = 0$. Then $u \in \mathcal{H}$ and $I(u) \le I(v)$ for all $v \in \mathcal{H}$.*

***Proof.*** By the definition of the IIP (Definition 6.4), $u \in H^1(\Omega)$. The PDE (6.18) implies that $\mathcal{L}_k u = 0$ and thus $\Delta u \in L^2(\Omega)$. Since $g \in L^2(\Gamma)$, the boundary condition (6.19) implies that $\partial_n u \in L^2(\Gamma)$; therefore $u \in \mathcal{H}$.

If $v \in \mathcal{H}$, then, by (6.96) with $D = \Omega$,

$$I(v) = \|\partial_n v - i\eta\gamma v\|^2_{L^2(\Gamma)} - 2\operatorname{Re}\big(\partial_n v - i\eta\gamma v, g\big)_{L^2(\Gamma)}.$$

Now, since $|a-b|^2 = |a|^2 + |b|^2 - 2\operatorname{Re}(a\overline{b})$,

$$\|\partial_n v - i\eta\gamma v - g\|^2_{L^2(\Gamma)} = \|\partial_n v - i\eta\gamma v\|^2_{L^2(\Gamma)} + \|g\|^2_{L^2(\Gamma)} - 2\operatorname{Re}\big(\partial_n v - i\eta\gamma v, g\big)_{L^2(\Gamma)},$$

so

$$I(v) = \|\partial_n v - i\eta\gamma v - g\|^2_{L^2(\Gamma)} - \|g\|^2_{L^2(\Gamma)}.$$

Therefore, the minimum of $I(v)$ is $-\|g\|^2_{L^2(\Gamma)}$ and this is reached when $\partial_n v - i\eta\gamma v = g$; by uniqueness of the solution of the IIP, this is when $v = u$. □

Define the *impedance trace operators* $h^{\pm}(\cdot)$ by

$$h^{\pm}(v) := \partial_n v \pm i\eta\gamma v$$

so that

$$I(v) = \frac{1}{2} \|h^-(v)\|^2_{L^2(\Gamma)} + \frac{1}{2} \|h^+(v)\|^2_{L^2(\Gamma)} - 2\operatorname{Re}\big(h^-(v), g\big)_{L^2(\Gamma)}.$$

Define the sesquilinear form $a(\cdot,\cdot)$ and antilinear functional $F(\cdot)$ by

$$a(v,w) := \frac{1}{2}\big(h^+(v), h^+(w)\big)_{L^2(\Gamma)} + \frac{1}{2}\big(h^-(v), h^-(w)\big)_{L^2(\Gamma)} \quad \text{and} \quad F(w) = \big(g, h^-(w)\big)_{L^2(\Gamma)}. \tag{6.106}$$

**Lemma 6.39.** *Given $g \in L^2(\Gamma)$ and $\eta \in \mathbb{R}\setminus\{0\}$, define $h^{\pm}(\cdot)$, $a(\cdot,\cdot)$, $F(\cdot)$, and $\mathcal{H}$ as above. If $u$ is the solution of the IIP (with $f=0$), then*

$$a(u,v) = F(v) \quad \textit{for all } v \in \mathcal{H}. \tag{6.107}$$

***Proof.*** Let $\varepsilon \in \mathbb{R}$. Theorem 6.38 implies that $I(u) \le I(v)$ for all $v \in \mathcal{H}$, and therefore

$$\frac{\mathrm{d}}{\mathrm{d}\varepsilon} I(u+\varepsilon v)\bigg|_{\varepsilon=0} = 0 \quad \text{and} \quad \frac{\mathrm{d}}{\mathrm{d}\varepsilon} I(u+\mathrm{i}\varepsilon v)\bigg|_{\varepsilon=0} = 0 \tag{6.108}$$

for all $v \in \mathcal{H}$. The first condition in (6.108) simplifies to the real part of (6.107), and the second condition in (6.108) simplifies to the imaginary part of (6.107). □

We now need to specify a norm on $\mathcal{H}$. We would like to let

$$\|w\|_{\mathcal{H}}^2 := \|h^+(w)\|_{L^2(\Gamma)}^2 + \|h^-(w)\|_{L^2(\Gamma)}^2, \tag{6.109}$$

since, if this is a norm, then $a(\cdot,\cdot)$ defined by (6.106) is continuous with $C_c = 1$ (via the inequality (6.12)) and coercive with $\alpha = 1/2$.

**Lemma 6.40.** *$\mathcal{H}$ is a Hilbert space with norm $\|\cdot\|_{\mathcal{H}}$ defined by* (6.109).

***Proof.*** Define

$$\mathcal{V} := \left\{ v \in H^1(\Omega,\Delta) : \partial_n v \in L^2(\Gamma), \gamma v \in H^1(\Gamma)\right\}. \tag{6.110}$$

A regularity result of Nečas (see [73, Lemma 4.24]) implies that if $v \in H^1(\Omega,\Delta)$, then the conditions $\partial_n v \in L^2(\Gamma)$ and $\gamma v \in H^1(\Gamma)$ are equivalent (i.e., if one holds, then so does the other, and vice versa). Therefore, we can append the condition $\gamma v \in H^1(\Gamma)$ to the definition of $\mathcal{H}$ (6.105) and thus see that $\mathcal{H} \subset \mathcal{V}$.

Now $\mathcal{V}$ is a Hilbert space with norm

$$\|v\|_{\mathcal{V}}^2 := k^2\|v\|_{L^2(\Omega)}^2 + \|\nabla v\|_{L^2(\Omega)}^2 + k^{-2}\|\Delta v\|_{L^2(\Omega)}^2 + k^2\|\gamma v\|_{L^2(\Gamma)}^2 + \|\nabla_\Gamma \gamma v\|_{L^2(\Gamma)}^2 + \|\partial_n v\|_{L^2(\Gamma)}^2,$$

where we have weighted the terms with $k$ similarly as in the norm (6.68).

It is straightforward to show that $\mathcal{H}$ is closed in $\mathcal{V}$ with the norm $\|\cdot\|_{\mathcal{V}}$; therefore if we can show that $\|\cdot\|_{\mathcal{V}}$ is equivalent to $\|\cdot\|_{\mathcal{H}}$, then we are done. From the definitions of the norms and the operators $h^{\pm}(\cdot)$, there exists a $C_1(k,\eta)$ such that $\|v\|_{\mathcal{H}} \le C_1(k,\eta)\|v\|_{\mathcal{V}}$ for all $v \in \mathcal{V}$ (if $\eta = k$, then $C_1(k,\eta) \sim 1$). Furthermore, by the well-posedness of the IIP, there exists a $C_2(k,\eta)$ such that

$$\|v\|_{\mathcal{V}} \le C_2(k,\eta)\,\|h^+(v)\|_{L^2(\Gamma)}$$

for all $v \in \mathcal{V}$. Because $\|h^+(v)\|_{L^2(\Gamma)} \le \|v\|_{\mathcal{H}}$, the other half of the norm equivalence follows. □

In summary, with $\mathcal{H}$ defined by (6.105), $a(\cdot,\cdot)$ and $F(\cdot)$ defined by (6.106), and $\|\cdot\|_{\mathcal{H}}$ defined by (6.109), the variational problem (6.1) is a continuous and coercive variational formulation of the Helmholtz IIP. The disadvantage of this formulation is that it is difficult to obtain piecewise-polynomial finite-dimensional subspaces of $\mathcal{H}$ (the difficulty arises from requiring that $\mathcal{L}_k v = 0$). For the analogous formulation of the exterior impedance problem (discussed briefly below), bypassing this difficulty is investigated in [34], [35], and [12]. In particular, from the variational formulation of the exterior impedance problem one can derive a system of integral equations (which can also be derived starting from Green's integral representation); see [12, Sections 3.2–3.4], [35, Section 3-4], [82, Section 3.4.4] for more details.

**Remark 37 (The analogous formulation for the exterior impedance problem).** *We consider the two-dimensional exterior impedance problem for the homogeneous Helmholtz equation with* $\eta = k$, *i.e., given* $g \in L^2(\Gamma)$, *find* $u$ *satisfying*

$$\mathscr{L}_k u = 0 \ \text{in}\ \Omega_+, \quad \partial_n^+ u + \mathrm{i}k\gamma_+ u = g \ \text{on}\ \Gamma,$$

*and the Sommerfeld radiation condition* (6.17) *(the solution to this problem is unique by, e.g.,* [22, Theorem 3.12] [18, Corollary 2.9]). *The space* $\mathscr{H}$ *now consists of solutions of* $\mathscr{L}_k v = 0$ *in* $\Omega_+$ *with* $\partial_n^+ v \pm \mathrm{i}k\gamma_+ v \in L^2(\Gamma)$ *and*

$$v(r,\theta) = \frac{\mathrm{e}^{\mathrm{i}kr} a(\theta)}{\sqrt{r}} + \frac{\mathrm{e}^{-\mathrm{i}kr} b(\theta)}{\sqrt{r}} + o\left(\frac{1}{\sqrt{r}}\right) \quad \text{as } r \to \infty$$

*for* $a, b \in L^2[0, 2\pi]$; *i.e., the space contains both outgoing and incoming solutions (and linear combinations of the two). (Note that we have chosen to use the notation of* [12] *over that of* [34] *and* [35], *since* [12] *shares our conventions for "outgoing" and "incoming.") The analogue of the isometry property* (6.96) *is now*

$$4k^2 \|a\|^2_{L^2[0,2\pi]} + \|\partial_n^+ v - \mathrm{i}k\gamma_+ v\|^2_{L^2(\Gamma)} = 4k^2 \|b\|^2_{L^2[0,2\pi]} + \|\partial_n^+ v + \mathrm{i}k\gamma_+ v\|^2_{L^2(\Gamma)}; \tag{6.111}$$

*see* [12, Lemma 3.3], [34, Lemma 3.1], *and* [35, Theorem 1]. *Just as* (6.96) *gives rise to the quadratic functional* (6.104) *(which is minimized at the solution of the IIP),* (6.111) *gives rise to a quadratic functional that is minimized at the solution of the exterior impedance problem.*

### 6.1.9 ▪ Green's Integral Representation and Boundary Integral Equations (BIEs)

**Summary:** G2 with $v$ equivalent to the fundamental solution of $\mathscr{L}$ and $D = \Omega_\pm$ gives Green's integral representation (from which BIEs can be obtained).

#### Green's Integral Representation

In this section we let $\mathscr{L}$ denote a general linear differential operator. Given such an $\mathscr{L}$, $E$ is a *fundamental solution* for $\mathscr{L}$ if

$$\mathscr{L}_{\mathbf{x}} E(\mathbf{x},\mathbf{y}) = -\delta(\mathbf{x}-\mathbf{y}), \tag{6.112}$$

where the subscript $\mathbf{x}$ on $\mathscr{L}_{\mathbf{x}}$ indicates that the differentiation is in the $\mathbf{x}$-variable. The $\delta$ on the right-hand side of (6.112) is the Dirac delta function, and thus equation (6.112) is understood in the sense of distributions. Note that $E$ is *a* fundamental solution, and not *the* fundamental solution, since there can be several different fundamental solutions for a given $\mathscr{L}$ (and we see examples of this below).

We now discuss fundamental solutions for the operator $\Delta + \lambda$ in two and three dimensions (the situation in higher dimensions is similar). We begin with the case $\lambda = -\mu^2$ for $\mu > 0$ (i.e., the modified Helmholtz equation). In thee dimensions, two solutions of (6.112) with $\mathscr{L} = \Delta - \mu^2$ are

$$\frac{\mathrm{e}^{-\mu|\mathbf{x}-\mathbf{y}|}}{4\pi|\mathbf{x}-\mathbf{y}|} \quad \text{and} \quad \frac{\mathrm{e}^{\mu|\mathbf{x}-\mathbf{y}|}}{4\pi|\mathbf{x}-\mathbf{y}|} \tag{6.113}$$

(see, e.g., [96, Section 5.8, Pages 53–55]), illustrating the fact that fundamental solutions are not unique. It is usually convenient to have the fundamental solution tending to zero as $|\mathbf{x}-\mathbf{y}| \to \infty$ (if this is possible), and therefore we choose the first fundamental solution in (6.113).

In two dimensions, solutions of (6.112) with $\mathscr{L} = \Delta - \mu^2$ are given by modified Bessel functions (see, e.g., [96, Section 5.8, Pages 53–55]), and we encounter a similar situation to that in three dimensions; one solution tends to infinity as $|\mathbf{x}-\mathbf{y}| \to \infty$, and the other tends to zero. Choosing the one that tends to zero, we therefore define $\Phi_\mu(\mathbf{x},\mathbf{y})$ by

$$\Phi_\mu(\mathbf{x},\mathbf{y}) := \begin{cases} \frac{1}{2\pi} K_0(\mu|\mathbf{x}-\mathbf{y}|), & d=2, \\ \dfrac{e^{-\mu|\mathbf{x}-\mathbf{y}|}}{4\pi|\mathbf{x}-\mathbf{y}|}, & d=3, \end{cases}$$

where $K_0$ is defined by, e.g., [36, Equations 10.25.3 and 10.27.4].

When $\mathscr{L} = \Delta$ (i.e., the PDE is Laplace's/Poisson's equation) we define $\Phi_0(\mathbf{x},\mathbf{y})$ by

$$\Phi_0(\mathbf{x},\mathbf{y}) := \begin{cases} \dfrac{1}{2\pi} \log\left(\dfrac{a}{|\mathbf{x}-\mathbf{y}|}\right), & d=2, \\ \dfrac{1}{4\pi|\mathbf{x}-\mathbf{y}|}, & d=3, \end{cases}$$

where $a \in \mathbb{R}$. We immediately see that $\Phi_0(\mathbf{x},\mathbf{y})$ does not tend to zero as $|\mathbf{x}-\mathbf{y}| \to \infty$ in two dimensions; this means that the theory for BIEs for the Laplace equation in two dimensions is more awkward than in three dimensions. Furthermore, $\Phi_0(\mathbf{x},\mathbf{y})$ contains the arbitrary parameter $a$ when $d = 2$. Usually one lets $a = 1$, but considering other values of $a$ can sometimes be useful (for an example in the theory of BIEs see [73, Theorem 8.16]).

Finally, moving on to the Helmholtz equation, we find that the two solutions of (6.112) in three dimensions when $\mathscr{L} = \Delta + k^2$ with $k > 0$ are

$$\frac{e^{ik|\mathbf{x}-\mathbf{y}|}}{4\pi|\mathbf{x}-\mathbf{y}|} \quad \text{and} \quad \frac{e^{-ik|\mathbf{x}-\mathbf{y}|}}{4\pi|\mathbf{x}-\mathbf{y}|}.$$

Both of these functions decay at the same rate as $|\mathbf{x}-\mathbf{y}| \to \infty$. The first, however, satisfies the outgoing radiation condition (6.17), whereas the second satisfies the incoming radiation condition (6.36), and we therefore choose the first (but see Remark 38). We find a similar situation in two dimensions; here there are two solutions of (6.112) given in terms of the Hankel functions $H_0^{(1)}$ and $H_0^{(2)}$ (defined by, e.g., [36, Equations 10.4.3, 10.2.5, and 10.2.6]). $H_0^{(1)}$ satisfies the outgoing radiation condition and $H_0^{(2)}$ satisfies the incoming radiation condition, and so we chose the first. We therefore define $\Phi_k(\mathbf{x},\mathbf{y})$ by

$$\Phi_k(\mathbf{x},\mathbf{y}) := \begin{cases} \frac{i}{4} H_0^{(1)}(k|\mathbf{x}-\mathbf{y}|), & d=2, \\ \dfrac{e^{ik|\mathbf{x}-\mathbf{y}|}}{4\pi|\mathbf{x}-\mathbf{y}|}, & d=3. \end{cases} \tag{6.114}$$

Green's integral representation arises from applying G2 with $v = \Phi$ and $D = \Omega_\pm$. From now on we restrict our attention to the homogeneous Helmholtz equation, but we note that analogous results hold for Laplace's equation and the modified Helmholtz equation, as well as for the inhomogeneous counterparts of all three equations (see, e.g., [73, Theorem 7.5], [41, Section 2.2.4]).

Recall the notation introduced in Section 6.1.2, namely that $\Omega_-$ is a bounded Lipschitz open set such that $\Omega_+ := \mathbb{R}^d \setminus \overline{\Omega_-}$ is connected.

**Theorem 6.41 (Green's integral representation for $\Omega_-$).** *If $u \in H^1(\Omega_-)\cap C^2(\Omega_-)$ satisfies $\mathscr{L}_k u = 0$, then*

$$\int_\Gamma \left( \Phi_k(\mathbf{x},\mathbf{y})\partial_n^- u(\mathbf{y}) - \frac{\partial \Phi_k(\mathbf{x},\mathbf{y})}{\partial n(\mathbf{y})}\gamma_- u(\mathbf{y}) \right) ds(\mathbf{y}) = \begin{cases} u(\mathbf{x}), & \mathbf{x}\in\Omega_-, \\ 0, & \mathbf{x}\in\Omega_+. \end{cases} \tag{6.115}$$

*Proof.* For $\mathbf{x}\in\Omega_+$, this is an immediate consequence of G2 (without the complex conjugate) (6.25) applied with $v(\cdot)=\Phi_k(\mathbf{x},\cdot)$ and $D=\Omega_-$. For $\mathbf{x}\in\Omega_-$ we apply G2 (6.25) with $D=\Omega_-\setminus B_\varepsilon(\mathbf{x})$ and then let $\varepsilon\to 0$; see [22, Theorem 3.1] for the details. □

Note that if $u\in H^1(\Omega_-)$ satisfies $\mathscr{L}_k u = 0$, then $u\in C^\infty(\Omega_-)$ by Remark 21, so the assumption in Theorem 6.41 that $u\in C^2(\Omega_-)$ is not restrictive.

**Theorem 6.42 (Green's integral representation for $\Omega_+$).** *If $u \in H^1_{\mathrm{loc}}(\Omega_+)\cap C^2(\Omega_+)$ satisfies $\mathscr{L}_k u = 0$ and the Sommerfeld radiation condition* (6.17)*, then*

$$-\int_\Gamma \left( \Phi_k(\mathbf{x},\mathbf{y})\partial_n^+ u(\mathbf{y}) - \frac{\partial \Phi_k(\mathbf{x},\mathbf{y})}{\partial n(\mathbf{y})}\gamma_+ u(\mathbf{y}) \right) ds(\mathbf{y}) = \begin{cases} 0, & \mathbf{x}\in\Omega_-, \\ u(\mathbf{x}), & \mathbf{x}\in\Omega_+. \end{cases} \tag{6.116}$$

*Proof.* This follows in a way similar to Theorem 6.41 (recalling that the normal vector $\mathbf{n}$ points into $\Omega_+$), with the integral at infinity vanishing by Lemma 6.13 (note that if we had used G2 *with* the complex conjugate, then this would not happen). □

**Remark 38 (Chase the complex conjugate).** *We chose the fundamental solution $\Phi_k(\mathbf{x},\mathbf{y})$* (6.114) *on the basis that it should satisfy the outgoing radiation condition* (6.17)*, but this meant that to obtain Green's integral representation in $\Omega_+$* (6.116) *we had to use* G2 *without the complex conjugate (because of Lemma* 6.13*). If we had chosen the fundamental solution satisfying the* incoming *radiation condition* (6.36) *and used this in $\Omega_+$ with the version of* G2 with *the complex conjugate, then we would have also obtained the representation formulae* (6.116).

*In $\Omega_-$ it doesn't matter which fundamental solution we use. This is because the difference between the two is a solution of the homogeneous Helmholtz equation, and then* G2 *implies that the two different representation formulae (one with each fundamental solution) are actually equivalent.*

The key point about Green's integral representation is that once you know $\gamma_\pm u$ and $\partial_n^\pm u$ on $\Gamma$, then you know $u$ in $\Omega_\pm$. Therefore, to find the solution of the IDP, one only needs to find $\partial_n^- u$ on $\Gamma$, thereby reducing the problem from one posed in a $d$-dimensional domain to one posed in a $(d-1)$-dimensional domain. The case of the EDP is similar, except that now there is the additional advantage that, whereas $\Omega_+$ is unbounded, $\Gamma$ is bounded.

It is convenient to introduce the following notation. The *single-layer potential* $\mathscr{S}_k$ is defined by

$$\mathscr{S}_k\phi(\mathbf{x}) := \int_\Gamma \Phi_k(\mathbf{x},\mathbf{y})\phi(\mathbf{y})\,ds(\mathbf{y}), \quad \mathbf{x}\in\mathbb{R}^d\setminus\Gamma, \tag{6.117}$$

and the *double-layer potential* $\mathscr{D}_k$ is defined by

$$\mathscr{D}_k\phi(\mathbf{x}) := \int_\Gamma \frac{\partial \Phi_k(\mathbf{x},\mathbf{y})}{\partial n(\mathbf{y})}\phi(\mathbf{y})\,ds(\mathbf{y}), \quad \mathbf{x}\in\mathbb{R}^d\setminus\Gamma.$$

(Observe that both these operators take functions defined on $\Gamma$ to functions defined on

$\mathbb{R}^d \setminus \Gamma$.) Green's integral representation for $\Omega_+$ (6.116) then reads as

$$-\mathscr{S}_k \partial_n^+ u(\mathbf{x}) + \mathscr{D}_k \gamma_+ u(\mathbf{x}) = \begin{cases} 0, & \mathbf{x} \in \Omega_-, \\ u(\mathbf{x}), & \mathbf{x} \in \Omega_+. \end{cases} \tag{6.118}$$

**Remark 39 (Kupradze's method).** *The following formulation of the EDP is based on viewing the first equation in* (6.118) *as an equation to be solved for* $\partial_n^+ u$. *That is, given* $g_D \in H^{1/2}(\Gamma)$, *we find a* $\psi \in H^{-1/2}(\Gamma)$ *such that*

$$\mathscr{S}_k \psi(\mathbf{x}) = \mathscr{D}_k g_D(\mathbf{x}) \quad \textit{for all } \mathbf{x} \in \Omega_- \tag{6.119}$$

*and then set* $\partial_n^+ u = \psi$. *Note that this is not a variational problem in the sense of Section* 6.1.5, *since the trial space is a space of functions on* $\Gamma$, *and the test space consists of all points in* $\Omega_-$.

*We now show that the equation* (6.119) *has a unique solution. Assume that* $\mathscr{S}_k \psi(\mathbf{x}) = 0$ *for all* $\mathbf{x} \in \Omega_-$. *Let* $u := \mathscr{S}\psi$. *Then* $u = 0$ *in* $\Omega_-$ *and thus* $\gamma_- u = 0$ *on* $\Gamma$. *The jump relations for the single-layer potential (*(6.120) *below) then imply that* $\gamma_+ u = 0$ *on* $\Gamma$. $u$ *is then a solution of the EDP for the Helmholtz equation with zero right-hand side and zero Dirichlet trace. By uniqueness of the solution to this BVP,* $u = 0$ *in* $\Omega_+$. *Then, by the jump relations* (6.120), $\psi = \partial_n^- u - \partial_n^+ u = 0$.

*Although* (6.119) *is uniquely solvable, nonuniqueness arises when one tries to discretize the equation; see* [72, Section 7.3] *and the references therein.*

### Boundary Integral Equations

We now focus on the Helmholtz EDP. We saw above that Green's integral representation means that we only need to find $\partial_n^+ u$, and we now show how to do this using BIEs.

We first need to understand how $\mathscr{S}_k\phi(\mathbf{x})$, $\mathscr{D}_k\phi(\mathbf{x})$, and their normal derivatives behave as $\mathbf{x} \to \Gamma$ from either $\Omega_-$ or $\Omega_+$ (in other words, we need to know what the interior and exterior Dirichlet and Neumann traces of $\mathscr{S}_k\phi$ and $\mathscr{D}_k\phi$ are). Define the operators $S_k$, $D_k$, $D_k'$, and $H_k$ by

$$S_k\phi(\mathbf{x}) = \int_\Gamma \Phi_k(\mathbf{x},\mathbf{y})\phi(\mathbf{y})\,ds(\mathbf{y}), \quad D_k\phi(\mathbf{x}) = \int_\Gamma \frac{\partial \Phi_k(\mathbf{x},\mathbf{y})}{\partial n(\mathbf{y})}\phi(\mathbf{y})\,ds(\mathbf{y}), \quad \mathbf{x} \in \Gamma,$$

and

$$D_k'\phi(\mathbf{x}) = \int_\Gamma \frac{\partial \Phi_k(\mathbf{x},\mathbf{y})}{\partial n(\mathbf{x})}\phi(\mathbf{y})\,ds(\mathbf{y}), \quad H_k\phi(\mathbf{x}) = \frac{\partial}{\partial n(\mathbf{x})}\int_\Gamma \frac{\partial \Phi_k(\mathbf{x},\mathbf{y})}{\partial n(\mathbf{y})}\phi(\mathbf{y})\,ds(\mathbf{y}), \quad \mathbf{x} \in \Gamma.$$

Observe that all these operators take functions defined on $\Gamma$ to functions defined on $\Gamma$. The notation $D_k$ and $D_k'$ expresses the fact that these two operators are adjoint with respect to the real-valued $L^2(\Gamma)$-inner product.

Lots of technical difficulties immediately arise. Indeed, if $\Gamma$ is not smooth, then the integrals defining $D_k$ and $D_k'$ must be understood as Cauchy principal values, and even when $\Gamma$ is smooth $H_k$ has to be understood as a limit (see, e.g., [18, Equation (2.36)], [73, Theorem 7.4(iii)], [86, Section 3.3.4]); we ignore all these difficulties here.

With the definitions of $S_k$, $D_k$, $D_k'$, and $H_k$ above, the *jump relations* for $\mathscr{S}_k$ are

$$\gamma_\pm \mathscr{S}_k = S_k, \quad \partial_n^\pm \mathscr{S}_k = \mp\frac{1}{2}I + D_k', \tag{6.120}$$

and those for $\mathscr{D}_k$ are

$$\gamma_{\pm}\mathscr{D}_k = \pm\frac{1}{2}I + D_k, \quad \partial_n^{\pm}\mathscr{D}_k = H_k. \tag{6.121}$$

Taking the exterior Dirichlet trace of the second equation in (6.118) (or, equivalently, taking the interior Dirichlet trace of the first equation in (6.118)) and using the jump relations (6.120) and (6.121), we obtain the integral equation

$$S_k\partial_n^+ u = \left(-\frac{1}{2}I + D_k\right)g_D \tag{6.122}$$

(where we have used the fact that $\gamma_+ u = g_D$).

Similarly, taking the exterior Neumann trace of the second equation in (6.118) (or, equivalently, taking the interior Neumann trace of the first equation in (6.118)), we obtain the integral equation

$$\left(\frac{1}{2}I + D_k'\right)\partial_n^+ u = H_k g_D. \tag{6.123}$$

In the case of Laplace's equation ($k = 0$), each of (6.122) and (6.123) is uniquely solvable. That is, the only solution to $S_0\phi = (-\frac{1}{2}I + D_0)g_D$ is $\phi = \partial_n^+ u$, and similarly for (6.123). However, in the Helmholtz case, neither of (6.122) and (6.123) is uniquely solvable for all $k > 0$. This nonuniqueness arises because the Neumann trace of the solution of the Helmholtz IDP also satisfies an integral equation of the form $S_k\partial_n^- u = \ldots$ (this can be seen by taking the interior Dirichlet trace of the first equation in (6.115)), and thus, since the Helmholtz IDP does not have a unique solution when $k^2$ is a Dirichlet eigenvalue of the Laplacian, neither do integral equations of the form $S_k\phi = \ldots$. Similarly, the Dirichlet trace of the solution of the Helmholtz interior Neumann problem satisfies an integral equation of the form $(\frac{1}{2}I + D_k)\gamma_- u = \ldots$ and since $(\frac{1}{2}I + D_k)$ is, roughly speaking, adjoint to $(\frac{1}{2}I + D_k')$, this implies that integral equations of the form $(\frac{1}{2}I + D_k')\phi = \ldots$ do not have a unique solution when $k^2$ is a Neumann eigenvalue of the Laplacian. (In the literature, the particular values of $k$ for which a BIE is not uniquely solvable are often called "spurious frequencies" or "spurious resonances" of that integral equation.)

There are at least four different ways around this nonuniqueness, with [72, Section 6.8] providing a good overview:

1. Supplement one of the integral equations with additional equations, such as the null-field equations (discussed in Section 6.1.10) or (6.119) evaluated at certain points in $\Omega_-$; see, e.g., [72, Section 6.11]

2. Modify the fundamental solution; see, e.g., [72, Section 6.9], [22, Section 3.6].

3. Modify the integral representation (this can be understood as using a particular *indirect* BIE method; see Remark 40).

4. Combine the two integral equations (6.122) and (6.123).

The third and fourth options have won out, at least from the point of view of mathematicians, and the relationship between *direct* and *indirect* methods (defined and discussed in Remark 40) means that these two options are, in some sense, equivalent. We now describe the fourth option.

Subtracting $i\eta$ times (6.122) from (6.123), we obtain

$$A_{k,\eta}'\partial_n^+ u = \left[H_k - i\eta\left(-\frac{1}{2}I + D_k\right)\right]g_D, \tag{6.124}$$

where

$$A'_{k,\eta} := \frac{1}{2}I + D'_k - i\eta S_k$$

is the so-called *combined field* or *combined potential* integral equation.

We saw above that the integral operators in each of the equations (6.122) and (6.123) could also be used to solve an interior BVP (for (6.122) the BVP was the IDP, and for (6.123) the BVP was the interior Neumann problem[5]). The interior BVP for $A'_{k,\eta}$ is the Helmholtz IIP; see [18, Theorem 2.30]. By Theorem 6.5, the solution of the IIP is unique for all $k > 0$ (when $\eta \in \mathbb{R}\setminus\{0\}$) and this fact can be used to prove that the integral operator $A'_{k,\eta}$ is invertible for all $k > 0$ when $\eta \in \mathbb{R}\setminus\{0\}$ [18, Theorem 2.27], thus overcoming the problem of nonuniqueness.

## Variational Formulations of the EDP Using BIEs

We consider two different variational formulations of the EDP, the first based on the integral equation (6.122) and the second based on the equation (6.124). In this section we slightly abuse notation, and use $\langle\cdot,\cdot\rangle_\Gamma$ to denote *both* the sesquilinear form on $H^{-1/2}(\Gamma)\times H^{1/2}(\Gamma)$ discussed in Remark 22 *and* its complex conjugate, which can be thought of as a sesquilinear form on $H^{1/2}(\Gamma)\times H^{-1/2}(\Gamma)$ after interchanging the arguments. Which one we're using will be clear from the arguments of the form, but the important point is that we always assume $\langle\cdot,\cdot\rangle_\Gamma$ is *sesquilinear*.

**Formulation Based on (6.122) (Involving $S_k$).** Define

$$a(\phi,\psi) := \langle S_k\phi,\psi\rangle_\Gamma \quad \text{and} \quad F(\psi) := \left\langle\left(-\frac{1}{2}I + D_k\right)g_D,\psi\right\rangle_\Gamma. \tag{6.125}$$

By the mapping properties $S_k : H^{-1/2}(\Gamma)\to H^{1/2}(\Gamma)$ and $D_k : H^{1/2}(\Gamma)\to H^{1/2}(\Gamma)$ [18, Theorem 2.17], [73, Theorem 7.1], $a(\cdot,\cdot)$ is a sesquilinear form on $H^{-1/2}(\Gamma)\times H^{-1/2}(\Gamma)$ and $F(\cdot)$ is an antilinear functional on $H^{-1/2}(\Gamma)$. The continuity of the sesquilinear form $\langle\cdot,\cdot\rangle_\Gamma : H^{1/2}(\Gamma)\times H^{-1/2}(\Gamma)\to\mathbb{C}$ and the continuity of $S_k : H^{-1/2}(\Gamma)\to H^{1/2}(\Gamma)$ imply that $a(\cdot,\cdot)$ is continuous with $\|a\|\le\|S_k\|_{H^{-1/2}(\Gamma)\to H^{1/2}(\Gamma)}$.

When $k=0$, $a(\cdot,\cdot)$ is coercive; indeed

$$\langle S_0\phi,\phi\rangle_\Gamma \gtrsim \|\phi\|^2_{H^{-1/2}(\Gamma)} \quad \text{for all } \phi\in H^{-1/2}(\Gamma) \tag{6.126}$$

by, e.g., [86, Theorem 3.5.3], [73, Corollary 8.13 and Theorem 8.16], [98, Theorems 6.22 and 6.23], [95, Section 1.4]. We saw above that, when $k^2$ is a Dirichlet eigenvalue of the Laplacian, $S_k$ is not invertible, and thus $a(\cdot,\cdot)$ cannot be coercive for these particular $k$. However, the difference $S_k - S_0$ is compact (this follows from the bounds in, e.g., [18, Equation (2.25)]), and thus $S_k$ is a compact perturbation of a coercive operator. Therefore, Theorem 6.29 implies that, when $k^2$ is not a Dirichlet eigenvalue of the Laplacian, the variational problem (6.1) with $a(\cdot,\cdot)$ and $F(\cdot)$ defined by (6.125) and $\mathcal{H} = H^{1/2}(\Gamma)$ has a unique solution that depends continuously on the data (i.e., the properties K1 and K2 hold).

---

[5]More precisely, we sketched above how the "adjoint" of $(\frac{1}{2}I + D'_k)$, namely $(\frac{1}{2}I + D_k)$, could be used to solve the interior Neumann problem, but the same is true for $(\frac{1}{2}I + D'_k)$ itself; see [18, Table 2.1].

**Formulation Based on (6.124) (Involving $A'_{k,\eta}$).** The operator $A'_{k,\eta} : H^s(\Gamma) \to H^s(\Gamma)$ for $|s| \leq 1/2$ [18, Theorem 2.27]. Since the unknown Neumann trace $\partial_n^+ u$ is in $H^{-1/2}(\Gamma)$, it is natural to pose equation (6.124) in the space $H^{-1/2}(\Gamma)$. That is, we define

$$a(\phi,\psi) := \left(A'_{k,\eta}\phi,\psi\right)_{H^{-1/2}(\Gamma)} \quad \text{and} \quad F(\psi) = \left(\left[H_k - \mathrm{i}\eta\left(-\frac{1}{2}I + D_k\right)\right] g_D, \psi\right)_{H^{-1/2}(\Gamma)}, \tag{6.127}$$

where $(\cdot,\cdot)_{H^{-1/2}(\Gamma)}$ denotes the $H^{-1/2}(\Gamma)$ inner product.

Since $A'_{k,\eta} : H^{-1/2}(\Gamma) \to H^{-1/2}(\Gamma)$, $a(\cdot,\cdot)$ is a continuous sesquilinear form on $H^{-1/2}(\Gamma)\times H^{-1/2}(\Gamma)$ (and similar mapping properties imply that $F(\cdot)$ is an antilinear functional on $H^{-1/2}(\Gamma)$). Turning to coercivity, we have that $A'_{0,0}$ is a compact perturbation of a coercive operator (see Remark 41 below), and since the difference $A'_{k,\eta} - A'_{0,0}$ is compact, $A'_{k,\eta}$ itself is then a compact perturbation of a coercive operator. Since $A'_{k,\eta}$ is injective for all $k > 0$ (as long as $\eta \in \mathbb{R}\setminus\{0\}$), Theorem 6.29 implies that the variational problem (6.1) with $a(\cdot,\cdot)$ and $F(\cdot)$ defined by (6.127) and $\mathcal{H} = H^{-1/2}(\Gamma)$ has a unique solution for all $k > 0$ (i.e., the properties K1 and K2 hold).

The drawback to posing (6.124) as an equation in $H^{-1/2}(\Gamma)$ is that the $H^{-1/2}(\Gamma)$-inner product is difficult to implement practically. If the Dirichlet trace $g_D$ is in $H^1(\Gamma)$ (as it is in the case of plane-wave or point-source scattering) the right-hand side of (6.124) is in $L^2(\Gamma)$ [18, Theorem 2.12]. We can therefore consider the integral equation (6.124) as an equation in $L^2(\Gamma)$; i.e., we define

$$a(\phi,\psi) := \left(A'_{k,\eta}\phi,\psi\right)_{L^2(\Gamma)} \quad \text{and} \quad F(\psi) = \left(\left[H_k - \mathrm{i}\eta\left(-\frac{1}{2}I + D_k\right)\right] g_D, \psi\right)_{L^2(\Gamma)}$$

and let $\mathcal{H} = L^2(\Gamma)$. The corresponding Galerkin method is then much easier to implement than the one in $H^{-1/2}(\Gamma)$.

Since $A'_{k,\eta} : L^2(\Gamma) \to L^2(\Gamma)$, $a(\cdot,\cdot)$ is continuous. Turning to coercivity, we have that, when $\Gamma$ is $C^1$, the operators $S_k$ and $D'_k$ are compact as mappings from $L^2(\Gamma)$ to itself (for $S_k$ this follows from the mapping properties given in, e.g., [18, Theorem 2.17], and for $D_k$ this is proved in [42, Theorem 1.2]). Therefore, when $\Gamma$ is $C^1$, $A'_{k,\eta}$ is a compact perturbation of a coercive operator, namely the identity. The question of whether $A'_{k,\eta}$ is a compact perturbation of a coercive operator in $L^2(\Gamma)$ when $\Gamma$ is Lipschitz is still open. However, under certain geometric restrictions $A'_{k,\eta}$ itself is coercive on $L^2(\Gamma)$; see [95, Theorem 1.2].

**Remark 40 (Indirect BIEs).** *We focused here on using Green's integral representation to obtain BIEs, and this is often called the* direct *method. Alternatively one can seek the solution of a BVP as either a single- or double-layer potential (or a linear combination of the two) and take appropriate traces of these potentials to reformulate the BVP as an integral equation on $\Gamma$; this is often called the* indirect *method. The BIEs obtained using the indirect method involve, roughly speaking, the adjoints of the operators in the BIEs obtained using the direct method; see, e.g.,* [18, Sections 2.5–2.6].

**Remark 41 (Coercivity up to a compact perturbation of the operator $A'_{k,\eta}$ on $H^{-1/2}(\Gamma)$).** *There are a few subtleties with considering $A'_{k,\eta}$ as a operator on $H^{-1/2}(\Gamma)$ that we glossed over above. The standard way to realize the $H^{-1/2}(\Gamma)$-inner product and norm is via the Laplace single-layer potential. Indeed, by the coercivity of $S_0$ (6.126), $\|\phi\|^2_{S_0} := \langle\phi, S_0\phi\rangle_\Gamma$*

*is an equivalent norm on* $H^{-1/2}(\Gamma)$. *We then let*

$$\widetilde{a}(\phi,\psi) := \langle A'_{k,\eta}\phi, S_0\psi\rangle_\Gamma \quad \text{and} \quad \widetilde{F}(\psi) = \left\langle \left[H_k - \mathrm{i}\eta\left(-\frac{1}{2}I + D_k\right)\right] g_D, S_0\psi\right\rangle_\Gamma,$$

*and these give rise to a variational problem on* $H^{-1/2}(\Gamma)$. *Using* G1 (6.24), *the single-layer jump relations* (6.120), *and the Poincaré-type inequality* (6.77), *Elschner showed that*

$$\left\langle \left(\frac{1}{2}I + D'_0\right)\phi, S_0\phi\right\rangle_\Gamma \gtrsim \|\phi\|^2_{H^{-1/2}(\Gamma)} - \|S_0\phi\|^2_{L^2(\Gamma)} \quad \text{for all } \phi \in H^{-1/2}(\Gamma) \tag{6.128}$$

[39, Proposition A1].[6] *Therefore, there exists a* $c > 0$ *such that*

$$\left\langle \left(\frac{1}{2}I + D'_0 + cS_0\right)\phi, S_0\phi\right\rangle_\Gamma \gtrsim \|\phi\|^2_{H^{-1/2}(\Gamma)} \quad \text{for all } \phi \in H^{-1/2}(\Gamma)$$

*and then, since* $S_0$, $D'_k - D'_0$, *and* $S_k$ *are compact operators on* $H^{-1/2}(\Gamma)$, *the operator associated with* $\widetilde{a}(\cdot,\cdot)$ *is a compact perturbation of a coercive operator.*

## 6.1.10 ▪ The Null-Field Method and the Fokas Transform Method

**Summary:** G2 with $v$ a separable solution of $\mathscr{L}v = 0$.

The null-field method is this idea applied to exterior Helmholtz problems with $v$ a separable solution in polar coordinates. The Fokas transform method can be understood as this idea applied with $v$ a separable solution in Cartesian coordinates; it turns out that the $v$'s in Cartesian coordinates are most suited to interior BVPs, and thus the Fokas transform method has been applied to interior BVPs for the Laplace, modified Helmholtz, and Helmholtz equations. An extension of this method has recently been successfully applied to exterior BVPs for the modified Helmholtz equation; see Remark 43 below.

We begin by introducing these ideas in their simplest possible setting, namely the IDP for the modified Helmholtz equation.

### Interior Dirichlet Problem for the Modified Helmholtz Equation

We consider the homogeneous IDP, i.e., Definition 6.2 with $\mathscr{L} = \mathscr{L}_\mu$ and $f = 0$. Let

$$\mathscr{R} := \left\{v \in H^1(\Omega,\Delta) : \mathscr{L}_\mu v = 0\right\}. \tag{6.129}$$

If $u$ and $v \in \mathscr{R}$, then G2 (6.26) with $D = \Omega$ implies that

$$\int_\Gamma \partial_n u\, \overline{\gamma v} = \int_\Gamma \gamma u\, \overline{\partial_n v}. \tag{6.130}$$

(Note that the integrals in (6.130), as well as all the other integrals over $\Gamma$ in this section, should be understood as duality pairings between $H^{-1/2}(\Gamma)$ and $H^{1/2}(\Gamma)$ unless we add the condition that $\partial_n v \in L^2(\Gamma)$ to the space $\mathscr{R}$.)

[6] More precisely, Elschner showed that (6.128) holds with $\|S_0\phi\|^2_{L^2(\Gamma)}$ replaced by $|\int_\Gamma S_0\phi|^2$, but (6.128) follows from this by using the Cauchy–Schwarz inequality (6.13).

If we define

$$a(u,v) := \int_\Gamma \partial_n u \, \overline{\gamma v} \quad \text{and} \quad F(v) := \int_\Gamma g_D \overline{\partial_n v}, \tag{6.131}$$

then the equation (6.130) gives rise to the following variational formulation of the IDP: given $g_D \in H^{1/2}(\Gamma)$,

$$\boxed{\text{find } u \in \mathscr{R} \text{ such that} \quad a(u,v) = F(v) \quad \text{for all } v \in \mathscr{R}.} \tag{6.132}$$

We now seek a variational problem where the Hilbert space consists of functions on $\Gamma$ (as opposed to functions in $\Omega\_$). Define $P_{\mathrm{DtN}} : H^{1/2}(\Gamma) \to H^{-1/2}(\Gamma)$ to be the Dirichlet-to-Neumann map. Of course, given $\phi \in H^{1/2}(\Gamma)$, finding $P_{\mathrm{DtN}}\phi$ is equivalent to solving the IDP, and so if the variational problem involves the operator $P_{\mathrm{DtN}}$, then we need to check at the end that it is practical to implement. We now define

$$b(\phi,\psi) := \int_\Gamma P_{\mathrm{DtN}} \phi \, \overline{\psi} \tag{6.133}$$

so that $a(u,v) = b(\gamma u, \gamma v)$. The variational problem (6.132) therefore becomes

$$\boxed{\text{find } u \in \mathscr{R} \text{ such that} \quad b(\gamma u, \gamma v) = F(v) \quad \text{for all } v \in \mathscr{R}.} \tag{6.134}$$

The next lemma concerns the properties of $b(\cdot,\cdot)$ as a sesquilinear form on $H^{1/2}(\Gamma) \times H^{1/2}(\Gamma)$. Note that everything so far in this section has only relied on G2, which holds for each of the Laplace, modified Helmholtz, and Helmholtz equations. The results of the lemma, however, are specific to the modified Helmholtz equation.

**Theorem 6.43 (Continuity and coercivity of $b(\cdot,\cdot)$ for modified Helmholtz).**
(i) *For any* $\phi, \psi \in H^{1/2}(\Gamma)$,

$$|b(\phi,\psi)| \leq \|P_{\mathrm{DtN}}\|_{H^{1/2}(\Gamma)\to H^{-1/2}(\Gamma)} \|\phi\|_{H^{1/2}(\Gamma)} \|\psi\|_{H^{1/2}(\Gamma)}.$$

(ii) *Given* $\mu_0 > 0$*, there exists a* $C$*, independent of* $\mu$ *(but dependent on* $\mu_0$*), such that*

$$|b(\psi,\psi)| \geq C \, \|\psi\|^2_{H^{1/2}(\Gamma)}$$

*for all* $\mu \geq \mu_0$.

***Proof.*** Continuity is straightforward (note that, given $\mu_0 > 0$, $\|P_{\mathrm{DtN}}\|_{H^{1/2}(\Gamma)\to H^{-1/2}(\Gamma)} \lesssim \mu$ for $\mu \geq \mu_0$; see [87, Proposition 2.5.2]). For coercivity, given $\psi \in H^{1/2}(\Gamma)$ there exists a unique $u \in \mathscr{R}$ such that $\gamma u = \psi$. Then, using G1 (6.24),

$$b(\psi,\psi) = \int_\Gamma \partial_n u \, \overline{\gamma u} = \int_\Omega |\nabla u|^2 + \mu^2 |u|^2,$$

and the result follows from the bound on the trace operator (6.9). □

We now consider the alternative variational problem, given $g_D \in H^{1/2}(\Gamma)$,

$$\boxed{\text{find } \phi \in H^{1/2}(\Gamma) \text{ such that } b(\phi,\psi) = G(\psi) \quad \text{for all } \psi \in H^{1/2}(\Gamma),} \tag{6.135}$$

where

$$G(\psi) := \int_\Gamma g_D \overline{P_{\mathrm{DtN}}\psi}. \tag{6.136}$$

Since $G(\gamma v) = F(v)$ when $v \in \mathscr{R}$, it is straightforward to show that if $u$ is a solution of the variational problem (6.134), then $\gamma u$ is a solution of the variational problem (6.135). Theorem 6.43 and the Lax–Milgram theorem (Theorem 6.25) imply that the solution of (6.135) exists and is unique, and therefore the *only* solution of (6.135) is $\phi = \gamma u$. We can therefore abandon the variational problem (6.134) and focus on (6.135).

Stepping back a moment, it may seem like we have achieved nothing. Indeed, the solution to the variational problem (6.135) is $\phi = \gamma u$, but we're given $\gamma u$ in the boundary condition and instead want to find $\partial_n u = P_{\mathrm{DtN}}(\gamma u)$. This shows us that when we discretize (6.135) we need to chose finite-dimensional subspaces where it is straightforward to apply $P_{\mathrm{DtN}}$, since after we have computed an approximation to $\phi$, $\phi_N$, we want to approximate $\partial_n u$ by $P_{\mathrm{DtN}}\phi_N$.

**Discretizing the Variational Problem (6.135).** The discussion in the previous paragraph indicated that, although we could in principle choose *any* finite-dimensional subspace of $H^{1/2}(\Gamma)$ when applying the Galerkin method to (6.135), we need to choose basis functions for which it is easy to apply $P_{\mathrm{DtN}}$. We also come to this conclusion when we observe that to compute $b(\phi,\psi)$ and $F(\psi)$ for given $\phi,\psi \in H^{1/2}(\Gamma)$ we need to be able to compute $P_{\mathrm{DtN}}\phi$ and $P_{\mathrm{DtN}}\psi$.

This situation is completely analogous to the situation encountered with the UWVF in Section 6.1.8, where we needed to choose subspaces in which it is easy to apply the operator $F_j$ (which involved calculating the "impedance-to-impedance" map).

Let $\mathscr{R}_N$ be a finite-dimensional subspace of $\mathscr{R}$ containing explicit solutions of $\mathscr{L}_\mu v = 0$ obtained by separation of variables. For example, the function

$$v(x,y) = \mathrm{e}^{m_1 x + m_2 y} \tag{6.137}$$

satisfies $\mathscr{L}_\mu v = 0$ in two dimensions if $m_1^2 + m_2^2 - \mu^2 = 0$. A natural parametrization of this last equation is $m_1 = \mu\cos\theta$, $m_2 = \mu\sin\theta$. Letting $\nu = \mathrm{e}^{\mathrm{i}\theta}$, we find that

$$m_1 = \frac{\mu}{2}\left(\nu + \frac{1}{\nu}\right) \quad \text{and} \quad m_2 = \frac{\mu}{2\mathrm{i}}\left(\nu - \frac{1}{\nu}\right),$$

and thus (6.137) becomes

$$v(x,y) = \exp\left[\frac{\mu}{2}\left(\nu + \frac{1}{\nu}\right)x + \frac{\mu}{2\mathrm{i}}\left(\nu - \frac{1}{\nu}\right)y\right], \tag{6.138}$$

where $\nu$ can be any complex number; we therefore write $v(x,y) = v(x,y;\nu)$.

With $\mathscr{R}_N$ such a subspace, let

$$\mathscr{Q}_N := \{\gamma v_N : v_N \in \mathscr{R}_N\}, \tag{6.139}$$

and note that $\mathscr{Q}_N \subset H^{1/2}(\Gamma)$ since $\mathscr{R}_N \subset \mathscr{R} \subset H^1(\Omega)$. Observe also that $\mathscr{Q}_N$ is a *global basis* of $H^{1/2}(\Gamma)$, i.e., each basis function has support on all of $\Gamma$. If $\phi_N \in \mathscr{Q}_N$, then there exists a $v_N \in \mathscr{R}_N$ such that $\phi_N = \gamma v_N$. Then $P_{\mathrm{DtN}}\phi_N = \partial_n v_N$, and this can be found easily from the explicit expression for $v_N$.

The Galerkin method for the variational problem (6.135) therefore reads as follows: given $g_D \in H^{1/2}(\Gamma)$,

$$\boxed{\text{find } \phi_N \in \mathcal{Q}_N \text{ such that } b(\phi_N, \psi_N) = G(\psi_N) \quad \text{for all } \psi_N \in \mathcal{Q}_N.} \tag{6.140}$$

Existence, uniqueness, and quasi optimality of $\phi_N$ (i.e., property K3) follow from using Theorem 6.43 and the Lax–Milgram theorem (Theorem 6.25). Once we have found $\phi_N$, an approximation of $\partial_n u$ is given by $P_{\mathrm{DtN}}\phi_N$ (which, from above, is straightforward to find).

The key point to take away from this discussion on discretizations is that (similar to the case of the UWVF) the Galerkin method applied to the variational formulation (6.135) is only practical when the finite-dimensional subspaces consist of (traces of) explicit solutions of $\mathcal{L}_\mu v = 0$.

### The Fokas Transform Method for the IDP for the Modified Helmholtz Equation

We now discuss the Fokas transform method. We focus on the IDP for the modified Helmholtz equation, but we also discuss how these ideas have been applied to interior BVPs for the Laplace and Helmholtz equations, and to exterior BVPs for the modified Helmholtz equation.

**The Global Relation.** The Fokas transform method for the IDP for the modified Helmholtz equation is based on the variational problem (6.134), i.e., the fact that if $u$ is the solution to the IDP, then

$$\int_\Gamma \partial_n u \, \overline{\gamma v} = \int_\Gamma g_D \, \overline{\partial_n v} \quad \text{for all } v \in \mathcal{R}.$$

Furthermore, one restricts attention to functions in $\mathcal{R}$ of the form (6.138), i.e.,

$$v(\mathbf{x}; \nu) := \exp\left[\frac{\mu}{2}\left(\nu + \frac{1}{\nu}\right)x + \frac{\mu}{2\mathrm{i}}\left(\nu - \frac{1}{\nu}\right)y\right] \tag{6.141}$$

for $\mathbf{x} \in \Omega_-$ and $\nu \in \mathbb{C} \setminus \{0\}$. The crux of the Fokas method is therefore the relation

$$\int_\Gamma \partial_n u \, \overline{\gamma v(\cdot;\nu)} = \int_\Gamma g_D \, \overline{\partial_n v(\cdot;\nu)} \quad \text{for all } \nu \in \mathbb{C} \setminus \{0\}, \tag{6.142}$$

with this one-parameter family of equations then called the *global relation*. (The word "global" reflects the fact that (6.142) contains information about certain integrals of the unknown Neumann trace, as opposed to information about the Neumann trace itself.)

With the notation introduced in the previous section (i.e., $b(\cdot,\cdot)$ and $G(\cdot)$ are defined by (6.133) and (6.136), respectively), solving the global relation (6.142) can be written as

$$\boxed{\text{find } \phi \in H^{1/2}(\Gamma) \text{ such that } b(\phi, \gamma v(\cdot;\nu)) = G(\gamma v(\cdot;\nu)) \quad \text{for all } \nu \in \mathbb{C}} \tag{6.143}$$

and then set $\partial_n u = P_{\mathrm{DtN}}\phi$.

A natural question is then, Is the set of equations (6.142) sufficient to determine $\partial_n u$? In [6] it is proved that if there exists a $\chi$ such that

$$\int_\Gamma \chi \, \overline{\gamma v(\cdot;\nu)} = \int_\Gamma g_D \, \overline{\partial_n v(\cdot;\nu)} \quad \text{for all } \nu \in \mathbb{C} \setminus \{0\}, \tag{6.144}$$

then there exists a solution of the IDP for the modified Helmholtz equation with $\partial_n u = \chi$; the proof considers smooth convex domains and continuous functions but can, in principle, be extended to nonsmooth (but still convex) domains and functions in appropriate Sobolev spaces. Since the solution of the IDP for the modified Helmholtz equation is unique, there is then only one $\chi$ satisfying (6.144) (and this is equal to the Neumann trace of the solution).

For certain domains one can solve global relation explicitly; i.e., one can *either* find the unknown Neumann trace $\partial_n u$ *or* find the solution $u$ in $\Omega_-$ (for this second option one also needs to use the analogue of Green's integral representation in the Fourier, or spectral, space); for more information see the review articles [53], [25] and the book [48]. For general domains, however, the global relation must be solved numerically.

**Solving the Global Relation Numerically (i.e., Discretizing the Variational Problem (6.143)).** The global relation (6.142) holds for all $\nu \in \mathbb{C}$, and the first step in solving this equation numerically is to enforce that (6.142) at a discrete set of points $(\nu_j)_{j=1}^N$ (we discuss the question of how to choose the points below).

Roughly speaking we then have two options.

1. The first option is to *define $\mathcal{Q}_N$ so that the Galerkin equations* (6.140) *are a discretization of* (6.143). Indeed, recalling that $\mathcal{R}_N$ denotes a finite-dimensional subspace of $\mathcal{R}$, and $\mathcal{Q}_N$ consists of Dirichlet traces of functions in $\mathcal{R}_N$, we let

$$\mathcal{R}_N := \{v(\cdot;\nu_j), j = 1,\dots,N\} \quad \text{and} \quad \mathcal{Q}_N := \{\gamma v(\cdot;\nu_j), j = 1,\dots,N\}, \qquad (6.145)$$

where $(\nu_j)_{j=1}^N$ are the points at which the global relation holds. With this particular $\mathcal{Q}_N$, the Galerkin equations (6.140) are then a discretization of the variational problem (6.143), and existence and uniqueness of a quasi-optimal Galerkin solution (property K3 of Section 6.1.5) can then be obtained using Theorem 6.43 and the Lax–Milgram theorem (Theorem 6.25).

2. The second option is to *use different trial and test spaces*; in particular we can use the "operator adapted" space $\mathcal{Q}_N$ in (6.145) for the test space but use a non operator adapted space for the trial space. (This allows us to use *local* trial spaces, i.e., trial spaces whose basis functions are supported only on parts of $\Gamma$.) The disadvantage of this option compared to the first is that to obtain K3 we would need to verify the discrete inf-sup condition (6.46), and this is difficult.

All the numerical implementations of the Fokas method in the literature have chosen the second option, and so we now discuss this option further.

Since the trial space no longer consists of traces of solutions of $\mathcal{L}_\mu v = 0$, it no longer makes sense to have $\gamma u$ as the unknown of the variational problem (since computing $P_{\mathrm{DtN}}\phi_N$, where $\phi_N$ is the approximation to $\phi$, will no longer be straightforward). We therefore consider the equivalent variational problem, given $g_D \in H^{1/2}(\Gamma)$,

$$\boxed{\text{find } \chi \in H^{-1/2}(\Gamma) \text{ such that } \langle \chi, \gamma v(\cdot;\nu)\rangle_\Gamma = G(\gamma v(\cdot;\nu)) \quad \text{for all } \nu \in \mathbb{C},} \qquad (6.146)$$

where $\langle\cdot,\cdot\rangle_\Gamma$ is the sesquilinear form on $H^{-1/2}(\Gamma) \times H^{1/2}(\Gamma)$ discussed in Remark 22 (i.e., when both arguments are in $L^2(\Gamma)$, $\langle\cdot,\cdot\rangle_\Gamma$ is the $L^2$-inner product). The result in [6] discussed above shows that if $\chi$ satisfies (6.146), then $\chi = \partial_n u$, where $u$ is the solution of the IDP for the modified Helmholtz equation.

To discretize (6.146) one needs to

1. choose a finite-dimensional subspace $\mathcal{V}_M \subset H^{-1/2}(\Gamma)$ (with dimension $M$),
2. choose points $(\nu_j)_{j=1}^N \in \mathbb{C}$, and
3. solve the problem

$$\boxed{\text{find } \chi_M \in \mathcal{V}_M \text{ such that } \langle \chi_M, \gamma v(\cdot;\nu_j)\rangle_\Gamma = G(\gamma v(\cdot;\nu_j)) \quad \text{for } j = 1,\ldots,N.}$$

Observe that, since the trial and test spaces are now independent, one can take the dimension of the test space, $N$, larger than the dimension of the trial space, $M$, i.e., create an overdetermined system that can be solved by, e.g., least squares.

Table 6.1 gives an overview of the different implementations in the literature of the preceding three steps. To understand this table we note the following:

- All the investigations consider BVPs in two-dimensional polygons.
- L stands for Laplace, MH stands for modified Helmholtz, and H stands for Helmholtz. L* means that the paper considers the $\overline{\partial}$-equation discussed in Remark 42 below (which can be understood as a reduction of Laplace's equation).
- Regarding the location of the $\{\nu_j\}_{j=1}^N$, the rays on which the $\{\nu_j\}_{j=1}^N$ are chosen in [55], [90], [93], and [51] are specified by the exponentials appearing in the global relation (which in turn are specified by the geometry of the polygon); see, e.g., [93, Equation 2.9 and Remark 2.3].
- For the definition of the *Halton notes* see [54, Appendix A], [24, Page 5] and the references therein. These are quasi-random points in the complex plane, with the particular feature that different points never get very close to each other.
- Regarding the basis functions in the trial space, "Fourier" means that a Fourier basis was used on each side of the polygon, and "polynomial" means that a polynomial basis (consisting of either Chebyshev or Legendre polynomials) was used on each side.
- The only works on the numerical implementation of the Fokas method not included in Table 6.1 are the five papers [92], [91], [85], [52], and [7]. The first three

Table 6.1: Overview of the different numerical implementations of the Fokas method (see the explanatory notes in the text).

| | PDEs considered | Dimension of test and trial spaces | Location of $\{\nu_j\}_{j=1}^N$ | Basis functions in trial space |
|---|---|---|---|---|
| [55] | L* | $N = M$ | on rays in $\mathbb{C}$ | Fourier |
| [90] | L* | $N = M$ | on rays in $\mathbb{C}$ | Fourier, polynomial |
| [93] | L, MH | $N = M$ | on rays in $\mathbb{C}$ | Fourier, polynomial |
| [54] | L | $N > M$ | at Halton nodes | polynomial |
| [24] | MH, H | $N > M$ | at Halton nodes | polynomial |
| [51] | MH | $N > M$ | on rays in $\mathbb{C}$ | polynomial |

of these are, in some sense, continuations of the method in [90] (they analyze the properties of the system of linear equations for specific geometries). The fourth paper [52] concerns the EDP for the modified Helmholtz equation and is discussed in Remark 43. The fifth paper [7] concerns interior problems, but uses a completely different variational formulation than those described above; see Remark 44.

**Remark 42 (The "Reduction" of Laplace's Equation to the $\overline{\partial}$-Equation).** *The papers* [55], [90], [92], [91], [85], *and* [7] *all effectively consider solving BVPs involving the operator $\partial/\partial\overline{z}$. The simplest such BVP is the following: let $\Omega$ be a bounded domain in $\mathbb{C}$ with boundary $\Gamma$. Given $g:\Gamma\to\mathbb{C}$, find $U:\mathbb{C}\to\mathbb{C}$ such that* $\operatorname{Re} U = g$ *on $\Gamma$, and*

$$\frac{\partial U}{\partial \overline{z}} = 0 \quad in\ \Omega \tag{6.147}$$

*(i.e., $U$ is analytic; see* [1, Section 2.6.3]*). Note that it is sufficient to find* $\operatorname{Im} U$ *on $\Gamma$, since once one knows $U$ on $\Gamma$ one can then find $U$ in $\Omega$ using Cauchy's integral formula. (Note the similarity with finding $u$ in $\Omega$ using Green's integral representation once one knows both $\gamma u$ and $\partial_n u$.)*

*How is this relevant to Laplace's equation? Identifying $\mathbb{R}^2$ with $\mathbb{C}$ (and thus writing $z = x + iy$), we have that*

$$\frac{\partial}{\partial z} = \frac{1}{2}\left(\frac{\partial}{\partial x} - \mathrm{i}\frac{\partial}{\partial y}\right) \quad and \quad \frac{\partial}{\partial \overline{z}} = \frac{1}{2}\left(\frac{\partial}{\partial x} + \mathrm{i}\frac{\partial}{\partial y}\right)$$

[1, Section 2.6.3], *and thus*

$$\Delta = \frac{\partial^2}{\partial x^2} + \frac{\partial^2}{\partial y^2} = 4\frac{\partial^2}{\partial z \partial \overline{z}}.$$

*Therefore, if $u$ satisfies Laplace's equation, then*

$$\frac{\partial}{\partial \overline{z}}\left(\frac{\partial u}{\partial z}\right) = 0.$$

*In some physical applications of Laplace's equation one is interested in finding the gradient of $u$, i.e., finding both $\partial u/\partial x$ and $\partial u/\partial y$ (for example, in fluid dynamics, the velocity potential for two-dimensional, irrotational, incompressible flow satisfies Laplace's equation, and then the velocity is given by the gradient of the potential; see, e.g.,* [1, Section 2.1.2]*). Since $\partial u/\partial z = (\partial u/\partial x - \mathrm{i}\partial u/\partial y)/2$, if $u$ is real, one can find $\nabla u$ from $\partial u/\partial z$.*

*In summary, for some applications it is sufficient to find the gradient of the solution of Laplace's equation, and this can be found by solving the $\overline{\partial}$-equation* (6.147).

**Remark 43 (Numerical implementation of the Fokas method for exterior problems).** *The paper* [52] *contains a preliminary numerical implementation of the Fokas transform method for the Dirichlet problem for the modified Helmholtz equation in the exterior of a square. The appropriate global relation is formed by applying* G2 *in the exterior domain with $v$ given by* (6.141)*. However, since these $v$'s do not tend to zero as $r := |\mathbf{x}| \to \infty$, there is a nonzero contribution from the integral at infinity (containing the coefficient of the first term in the asymptotic expansion of the solution as $r\to\infty$). The presence of this extra unknown means that one must supplement the global relation with an additional equation, and* [52] *uses the analogue of the BIE* (6.122) *in the transform space (i.e., the equation involves the* *transform of the unknown Neumann trace as opposed to the Neumann trace itself). Indeed, equation* (4.1) *in* [52] *is the transform-space analogue of the first equation in* (6.116) *(but now for modified Helmholtz instead of Helmholtz), and taking the limit as* $\mathbf{x} \to \Gamma$ *in equation* (4.1) *in* [52] *then yields the analogue of* (6.122) *in the transform space.*

**Remark 44 (The variation formulation in [7]).** *The paper* [7] *introduces a new variational formulation of the Dirichlet problem for Laplace's equation in a convex polygon (after reducing Laplace's equation to the* $\overline{\partial}$*-equation as discussed in Remark* 42*), with this variational formulation based on the analogue of the global relation* (6.142) *for this particular problem. The difference between the variational formulation in* [7] *and the variational formulations based on* (6.142) *discussed above is that, in* [7]*, the unknowns are a set of transforms of the Neumann trace* $\partial_n u$ *(as opposed to* $\partial_n u$ *itself), and therefore the Hilbert space is a space of analytic functions (as opposed to a space of functions on* $\Gamma$*).*

### Interior Dirichlet Problem for the Helmholtz Equation

Having discussed the Fokas transform method, our next goal is to discuss the null-field method. Although this second method is designed to solve exterior Helmholtz BVPs, it is helpful to first consider interior Helmholtz BVPs, and thus in this subsection we consider the Helmholtz IDP.

Repeating the steps we performed for the modified Helmholtz IDP earlier in this section, we see that everything up to Theorem 6.43 is the same for the Helmholtz equation, i.e., the space $\mathscr{R}$ and sesquilinear forms $a(\cdot,\cdot)$ and $b(\cdot,\cdot)$ are defined by (6.129), (6.131), and (6.133), respectively. However, the analogue of Theorem 6.43 is now the following.

**Theorem 6.44 (Continuity, Gårding inequality, and injectivity for $b(\cdot,\cdot)$ for Helmholtz).**
(i) *For any* $\phi, \psi \in H^{1/2}(\Gamma)$,

$$|b(\phi,\psi)| \leq \|P_{\mathrm{DtN}}\|_{H^{1/2}(\Gamma)\to H^{-1/2}(\Gamma)} \|\phi\|_{H^{1/2}(\Gamma)} \|\psi\|_{H^{1/2}(\Gamma)}.$$

(ii) *Assume that* $k^2$ *is not a Dirichlet eigenvalue of the Laplacian in* $\Omega$ *and also that, given* $\theta \in (1/2, 1)$*, there exists a* $C_1 > 0$ *(depending on* $k$ *and* $\theta$*) such that if* $u \in H^1(\Omega)$ *satisfies* $\mathscr{L}_k u = 0$*, then*

$$\|u\|_{H^{\theta}(\Omega)} \leq C_1 \|\gamma u\|_{H^{\theta-1/2}(\Gamma)}. \tag{6.148}$$

*Then, there exists a* $C_2 > 0$ *(independent of* $k$*) such that*

$$b(\psi,\psi) \geq \frac{1}{C_2^2}\|\psi\|^2_{H^{1/2}(\Gamma)} - (1+k^2)C_1^2 \|\psi\|^2_{H^{\theta-1/2}(\Gamma)}$$

*for all* $\psi \in H^{1/2}(\Gamma)$.
(iii) *If* $k^2$ *is not a Dirichlet eigenvalue of the Laplacian in* $\Omega$ *and* $b(\phi,\psi) = 0$ *for all* $\psi \in H^{1/2}(\Gamma)$*, then* $\phi = 0$ *(and thus the operator associated with* $b(\cdot,\cdot)$ *is injective).*

***Proof.*** The proof of (i) is identical to the modified Helmholtz case. For (ii), given $\psi \in H^{1/2}(\Gamma)$ there exists a unique $u \in \mathscr{R}$ such that $\gamma u = \psi$ ($u$ is unique because $k^2$ is not a Dirichlet eigenvalue). Then, by G1 (6.24),

$$b(\psi,\psi) = \int_\Gamma \partial_n u \overline{\gamma u} = \int_\Omega |\nabla u|^2 - k^2|u|^2 = \|u\|^2_{H^1(\Omega)} - (1+k^2)\|u\|^2_{L^2(\Omega)}. \tag{6.149}$$

Let $C_2$ be the constant in the inequality (6.9) (concerning the trace theorem). Then, using the inequalities (6.9) and (6.148) in (6.149), we obtain the result.

For (iii), the hypothesis and the definition of $b(\cdot,\cdot)$ (6.133) imply that

$$\langle P_{\mathrm{DtN}}\phi, \psi\rangle_\Gamma = 0 \quad \text{for all } \psi \in H^{1/2}(\Gamma),$$

which implies that $P_{\mathrm{DtN}}\phi = 0$ as an element of $H^{-1/2}(\Gamma)$ (if $F(\cdot)$ is an antilinear functional on $H^{1/2}(\Gamma)$ and $F(\psi) = 0$ for all $\psi \in H^{1/2}(\Gamma)$, then $F = 0$ as an element of $(H^{1/2}(\Gamma))^* \cong H^{-1/2}(\Gamma)$). When $k^2$ is not a Dirichlet eigenvalue of the Laplacian in $\Omega_-$, the operator $P_{\mathrm{DtN}} : H^{1/2}(\Gamma) \to H^{-1/2}(\Gamma)$ is invertible (this is perhaps most easily proved via BIEs; see [86, Remark 3.7.5]), and thus $\phi = 0$. □

**Remark 45 (The bound (6.148)).** *In* [9] *it is stated that the bound* (6.148) *follows "by a standard inequality for elliptic problems in bounded, smooth domains"; however, we have been unable to find a reference. Note that, when the solution of the IDP for $\mathscr{L}$ is unique, the bound with $\theta = 1$ holds by Fredholm theory (see, e.g.,* [86, Theorem 2.10.4]), *and the bound with $\theta = 1/2$ holds for the Laplace equation by* [67, Corollary 5.5] *(which is also stated as* [18, Theorem A.6]*).*

The results of Theorem 6.44 combined with Theorem 6.31 and part (i) of Theorem 6.29 then imply that the variational problem (6.135) has a unique solution (i.e., the properties K1 and K2 hold) when $k^2$ is not a Dirichlet eigenvalue of the Laplacian in $\Omega$. Similar to the case of the modified Helmhholtz equation we can then abandon the variational problem (6.134) and focus on (6.135).

As before, we need to use the finite-dimensional subspaces $\mathscr{Q}_N$ defined by (6.139) for the Galerkin method to be practical (with these subspaces now consisting of traces of explicit solutions of $\mathscr{L}_k v = 0$). Part (ii) of Theorem 6.29 can then be used to prove that the Galerkin equations (6.140) have a unique, quasi-optimal solution (i.e., property K3 holds), provided that $N$ is large enough and that $(\mathscr{Q}_N)_{N\in\mathbb{Z}^+}$ is a sequence of finite-dimensional nested subspaces of $H^{1/2}(\Gamma)$ whose union is dense in $H^{1/2}(\Gamma)$ (i.e., (6.43) holds with $\mathscr{H}_i = H^{1/2}(\Gamma)$ and $\mathscr{H}_N^i = \mathscr{Q}_N$). Since $\mathscr{Q}_N$ consists of traces of particular solutions of $\mathscr{L}_k v = 0$, this density property is not immediate and needs to be checked; we discuss this more for the case of the EDP in the section entitled "The Method of Aziz, Dorr, and Kellogg for the Helmholtz Exterior Dirichlet Problem" below, and we refer the reader to [19, Section 4.1] for discussion of this density property for the IDP. Indeed, [19] (which appears in this book) also discusses the Fokas transform method for the Helmholtz IDP, with [19, Lemma 4.1] proving an appropriate density result and then [19, Theorem 4.2] applying this to the Galerkin method.

### Exterior Dirichlet Problem for the Helmholtz Equation

We now let

$$\mathscr{R} := \left\{ u \in H^1_{\mathrm{loc}}(\Omega_+, \Delta) : \mathscr{L}_k u = 0 \text{ in } \Omega_+ \text{ and } u \text{ satisfies the radiation condition (6.17)} \right\}. \tag{6.150}$$

For $u, v \in \mathscr{R}$, applying G2 in $\Omega_R$, letting $R \to \infty$, and using Lemma 6.13, we find that

$$\int_\Gamma \partial_n^+ u\, \gamma_+ v = \int_\Gamma \gamma_+ u\, \partial_n^+ v, \tag{6.151}$$

but

$$\int_\Gamma \partial_n^+ u\, \overline{\gamma_+ v} \neq \int_\Gamma \gamma_+ u\, \overline{\partial_n^+ v}$$

(since the integral over $\Gamma_R$ does not tend to zero in the second case).

One can use (6.151) as the basis of a variational formulation of the Helmholtz EDP. Indeed, if

$$a(u,v) := \int_\Gamma \partial_n u \, \gamma v \quad \text{and} \quad F(v) = \int_\Gamma g_D \partial_n v,$$

then the equation (6.151) gives rise to the following variational formulation of the EDP: given $g_D \in H^{1/2}(\Gamma)$,

$$\boxed{\text{find } u \in \mathcal{R} \text{ such that} \quad a(u,v) = F(v) \quad \text{for all } v \in \mathcal{R}.} \tag{6.152}$$

Defining $b(\cdot,\cdot)$ by

$$b(\phi,\psi) := \int_\Gamma P_{\mathrm{DtN}} \phi \, \psi, \tag{6.153}$$

where $P_{\mathrm{DtN}} : H^{1/2}(\Gamma) \to H^{-1/2}(\Gamma)$ is now the exterior Dirichlet-to-Neumann map for the Helmholtz equation, we see that the variational problem (6.152) becomes

$$\boxed{\text{find } u \in \mathcal{R} \text{ such that} \quad b(\gamma u, \gamma v) = F(v) \quad \text{for all } v \in \mathcal{R}.} \tag{6.154}$$

The difficulty is that $b(\cdot,\cdot)$ is now a bilinear form on a complex-valued space (and $F(\cdot)$ is a linear functional). This is not a problem in and of itself, because the Lax–Milgram theorem, for example, still holds for such $b(\cdot,\cdot)$ and $F(\cdot)$. However, the lack of complex-conjugate in $b(\cdot,\cdot)$ makes it difficult to prove coercivity or a Gårding inequality. Indeed, with $R > \sup_{\mathbf{x}\in\Omega_-} |\mathbf{x}|$, applying G1 (6.23) in $\Omega_R$ implies that

$$b(\psi,\psi) = \int_{\Gamma_R} u \frac{\partial u}{\partial r} - \int_{\Omega_R} (\nabla u \cdot \nabla u - k^2 u^2), \tag{6.155}$$

where $u \in \mathcal{R}$ is such that $\gamma_+ u = \psi$ ($u$ is unique by the uniqueness of the solution to the Helmholtz EDP). The lack of complex conjugate in the bilinear form means that one cannot get a norm of $u$ from the terms in $\Omega_R$ on the right-hand side of (6.155). (Note that this difficulty can be understood as a consequence of the fact that the Helmholtz EDP is not self-adjoint; see Example 6.12.)

We now have two options: (i) continue with the variational problem (6.154) and accept that it will be difficult to prove that the properties K1, K2, and K3 hold, or (ii) modify $b(\cdot,\cdot)$. The first option is the null-field method, and we discuss this next. The second option was introduced by Aziz, Dorr, and Kellogg [9], and we discuss this in the section entitled "The Method of Aziz, Dorr, and Kellogg for the Helmholtz Exterior Dirichlet Problem."

### The Null-Field Method for the Helmholtz Exterior Dirichlet Problem

We start from (6.151) and choose a family of solutions of $\mathcal{L}_k v = 0$ that we hope span (in some sense) $\mathcal{R}$. In two dimensions we choose

$$v_n(\mathbf{x}) = H_n^{(1)}(kr)\mathrm{e}^{\mathrm{i}n\theta} \quad \text{for } n \in \mathbb{Z}, \text{ where } \mathbf{x} = (r,\theta), \tag{6.156}$$

and in three dimensions

$$v_{n,m}(\mathbf{x}) = h_n^{(1)}(kr) Y_{n,m}(\theta,\phi) \quad \text{for } n \in \mathbb{Z} \text{ and } m = -n,\dots,n, \text{ where } \mathbf{x} = (r,\theta,\phi), \tag{6.157}$$

where $H_n^{(1)}$ are the Hankel functions defined by, e.g., [36, Equations 10.4.3 and 10.2.5], $h_n^{(1)}$ are the spherical Bessel functions defined by [36, Equation 10.47.5], and $Y_{n,m}$ are the spherical harmonics defined by [36, Equation 14.30.1].

In what follows we consider the two-dimensional case, but everything follows in the same way for the three-dimensional case. Choosing $v$ in (6.151) to be $v_n$ defined by (6.156), we obtain the *null-field equations*

$$\int_\Gamma \partial_n^+ u \, \gamma_+ v_n = \int_\Gamma \gamma_+ u \, \partial_n^+ v_n \quad \text{for all } n \in \mathbb{Z}. \tag{6.158}$$

The null-field method for the EDP is then, given $g_D \in H^{1/2}(\Gamma)$,

$$\boxed{\text{find } \phi \in H^{-1/2}(\Gamma) \text{ such that } \quad \int_\Gamma \phi \, \gamma_+ v_n = \int_\Gamma g_D \, \partial_n^+ v_n \quad \text{for all } n \in \mathbb{Z},} \tag{6.159}$$

and then set $\partial_n^+ u = \phi$. (We can think of (6.159) as a moment problem: we know the moments of $\phi$ with respect to $v_n$ and the task is to find $\phi$ itself.) We prove below that the solution to (6.159) is unique for every $k > 0$. The unique solvability for every $k > 0$ of this boundary-based formulation of the EDP (in contrast to the "spurious frequencies" of the integral equations (6.122) and (6.123)) was the main novelty of the null-field equations when they were first introduced.

To practically implement (6.159), one chooses a family of functions $\{\psi_j\}_{j\in\mathbb{Z}}$ : $H^{-1/2}(\Gamma) \to \mathbb{C}$, approximates $\phi$ by

$$\phi_N := \sum_{j=1}^N a_j \psi_j,$$

and then solves the linear system

$$\sum_{j=1}^N \left( \int_\Gamma \psi_j \, \gamma_+ v_n \right) a_j = \int_\Gamma g_D \, \partial_n^+ v_n \quad \text{for } n = 1, \ldots, N. \tag{6.160}$$

Common choices of the functions $\psi_j$ are then $\gamma_{\scriptscriptstyle |} v_j$ and $\partial_n^+ v_j$; see [72, Section 7.7.2] for more discussion.

**Remark 46 (Connection with least squares [72, Section 7.8.2]).** *If we seek to approximate the solution of the EDP by*

$$u_N(\mathbf{x}) := \sum_{j=1}^N a_j \, v_j(\mathbf{x}),$$

*with $v_j$ defined by* (6.156) *and $a_j \in \mathbb{C}$, and impose the Dirichlet boundary condition $\gamma_+ u = g_D$ in a least squares sense, then we are led to minimizing the functional*

$$J(a_1, \ldots, a_N) = \int_\Gamma \left| \sum_{j=1}^N a_j \gamma_+ v_j - g_D \right|^2 .$$

*Taking the derivatives of $J(a_1, \ldots, a_N)$ with respect to the real and imaginary parts of $a_j$ for $j = 1, \ldots, N$ we arrive at the linear system*

$$\sum_{j=1}^N \left( \int_\Gamma \overline{\gamma_+ v_n} \, \gamma_+ v_j \, \mathrm{d}s \right) a_j = \int_\Gamma g_D \, \overline{\gamma_+ v_n} \quad \textit{for } n = 1, \ldots, N. \tag{6.161}$$

*If we choose* $\psi_j = \overline{\gamma_+ v_j}$ *in the discretized null-field equations* (6.160), *then the corresponding (Hermitian) matrix is the transpose (or complex conjugate) of the matrix in* (6.161).

**Alternative Derivation of the Null-Field Equations and a Proof of Uniqueness.** We now give an alternative derivation of the null-field equations that can be "reversed" to give a proof of uniqueness.

We start from the first equation in (6.116), i.e.,

$$-\int_\Gamma \left( \Phi_k(\mathbf{x},\mathbf{y})\partial_n^+ u(\mathbf{y}) - \frac{\partial \Phi_k(\mathbf{x},\mathbf{y})}{\partial n(\mathbf{y})} \gamma_+ u(\mathbf{y}) \right) ds(\mathbf{y}) = 0, \quad \mathbf{x} \in \Omega_-. \tag{6.162}$$

We introduce polar coordinates $\mathbf{x} = (r,\theta)$ and $\mathbf{y} = (\rho,\phi)$, and recall the Fourier series expansion of the fundamental solution

$$\Phi_k(\mathbf{x},\mathbf{y}) := \frac{i}{4} H_0^{(1)}(k|\mathbf{x}-\mathbf{y}|) = \frac{i}{4} \sum_{n=-\infty}^{\infty} H_n^{(1)}(kr_>) J_n(kr_<) e^{in(\phi-\theta)}, \tag{6.163}$$

where $r_> = \max(r,\rho)$ and $r_< = \min(r,\rho)$; see, e.g., [72, Equation (2.29)].[7]

Assume that $\Omega_-$ contains the ball of radius $a$ centered at the origin, $B_a$, and let $\mathbf{x}$ be in this ball. Substituting (6.163) into (6.162) and noting that $r_< = r$ and $r_> = \rho$, we obtain that

$$\sum_{n=-\infty}^{\infty} J_n(kr) e^{-in\theta} \int_\Gamma \left( -H_n^{(1)}(k\rho) e^{in\phi} \partial_n^+ u(\mathbf{y}) + \frac{\partial}{\partial n(\mathbf{y})} (H_n^{(1)}(k\rho) e^{in\phi}) \gamma_+ u(\mathbf{y}) \right) ds(\mathbf{y}) = 0 \quad \text{for } r < a. \tag{6.164}$$

(Interchanging the sum and integral is justified since the series (6.163) converges absolutely due to the asymptotics

$$H_n^{(1)}(kr_>) J_n(kr_<) \sim \frac{1}{i\pi n} \left( \frac{r_<}{r_>} \right)^n \quad \text{as } n \to \infty$$

[3, Equation 9.31], [36, Section 10.19]).

We now claim that

$$\int_\Gamma \left( -H_n^{(1)}(k\rho) e^{in\phi} \partial_n^+ u(\mathbf{y}) + \frac{\partial}{\partial n(\mathbf{y})} (H_n^{(1)}(k\rho) e^{in\phi}) \gamma_+ u(\mathbf{y}) \right) ds(\mathbf{y}) = 0 \quad \text{for all } n \in \mathbb{Z}. \tag{6.165}$$

Indeed, this is a consequence of applying the following lemma to (6.164).

**Lemma 6.45.** *If* $(\alpha_n)_{n\in\mathbb{Z}}$ *are such that* (i) *the series* $\sum_{n=-\infty}^{\infty} |\alpha_n||J_n(kr)|$ *converges and* (ii) *there exists an* $a > 0$ *such that*

$$\sum_{n=-\infty}^{\infty} \alpha_n J_n(kr) e^{-in\theta} = 0 \quad \textit{for } r < a, \tag{6.166}$$

*then* $\alpha_n = 0$ *for all* $n \in \mathbb{Z}$.

---

[7] To convert [72, Equation (2.29)] into an expression equivalent to (6.163) (after relabeling variables), let $\widetilde{\mathbf{b}} = -\mathbf{b}$ and observe that (i) $\mathbf{r}_2 = \mathbf{r}_1 - \widetilde{\mathbf{b}}$, and (ii) the angular co-ordinate of $\widetilde{\mathbf{b}}$ is $\pi + \beta$.

***Proof.*** Multiplying the sum (6.166) by $\mathrm{e}^{im\theta}$ and integrating over $\theta \in (0,2\pi)$, we find that

$$J_m(kr)\alpha_m = 0 \quad \text{for all } m \in \mathbb{Z} \text{ and for all } r < a.$$

If $J_m(kr) \neq 0$, then $\alpha_m = 0$. However, if $k = \beta_{m,l}/r$, where $\beta_{m,l}$ is the $l$th zero of $J_m$, then $J_m(kr) = 0$. (Recall that the Dirichlet eigenvalues of the negative Laplacian in $B_r$ are $(\beta_{m,l}/r)^2$). We therefore need to choose an $r < a$ such that $J_m(kr) \neq 0$ for all $m \in \mathbb{Z}$.

Recall that $0 < \beta_{0,1} \leq \beta_{m,l}$ for all $m = 0,1,\dots$ and $l = 1,\dots$ [36, Equation 10.21.2]. Therefore, given $k > 0$, choose $r^* < \min(\beta_{0,1}/k, a)$. Then $J_m(kr^*) \neq 0$ for all $m \in \mathbb{Z}$ and we are done. □

The definition of $v_n$ (6.156) implies that (6.165) can be rewritten as

$$\int_\Gamma \big(-\gamma_+ v_n(\mathbf{y})\partial_n^+ u(\mathbf{y}) + \partial_n^+ v_n(\mathbf{y})\gamma_+ u(\mathbf{y})\big)\,\mathrm{d}s(\mathbf{y}) = 0 \quad \text{for all } n \in \mathbb{Z},$$

which is (6.158). (This is how the null-field equations were originally derived by Waterman in [99] for the time-harmonic Maxwell equations, and in [100] for the Helmholtz equation.)

For uniqueness, we need to show that if $\phi \in H^{-1/2}(\Gamma)$ is such that

$$\int_\Gamma \phi\, \gamma_+ v_n = 0 \quad \text{for all } n \in \mathbb{Z}, \tag{6.167}$$

then $\phi = 0$. To begin, we assume that $\Omega_-$ contains $B_a$ and rewrite (6.167) as

$$\int_\Gamma \phi(\mathbf{y}) H_n^{(1)}(k\rho)\mathrm{e}^{in\phi}\,\mathrm{d}s(\mathbf{y}) = 0 \quad \text{for all } n \in \mathbb{Z}$$

(where, as above, $\mathbf{y} = \rho \mathrm{e}^{i\phi}$). Multiplying each of these equations by $J_n(kr)\mathrm{e}^{-in\theta}$, with $r < a$, and summing them up, we obtain that

$$0 = \sum_{n=-\infty}^{\infty} J_n(kr)\mathrm{e}^{-in\theta} \int_\Gamma \phi(\mathbf{y}) H_n^{(1)}(k\rho)\mathrm{e}^{in\phi}\,\mathrm{d}s(\mathbf{y}) \quad \text{for } r < a.$$

However, the expansion of the fundamental solution (6.163) then implies that

$$\sum_{n=-\infty}^{\infty} J_n(kr)\mathrm{e}^{-in\theta} \int_\Gamma \phi(\mathbf{y}) H_n^{(1)}(k\rho)\mathrm{e}^{in\phi}\,\mathrm{d}s(\mathbf{y}) = \int_\Gamma \Phi_k(\mathbf{x},\mathbf{y})\phi(\mathbf{y})\,\mathrm{d}s(\mathbf{y}) \quad \text{for } \mathbf{x} \in B_a \tag{6.168}$$

(i.e., we have "reversed" the original derivation of the null-field equations).

Recalling the definition of the single-layer potential $\mathcal{S}_k$ (6.117), we see that if $\phi$ is such that (6.168) holds, then $u := \mathcal{S}_k\phi$ is zero in $B_a$. Now $u$ is in $C^2(\Omega_-)$ and satisfies $\mathcal{L}_k u = 0$ and, by Green's integral representation, $C^2$ solutions of the Helmholtz equation are analytic (see, e.g., [22, Theorem 3.5]). Therefore, if $u = 0$ in $B_a$, then $u = 0$ in $\Omega_-$. By the jump relations (6.120), $u$ is continuous across $\Gamma$, and so $\gamma_+ u = \gamma_- u = 0$. By uniqueness of the Helmholtz EDP, $u = 0$ in $\Omega_+$. The jump relations (6.120) imply that $\phi = \partial_n^- u - \partial_n^+ u$, and then (since $u = 0$ in both $\Omega_-$ and $\Omega_+$) $\phi = 0$ and we have uniqueness.

### The Method of Aziz, Dorr, and Kellogg for the Helmholtz Exterior Dirichlet Problem

The method introduced in [9] reformulates the Helmholtz EDP as the following variational problem: given $g_D \in H^{1/2}(\Gamma)$,

$$\boxed{\text{find } u \in \mathcal{R} \text{ such that } \int_\Gamma \gamma_+ u\, \overline{\partial_n^+ v} = \int_\Gamma g_D\, \overline{\partial_n^+ v} \quad \text{for all } v \in \mathcal{R},} \tag{6.169}$$

where $\mathscr{R}$ is defined by (6.150). This variational problem can be understood as imposing the Dirichlet boundary condition on $\Gamma$ in a weak sense, i.e., there is no use of Green's identities.

Define the sesquilinear form $\widetilde{b}(\cdot,\cdot)$ by

$$\widetilde{b}(\phi,\psi):=\int_\Gamma \phi\,\overline{P_{\rm DtN}\psi},$$

where $P_{\rm DtN}: H^{1/2}(\Gamma)\to H^{-1/2}(\Gamma)$ is the exterior Dirichlet-to-Neumann map for the Helmholtz equation. With this definition, the variational problem (6.169) can be written as

$$\boxed{\text{find } u\in\mathscr{R} \text{ such that } \widetilde{b}(\gamma_+u,\gamma_+v)=\int_\Gamma g_D\,\overline{\partial_n^+ v} \quad \text{for all } v\in\mathscr{R}.} \tag{6.170}$$

The next theorem allows us to reduce the variational problem (6.170) to the following one:

$$\boxed{\text{find } \phi\in H^{1/2}(\Gamma) \text{ such that } \widetilde{b}(\phi,\psi)=G(\psi) \quad \text{for all } \psi\in H^{1/2}(\Gamma),} \tag{6.171}$$

where $G(\cdot)$ is defined by (6.136).

**Theorem 6.46 (Continuity, Gårding inequality, and injectivity for $\widetilde{b}(\cdot,\cdot)$).**
(i) *For any* $\phi,\psi\in H^{1/2}(\Gamma)$,

$$\left|\widetilde{b}(\phi,\psi)\right|\le \|P_{\rm DtN}\|_{H^{1/2}(\Gamma)\to H^{-1/2}(\Gamma)}\|\phi\|_{H^{1/2}(\Gamma)}\|\psi\|_{H^{1/2}(\Gamma)}.$$

(ii) *Assume that, given* $\theta\in(1/2,1)$ *and* $R>\sup_{\mathbf{x}\in\Omega_-}|\mathbf{x}|$, *there exists a* $C_1>0$ *(depending on* $k$ *and* $\theta$*) such that if* $u\in H^1_{\rm loc}(\Omega_+)$ *satisfies* $\mathscr{L}_k u=0$, *then*

$$\|u\|_{H^\theta(\Omega_R)}\le C_1\left\|\gamma_+u\right\|_{H^{\theta-1/2}(\Gamma)}. \tag{6.172}$$

*Then, there exists a* $C_2>0$ *(independent of* $k$*) such that*

$$\left|\widetilde{b}(\psi,\psi)\right|\ge \frac{1}{C_2^2}\|\psi\|^2_{H^{1/2}(\Gamma)}-(1+k^2)C_1^2\|\psi\|^2_{H^{\theta-1/2}(\Gamma)} \tag{6.173}$$

*for all* $\psi\in H^{1/2}(\Gamma)$.

(iii) *If* $\widetilde{b}(\phi,\psi)=0$ *for all* $\psi\in H^{1/2}(\Gamma)$, *then* $\phi=0$ *(and thus the operator associated with* $\widetilde{b}(\cdot,\cdot)$ *is injective).*

***Proof.*** Note that part (ii) of this theorem was first proved in [9, Theorem 3.1].

The proof of (i) is straightforward (note that if $\Omega_+$ is nontrapping, then, given $k_0>0$, $\|P_{\rm DtN}\|_{H^{1/2}(\Gamma)\to H^{-1/2}(\Gamma)}\lesssim k$ for all $k\ge k_0$ by [13]). The proof of (ii) is very similar to the proof of part (ii) of Theorem 6.44. Indeed, given $\psi\in H^{1/2}(\Gamma)$ there exists a unique $u\in\mathscr{R}$ such that $\gamma_+u=\psi$. Then, by G1 (6.24),

$$\left|\widetilde{b}(\psi,\psi)\right|\ge -\operatorname{Re}\widetilde{b}(\psi,\psi)=\int_{\Omega_R}\left(|\nabla u|^2-k^2|u|^2\right)-\operatorname{Re}\int_{\Gamma_R}\overline{u}\frac{\partial u}{\partial r}. \tag{6.174}$$

Now, if $u\in\mathscr{R}$, then $\operatorname{Re}\int_{\Gamma_R}\overline{u}\,\partial u/\partial r\,{\rm d}s\le 0$ [82, Theorem 2.6.4]. After using this inequality in (6.174), we find that the proof of (6.173) is identical to the proof of part (ii) of Theorem 6.44.

The proof of (iii) is also very similar to the proof of part (iii) of Theorem 6.44. The exterior Dirichlet-to-Neumann map $P_{\mathrm{DtN}} : H^{1/2}(\Gamma) \to H^{1/2}(\Gamma)$ is invertible for all $k > 0$ by uniqueness of the solution to the Helmholtz EDP (although the invertibility is perhaps most easily proved via integral equations; see [18, Theorem 2.31]). □

Similar to the case of the Helmholtz IDP above, Theorems 6.46 and 6.29 show that the variational problem (6.171) has a unique solution. It is then straightforward to show that this solution is $\gamma_+ u$, where $u$ is the solution to (6.170). We can therefore forget about the variational problem (6.170) and instead concentrate on (6.171).

Similar to the interior problems in the sections entitled "Interior Dirichlet Problem for the Modified Helmholtz Equation" and "The Fokas Transform Method for the IDP for the Modified Helmholtz Equation," given $\phi, \psi \in H^{1/2}(\Gamma)$, to find $\widetilde{b}(\phi,\psi)$ and $G(\psi)$ one needs to find $P_{\mathrm{DtN}}\psi$. Moreover (and exactly as before), the unknown $\phi$ in the variational problem (6.171) equals $\gamma_+ u$, which is already given by the boundary condition. Therefore, once we have found the Galerkin approximation to $\phi$, $\phi_N$, we need to be able to work out $P_{\mathrm{DtN}}\phi_N$ easily.

Therefore, for the Galerkin method applied to (6.171) to be practical, we need to choose finite-dimensional subspaces where it is easy to apply the Dirichlet-to-Neumann map (for example, subspaces consisting of traces of explicit solutions to the Helmholtz equation in $\Omega_+$). A natural choice in two dimensions is

$$\mathcal{Q}_N := \operatorname{span}\left\{\gamma_+ v_n : n = -N, \dots, N\right\}, \tag{6.175}$$

where $v_n$ are defined by (6.156). (Note that $\mathcal{Q}_N \subset \mathcal{Q}_{N+1}$ by definition.)

**Theorem 6.47 (Asymptotic density of $\mathcal{Q}_N$ in $H^{1/2}(\Gamma)$).** *With $\mathcal{Q}_N$ defined by* (6.175), *their union over $N \in \mathbb{Z}^+$ is dense in $H^{1/2}(\Gamma)$ (i.e.,* (6.43) *holds with $\mathcal{H}_i = H^{1/2}(\Gamma)$ and $\mathcal{H}_N^i = \mathcal{Q}_N$).*

***Proof.*** The analogue of this result in three dimensions (with $v_n$ replaced by $v_{n,m}$ defined by (6.157)) is proved in [9, Theorem 5.1]; the proof for the two-dimensional case is completely analogous. □

This density result and part (ii) of Theorem 6.29 then imply that, when $N$ is sufficiently large, the Galerkin method applied to the variational problem (6.171) has a unique, quasi-optimal solution (i.e., the property K3 holds).

**Remark 47 (The method of DeSanto for scattering by diffraction gratings).** *In this paper, the only BVP for the Helmholtz equation posed on an* unbounded *domain that we considered was the EDP. Another such BVP is the Helmholtz equation posed above an infinite rough surface, and in the case when the surface is periodic the surface is known as a* diffraction grating. *A method for solving scattering by diffraction gratings was introduced by DeSanto in* [30] *and further developed by DeSanto and co-workers in* [32], [33], [31], *and* [4]. *In* [19, Section 4.2] *it is shown that this method can be viewed as an implementation of the Fokas transform method to diffraction grating problems.*

### 6.1.11 ▪ Concluding Remarks

This paper had the following two goals:

> **Goal 1**: To give an overview of variational formulations for second-order linear elliptic PDEs based on multiplying by a test function and integrating by parts (or, equivalently, based on Green's identities).

**Goal 2**: To show how the Fokas transform method applied to second-order linear elliptic PDEs can be placed into the framework established in Goal 1.

Stepping back to look at the bigger picture around Goal 1, we might ask the following two questions:

1. Goal 1 has assumed that all variational formulations based on multiplying by a test function and integrating by parts are equivalent to using Green's identities, but are there any formulations based on other identities?
2. What about variational formulations that are not based on any identities?

Similarly, concerning the wider context of Goal 2 we might ask the following question:

3. Since the Fokas transform method arose from investigations of certain nonlinear PDEs (the so-called *integrable* PDEs), what have integrable nonlinear PDEs got to do with Green's identities?

In this final section we answer these three questions.

### Variational Formulations Based Identities Other than Green's Identities

New variational formulations of the interior impedance and EDPs for the Helmholtz equation were introduced in [76], and these variational formulations are based on identities other than Green's identities. The idea behind these variational formulations is the following.

We saw in Section 6.1.6 that the standard variational formulations of the interior Dirichlet and impedance problems for the Helmholtz equation, (6.64) and (6.75), respectively, arose from integrating over $\Omega$ the identity

$$\overline{v}\,\mathcal{L}_k u = \nabla\cdot\left[\overline{v}\nabla u\right] - \nabla u\cdot\overline{\nabla v} + k^2 u\overline{v}, \tag{6.176}$$

i.e., G1 (6.24). Recall that the sesquilinear forms in these variational formulations, $a_D(\cdot,\cdot)$ and $a_I(\cdot,\cdot)$, are not coercive when $k$ is sufficiently large (see Lemmas 6.34 and 6.36). On one level, the reason for this is that, when $u = v$, the nondivergence terms on the right-hand side of (6.176) equal $-|\nabla v|^2 + k^2|v|^2$, and this expression is *not* single-signed (i.e., for some $v$ it will be positive, and for some $v$ it will be negative).

This observation motivates the following question: if the identity (6.176) is replaced by a different identity with the property that when $u = v$ the nondivergence terms on the right-hand side *are* single-signed, will this lead to coercive variational formulations? The paper [76] shows that the answer is yes for the interior impedance and exterior Dirichlet problems. (Note that there cannot exist a formulation of the IDP that is coercive for all $k > 0$ because the solution of this problem is not unique for all $k > 0$.)

The formulations in [76] are based on a class of identities for the Helmholtz equation introduced by Morawetz in [80] and [78] (building on the earlier work [79] concerning the wave equation). The simplest such identity involving two functions $u$ and $v$ is

$$\overline{\mathcal{M}v}\,\mathcal{L}u + \mathcal{M}u\,\overline{\mathcal{L}v} = \nabla\cdot\left[\overline{\mathcal{M}v}\,\nabla u + \mathcal{M}u\,\overline{\nabla v} + \mathbf{x}(k^2 u\overline{v} - \nabla u\cdot\overline{\nabla v})\right] - \nabla u\cdot\overline{\nabla v} - k^2 u\overline{v}, \tag{6.177}$$

where the multiplier $\mathcal{M}$ is defined by

$$\mathcal{M}v := \mathbf{x}\cdot\nabla v - \mathrm{i}k\beta v + \frac{d-1}{2}v, \tag{6.178}$$

and $\beta$ is an arbitrary real number. We see that, when $u = v$, the nondivergence terms of

(6.177) equal $-|\nabla v|^2 - k^2|v|^2$ and this expression is single-signed. Integrating the identity (6.177) over $\Omega$ and using the PDE (6.18) and boundary conditions (6.19) gives rise to a variational formulation of the Helmholtz IIP that is coercive for all $k > 0$ when $\Omega$ is star-shaped with respect to a ball [76, Theorem 1.1 and Section 3].

The only disadvantage of the new formulation compared to the standard variational formulation is that the Hilbert space of the new formulation is $\mathcal{V}$ defined by (6.110) (we need to work in this space because of the $\mathcal{M}u\overline{\mathcal{L}v}$ term on the left-hand side of (6.177)). Because the space $\mathcal{V}$ is smaller than $H^1(\Omega)$ it is harder to create piecewise-polynomial finite-dimensional subspaces of $\mathcal{V}$ than it is for $H^1(\Omega)$ (see [76, Section 5] for more details).

### Variational Formulations Not Based on Any Identities

The main class of variational formulations that are not based on integrating identities over the domain (or over elements of a triangulation of the domain) are *least squares methods*.

Given a linear PDE, $\mathcal{L}u = f$, on a domain $\Omega$, with boundary conditions $\mathcal{B}u = g$ on $\Gamma := \partial\Omega$, the simplest least squares method consists of minimizing the functional

$$J(v) = \|\mathcal{L}v - f\|_\Omega^2 + \|\mathcal{B}v - g\|_\Gamma^2$$

(where $\|\cdot\|_\Omega$ is an appropriate norm on $\Omega$ arising from an inner product $(\cdot,\cdot)_\Omega$, and $\|\cdot\|_\Gamma$ is an appropriate norm on $\Gamma$ arising from an inner product $(\cdot,\cdot)_\Gamma$).

The problem of minimizing $J(\cdot)$ can be shown to be equivalent to the variational problem (6.1) with

$$a(u,v) = (\mathcal{L}u, \mathcal{L}v)_\Omega + (\mathcal{R}u, \mathcal{R}v)_\Gamma \quad \text{and} \quad F(v) = (f, \mathcal{L}v)_\Omega + (g, \mathcal{R}v)_\Gamma,$$

and this formulation does not (in general) arise from integrating an identity over $\Omega$.

We refer the reader to [14] for a good introduction to least squares methods in general and to [40, Section 6] for a review of least squares methods for the Helmholtz equation that use subspaces satisfying the Trefftz property (6.92).

### From Green to Lax

The Fokas transform method arose from attempts to solve BVPs for integrable nonlinear PDEs, where we say that a PDE is *integrable* if it possesses a Lax pair formulation. The reason that the Fokas transform method is also applicable to linear PDEs is that linear PDEs possess Lax pair formulations; this fact was first noted by Fokas and Gelfand in [50].

The second goal of this paper was to place the Fokas transform method into the framework of variational formulations arising from Green's identities. The reason that we have been able to do this is that *Lax pairs for linear PDEs arise naturally from the differential (as opposed to integrated) form of Green's second identity.* Indeed, a Lax pair formulation for the second-order linear elliptic PDE (6.2) can be obtained from (6.22), and a Lax pair formulation for a general linear PDE can be obtained from (6.29); see [53, Section 7] for more details.

## Bibliography

[1] M. J. Ablowitz and A. S. Fokas, Complex Variables: Introduction and Applications, Cambridge University Press, Cambridge, UK (2003).

[2] Y. A. Abramovich and C. D. Aliprantis, An Invitation to Operator Theory, AMS, Providence, RI (2002).

[3] M. Abramowitz and I. A. Stegun, Handbook of Mathematical Functions with Formulas, Graphs, and Mathematical Tables, Dover, New York (1964).

[4] T. Arens, S. N. Chandler-Wilde, and J. A. DeSanto, On integral equation and least squares methods for scattering by diffraction gratings, Commun. Comput. Phys. **1** (2006), 1010–1042.

[5] D. N. Arnold, F. Brezzi, B. Cockburn, and L. D. Marini, Unified analysis of discontinuous Galerkin methods for elliptic problems, SIAM J. Numer. Anal. **39** (2002), 1749–1779.

[6] A. C. L. Ashton, On the rigorous foundations of the Fokas method for linear elliptic partial differential equations, Proc. R. Soc. Lond. Ser. A **468** (2012), 1325–1331.

[7] A. C. L. Ashton, The spectral Dirichlet–Neumann map for Laplace's equation in a convex polygon, SIAM J. Math. Anal. **45** (2013), 3575–3592.

[8] J. P. Aubin, Behavior of the error of the approximate solutions of boundary value problems for linear elliptic operators by Galerkin's and finite difference methods, Ann. Sc. Norm. Super. Pisa Cl. Sci. **21** (1997), 599–637.

[9] A. K. Aziz, M. R. Dorr, and R. B. Kellogg, A new approximation method for the Helmholtz equation in an exterior domain, SIAM J. Numer. Anal. **19** (1982), 899–908.

[10] I. Babuška and J. M. Melenk, The partition of unity method, Internat. J. Numer. Methods Engrg. **40** (1997), 727–758.

[11] T. P. Barrios, R. Bustinza, and V. Domínguez, On the Discontinuous Galerkin Method for Solving Boundary Value Problems for the Helmholtz Equation: A Priori and A Posteriori Error Analyses, arXiv:1310.2847 (2013).

[12] N. Bartoli and F. Collino, Integral equations via saddle point problem for 2d electromagnetic problems, ESAIM Math. Model. Numer. Anal. **34** (2000), 1023–1049.

[13] D. Baskin, E. A. Spence, and J. Wunsch, Sharp high-frequency estimates for the Helmholtz equation and applications to boundary integral equations, preprint (2014)

[14] P. B. Bochev and M. D. Gunzburger, Least-Squares Finite Element Methods, Springer, New York (2009).

[15] S. C. Brenner and L. R. Scott, The Mathematical Theory of Finite Element Methods, Springer, New York (2000).

[16] A. Buffa and P. Monk, Error estimates for the ultra weak variational formulation of the Helmholtz equation, M2AN Math. Model. Numer. Anal. **42** (2008), 925–940.

[17] O. Cessenat and B. Despres, Application of an Ultra Weak Variational Formulation of Elliptic PDEs to the Two-Dimensional Helmholtz Problem, SIAM J. Numer. Anal. **35** (1998), 255–299.

[18] S. N. Chandler-Wilde, I. G. Graham, S. Langdon, and E. A. Spence, Numerical-asymptotic boundary integral methods in high-frequency acoustic scattering, Acta Numer. **21** (2012), 89–305.

[19] S. N. Chandler-Wilde and S. Langdon, Acoustic Scattering: High Frequency Boundary Element Methods and Unified Transform Methods in Unified Transform Method for Boundary Value Problems: Applications and Advances, SIAM (2015).

[20] P. G. Ciarlet, Basic Error Estimates for Elliptic Problems, Handb. Numer. Anal., Vol. II, North–Holland, Amsterdam (1991).

[21] F. Collino and B. Després, Integral equations via saddle point problems for time-harmonic Maxwell's equations, J. Comput. Appl. Math. **150** (2003), 157–192.

[22] D. Colton and R. Kress, Integral Equation Methods in Scattering Theory, Wiley, New York (1983).

[23] P. Cummings and X. Feng, Sharp regularity coefficient estimates for complex-valued acoustic and elastic Helmholtz equations, Math. Models Methods Appl. Sci. **16** (2006), 139–160.

[24] C-I. R. Davis and B. Fornberg, A spectrally accurate numerical implementation of the Fokas transform method for Helmholtz-type PDEs, Complex Var. Elliptic Equ. **59** (2014), 564–577.

[25] B. Deconinck, T. Trogdon, and V. Vasan, The Method of Fokas for Solving Linear Partial Differential Equations, SIAM Rev. **56** (2014), 159–186.

[26] L. Demkowicz, Babuška ⟷ Brezzi, University of Texas at Austin, http://www.ices.utexas.edu/media/reports/2006/0608.pdf (2006).

[27] L. Demkowicz and J. Gopalakrishnan, Analysis of the DPG method for the Poisson equation, SIAM J. Numer. Anal. **49** (2011), 1788–1809.

[28] L. Demkowicz, J. Gopalakrishnan, I. Muga, and J. Zitelli, Wavenumber explicit analysis for a DPG method for the multidimensional Helmholtz equation, Comput. Methods Appl. Mech. Engrg. **213/216** (2012), 126–138.

[29] L. F. Demkowicz and J. Gopalakrishnan, An Overview of the Discontinuous Petrov Galerkin Method, in Recent Developments in Discontinuous Galerkin Finite Element Methods for Partial Differential Equations, 149–180, Springer, New York (2014).

[30] J. A. DeSanto, Scattering from a perfectly reflecting arbitrary periodic surface: An exact theory, Radio Sci. **16** (1981), 1315–1326.

[31] J. DeSanto, G. Erdmann, W. Hereman, B. Krause, M. Misra, and E. Swim, Theoretical and computational aspects of scattering from periodic surfaces: Two-dimensional perfectly reflecting surfaces using the spectral-coordinate method, Waves Random Media **11** (2001), 455–488.

[32] J. DeSanto, G. Erdmann, W. Hereman, and M. Misra, Theoretical and computational aspects of scattering from rough surfaces: One-dimensional perfectly reflecting surfaces, Waves Random Media **8** (1998), 385–414.

[33] J. DeSanto, G. Erdmann, W. Hereman, and M. Misra, Theoretical and computational aspects of scattering from periodic surfaces: One-dimensional transmission interface, Waves Random Media **11** (2001), 425–451.

[34] B. Després, Fonctionnelle quadratique et equations integrales pour les problemes d'onde harmonique en domaine exterieur, Modél. Math. Anal. Numér. **31** (1997), 679–732.

[35] B. Després, Quadratic functional and integral equations for harmonic wave equations, in Proceedings of the 4th International Conference on Mathematical and Numerical Aspects of Wave Propagation, SIAM (1998).

[36] Digital Library of Mathematical Functions, http://dlmf.nist.gov/ (2014).

[37] S. C. Eisenstat, H. C. Elman, and M. H. Schultz, Variational iterative methods for nonsymmetric systems of linear equations, SIAM J. Numer. Anal. **20** (1983), 345–357.

[38] H. C. Elman, Iterative Methods for Sparse Nonsymmetric Systems of Linear Equations, PhD thesis, Yale University, New Haven, CT (1982).

[39] J. Elschner, The double layer potential operator over polyhedral domains I: Solvability in weighted Sobolev spaces, Appl. Anal. **45** (1992), 117–134.

[40] S. Esterhazy and J. M. Melenk, On stability of discretizations of the Helmholtz equation, Lecture Notes in Comput. Sci. Eng. 83, Springer, Berlin (2012).

[41] L. C. Evans, Partial Differential Equations, AMS, Providence, RI (1998).

[42] E. B. Fabes, M. Jodeit, and N. H. Riviere, Potential techniques for boundary value problems on $C^1$ domains, Acta Math. **141** (1978), 165–186.

[43] C. Farhat, I. Harari, and L. P. Franca, The discontinuous enrichment method, Comput. Methods Appl. Mech. Engrg. **190** (2001), 6455–6479.

[44] C. Farhat, I. Harari, and V. Hetmaniuk, A discontinuous Galerkin method with Lagrange multipliers for the solution of Helmholtz problems in the mid-frequency regime, Comput. Methods Appl. Mech. Engrg. **192** (2003), 1389–1419.

[45] X. Feng and H. Wu, Discontinuous Galerkin methods for the Helmholtz equation with large wave number, SIAM J. Numer. Anal. **47** (2009), 2872–2896.

[46] X. Feng and H. Wu, $hp$-Discontinuous Galerkin methods for the Helmholtz equation with large wave number, Math. Comp. **80** (2011), 1997–2024.

[47] X. Feng and Y. Xing, Absolutely stable local discontinuous Galerkin methods for the Helmholtz equation with large wave number, Math. Comp. **82** (2013), 1269–1296.

[48] A. S. Fokas, A Unified Approach to Boundary Value Problems, SIAM (2008).

[49] A. S. Fokas, A unified transform method for solving linear and certain nonlinear PDEs, Proc. R. Soc. Lond. Ser. A **453** (1997), 1411–1443.

[50] A. S. Fokas and I. M. Gelfand, Integrability of linear and nonlinear evolution equations and the associated nonlinear Fourier transforms, Lett. Math. Phys. **32** (1994), 189–210.

[51] A. S. Fokas, A. Iserles, and S. A. Smitheman, The Unified Method in Polygonal Domains via the Explicit Fourier Transform of Legendre Polynomials, in Unified Transform for Boundary Value Problems: Applications and Advances, A. S. Fokas and B. Pelloni, eds., SIAM, 2015.

[52] A. S. Fokas and J. Lenells, The Unified Transform for the Modified Helmholtz Equation in the Exterior of a Square, in Unified Transform for Boundary Value Problems: Applications and Advances, A. S. Fokas and B. Pelloni, eds., SIAM, 2015.

[53] A. S. Fokas and E. A. Spence, Synthesis, as opposed to separation, of variables, SIAM Rev. **54** (2012), 291–324.

[54] B. Fornberg and N. Flyer, A numerical implementation of Fokas boundary integral approach: Laplace's equation on a polygonal domain, Proc. R. Soc. A **467** (2011), 2983–3003.

[55] S. R. Fulton, A. S. Fokas, and C. A. Xenophontos, An analytical method for linear elliptic PDEs and its numerical implementation, J. Comput. Appl. Math. **167** (2004), 465–483.

[56] G. Gabard, Discontinuous Galerkin methods with plane waves for time-harmonic problems, J. Comput. Phys. **225** (2007), 1961–1984.

[57] M. J. Gander, I. G. Graham, and E. A. Spence, Applying GMRES to the Helmholtz equation with shifted Laplacian preconditioning: What is the largest shift for which wavenumber-independent convergence is guaranteed?, preprint (2014).

[58] C. J. Gittelson, R. Hiptmair, and I. Perugia, Plane wave discontinuous Galerkin methods: analysis of the $h$-version, M2AN Math. Model. Numer. Anal. **2** (2009), 297–331.

[59] J. Gopalakrishnan, Five Lectures on DPG Methods, arXiv:1306.0557 (2013).

[60] I. G. Graham, Advanced Finite Element methods, http://www.maths.bath.ac.uk/~masigg/ma60202/lectures.pdf

[61] A. Greenbaum, Iterative Methods for Solving Linear Systems, SIAM (1997).

[62] P. Grisvard, Elliptic Problems in Nonsmooth Domains, Pitman, Boston (1985).

[63] K. E. Gustafson and D. K. M. Rao, Numerical range: The Field of Values of Linear Operators and Matrices, Springer, New York (1997).

[64] R. Hiptmair, A. Moiola, and I. Perugia, Stability results for the time-harmonic Maxwell equations with impedance boundary conditions, Math. Models Methods Appl. Sci. **21** (2011), 2263–2287.

[65] R. Hiptmair, A. Moiola, and I. Perugia, Plane wave discontinuous Galerkin methods for the 2D Helmholtz equation: Analysis of the $p$-version, J. Comput. Phys. **182** (2002), 27–46.

[66] G. C. Hsiao and W. L. Wendland, Boundary Integral Equations, Appl. Math. Sci. 164, Springer, Berlin (2008).

[67] D. Jerison and C. E. Kenig, The Inhomogeneous Dirichlet Problem in Lipschitz Domains, J. Funct. Anal. **130** (1995), 161–219.

[68] T. Kato, Estimation of iterated matrices, with application to the von Neumann condition, Numer. Math. **2** (1960), 22–29.

[69] J. P. Keener, Principles of Applied Mathematics, Perseus, New York (1995).

[70] T. Luostari, Non-Polynomial Approximation Methods in Acoustics and Elasticity, PhD thesis, University of Eastern Finland, Joensuu, Finland (2013).

[71] T. Luostari, T. Huttunen, and P. Monk, Plane Wave Methods for Approximating the Time Harmonic Wave Equation, in Highly Oscillatory Problems: Computation, Theory and Applications, Cambridge University Press, Cambridge, UK (2009).

[72] P. A. Martin, Multiple Scattering: Interaction of Time-Harmonic Waves with N Obstacles, Cambridge University Press, Cambridge, UK (2006).

[73] W. McLean, Strongly Elliptic Systems and Boundary Integral Equations, Cambridge University Press, Cambridge, UK (2000).

[74] J. M. Melenk, On generalized finite element methods, PhD thesis, University of Maryland, College Park, MD (1995).

[75] J. M. Melenk and I. Babuška, The partition of unity finite element method: Basic theory and applications, Comput. Methods Appl. Mech. Engrg. **139** (1996), 289–314.

[76] A. Moiola and E. A. Spence, Is the Helmholtz equation really sign-indefinite?, SIAM Rev. **56** (2014), 274–313.

[77] P. Monk and D. Q. Wang, A least-squares method for the Helmholtz equation, Comput. Methods Appl. Mech. Engrg. **175** (2013), 1269–1296.

[78] C. S. Morawetz, Decay for solutions of the exterior problem for the wave equation, Comm. Pure Appl. Math. **28** (1975), 229–264.

[79] C. S. Morawetz, The decay of solutions of the exterior initial-boundary value problem for the wave equation, Comm. Pure Appl. Math. **14** (1961), 561–568.

[80] C. S. Morawetz and D. Ludwig, An inequality for the reduced wave operator and the justification of geometrical optics, Comm. Pure Appl. Math. **21** (1968), 187–203.

[81] M. A. Naimark, Linear Differential Operators. Part I: Elementary Theory of Linear Differential Operators, Frederick Ungar, New York (1967).

[82] J. C. Nédélec, Acoustic and Electromagnetic Equations: Integral Representations for Harmonic Problems, Springer, New York (2001).

[83] J. Nitsche, Ein kriterium für die quasi-optimalität des ritzschen verfahrens, Numer. Math. **11** (1968), 346–348.

[84] I. Perugia and D. Schötzau, An $hp$-analysis of the local discontinuous Galerkin method for diffusion problems, J. Sci. Comput. **17** (2002), 561–571.

[85] Y. G. Saridakis, A. G. Sifalakis, and E. P. Papadopoulou, Efficient numerical solution of the generalized Dirichlet–Neumann map for linear elliptic PDEs in regular polygon domains, J. Comput. Appl. Math. **236** (2012), 2515–2528.

[86] S. A. Sauter and C. Schwab, Boundary Element Methods, Springer, New York (2011).

[87] F. J. Sayas, Retarded Potentials and Time Domain Boundary Integral Equations: A Road-Map, http://www.math.udel.edu/~fjsayas/TDBIEclassnotes2012.pdf.

[88] A. H. Schatz, An observation concerning Ritz-Galerkin methods with indefinite bilinear forms, Math. Comp. **28** (1974), 959–962.

[89] L. R. Scott and S. Zhang, Finite element interpolation of nonsmooth functions satisfying boundary conditions, Math. Comp. **54** (1998), 483–493.

[90] A. G. Sifalakis, A. S. Fokas, S. R. Fulton, and Y. G. Saridakis, The generalized Dirichlet-Neumann map for linear elliptic PDEs and its numerical implementation, J. Comput. Appl. Math. **219** (2008), 9–34.

[91] A. G. Sifalakis, S. R. Fulton, E. P. Papadopoulou, and Y. G. Saridakis, Direct and iterative solution of the generalized Dirichlet–Neumann map for elliptic PDEs on square domains, J. Comput. Appl. Math. **227** (2009), 171–184.

[92] A. G. Sifalakis, E. P. Papadopoulou, and Y. G. Saridakis, Numerical study of iterative methods for the solution of the Dirichlet-Neumann map for linear elliptic PDEs on regular polygon domains, Int. J. Appl. Math. Comput. Sci. **4** (2007), 173–178.

[93] S. A. Smitheman, E. A. Spence, and A. S. Fokas, A spectral collocation method for the Laplace and modified Helmholtz equations in a convex polygon, IMA J. Numer. Anal. **30** (2010), 1184–1205.

[94] E. A. Spence, Wavenumber-Explicit Bounds in Time-Harmonic Acoustic Scattering, SIAM J. Math. Anal. **46** (2014), 2987–3024.

[95] E. A. Spence, I. V. Kamotski, and V. P. Smyshlyaev, Coercivity of combined boundary integral equations in high frequency scattering, Comm. Pure Appl. Math., to appear.

[96] I. Stakgold, Boundary Value Problems of Mathematical Physics, Volume II, Macmillan, New York (1968).

[97] I. Stakgold, Green's Functions and Boundary Value Problems, Wiley, New York (1979).

[98] O. Steinbach, Numerical Approximation Methods for Elliptic Boundary Value Problems: Finite and Boundary Elements, Springer, New York (2008).

[99] P. C. Waterman, Matrix formulation of electromagnetic scattering, Proc. IEEE **53** (1965), 805–812.

[100] P. C. Waterman, New formulation of acoustic scattering, J. Acoust. Soc. Amer. **45** (1969), 1417.

[101] H. Wu, Pre-asymptotic error analysis of CIP-FEM and FEM for the Helmholtz equation with high wave number. Part I: Linear version, IMA J. Numer. Anal. **34** (2014), 1266–1288.

[102] J. Xu and L. Zikatanov, Some observations on Babuška and Brezzi theories, Numer. Math. **94** (2003), 195–202.

# Part III

# Applications

Chapter 7

# Numerical Methods for Elliptic Problems

## 7.1 ▪ The Unified Transform in Polygonal Domains via the Explicit Fourier Transform of Legendre Polynomials

*A. S. Fokas, A. Iserles, and S. A. Smitheman*

The authors are grateful to EPSRC. ASF thanks the Onassis Foundation (USA) for partial support. SAS wishes to thank Prof. B. Fornberg for many useful discussions.

### 7.1.1 ▪ Introduction

A novel method, often called the unified transform (or Fokas transform), for analyzing boundary value problems for linear and integrable nonlinear PDEs has been introduced by one of the authors [5], [6], [7] and used extensively in the literature. Recently, Fornberg and collaborators (see [4], [9] have introduced a powerful numerical implementation of the unified transform to linear elliptic PDEs formulated in the interior of a polygon. In this case, the unified transform yields two simple algebraic equations, the so-called global relations, which couple the finite Fourier transforms of the given boundary data with the finite Fourier transforms of the unknown boundary values. For the determination of these boundary values one has to choose (a) appropriate basis functions and (b) suitable collocation points in the Fourier space. Several such choices have been featured in the literature [4], [8], [9], [10], [11], [12], [13], [14], [15], [16]; it appears that the best choice is the following [15]: (a) the unknown boundary values are expanded in terms of Legendre polynomials, and (b) the collocation points are chosen on the rays introduced in [8], [16]. We note that Chebyshev and Legendre polynomials give rise to equivalent finite bases of $L^2$, and hence either choice will result in the same numerical method (however, the conditioning of the resulting linear systems will differ, as conditioning is not invariant under matrix column operations).

The above numerical implementation requires the computation of the finite Fourier transform of Legendre polynomials. For this purpose, Fornberg and collaborators have used the fact that the finite Fourier transform of the Legendre polynomials can be expressed in terms of $J_{n+\frac{1}{2}}$, Bessel functions of order half integer. Here we employ an explicit formula for the Fourier transform of the Legendre polynomials. We have not been able to find such a formula in the literature, so we include the relevant derivation. Furthermore, for completeness, we include the corresponding formula for all Jacobi polynomials.

As an example of the applicability of the analytic formula for the Fourier transform of the Legendre polynomials, we implement the unified transform to the modified Helmholtz equation in the simplest possible polygon, namely a square.

Taking into consideration that the finite Fourier transform of the Legendre polynomials can be computed either explicitly or in terms of Bessel functions, the explicit formula presented in Section 7.1.3 implies a new explicit expression for Bessel functions of order half integer.

### 7.1.2 ▪ The Modified Helmholtz Equation in the Interior of a Square

Let $u(x,y)$ satisfy the modified Helmholtz equation

$$\frac{\partial^2 u}{\partial x^2}+\frac{\partial^2 u}{\partial y^2}-\kappa^2 u=0, \quad \kappa>0, \quad (x,y)\in\mathbb{R}, \tag{7.1}$$

in the interior of a domain $\Omega\subset\mathbb{R}^2$. It is straightforward to verify that equation (7.1) can be rewritten as the following one-parameter family of divergence forms:

$$\begin{aligned}&\left\{\exp\left[\frac{ik}{2}\left(\frac{1}{\lambda}-\lambda\right)x+\frac{k}{2}\left(\frac{1}{\lambda}+\lambda\right)y\right]\left[u_x+\frac{ik}{2}\left(\lambda-\frac{1}{\lambda}\right)u\right]\right\}_x\\ &-\left\{\exp\left[\frac{ik}{2}\left(\frac{1}{\lambda}-\lambda\right)x+\frac{k}{2}\left(\frac{1}{\lambda}+\lambda\right)y\right]\left[-u_y+\frac{k}{2}\left(\lambda+\frac{1}{\lambda}\right)u\right]\right\}_y=0.\end{aligned}$$

Hence if $\partial\Omega$ denotes the boundary of $\Omega$ and if $s$ is a smooth parametrization of the boundary, then the following equation is valid:

$$\int_{\partial\Omega} W(s,\lambda)ds=0, \quad \lambda\in\mathbb{C}\setminus\{0\}, \tag{7.2}$$

where

$$W(s,\lambda)=e^{\frac{i\kappa}{2}\left(\frac{1}{\lambda}-\lambda\right)x(s)+\frac{\kappa}{2}\left(\frac{1}{\lambda}+\lambda\right)y(s)}\left[u_x(s)+\frac{i\kappa}{2}\left(\lambda-\frac{1}{\lambda}\right)u(s)\right]\frac{dy}{ds}+\left[-u_y(s)+\frac{\kappa}{2}\left(\lambda+\frac{1}{\lambda}\right)u(s)\right]\frac{dx}{ds}.$$

Equation (7.2), called in [5] *the global relation,* is valid for all $\lambda$; thus it provides a *family* of equations which can be used to characterize the Dirichlet-to-Neumann map. Actually, for elliptic PDEs involving second-order derivatives it is necessary to employ *two* global relations. The second global relation is

$$\int_{\partial\Omega} \tilde{W}(s,\lambda)ds=0, \quad \lambda\in\mathbb{C}\setminus\{0\}, \tag{7.3}$$

where

$$\tilde{W}(s,\lambda)=W\left(s,\frac{1}{\lambda}\right). \tag{7.4}$$

For polygonal domains, equations (7.2) and (7.3) characterize the Dirichlet-to-Neumann map. Rigorous aspects of this method, often referred to as the *unified transform*, are discussed in [1], [2], [3].

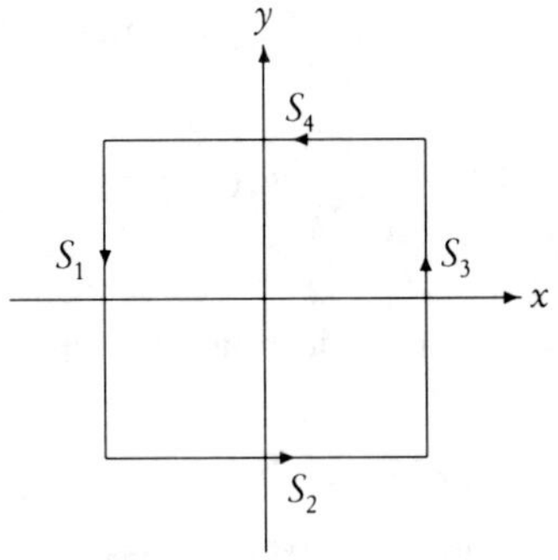


**Figure 7.1.** *The square with sides of length* 2.

In what follows we will use the simplest possible polygon, namely a square. Consider the square in Figure 7.1 with corners at the points

$$(-1,1), \quad (-1,-1), \quad (1,-1), \quad (1,1). \tag{7.5}$$

For the sides $S_1$, $S_2$, $S_3$, $S_4$, we have, respectively,

$$x=-1, \quad y=-1, \quad x=1, \quad y=1. \tag{7.6}$$

In order to analyze the global relations, taking into account the orientations of the sides, we introduce the following expressions:

$$\hat{u}_1(\lambda)=e^{\frac{\varkappa}{2}(i\lambda+\frac{1}{i\lambda})}\int_{+1}^{-1} e^{\frac{\varkappa}{2}(\lambda+\frac{1}{\lambda})y}\left[u_x^{(1)}+\frac{\varkappa}{2}\left(i\lambda+\frac{1}{i\lambda}\right)u^{(1)}\right]Dy, \tag{7.7a}$$

$$\hat{u}_2(\lambda)=e^{-\frac{\varkappa}{2}(\lambda+\frac{1}{\lambda})}\int_{-1}^{+1} e^{\frac{\varkappa}{2}(-i\lambda-\frac{1}{i\lambda})x}\left[-u_y^{(2)}+\frac{\varkappa}{2}\left(\lambda+\frac{1}{\lambda}\right)u^{(2)}\right]Dx, \tag{7.7b}$$

$$\hat{u}_3(\lambda)=e^{-\frac{\varkappa}{2}(i\lambda+\frac{1}{i\lambda})}\int_{-1}^{+1} e^{\frac{\varkappa}{2}(\lambda+\frac{1}{\lambda})y}\left[u_x^{(3)}+\frac{\varkappa}{2}\left(i\lambda+\frac{1}{i\lambda}\right)u^{(3)}\right]Dy, \tag{7.7c}$$

$$\hat{u}_4(\lambda)=e^{\frac{\varkappa}{2}(\lambda+\frac{1}{\lambda})}\int_{+1}^{-1} e^{\frac{\varkappa}{2}(-i\lambda-\frac{1}{i\lambda})x}\left[-u_y^{(4)}+\frac{\varkappa}{2}\left(\lambda+\frac{1}{\lambda}\right)u^{(4)}\right]Dx. \tag{7.7d}$$

Let $\hat{D}_j$ and $\hat{N}_j$ denote the parts of $\hat{u}_j$ corresponding to Dirichlet and Neumann boundary values, respectively. Then

$$\hat{u}_1(\lambda)=-e^{\frac{\varkappa}{2}(i\lambda+\frac{1}{i\lambda})}\hat{N}_1(\lambda)-\frac{\varkappa}{2}\left(i\lambda+\frac{1}{i\lambda}\right)e^{\frac{\varkappa}{2}(i\lambda+\frac{1}{i\lambda})}\hat{D}_1(\lambda), \tag{7.8a}$$

$$\hat{u}_2(\lambda)=-e^{-\frac{\varkappa}{2}(\lambda+\frac{1}{\lambda})}\hat{N}_2(-i\lambda)+\frac{\varkappa}{2}\left(\lambda+\frac{1}{\lambda}\right)e^{-\frac{\varkappa}{2}(\lambda+\frac{1}{\lambda})}\hat{D}_2(-i\lambda), \tag{7.8b}$$

$$\hat{u}_3(\lambda)=e^{-\frac{\varkappa}{2}(i\lambda+\frac{1}{i\lambda})}\hat{N}_3(\lambda)+\frac{\varkappa}{2}\left(i\lambda+\frac{1}{i\lambda}\right)e^{-\frac{\varkappa}{2}(i\lambda+\frac{1}{i\lambda})}\hat{D}_3(\lambda), \tag{7.8c}$$

$$\hat{u}_4(\lambda)=e^{\frac{\varkappa}{2}(\lambda+\frac{1}{\lambda})}\hat{N}_4(-i\lambda)-\frac{\varkappa}{2}\left(\lambda+\frac{1}{\lambda}\right)e^{\frac{\varkappa}{2}(\lambda+\frac{1}{\lambda})}\hat{D}_4(-i\lambda). \tag{7.8d}$$

The first of the global relations (7.2) becomes

$$\sum_{j=1}^{n} \hat{u}_j(\lambda) = 0, \quad \lambda \in \mathbb{C} \setminus \{0\}. \tag{7.9}$$

For simplicity, we consider the following *symmetric* Dirichlet boundary value problem:

$$u^{(1)} = u(-1,y) = \cosh(1)\cosh(\sqrt{3}y) + \cosh(\sqrt{3})\cosh(y), \quad -1 < y < 1; \tag{7.10a}$$
$$u^{(3)} = u(1,y) = \cosh(1)\cosh(\sqrt{3}y) + \cosh(\sqrt{3})\cosh(y), \quad -1 < y < 1; \tag{7.10b}$$
$$u^{(2)} = u(x,-1) = \cosh(1)\cosh(\sqrt{3}x) + \cosh(\sqrt{3})\cosh(x), \quad -1 < x < 1; \tag{7.10c}$$
$$u^{(4)} = u(x,1) = \cosh(1)\cosh(\sqrt{3}x) + \cosh(\sqrt{3})\cosh(x), \quad -1 < x < 1. \tag{7.10d}$$

Symmetry implies

$$u(x,y) = u(-x,y), \quad u(x,y) = u(x,-y), \quad u(x,y) = u(y,x). \tag{7.11}$$

Thus,

$$u_x^{(3)} = -u_x^{(1)}, \quad u_y^{(4)} = -u_y^{(2)}, \quad u_y^{(2)} = u_x^{(1)}\Big|_{(x,y)\leftrightarrow(y,x)}. \tag{7.12}$$

Hence, the first of the global relations, namely equation (7.9), becomes

$$\begin{aligned}
&\cos\left[\frac{x}{2}\left(\lambda - \frac{1}{\lambda}\right)\right]\hat{N}_1(\lambda) + \cos\left[\frac{x}{2}\left(\mathrm{i}\lambda - \frac{1}{\mathrm{i}\lambda}\right)\right]\hat{N}_1(-\mathrm{i}\lambda) \\
&\quad = \frac{x}{2}\left(\lambda - \frac{1}{\lambda}\right)\sin\left[\frac{x}{2}\left(\lambda - \frac{1}{\lambda}\right)\right]\hat{D}_1(\lambda) \\
&\qquad + \frac{x}{2}\left(\mathrm{i}\lambda - \frac{1}{\mathrm{i}\lambda}\right)\sin\left[\frac{x}{2}\left(\mathrm{i}\lambda - \frac{1}{\mathrm{i}\lambda}\right)\right]\hat{D}_1(-\mathrm{i}\lambda), \quad \lambda \in \mathbb{C}\setminus\{0\}.
\end{aligned} \tag{7.13a}$$

The simplest way to obtain the second of the global relations, namely equation (7.3), is to take the Schwartz conjugate of equation (7.13a) (i.e., to take the complex conjugate of (7.13a), subsequently replacing $\bar{\lambda}$ with $\lambda$). This yields the equation

$$\begin{aligned}
&\cos\left[\frac{x}{2}\left(\lambda - \frac{1}{\lambda}\right)\right]\hat{N}_1(\lambda) + \cos\left[\frac{x}{2}\left(\mathrm{i}\lambda - \frac{1}{\mathrm{i}\lambda}\right)\right]\hat{N}_1(\mathrm{i}\lambda) \\
&\quad = \frac{x}{2}\left(\lambda - \frac{1}{\lambda}\right)\sin\left[\frac{x}{2}\left(\lambda - \frac{1}{\lambda}\right)\right]\hat{D}_1(\lambda) \\
&\qquad + \frac{x}{2}\left(\mathrm{i}\lambda - \frac{1}{\mathrm{i}\lambda}\right)\sin\left[\frac{x}{2}\left(\mathrm{i}\lambda - \frac{1}{\mathrm{i}\lambda}\right)\right]\hat{D}_1(\mathrm{i}\lambda), \quad \lambda \in \mathbb{C}\setminus\{0\}.
\end{aligned} \tag{7.13b}$$

Equations (7.13) are coupling the finite Fourier transform of the unknown Neumann boundary value $u_x^{(1)}$ to the finite Fourier transform of the given Dirichlet datum $u^{(1)}$. In order to solve these equations we have to make two choices: (a) expand $u_x^{(1)}$ in terms of appropriate basis functions; (b) evaluate (7.13) at appropriate collocation points $\{\lambda_n\}_{n=1}^{M}$.

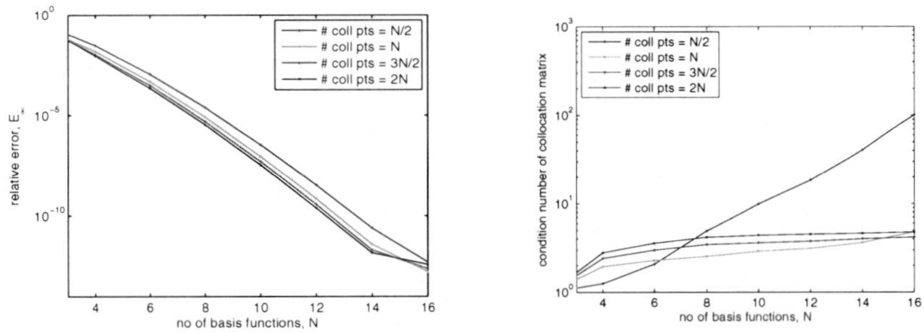


**Figure 7.2.** *Numerical solution of the global relations* (7.13) *for the symmetric Dirichlet boundary value problem* (7.10).

Regarding (a), Fourier basis functions [8], [10], [11], [12], [13], [14], [15], [16], as well as Chebyshev and Legendre polynomials [4], [9], [12], have been used earlier. Regarding (b), in most of the earlier papers the collocation points $\lambda \in \mathbb{C} \setminus \{0\}$ were chosen to lie on the rays in the complex $\lambda$-plane which are parallel to the edges of the polygon and its reflection in the imaginary axis. Recently, Fornberg and collaborators introduced the use of the so-called Halton nodes (see [4], [9]).

It appears that the most efficient numerical method involves the following [15]: (a) approximating the unknown boundary values in terms of Legendre polynomials (following Fornberg) and (b) using the collocation points employed in our earlier work [16], where ideas of Sifalakis and collaborators [12] for the Laplace equation were extended to the modified Helmholtz equation. These points are on certain rays of the complex $\lambda$-plane, determined by the exponential appearing in the global relation. Furthermore, in order to ensure that the collocation matrix remains well conditioned as the number $N$ of basis functions increases, it is important following Fornberg (i) to normalize each row, as well as each column, of the collocation matrix by its $l^1$-norm [4] and (ii) to "overdetermine" the linear system by choosing the number of collocation points to be about the same as the number of unknowns.

Numerical experiments suggest that Legendre polynomials yield spectral accuracy rather than the algebraic accuracy found with a Fourier basis. Furthermore, as a result of choosing the collocation points to be on the above rays, the semiblock circulant structure of the collocation matrix for regular polygons, demonstrated for the Laplace equation in [11], is preserved in the modified Helmholtz equation as well.

Plots of the relative error $E_\infty$ (defined in [15]), as well as of the matrix condition number as a function of $N$, for $N/2$, $N$, $3N/2$, and $2N$ collocation points, are presented in Figure 7.2. The rectangular collocation matrix was inverted by using the "backslash" command in MATLAB®. It is clear that overdetermining the linear system by a factor of 2 is sufficient to achieve very good matrix conditioning.

## 7.1.3 ▪ Fourier Expansions of Polynomials

### A General Theory

Let $p_m \in \mathbb{P}_m[x]$, the linear space of $m$th degree polynomials. We set

$$I[p_m] = \int_{-1}^{1} e^{-i\lambda x} p_m(x)dx, \qquad m \in \mathbb{Z}_+. \tag{7.14}$$

Repeatedly integrating by parts we obtain the following:

$$\begin{aligned}
I[p_m] &= -\frac{1}{\mathrm{i}\lambda}\int_{-1}^{1} p_m(x)\frac{d\mathrm{e}^{-\mathrm{i}\lambda x}}{dx}dx = -\frac{1}{\mathrm{i}\lambda}[p_m(1)\mathrm{e}^{-\mathrm{i}\lambda} - p_m(-1)\mathrm{e}^{\mathrm{i}\lambda}] \\
&\quad + \frac{1}{\mathrm{i}\lambda}I[p_m'] \\
&= -\frac{1}{\mathrm{i}\lambda}[p_m(1)\mathrm{e}^{-\mathrm{i}\lambda} - p_m(-1)\mathrm{e}^{\mathrm{i}\lambda}] - \frac{1}{(\mathrm{i}\lambda)^2}[p_m'(1)\mathrm{e}^{-\mathrm{i}\lambda} - p_m'(-1)\mathrm{e}^{\mathrm{i}\lambda}] \\
&\quad + \frac{1}{(\mathrm{i}\lambda)^2}I[p_m''] \\
&= \cdots \\
&= -\sum_{n=0}^{k}\frac{1}{(\mathrm{i}\lambda)^{n+1}}[p_m^{(n)}(1)\mathrm{e}^{-\mathrm{i}\lambda} - p_m^{(n)}(-1)\mathrm{e}^{\mathrm{i}\lambda}] + \frac{1}{(\mathrm{i}\lambda)^{k+1}}I[p_m^{(k+1)}], \\
&\qquad\qquad k \in \mathbb{Z}_+.
\end{aligned}$$

Using $p_m^{(m+1)} \equiv 0$, we deduce the explicit representation

$$I[p_m] = -\mathrm{e}^{-\mathrm{i}\lambda}\sum_{n=0}^{m}\frac{p_m^{(n)}(1)}{(\mathrm{i}\lambda)^{n+1}} + \mathrm{e}^{\mathrm{i}\lambda}\sum_{n=0}^{m}\frac{p_m^{(n)}(-1)}{(\mathrm{i}\lambda)^{n+1}}, \qquad m \in \mathbb{Z}_+. \tag{7.15}$$

#### Jacobi Polynomials

Let $p_m = \mathrm{P}_m^{(\alpha,\beta)}$, the $m$th *Jacobi polynomial,* where $\alpha, \beta > -1$. Then

$$\begin{aligned}
\mathrm{P}_m^{(\alpha,\beta)}(1) &= \frac{(1+\alpha)_m}{m!}, \\
\mathrm{P}_m^{(\alpha,\beta)}(-1) &= (-1)^m\mathrm{P}_m^{(\beta,\alpha)}(1) = (-1)^m\frac{(1+\beta)_m}{m!}
\end{aligned}$$

(this can be easily deduced from http://dlmf.nist.gov/18.6). Moreover,

$$\partial_x\mathrm{P}_m^{(\alpha,\beta)}(x) = \frac{1}{2}(\alpha+\beta+m+1)\mathbb{P}_{m-1}^{(\alpha+1,\beta+1)}(x) \tag{7.16}$$

(http://dlmf.nist.gov/18.9) and, by induction,

$$\partial_x^m\mathrm{P}_m^{(\alpha,\beta)}(x) = \frac{(\alpha+\beta+m+1)_n}{2^n}\mathrm{P}_{m-n}^{(\alpha+n,\beta+n)}(x), \qquad n = 0,\ldots,m. \tag{7.17}$$

Therefore

$$\begin{aligned}
\partial_x^n\mathrm{P}_m^{(\alpha,\beta)}(1) &= \frac{(\alpha+\beta+m+1)_n}{2^n}\,\frac{(\alpha+n+1)_{m-n}}{(m-n)!} \\
&= \frac{(\alpha+\beta+1)_{m+n}}{(\alpha+\beta+1)_m}\,\frac{(\alpha+1)_m}{(\alpha+1)_n}\,\frac{1}{2^n(m-n)!} \\
&= \frac{(\alpha+1)_m}{(\alpha+\beta+1)_m}\,\frac{(\alpha+\beta+1)_{m+n}}{2^n(m-n)!(\alpha+1)_n}.
\end{aligned}$$

Similarly

$$\partial_x^n \mathrm{P}_m^{(\alpha,\beta)}(-1) = (-1)^n \partial_x^n \mathrm{P}_m^{(\beta,\alpha)}(1)$$
$$= (-1)^{m+n} \frac{(\beta+1)_m}{(\alpha+\beta+1)_m} \frac{(\alpha+\beta+1)_{m+n}}{2^n(m-n)!(\beta+1)_n}.$$

Substitution in (7.35) implies the following formula for the finite Fourier transform of the $m$th Jacobi polynomial:

$$I[\mathrm{P}_m^{(\alpha,\beta)}] = (-1)^m \frac{(\beta+1)_m}{(\alpha+\beta+1)_m} \mathrm{e}^{\mathrm{i}\lambda} \sum_{n=0}^{m} (-1)^n \frac{(\alpha+\beta+1)_{m+n}}{2^n(\beta+1)_n(m-n)!} \frac{1}{(\mathrm{i}\lambda)^{n+1}}$$
$$- \frac{(\alpha+1)_m}{(\alpha+\beta+1)_m} \mathrm{e}^{-\mathrm{i}\lambda} \sum_{n=0}^{m} \frac{(\alpha+\beta+1)_{m+n}}{2^n(\alpha+1)_n(m-n)!} \frac{1}{(\mathrm{i}\lambda)^n}. \tag{7.18}$$

### Legendre Polynomials

Since $\mathrm{P}_m = \mathrm{P}_m^{(0,0)}$, (7.36) reduces to

$$I[\mathrm{P}_m] = (-1)^m \mathrm{e}^{\mathrm{i}\lambda} \sum_{n=0}^{m} (-1)^n \frac{(m+n)!}{2^n n!(m-n)!} \frac{1}{(\mathrm{i}\lambda)^{n+1}}$$
$$- \mathrm{e}^{-\mathrm{i}\lambda} \sum_{n=0}^{m} \frac{(m+n)!}{2^n n!(m-n)!} \frac{1}{(\mathrm{i}\lambda)^{n+1}}.$$

Hence, we obtain the following formula for the finite Fourier transform of Legendre polynomials:

$$I[\mathrm{P}_m] = \sum_{n=0}^{m} \frac{(m+n)!}{2^n n!(m-n)!} \frac{1}{(\mathrm{i}\lambda)^{n+1}} [(-1)^{m+n}\mathrm{e}^{\mathrm{i}\lambda} - \mathrm{e}^{-\mathrm{i}\lambda}], \qquad m \in \mathbb{Z}_+. \tag{7.19}$$

### Chebyshev Polynomials

We have

$$\mathrm{T}_m(x) = \frac{m!}{(\frac{1}{2})_m} \mathrm{P}_m^{(-1/2,-1/2)}(x); \tag{7.20}$$

therefore

$$\mathrm{T}_m^{(n)}(1) = \frac{m!}{(\frac{1}{2})_m} \frac{(m)_n}{2^n} \frac{(n+\frac{1}{2})_{m-n}}{(m-n)!} = \frac{m(m+n-1)!}{2^n(m-n)!(\frac{1}{2})_n}, \qquad m \in \mathbb{N}, \tag{7.21}$$

and

$$\mathrm{T}_m^{(n)}(-1) = (-1)^{m+n} \frac{m(m+n-1)!}{2^n(m-n)!(\frac{1}{2})_n}, \qquad m \in \mathbb{N}. \tag{7.22}$$

Thus, substituting in (7.35), we obtain the following formula for the finite Fourier transform of the Chebyshev polynomials:

$$I[\mathrm{T}_m] = \sum_{n=0}^{m} \frac{1}{(\mathrm{i}\lambda)^{n+1}} \frac{m(m+n-1)!}{2^n(m-n)!(\frac{1}{2})_n} [(-1)^{m+n}\mathrm{e}^{\mathrm{i}\lambda} - \mathrm{e}^{-\mathrm{i}\lambda}], \qquad m \in \mathbb{N}. \tag{7.23}$$

### Computations

It is slightly unwieldy to compute repeatedly Pochhammer symbols, and also unnecessary. Indeed, the following recursive formulae are valid:

$$I[\mathrm{P}_m] = \sum_{n=0}^{m} p_{m,n} \frac{(-1)^{m-n}\mathrm{e}^{\mathrm{i}\lambda} - \mathrm{e}^{-\mathrm{i}\lambda}}{(\mathrm{i}\lambda)^{n+1}}, \tag{7.24}$$

where

$$p_{m,0} = 1, \quad p_{m,n} = \frac{(m+\frac{1}{2})^2 - (n-\frac{1}{2})^2}{2n} p_{m,n-1}, \; n = 1, \ldots, m. \tag{7.25}$$

Similarly,

$$I[\mathrm{T}_m] = \sum_{n=0}^{m} t_{m,n} \frac{(-1)^{m-n}\mathrm{e}^{\mathrm{i}\lambda} - \mathrm{e}^{-\mathrm{i}\lambda}}{(\mathrm{i}\lambda)^{n+1}}, \tag{7.26}$$

where

$$t_{m,0} = 1, \quad t_{m,n} = \frac{m^2 - (n-1)^2}{2n-1} t_{m,n-1}, \; n = 1, \ldots, m. \tag{7.27}$$

### An Explicit Representation for the Half-Order Bessel Functions

Let $\mathrm{J}_{m+\frac{1}{2}}(\lambda)$ denote the half-order Bessel function, i.e.,

$$\begin{aligned} \mathrm{J}_{m+\frac{1}{2}}(\lambda) = {} & \frac{1}{\pi}\int_0^{\pi} \cos\left(\left(m+\frac{1}{2}\right)\tau - \lambda \sin\tau\right) d\tau \\ & + \frac{(-1)^{m+1}}{\pi}\int_0^{\infty} \exp\left(-\lambda \sinh\tau - \left(m+\frac{1}{2}\right)\tau\right) d\tau, \\ & \lambda \in \mathbb{C}, \quad m \in \mathbb{Z}_+. \end{aligned} \tag{7.28}$$

Then equation (7.39) together with the formula for the Fourier transform of the Legendre polynomials employed in [9], namely

$$I[P_m](\lambda) = \frac{1}{\mathrm{i}^m}\sqrt{\frac{2\pi}{\lambda}}\,\mathrm{J}_{m+\frac{1}{2}}(\lambda), \tag{7.29}$$

gives the following new explicit representation for Bessel functions of order half integer:

$$\mathrm{J}_{m+\frac{1}{2}}(\lambda) = \frac{1}{\sqrt{2\pi}} \sum_{n=1}^{m+1} p_{m,n-1} \frac{(-1)^{m-n-1}\mathrm{e}^{\mathrm{i}\lambda} - \mathrm{e}^{-\mathrm{i}\lambda}}{\mathrm{i}^{n-m}\lambda^{n-\frac{1}{2}}}, \qquad \lambda \in \mathbb{C}\setminus\{0\}, \quad m = 0, 1, 2, \ldots, \tag{7.30}$$

where $p_{m,n}$ are defined in (7.25) (see also (7.37)). Since Bessel functions of order half integer can be easily converted to *spherical Bessel functions,*

$$\mathrm{J}_{n+\frac{1}{2}}(\lambda) = \sqrt{\frac{2\lambda}{\pi}}\,\mathrm{j}_n(\lambda), \tag{7.31}$$

(7.30) is also available for the representation and rapid computation of the latter.

# Bibliography

[1] A. C. L. Ashton, On the rigorous foundations of the Fokas method for linear elliptic partial differential equations, Proc. R. Soc. A **468** (1997), 1325–1331.

[2] A. C. L. Ashton, The spectral Dirichlet–Neumann map for Laplace's equation in a convex polygon, SIAM J. Math. Anal. **45** (2013), 3575–3591.

[3] A. C. L. Ashton and A. S. Fokas, Elliptic boundary value problems in convex polygons with low regularity boundary data via the unified method (submitted).

[4] C.-I. Davis, Numerical Tests of the Fokas Method for Helmholtz-Type Partial Differential Equations: Dirichlet to Neumann Maps, Masters thesis, University of Colorado at Boulder, Boulder, CO (2008).

[5] A. S. Fokas, A unified transform method for solving linear and certain nonlinear PDEs, Proc. R. Soc. A **453** (1997), 1411–1443.

[6] A. S. Fokas, On the integrability of linear and nonlinear partial differential equations, J. Math. Phys. **41** (2000), 4188-4237.

[7] A. S. Fokas, Two-dimensional linear PDEs in a convex polygon, Proc. R. Soc. Lond. Ser. A **457** (2001), 371–393.

[8] A. S. Fokas, N. Flyer, S. A. Smitheman, and E. A. Spence, A semi-analytical numerical method for solving evolution and elliptic partial differential equations, J. Comput. Appl. Math. **227** (2009), 59–74.

[9] B. Fornberg and N. Flyer, A numerical implementation of Fokas boundary integral approach: Laplace's equation on a polygonal domain, Proc. R. Soc. A **467** (2011), 2983–3003.

[10] S. R. Fulton, A. S. Fokas, and C. A. Xenophontos, An analytical method for linear elliptic PDEs and its numerical implementation, J. Comput. Appl. Math. **167** (2004), 465–483.

[11] Y. G. Saridakis, A. G. Sifalakis, and E. P. Papadopoulou, Efficient numerical solution of the generalized Dirichlet–Neumann map for linear elliptic PDEs in regular polygon domains, J. Comput. Appl. Math. **236** (2012), 2515–2528.

[12] A. G. Sifalakis, A. S. Fokas, S. R. Fulton, and Y. G. Saridakis, The generalized Dirichlet-Neumann map for linear elliptic PDEs and its numerical implementation, J. Comput. Appl. Math. **219** (2008), 9–34.

[13] A. G. Sifalakis, S. R. Fulton, E. P. Papadopoulou, and Y. G. Saridakis, Direct and iterative solution of the generalized Dirichlet-Neumann map for elliptic PDEs on square domains, J. Comput. Appl. Math. **227** (2009), 171–184.

[14] A. G. Sifalakis, E. P. Papadopoulou, and Y. G. Saridakis, Numerical study of iterative methods for the solution of the Dirichlet-Neumann map for linear elliptic PDEs on regular polygon domains, Int. J. Appl. Math. Comput. Sci. **4** (2007), 173–178.

[15] A. S. Fokas and Hashemzadeh, Linear Elliptic PDEs in Polygonal Domains (preprint).

[16] S. A. Smitheman, E. A. Spence, and A. S. Fokas, A spectral collocation method for the Laplace and modified Helmholtz equations in a convex polygon, IMA J. Numer. Anal. **30** (2010), 1184–1205.

## 7.2 ▪ The Unified Transform for the Modified Helmholtz Equation in the Exterior of a Square

*A. S. Fokas and J. Lenells*

The authors acknowledge support from EPSRC, UK.

### 7.2.1 ▪ Introduction

The so-called unified transform, introduced by the first author in the late nineties [3, 6], has been implemented to a variety of boundary value problems for linear elliptic PDEs. In particular, it has led to the emergence of a new numerical technique for the Laplace, the Helmholtz, and the modified Helmholtz equations formulated in the *interior* of a polygon [1, 2, 4, 7, 8, 10, 11, 13, 14, 15, 16, 17]. This technique is illustrated for the particular case of a square in this book (see Section 7.1). A novel numerical implementation of the unified transform to the Helmholtz and modified Helmholtz equations formulated in the *exterior* of a polygon is presented in [9]. Here, we illustrate this implementation in the particular case of a square.

### 7.2.2 ▪ The Unified Transform

The starting point of the unified transform is rewriting the given PDE as a one-parameter family of equations in divergence form. In particular, the modified Helmholtz equation

$$\frac{\partial^2 u}{\partial x^2}+\frac{\partial^2 u}{\partial y^2}-\varkappa^2 u=0, \qquad \varkappa>0, \quad (x,y)\in D\subset\mathbb{R}^2, \tag{7.32}$$

can be written in the form

$$\left\{e^{\frac{i\varkappa}{2}(\frac{1}{\lambda}-\lambda)x+\frac{\varkappa}{2}(\frac{1}{\lambda}+\lambda)y}\left[u_x+\frac{i\varkappa}{2}\left(\lambda-\frac{1}{\lambda}\right)u\right]\right\}_x$$
$$+\left\{e^{\frac{i\varkappa}{2}(\frac{1}{\lambda}-\lambda)x+\frac{\varkappa}{2}(\frac{1}{\lambda}+\lambda)y}\left[u_y-\frac{\varkappa}{2}\left(\lambda+\frac{1}{\lambda}\right)u\right]\right\}_y=0, \qquad (x,y)\in D, \quad \lambda\in\mathbb{C}\setminus\{0\}.$$

Let $s$ be a smooth parametrization of the boundary of $D$, which is denoted by $\partial D$, and let $u_{\mathcal{N}}$ denote the derivative along the outward normal of the boundary. The identities

$$u_x dx+u_y dy=u_s ds, \qquad u_x dy-u_y dx=u_{\mathcal{N}} ds,$$

together with Green's theorem applied in $D$, imply

$$\int_{\partial D} W(s,\lambda)ds=0, \qquad \lambda\in\mathbb{C}\setminus\{0\}, \tag{7.33}$$

where $W(s,\lambda)$ is defined by

$$W(s,\lambda)=e^{-\frac{i\varkappa}{2}(\lambda z(s)-\frac{\overline{z(s)}}{\lambda})}\left[u_{\mathcal{N}}(s)+\frac{\varkappa}{2}\left(\lambda\frac{dz(s)}{ds}+\frac{1}{\lambda}\frac{d\overline{z(s)}}{ds}\right)u(s)\right],$$

with

$$z(s)=x(s)+iy(s).$$

### 7.2.3 ▪ The Global Relation for Exterior Domains

Suppose that the domain $D$ is exterior to some bounded domain $D_B$. Then equation (7.33) becomes

$$\int_{\partial D_B} W(s,\lambda)ds = I(\lambda), \qquad \lambda \in \mathbb{C}\setminus\{0\}, \tag{7.34}$$

where $I(\lambda)$ is defined by

$$I(\lambda) = \lim_{r\to\infty} \int_{|z|=r} W(s,\lambda)ds, \qquad \lambda \in \mathbb{C}\setminus\{0\}. \tag{7.35}$$

Assume that $u$ satisfies the Sommerfeld radiation condition, i.e.,

$$\lim_{r\to\infty} \sqrt{r}\left(\frac{\partial}{\partial r} + x\right)u = 0. \tag{7.36}$$

It is well known (see [12]) that as $r \to \infty$, $u$ admits the asymptotic expansion

$$u \sim \sqrt{\frac{2}{\pi i x r}}\, e^{i(ixr - \frac{\pi}{4})}\left(f_0(\varphi) + \frac{f_1(\varphi)}{r} + \frac{f_2(\varphi)}{r^2} + \cdots\right), \qquad r \to \infty. \tag{7.37}$$

Employing the above representation in (7.35), it is shown in [9] that $I(\lambda)$ is given by

$$I(\lambda) = 4 i f_0(i \ln(-i\lambda)), \qquad \lambda \in \mathbb{C}\setminus\{0\}. \tag{7.38}$$

It is shown in [9] that $f_0(\varphi)$ admits an analytic continuation to complex values of $\varphi$ and hence the right-hand side of (7.38) is well defined.

We recall that if the modified Helmholtz equation (7.32) is defined in the *interior* domain $D_B$, then $W$ satisfies the equation

$$\int_{\partial D_B} W(s,\lambda)ds = 0, \qquad \lambda \in \mathbb{C}\setminus\{0\}. \tag{7.39}$$

It turns out that if $D_B$ is the interior of a polygon, then equation (7.39) provides an efficient way for computing the generalized Dirichlet-to-Neumann map (see [10]). Indeed, for a well-posed problem, one specifies a relationship between the Dirichlet and the Neumann boundary values. Thus, if the relevant polygon consists of $n$ sides, then equation (7.39) contains $n$ unknowns. For example, in the case of the Dirichlet problem, equation (7.39) contains the $n$ unknown Neumann boundary values. However, equation (7.39) is valid for *all* complex values of $\lambda$. Thus, by expanding the unknown boundary values in an appropriate basis, and by evaluating equation (7.39) at a sufficiently large number of collocation points, we can obtain the unknown coefficients appearing in the above expansions. Fornberg and colleagues have shown that by choosing Legendre polynomials as base functions, and by "overdetermining" the relevant linear system (i.e., by choosing the number of collocation points much larger than the number of unknown coefficients), the above technique yields a linear system with a small condition number. Furthermore, it is shown in [8] that, by choosing the collocation points to be on certain curves, it is possible to obtain a condition number which is *independent* of $x$.

Equation (7.34) differs from equation (7.39) only by the presence of the explicit term defined in (7.38). Thus, we can still apply the techniques developed for analyzing equation (7.39), but we must supplement equation (7.34) with an *additional* equation in order to compensate for the existence of the additional *unknown* function $f_0$.

### 7.2.4 ▪ A Supplement to the Global Relation

Suppose that $D$ is the exterior of the $n$-gon with vertices at $\{z_j\}_1^n$.

It is shown in [18] and [9] that in this case the following relation is valid:

$$\sum_{j=1}^{n} \int_{l_j} e^{i\beta(\lambda z - \frac{\bar{z}}{\lambda})} \hat{u}_j(\lambda) \frac{d\lambda}{\lambda} = 0, \qquad z \in D_B, \tag{7.40}$$

where $D_B$ denotes the interior of the polygon, the rays $l_j$ oriented from the origin to infinity are defined by

$$l_j = \{\lambda \in \mathbb{C} \mid \arg \lambda = -\arg(z_{j+1} - z_j)\}, \qquad j = 1, \dots, n, \tag{7.41}$$

and the functions $\{\hat{u}_j(\lambda)\}_1^n$ are defined by

$$\hat{u}_j(\lambda) = \int_{z_j}^{z_{j+1}} e^{-i\beta(\lambda z - \frac{\bar{z}}{\lambda})} \left[ i u_{\mathcal{N}} + i\beta\left( \frac{1}{\lambda}\frac{d\bar{z}}{dt} + \lambda \frac{dz}{dt} \right) u \right] dt, \tag{7.42}$$

where $u_{\mathcal{N}}$ denotes the Neumann boundary value. Let us parametrize the side $[z_j, z_{j+1}]$ by

$$z(t) = m_j + t h_j, \qquad -1 \le t \le 1, \tag{7.43}$$

where

$$m_j = \frac{z_j + z_{j+1}}{2}, \qquad h_j = \frac{z_{j+1} - z_j}{2}. \tag{7.44}$$

Then, the expression $\hat{u}_j(\lambda)$ defined in (7.42) can be written in the form

$$\hat{u}_j(\lambda) = \hat{u}_j^{\mathcal{D}}(\lambda) + \hat{u}_j^{\mathcal{N}}(\lambda), \qquad j = 1, \dots, n, \tag{7.45}$$

where

$$\hat{u}_j^{\mathcal{D}}(\lambda) = i\beta\left( \frac{1}{\lambda}\bar{h}_j + \lambda h_j \right) e^{-i\beta(\lambda m_j - \frac{\bar{m}_j}{\lambda})} \int_{-1}^{1} e^{-i\beta t(\lambda h_j - \frac{\bar{h}_j}{\lambda})} u\, dt, \tag{7.46}$$

$$\hat{u}_j^{\mathcal{N}}(\lambda) = i e^{-i\beta(\lambda m_j - \frac{\bar{m}_j}{\lambda})} \int_{-1}^{1} e^{-i\beta t(\lambda h_j - \frac{\bar{h}_j}{\lambda})} u_{\mathcal{N}}\, dt. \tag{7.47}$$

### 7.2.5 ▪ The Dirichlet Problem

For the Dirichlet problem the functions $\{\hat{u}_j^{\mathcal{D}}(\lambda)\}_1^n$ are known (see (7.46)). In order to compute $\hat{u}_j^{\mathcal{N}}(\lambda)$, we approximate the Neumann boundary value on the side $j$ as follows:

$$u_j^{\mathcal{N}}(t) \approx \sum_{m=0}^{M} c_m^{(j)} P_m(t), \qquad j = 1, \dots, n, \tag{7.48}$$

where $P_m$ is the Legendre polynomial of degree $m$, and the unknown constants $c_m^{(j)}$ are to be determined.

Equation (7.47) yields

$$\hat{u}_j^{\mathcal{N}}(\lambda) = i e^{-i\beta(\lambda m_j - \frac{\bar{m}_j}{\lambda})} \sum_{m=0}^{M} c_m^{(j)} \int_{-1}^{1} e^{-i\beta t(\lambda h_j - \frac{\bar{h}_j}{\lambda})} P_m(t) dt. \tag{7.49}$$

Thus, using the expression [8]

$$\int_{-1}^{1} e^{-i\Lambda x} P_m(x) dx = \sum_{p=0}^{m} \frac{A_{mp} e^{i\Lambda} + B_{mp} e^{-i\Lambda}}{\Lambda^{p+1}}, \qquad m \geq 0, \quad \Lambda \in \mathbb{C} \setminus \{0\}, \tag{7.50}$$

where

$$A_{mp} = \frac{(m+p)!(-1)^{m+p}}{2^p p!(m-p)! i^{p+1}}, \qquad B_{mp} = -\frac{(m+p)!}{2^p p!(m-p)! i^{p+1}}, \tag{7.51}$$

equation (7.49) becomes

$$\hat{u}_j^{\mathcal{N}}(\lambda) = i e^{-i\beta(\lambda m_j - \frac{\bar{m}_j}{\lambda})} \sum_{m=0}^{M} c_m^{(j)} \sum_{p=0}^{m} \frac{A_{mp} e^{i\Lambda} + B_{mp} e^{-i\Lambda}}{\Lambda^{p+1}} \bigg|_{\Lambda = \beta(\lambda h_j - \frac{\bar{h}_j}{\lambda})}. \tag{7.52}$$

Thus, equation (7.40) yields

$$u^{\mathcal{D}}(z) + i \sum_{j=1}^{n} \sum_{m=0}^{M} c_m^{(j)} \sum_{p=0}^{m} \frac{A_{mp} Q_{pj}^{+}(z) + B_{mp} Q_{pj}^{-}(z)}{\beta^{p+1}} = 0, \qquad z \in D_B, \tag{7.53}$$

where

$$u^{\mathcal{D}}(z) = \sum_{j=1}^{n} \int_{\hat{l}_j} e^{i\beta(\lambda z - \frac{\bar{z}}{\lambda})} \hat{u}_j^{\mathcal{D}}(\lambda) \frac{d\lambda}{\lambda}, \tag{7.54}$$

and the functions $Q_{pj}^{\pm}(z)$ are defined by

$$Q_{pj}^{\pm}(z) = \int_{\hat{l}_j} \frac{e^{i\beta(\lambda z - \frac{\bar{z}}{\lambda} - \lambda m_j + \frac{\bar{m}_j}{\lambda} \pm (\lambda h_j - \frac{\bar{h}_j}{\lambda}))}}{(\lambda h_j - \frac{\bar{h}_j}{\lambda})^{p+1}} \frac{d\lambda}{\lambda}, \tag{7.55}$$

with $\hat{l}_j$ denoting the contour $l_j$ with small indentations inserted appropriately so that $\hat{l}_j$ passes around the points $\lambda = \pm\exp(-i \arg h_j)$ (these points are *removable* singularities; thus we deform $l_j$ to $\hat{l}_j$ *before* splitting the relevant integrals).

Suppose we take the limit as $z$ approaches the $j_0$th side. Then, using for this side the parametrization

$$z_0 = m_{j_0} + t_0 h_{j_0}, \qquad -1 \leq t_0 \leq 1, \tag{7.56}$$

we find

$$Q_{pj_0}^{\sigma}(z_0) = \int_{\hat{l}_{j_0}} \frac{e^{i\beta(t_0+\sigma)[\lambda h_{j_0} - \frac{\bar{h}_{j_0}}{\lambda}]}}{(\lambda h_{j_0} - \frac{\bar{h}_{j_0}}{\lambda})^{p+1}} \frac{d\lambda}{\lambda}, \qquad \sigma = \pm 1, \tag{7.57}$$

$$Q_{pj}^{\sigma}(z_0) = \int_{\hat{l}_j} \frac{e^{i\beta[\lambda(m_{j_0} - m_j) + \lambda(t_0 h_{j_0} + \sigma h_j) - \frac{\bar{m}_{j_0} - \bar{m}_j}{\lambda} - \frac{t_0 \bar{h}_{j_0} + \sigma \bar{h}_j}{\lambda}]}}{(\lambda h_j - \frac{\bar{h}_j}{\lambda})^{p+1}} \frac{d\lambda}{\lambda}, \qquad j \neq j_0, \quad \sigma = \pm 1. \tag{7.58}$$

By employing the above limiting procedure, we can obtain $n$ additional equations.

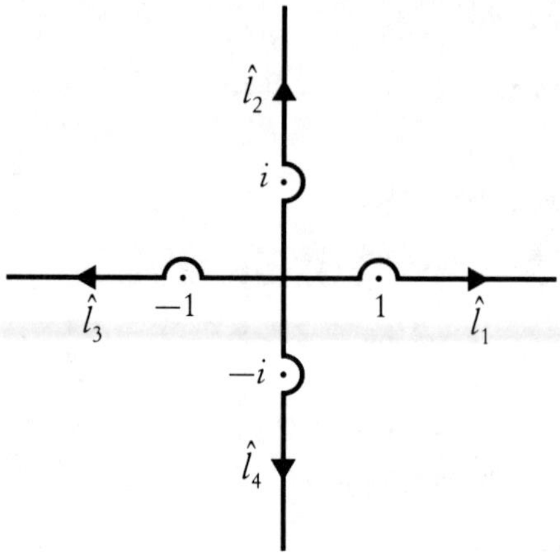


**Figure 7.3.** *The indented contours $\hat{l}_j$, $j=1,\dots,4$, utilized for the exterior of the square.*

### 7.2.6 ▪ The Exterior of a Square

Let $D_B$ be the square with corners at

$$z_1 = 1+i, \qquad z_2 = -1+i, \qquad z_3 = -1-i, \qquad z_4 = 1-i.$$

Then, the four midpoints of the sides are

$$m_1 = i, \qquad m_2 = -1, \qquad m_3 = -i, \qquad m_4 = 1,$$

with the corresponding tangential vectors

$$h_1 = -1, \qquad h_2 = -i, \qquad h_3 = 1, \qquad h_4 = i.$$

Also,

$$l_1 = (0,-\infty), \qquad l_2 = (0, i\infty), \qquad l_3 = (0,\infty), \qquad l_4 = (0,-i\infty).$$

Thus, if $j_0 = 1$ so that $z_0 = i - t_0$, $-1 \le t_0 \le 1$ belongs to the top side, then

$$Q_{p1}^{\sigma}(z_0) = \int_{\hat{l}_1} \frac{e^{-i\beta(t_0+\sigma)[\lambda-\frac{1}{\lambda}]}}{(-\lambda+\frac{1}{\lambda})^{p+1}} \frac{d\lambda}{\lambda}, \qquad \sigma = \pm 1,$$

$$Q_{p2}^{\sigma}(z_0) = \int_{\hat{l}_2} \frac{e^{i\beta[i(1-\sigma)(\lambda+\frac{1}{\lambda})+(1-t_0)(\lambda-\frac{1}{\lambda})]}}{(-i\lambda-\frac{i}{\lambda})^{p+1}} \frac{d\lambda}{\lambda}, \qquad \sigma = \pm 1,$$

$$Q_{p3}^{\sigma}(z_0) = \int_{\hat{l}_3} \frac{e^{i\beta[2i(\lambda+\frac{1}{\lambda})+(\sigma-t_0)(\lambda-\frac{1}{\lambda})]}}{(\lambda-\frac{1}{\lambda})^{p+1}} \frac{d\lambda}{\lambda}, \qquad \sigma = \pm 1,$$

$$Q_{p4}^{\sigma}(z_0) = \int_{\hat{l}_4} \frac{e^{i\beta[i(1+\sigma)(\lambda+\frac{1}{\lambda})-(1+t_0)(\lambda-\frac{1}{\lambda})]}}{(\lambda i+\frac{i}{\lambda})^{p+1}} \frac{d\lambda}{\lambda}, \qquad \sigma = \pm 1.$$

We choose the indented contours $\hat{l}_j$, $j=1,\dots,4$, displayed in Figure 7.3.

The exponential $e^{-i\beta(t_0+\sigma)[\lambda-\frac{1}{\lambda}]}$ has decay for $\operatorname{Im}\lambda < 0$ if $\sigma = 1$ and for $\operatorname{Im}\lambda > 0$ if $\sigma = -1$. Using that

$$\operatorname*{Res}_{\lambda=-1} \frac{e^{-i\beta(t_0+\sigma)[\lambda-\frac{1}{\lambda}]}}{(-\lambda+\frac{1}{\lambda})^{p+1}} \frac{1}{\lambda} = \frac{1}{p!} \frac{d^p}{d\lambda^p}\bigg|_{\lambda=-1} \left(\frac{\lambda}{1-\lambda}\right)^{p+1} e^{-i\beta(t_0+\sigma)[\lambda-\frac{1}{\lambda}]}\frac{1}{\lambda},$$

we find

$$Q^{\sigma}_{p1}(z_0) = \int_0^{e^{-\frac{3i\pi}{4}}\infty} \frac{e^{-i\beta(t_0+\sigma)[\lambda-\frac{1}{\lambda}]}}{(-\lambda+\frac{1}{\lambda})^{p+1}} \frac{d\lambda}{\lambda} + \frac{2\pi i}{p!}\frac{d^p}{d\lambda^p}\bigg|_{\lambda=-1}\left(\frac{\lambda}{1-\lambda}\right)^{p+1} e^{-i\beta(t_0+\sigma)[\lambda-\frac{1}{\lambda}]}\frac{1}{\lambda}, \qquad \sigma = 1,$$

and

$$Q^{\sigma}_{p1}(z_0) = \int_0^{e^{\frac{3i\pi}{4}}\infty} \frac{e^{-i\beta(t_0+\sigma)[\lambda-\frac{1}{\lambda}]}}{(-\lambda+\frac{1}{\lambda})^{p+1}} \frac{d\lambda}{\lambda}, \qquad \sigma = -1.$$

Similarly, we find

$$Q^{\sigma}_{p2}(z_0) = \int_0^{\infty e^{\frac{i\pi}{4}}} \frac{e^{i\beta[i(1-\sigma)(\lambda+\frac{1}{\lambda})+(1-t_0)(\lambda-\frac{1}{\lambda})]}}{(-i\lambda-\frac{i}{\lambda})^{p+1}} \frac{d\lambda}{\lambda}, \qquad \sigma = \pm 1,$$

$$Q^{\sigma}_{p3}(z_0) = \int_0^{\infty e^{\frac{i\pi}{4}}} \frac{e^{i\beta[2i(\lambda+\frac{1}{\lambda})+(\sigma-t_0)(\lambda-\frac{1}{\lambda})]}}{(\lambda-\frac{1}{\lambda})^{p+1}} \frac{d\lambda}{\lambda}, \qquad \sigma = 1,$$

$$Q^{\sigma}_{p3}(z_0) = \int_0^{\infty e^{-\frac{i\pi}{4}}} \frac{e^{i\beta[2i(\lambda+\frac{1}{\lambda})+(\sigma-t_0)(\lambda-\frac{1}{\lambda})]}}{(\lambda-\frac{1}{\lambda})^{p+1}} \frac{d\lambda}{\lambda} - \frac{2\pi i}{p!}\frac{d^p}{d\lambda^p}\bigg|_{\lambda=1}\left(\frac{\lambda}{\lambda+1}\right)^{p+1} e^{i\beta[2i(\lambda+\frac{1}{\lambda})+(\sigma-t_0)(\lambda-\frac{1}{\lambda})]}\frac{1}{\lambda}, \qquad \sigma = -1,$$

$$Q^{\sigma}_{p4}(z_0) = \int_0^{\infty e^{-\frac{i\pi}{4}}} \frac{e^{i\beta[-(1+t_0)(\lambda-\frac{1}{\lambda})+i(1+\sigma)(\lambda+\frac{1}{\lambda})]}}{(\lambda i+\frac{i}{\lambda})^{p+1}} \frac{d\lambda}{\lambda}, \qquad \sigma = \pm 1.$$

The above integrals have exponential decay and are suitable for numerical evaluation. Similar considerations lead to analogous expressions for $Q^{\pm}_{pj}(z_0)$ for $z_0$ belonging to the left, bottom, and right sides of the square.

Substituting the above expressions into relation (7.53) evaluated at $z = z_0$, where $z_0$ belongs to one of the sides of the square, we find relations for the $c^{(j)}_m$'s which can be used to supplement the exterior global relation.

It was noted earlier that the term $u^{\mathcal{D}}$ defined by (7.54) is known. However, the question of computing this term numerically in the most efficient way remains open. In the examples below we compute the known expression (7.54) by approximating the given Dirichlet data in terms of Legendre polynomials.

## Examples

Let $D$ be the domain exterior to the square with vertices at $\pm 1 \pm i$. The function

$$u(r,\varphi) = H^{(1)}_n(2i\beta r)(Ae^{in\varphi} + Be^{-in\varphi}) \tag{7.59}$$

satisfies the modified Helmholtz equation (7.32) in $D$ for any $n \geq 1$ and any complex constants $A$ and $B$.

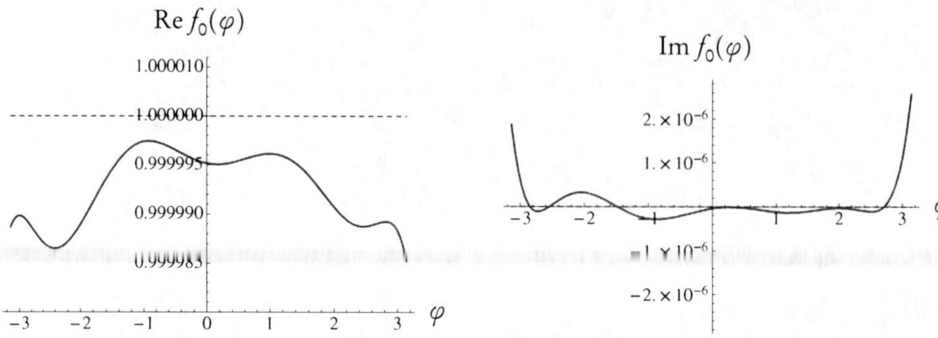


**Figure 7.4.** *The real and imaginary parts of the scattering amplitude $f_0(\varphi)$ for Example* 1. *Numerical values (solid) compared with true values (dashed).*

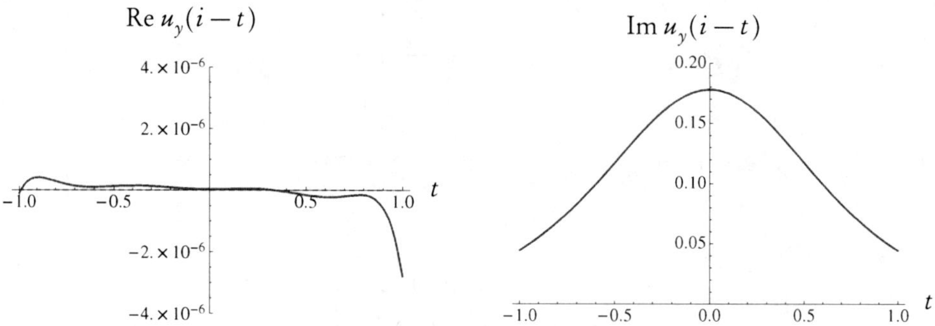


**Figure 7.5.** *The real and imaginary parts of the Neumann value on the top side for Example* 1.

**Example 1**

For the first example, we let $u(r,\varphi) = H_0^{(1)}(2i\beta r)$. This solution satisfies

$$u(r,\varphi) = \sqrt{\frac{1}{\pi i \beta r}}\, e^{i(2i\beta r - \frac{\pi}{4})}\left(1 - \frac{1}{16\beta r} + \frac{9}{512\beta^2 r^2} + O(r^{-3})\right), \qquad r \to \infty.$$

Hence equation (7.37) is satisfied with $f_0(\varphi) \equiv 1$. We use the values of $u$ on the four sides of the square as the given Dirichlet data. Letting $\beta = 1$, we apply the above technique with $M = 8$ and with 54 collocation points. The numerical scheme yields the scattering amplitude $f_0(\varphi)$ displayed in Figure 7.4. The numerically obtained and true Neumann values on the top side are shown in Figure 7.5; the results for the other three sides are similar. The numerics gives the correct profiles with a maximal error of approximately $10^{-4}$.

**Example 2**

For the second example, we let $u(r,\varphi) = H_1^{(1)}(2i\beta r)e^{i\varphi}$. This solution satisfies

$$u(r,\varphi) = \sqrt{\frac{1}{\pi i \beta r}}\, e^{i(2i\beta r - \frac{\pi}{4})}\left(-ie^{i\varphi} + \frac{3e^{i\varphi}}{16i\beta r} + \frac{15ie^{i\varphi}}{512\beta^2 r^2} + O(r^{-3})\right), \qquad r \to \infty.$$

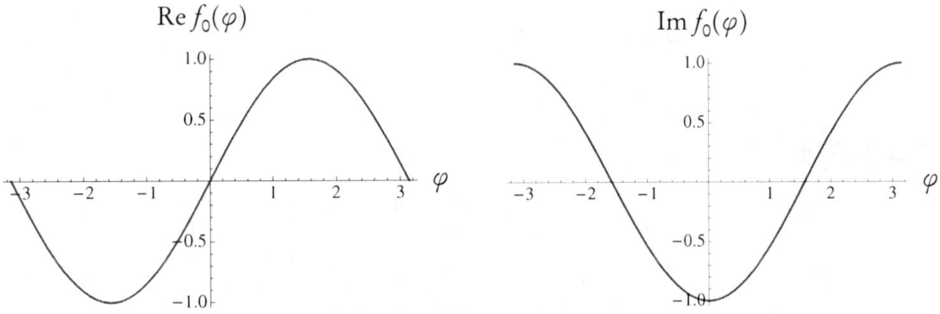


**Figure 7.6.** *The real and imaginary parts of the scattering amplitude $f_0(\varphi)$ for Example 2.*

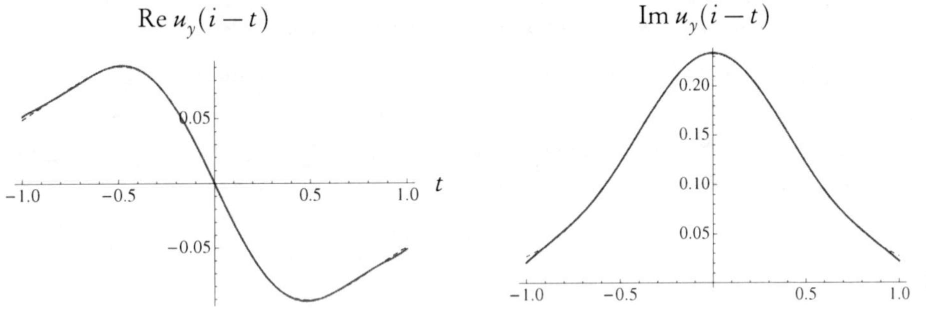


**Figure 7.7.** *The real and imaginary parts of the Neumann value on the top side for Example 2.*

Hence equation (7.37) is satisfied with $f_0(\varphi) = -ie^{i\varphi}$. Letting $\beta = 1$, the numerical scheme with $M = 8$ and with 54 collocation points yields the scattering amplitude $f_0(\varphi)$ displayed in Figure 7.6. The numerically obtained and true Neumann values on the top side are shown in Figure 7.7. The numerics gives the correct profiles with a maximal error of approximately 0.002.

## 7.2.7 ▪ Conclusions

We have illustrated a novel numerical technique for linear elliptic PDEs formulated in the *exterior* of a polygon. This technique, which is based on the unified transform, uses two main ingredients. The first is the global relation; in the case of the modified Helmholtz equation, the global relation is given by (7.34), where $I(\lambda)$ is defined by equation (7.38). This equation can be analyzed numerically in the same way that the global relation is analyzed in the case of linear elliptic PDEs formulated in the *interior* of a polygon. However, now the global relation, in addition to the unknown boundary values, also involves the unknown function $f_0$ (the scattering amplitude). Hence, it must be supplemented with an additional equation. Second, additional equations can be obtained by taking the limit of a certain equation which is valid for $z$ in the interior of the given polygon, as $z$ approaches the boundary; in the case of the modified Helmholtz equation, this equation is given by (7.53). Actually, by taking the limit as $z$ approaches each side of an $n$-gon, one obtains $n$ additional equations. These equations provide the formulation in the complex Fourier plane (complex extension of $\lambda$) of the classical boundary integral method. By exploiting the analyticity structure in the complex Fourier plane, it is possible to perform contour deformations and to obtain integrals which decay exponentially for large $\lambda$. This yields a most efficient numerical evaluation of these integrals. Detailed comparisons will be presented elsewhere.

## Bibliography

[1] A. C. L. Ashton, On the rigorous foundations of the Fokas method for linear elliptic partial differential equations, Proc. R. Soc. A **468** (2012), 1325–1331.

[2] A. C. L. Ashton, The spectral Dirichlet–Neumann map for the Laplace equation in a convex polygon, SIAM J. Math. Anal. **45** (2013), 3575–3591.

[3] A. S. Fokas, A unified transform method for solving linear and certain nonlinear PDEs, Proc. Roy. Soc. Lond. Ser. A **453** (1997), 1411–1443.

[4] A. S. Fokas, Two-dimensional linear partial differential equations in a convex polygon, R. Soc. Lond. Proc. Ser. A Math. Phys. Eng. Sci. **457** (2001), 371–393.

[5] A. S. Fokas, Integrable nonlinear evolution equations on the half-line, Comm. Math. Phys. **230** (2002), 1–39.

[6] A. S. Fokas, A Unified Approach to Boundary Value Problems, CBMS-NSF Regional Conf. Ser. in Appl. Math., SIAM (2008).

[7] A. S. Fokas, N. Flyer, S. A. Smitheman, and E. A. Spence, A semi-analytical numerical method for solving evolution and elliptic partial differential equations, J. Comput. Appl. Math. **227** (2009), 59–74.

[8] A. S. Fokas, P. Hashemzadeh and S. A. Smitheman, A Numerical Technique for Linear Elliptic PDEs in Polygonal Domains (preprint).

[9] A. S. Fokas and J. Lenells, The Unified Transform for the Helmholtz and modified Helmholtz equations in the exterior of a polygon (in preparation).

[10] B. Fornberg and N. Flyer, A numerical implementation of Fokas boundary integral approach: Laplace's equation on a polygonal domain, Proc. R. Soc. Lond. Ser. A Math. Phys. Eng. Sci. **467** (2011), 2983–3003.

[11] S. R. Fulton, A. S. Fokas, and C. A. Xenophontos, An analytical method for linear elliptic PDEs and its numerical implementation, J. Comput. Appl. Math. **167** (2004), 465–483.

[12] S. N. Karp, A convergent "farfield" expansion for two-dimensional radiation functions, Comm. Pure Appl. Math. **14** (1961), 427–434.

[13] Y. G. Saridakis, A. G. Sifalakis, and E. P. Papadopoulou, Efficient numerical solution of the generalized Dirichlet–Neumann map for linear elliptic PDEs in regular polygon domains, J. Comput. Appl. Math. **236** (2012), 2515–2528.

[14] A. G. Sifalakis, A. S. Fokas, S. R. Fulton, and Y. G. Saridakis, The generalized Dirichlet-Neumann map for linear elliptic PDEs and its numerical implementation, J. Comput. Appl. Math. **219** (2008), 9–34.

[15] A. G. Sifalakis, S. R. Fulton, E. P. Papadopoulou, and Y. G. Saridakis, Direct and iterative solution of the generalized Dirichlet-Neumann map for elliptic PDEs on square domains, J. Comput. Appl. Math. **227** (2009), 171–184.

[16] A. G. Sifalakis, E. P. Papadopoulou, and Y. G. Saridakis, Numerical study of iterative methods for the solution of the Dirichlet-Neumann map for linear elliptic PDEs on regular polygon domains, Int. J. Appl. Math. Comput. Sci. **4** (2007), 173–178.

[17] S. A. Smitheman, E. A. Spence, and A. S. Fokas, A spectral collocation method for the Laplace and modified Helmholtz equations in a convex polygon, IMA J. Numer. Anal. 30 (2010), 1184–1205.

[18] E. A. Spence, Boundary Value Problems for Linear Elliptic PDEs, PhD thesis, University of Cambridge, Cambridge, UK (2010).

## 7.3 ▪ Acoustic Scattering: High-Frequency Boundary Element Methods and Unified Transform Methods

*S. N. Chandler-Wilde and S. Langdon*

SL was supported by EPSRC grants EP/K000012/1 and EP/K037862/1, and both authors by EPSRC grant EP/F067798/1.

### 7.3.1 ▪ Introduction

The reliable simulation of processes in which acoustic waves are scattered by obstacles is of great practical interest, with applications including the modeling of sonar and other methods of acoustic detection and the study of problems of outdoor noise propagation and noise control, associated, for example, with road, rail, or aircraft noise. Unless the geometry of the scattering obstacle is particularly simple, analytical solution of scattering problems is usually impossible, and hence in general numerical schemes are required.

Most problems of acoustic scattering can be formulated, in the frequency domain, as linear elliptic boundary value problems (BVPs). In this section we describe how the boundary element method (BEM) can be used to solve such problems, and we summarize in particular recent progress in tackling high-frequency scattering problems by combining classical BEMs with insights from ray tracing methods and high-frequency asymptotics. We also make connections with a newer method for numerical solution of elliptic BVPs, the unified transform method due to Fokas and co-authors, and we detail the independent development of this method for acoustic scattering problems by DeSanto and co-authors.

In a homogeneous medium at rest, the acoustic pressure satisfies the linear wave equation. Under the further assumption of harmonic ($\mathrm{e}^{-\mathrm{i}\omega t}$) time dependence, the problem of computing perturbations in acoustic pressure reduces to the solution of the Helmholtz equation

$$\Delta u + k^2 u = 0, \quad \text{in } D \subset \mathbb{R}^d, \tag{7.60}$$

where $d = 2$ or 3 is the dimension of the problem under consideration, and $D$ denotes the domain of propagation, the region in which the wave propagates. The positive constant $k := \omega/c > 0$, where $\omega$ is angular frequency and $c$ is the speed of sound in $D$, is known as the wavenumber.

We will describe below how to compute, by the BEM, solutions to (7.60) that also satisfy certain boundary conditions on the boundary of the domain of propagation. The most commonly relevant boundary condition is the impedance boundary condition

$$\frac{\partial u}{\partial \nu} - \mathrm{i}k\beta u = h \quad \text{on } \Gamma, \tag{7.61}$$

where $\Gamma$ denotes the boundary of $D$ and $\partial/\partial\nu$ denotes the normal derivative on the boundary, where $\nu(\mathbf{x})$ denotes the unit normal at $\mathbf{x} \in \Gamma$ directed out of $D$. The function $h$ is identically zero in acoustic scattering problems (problems where we are given an incident wave and a stationary scatterer and have to compute the resulting acoustic field) but is nonzero for radiation problems (where the motion of a radiating structure is given, and we have to compute the radiated acoustic field). Finally, $\beta$ is the relative surface admittance which, in general, is a function of position on the boundary and of frequency $\omega$. When $\beta = 0$ the boundary is described as being acoustically rigid, or sound-hard (this is the Neumann boundary condition). More generally, $\beta \in \mathbb{C}$, with $\mathrm{Re}[\beta] \geq 0$ to satisfy the physical condition that the surface absorbs (rather than emits) energy. When $|\beta|$ is very large, one can consider as an approximation to (7.61) the sound-soft (Dirichlet) boundary condition

$$u = h \quad \text{on } \Gamma. \tag{7.62}$$

Equation (7.60) models acoustic propagation in a homogeneous medium at rest. We are often interested in applications in propagation through a medium with variable wave speed. The BEM is well adapted to compute solutions in the case when the wave speed $c$ is piecewise constant. In particular, when a homogeneous region with a different wave speed is embedded in a larger homogeneous medium, acoustic waves are transmitted across the boundary $\Gamma$ between the two media, (7.60) holds on either side of $\Gamma$ with different values of $k = \omega/c$, and, at least in the simplest case when the density of the two media is the same, the boundary conditions on $\Gamma$ (so-called *transmission conditions*) are that $u$ and $\partial u/\partial \nu$ are continuous across $\Gamma$.

The domain $D$ can be a bounded domain (e.g., for applications in room acoustics), but in many practical applications it may be unbounded (e.g., for outdoor noise propagation). In this case, the complete mathematical formulation must also include a condition to represent the idea that the acoustic field (or at least some part of it, e.g., the part reflected by a scattering obstacle) is traveling outward. The usual condition imposed is the Sommerfeld radiation condition,

$$\frac{\partial u}{\partial r}(\mathbf{x}) - \mathrm{i}ku(\mathbf{x}) = o(r^{-(d-1)/2}), \tag{7.63}$$

as $r := |\mathbf{x}| \to \infty$, uniformly in $\hat{\mathbf{x}} := \mathbf{x}/r$.

Numerical solution of (7.60) together with a boundary condition (and a radiation condition if the problem is posed on an unbounded domain) can be achieved by many means. The key idea of the BEM is to reformulate the BVP as a boundary integral equation (BIE) (i.e., an integral equation that holds on the boundary $\Gamma$) and then to solve that numerically. Under the assumptions above (homogeneous medium, time-harmonic waves) the advantages of BEM over domain-based methods such as finite difference or finite element methods are twofold. First, as the integral equation to be solved holds only on the boundary of the domain, the dimensionality of the problem is reduced; this has the obvious advantage that it is easier to construct an approximation space in a lower dimension, but moreover we will see below (in the section entitled "Hybrid Numerical-Asymptotic BEMs for High-Frequency Problems") that it can be easier to understand the oscillatory behavior of the solution on the boundary than throughout the domain, which can assist greatly in designing a good approximation space at high frequencies. Second, for problems in unbounded domains, reformulating as an integral equation on the boundary removes the need to truncate the unbounded domain in order to be able to construct a discretization space.

We describe the reformulation of our BVPs as BIEs in Section 7.3.2. Following [19] we consider the case that $D$ is Lipschitz and thus work throughout in a Sobolev space setting (see, e.g., [65]; for the simpler case of smooth boundaries we refer to [28]). We then consider the numerical solution of these BIEs in Section 7.3.3, focusing in particular, in the section entitled "Hybrid Numerical-Asymptotic BEMs for High-Frequency Problems," on schemes that are well adapted to the case when the wavenumber $k$ is large. As we report, for many scattering problems these methods provably compute solutions of any desired accuracy with a cost, in terms of numbers of degrees of freedom and size of matrix to be inverted, that is close to frequency independent.

There is a wide literature on BIE formulations for acoustic scattering problems and BEMs for linear elliptic BVPs: see, e.g., [6, 28, 57, 60, 65, 68, 72, 73, 74], and see also, e.g., [59] for a comparison with finite element methods. The question of how to develop schemes efficient for large $k$ has been considered fully in the review article [19], where a very complete literature review of the topic can be found. Our presentation in this section summarizes some of the key ideas found therein, but primarily focuses, in the section entitled "Hybrid Numerical-Asymptotic BEMs for High-Frequency Problems," on more recent developments.

This section is one of a collection of articles in significant part focused on the so-called *unified transform* or *Fokas transform*, introduced by Fokas in 1997 [44]. This method is on the one hand an analytical transform technique which can be employed to solve linear BVPs with constant coefficients in canonical domains. But this method also, when applied in general domains, has many aspects in common with BIE and BEMs; in particular it reduces solution of the BVP to solution of an equation on the boundary $\Gamma$ for the unknown part of the Cauchy data ($u$ or its normal derivative on $\Gamma$), just as in the BIE method.

We discuss this methodology in Section 7.3.4. In particular, we describe how this method can be used to solve acoustic problems in interior domains (see also Spence [76] in this book), proposing a new version of this method which computes the best approximation to the unknown boundary data from a space of restrictions of (generalized) plane waves to the boundary. We also discuss the application of this method to acoustic scattering problems, noting that the methodology applies in particular to so-called *rough surface scattering problems*; indeed it has been developed independently as a numerical method for these problems by DeSanto and co-authors, in articles from 1981 onward [33, 35, 36, 34, 5]. The numerical implementations we focus on have in common with the high-frequency BEMs of the section entitled "Hybrid Numerical-Asymptotic BEMs for High-Frequency Problems" that they utilize oscillatory basis functions (restrictions of plane waves to $\Gamma$). The results reported in Section 7.3.4 from [5] suggest that, in terms of numbers of degrees of freedom required to achieve accurate approximations, these methods outperform standard BEMs for certain scattering problems.

## 7.3.2 ▪ BIE Formulations

In this section we state the BVPs introduced in Section 7.3.1 more precisely, and we reformulate them as BIEs. Given that many scatterers in applications have corners and edges, we will consider throughout domains $D \subset \mathbb{R}^d$ (with $d = 2$ or 3) that are Lipschitz, usually denoting the boundary of our domain $D$ by $\Gamma$. We denote the trace operator by $\gamma$, so that $\gamma u$ is just the restriction of $u$ to $\Gamma$ when $u$ is sufficiently regular, and the normal derivative operator by $\partial_\nu$, noting that $\partial_\nu u$ coincides with the classical definition of the normal derivative $\partial u/\partial \nu$ when $u$ is sufficiently regular (see [19, Appendix A] for more details). For further mathematical details on the results in this section, we refer particularly to [19, Section 2]. Our Sobolev space notation is defined precisely in [19, 65]. This notation is standard and we will not repeat it here except to note that, for a general open set $D$, following [65], $H^s(D)$, for $s \in \mathbb{R}$, will denote the space of restrictions to $D$ of elements of $H^s(\mathbb{R}^d)$, while, for $n \in \mathbb{N}$, $W^n(D)$ will denote those $u \in L^2(D)$ whose partial derivatives of order $\leq n$ are also in $L^2(D)$: in particular $W^1(D) = \{u \in L^2(D) : \nabla u \in L^2(D)\}$. These notions coincide, that is, $H^n(D) = W^n(D)$ (with equivalence of norms), when $D$ is Lipschitz [65], but we need to tread carefully where $D$ is not Lipschitz, for example when studying screen problems.

### Acoustic BVPs

This paper is about scattering problems, which are predominantly BVPs in exterior domains. However, the BIE methods we use intimately make connections between problems in exterior and interior domains, so that we will also make mention of interior problems. (And we flag that, to introduce the unified transform in Section 7.3.4, we study its application first to interior problems in the section entitled "The Unified Transform Method for the Interior Dirichlet Problem".) As alluded to above, we will formulate all of our BVPs in Sobolev space settings. We will consider as BVPs interior and exterior Dirichlet and impedance problems for the Helmholtz equation (7.60) in the interior and exterior of Lipschitz domains, a Dirichlet problem in the exterior of a planar screen, and a particular Helmholtz transmission problem.

We state first, for $D$ a bounded Lipschitz domain, the *interior Dirichlet problem*:

$$\begin{array}{l}\text{Given } h \in H^{1/2}(\Gamma),\ \text{find } u \in C^2(D)\cap H^1(D)\\ \text{such that (7.60) holds in } D \text{ and } \gamma u = h \text{ on } \Gamma.\end{array} \tag{7.64}$$

Next, for $D$ a bounded Lipschitz domain, the *interior impedance problem*:

$$\begin{array}{l}\text{Given } h \in H^{-1/2}(\Gamma),\ \text{and } \beta \in L^\infty(\Gamma) \text{ (with } \mathrm{Re}[\beta] \geq 0),\\ \text{find } u \in C^2(D)\cap H^1(D) \text{ such that (7.60) holds in } D\\ \text{and } \partial_\nu u - \mathrm{i}k\beta\gamma u = h \text{ on } \Gamma.\end{array} \tag{7.65}$$

Solvability results for (7.64) and (7.65) are well known (see, e.g., [28], [65, Page 286], [19, Theorems 2.1, 2.3]). For the interior Dirichlet problem (7.64) there exists a sequence $0 < k_1 < k_2 < \cdots$ of positive wavenumbers, with $k_m \to \infty$ as $m \to \infty$, such that (7.64) with $h = 0$ has a nontrivial solution (so $-k_m^2$ is a Dirichlet eigenvalue of the Laplacian in $D$). For all other values of $k > 0$, (7.64) has exactly one solution. The same result holds for the Neumann problem ((7.65) with $\beta = 0$), with a different sequence of wavenumbers. The interior impedance problem (7.65) with $\mathrm{Re}[\beta]$ not identically zero has exactly one solution.

Next we state the exterior BVPs. Suppose that $\Omega_+ \subset \mathbb{R}^d$ is an unbounded Lipschitz domain with boundary $\Gamma$ such that $\Omega_- := \mathbb{R}^d \setminus \overline{\Omega_+}$, $d = 2$ or $3$, is a bounded Lipschitz open set. The *exterior Dirichlet problem* is

$$\begin{array}{l}\text{Given } h \in H^{1/2}(\Gamma),\ \text{find } u \in C^2(\Omega_+)\cap H^1_{\mathrm{loc}}(\Omega_+)\\ \text{such that (7.60) holds in } \Omega_+, \gamma u = h \text{ on } \Gamma,\\ \text{and } u \text{ satisfies the radiation condition (7.63).}\end{array} \tag{7.66}$$

The *exterior impedance problem* is (where the normal $\nu$ here points out of $\Omega_-$ *into* the domain of propagation $\Omega_+$)

$$\begin{array}{l}\text{Given } h \in H^{-1/2}(\Gamma),\ \text{and } \beta \in L^\infty(\Gamma) \text{ (with } \mathrm{Re}[\beta] \geq 0),\\ \text{find } u \in C^2(\Omega_+)\cap H^1_{\mathrm{loc}}(\Omega_+) \text{ such that (7.60) holds in } \Omega_+,\\ \partial_\nu u + \mathrm{i}k\beta u = h \text{ on } \Gamma,\\ \text{and } u \text{ satisfies the radiation condition (7.63).}\end{array} \tag{7.67}$$

Both problems (7.66) and (7.67) have exactly one solution (see, e.g., [28], [19, Theorem 2.10]).

In the above exterior Dirichlet problem the boundary $\Gamma$ is a closed surface, separating two Lipschitz open sets, $\Omega_+$ and $\Omega_-$. A variant, which models acoustic scattering by infinitely thin screens, is to consider the case where $\Gamma$ is an open surface, and the domain in which the BVP is to be solved lies on both sides of $\Gamma$. We will show results in the section entitled "Hybrid Numerical-Asymptotic BEMs for High-Frequency Problems" for the special case of a planar screen, when, for some bounded $C^0$ open subset $S \subset \mathbb{R}^{d-1}$,

$$\Gamma := \{(\tilde{x}, 0) : \tilde{x} \in S\}. \tag{7.68}$$

With $\Gamma$ given by (7.68) and $D := \mathbb{R}^d \setminus \overline{\Gamma}$, the *Dirichlet screen problem* is

$$\begin{aligned}&\text{Given } h \in H^{1/2}(\Gamma), \text{ find } u \in C^2(D) \cap W^1_{\text{loc}}(D)\\ &\text{such that (7.60) holds in } D, \gamma_\pm u = h \text{ on } \Gamma,\\ &\text{and } u \text{ satisfies the radiation condition (7.63).}\end{aligned} \tag{7.69}$$

In the above formulation $\gamma_\pm$ are the trace operators onto the plane $x_d = 0$ containing $\Gamma$ from the upper and lower half planes, respectively. That this problem is well-posed dates back at least to Stephan [81] in the case that $S$ is a $C^\infty$ open set; this result is extended to the case that $S$ is merely $C^0$ in [20].

Finally we formulate a transmission problem for the Helmholtz equation. Let $\Omega_-$ be a bounded Lipschitz open set, let $\Omega_+ := \mathbb{R}^d \setminus \overline{\Omega_-}$, $\Gamma$ be the common boundary of $\Omega_+$ and $\Omega_-$, and suppose that $k$ in (7.60) is a function of position rather than constant, precisely that, for some constants $k_+ > 0$ and $k_- \in \mathbb{C}$ with $\text{Re}[k_-] > 0$ and $\text{Im}\, k_- \geq 0$, $k(x) = k_+$ in $\Omega_+$ and $k(x) = k_-$ in $\Omega_-$. The *Helmholtz transmission problem* we consider is

$$\begin{aligned}&\text{Given } h \in H^{1/2}(\Gamma),\ g \in H^{-1/2}(\Gamma), \text{ find } u \in C^2(\Omega_+ \cup \Omega_-) \cap\\ &W^1_{\text{loc}}(\Omega_+ \cup \Omega_-) \text{ such that (7.60) holds in } \Omega_+ \cup \Omega_-, \gamma_+ u - \gamma_- u = h \text{ and}\\ &\partial_\nu^+ u - \partial_\nu^- u = g \text{ on } \Gamma, \text{ and } u \text{ satisfies the radiation condition (7.63).}\end{aligned} \tag{7.70}$$

Here $\gamma_+ u$ and $\gamma_- u$ are the traces of $u$, and $\partial_\nu^\pm u$ the normal derivatives, taken from the $\Omega_+$ and $\Omega_-$ sides, respectively. That this BVP has exactly one solution is shown for the case in which $\Omega_-$ is a $C^2$ domain in [28], and the argument there easily extends to the Lipschitz case. We recall from the introduction that the different values for $k$ in $\Omega_\pm$ correspond physically to different wave speeds; allowing $k_-$ to have a positive imaginary part models an interior medium which dissipates energy as the wave propagates. We remark that [28] (and many other authors) consider a more general case with $\gamma_+ u - \gamma_- u = h$ replaced by $\mu_+ \gamma_+ u - \mu_- \gamma_- u = h$, for some constants $\mu_\pm \in \mathbb{C}$, which models a jump in density in addition to wave speed across $\Gamma$.

### Acoustic Scattering Problems

In acoustic scattering problems one is interested in computing the scattered acoustic field $u^S$ produced when an incident field $u^I \in L^1_{\text{loc}}(\mathbb{R}^d)$ interacts with some obstacle or obstacles (the *scatterer*) occupying the closed set $\Omega$ (with surface $\Gamma$), such that $\Omega_+ := \mathbb{R}^d \setminus \Omega$ is an unbounded domain. One assumes that the incident field itself is a solution of the Helmholtz equation, precisely that it satisfies (7.60) in some neighborhood $G$ of $\Omega$, in which case also $u^I|_G \in C^\infty(G)$. We will refer throughout to the sum $u := u^I + u^S$ as the *total acoustic field* (total field for short).

In many applications, the incident field is generated by a point source, i.e., for some $\mathbf{z} \in \mathbb{R}^d \setminus \Omega$,

$$u^I(\mathbf{x}) = \Phi_k(\mathbf{x}, \mathbf{z}), \quad \mathbf{x} \in \mathbb{R}^d \setminus \{\mathbf{z}\},$$

where $\Phi_k$ is the fundamental solution of the Helmholtz equation, given, in the two-dimensional (2D) and three-dimensional (3D) cases, by

$$\Phi_k(\mathbf{x},\mathbf{y}) := \begin{cases} \frac{\mathrm{i}}{4} H_0^{(1)}(k|\mathbf{x}-\mathbf{y}|), & d=2, \\ \frac{\exp(\mathrm{i}k|\mathbf{x}-\mathbf{y}|)}{4\pi|\mathbf{x}-\mathbf{y}|}, & d=3, \end{cases} \tag{7.71}$$

for $\mathbf{x},\mathbf{y} \in \mathbb{R}^d$, $\mathbf{x} \neq \mathbf{y}$, where $H_\nu^{(1)}$ denotes the Hankel function of the first kind of order $\nu$. To represent a source far from the scatterer, the incident field can be taken to be a plane wave; i.e., for some $\hat{\mathbf{a}} \in \mathbb{R}^d$ with $|\hat{\mathbf{a}}| = 1$,

$$u^I(\mathbf{x}) = \exp(\mathrm{i}k\mathbf{x}\cdot\hat{\mathbf{a}}), \quad \mathbf{x} \in \mathbb{R}^d. \tag{7.72}$$

The scattering problems we consider are then particular cases of some of the BVPs above, namely the exterior Dirichlet and impedance problems, the Dirichlet screen problem, and the transmission problem. The *sound-soft scattering problem* is

$$\begin{aligned} &\text{Find } u^S \in C^2(\Omega_+)\cap H^1_{\mathrm{loc}}(\Omega_+) \text{ such that (7.60) holds in } \Omega_+, \\ &\gamma u = 0 \text{ on } \Gamma \; (\text{so } \gamma u^S = -u^I|_\Gamma), \\ &\text{and } u^S \text{ satisfies the radiation condition (7.63).} \end{aligned} \tag{7.73}$$

The *impedance scattering problem* is (where $\beta \in L^\infty(\Gamma)$ and $\mathrm{Re}[\beta] \geq 0$ on $\Gamma$)

$$\begin{aligned} &\text{Find } u^S \in C^2(\Omega_+)\cap H^1_{\mathrm{loc}}(\Omega_+) \text{ such that (7.60) holds in } \Omega_+, \\ &\partial_\nu u + \mathrm{i}k\beta u = 0 \text{ on } \Gamma \\ &(\text{so } (\partial_\nu + \mathrm{i}k\beta\gamma)u^S|_\Gamma = -(\partial_\nu + \mathrm{i}k\beta\gamma)u^I|_\Gamma), \\ &\text{and } u^S \text{ satisfies the radiation condition (7.63).} \end{aligned} \tag{7.74}$$

The *screen scattering problem* we consider is (where $\Gamma$ is given by (7.68) and $D = \mathbb{R}^d \setminus \overline{\Gamma}$)

$$\begin{aligned} &\text{Find } u^S \in C^2(D)\cap W^1_{\mathrm{loc}}(D) \text{ such that (7.60) holds in } D, \\ &\gamma_\pm u = 0 \text{ on } \Gamma \; (\text{so } \gamma_\pm u^S = -u^I|_\Gamma), \\ &\text{and } u^S \text{ satisfies the radiation condition (7.63).} \end{aligned} \tag{7.75}$$

Finally the *transmission scattering problem* we consider is (where $k(x) = k_+$ in $\Omega_+$ and $k(x) = k_-$ in $\Omega_-$)

$$\begin{aligned} &\text{Find } u \in C^2(\Omega_+\cup\Omega_-)\cap W^1_{\mathrm{loc}}(\Omega_+\cup\Omega_-) \text{ such that (7.60)} \\ &\text{holds in } \Omega_+\cup\Omega_-, \gamma_+ u = \gamma_- u \text{ and } \partial_\nu^+ u = \partial_\nu^- u \text{ on } \Gamma, \\ &\text{and } u^S \text{ satisfies the radiation condition (7.63).} \end{aligned} \tag{7.76}$$

### BIE Formulations

Suppose that $\Omega_-$ is a bounded Lipschitz open set with boundary $\Gamma$ such that $\Omega_+ := \mathbb{R}^d \setminus \overline{\Omega_-}$ is a Lipschitz domain, and let $\nu(\mathbf{x})$ denote the unit normal at $\mathbf{x} \in \Gamma$ directed into $\Omega_+$. To formulate BIEs for (7.60), we introduce the single-layer potential operator $\mathcal{S}_k : H^{-1/2}(\Gamma) \to H^1_{\mathrm{loc}}(\mathbb{R}^d)$ and the double-layer potential operator $\mathcal{D}_k : H^{1/2}(\Gamma) \to H^1_{\mathrm{loc}}(\Omega_\pm)$, defined by

$$\mathcal{S}_k\phi(\mathbf{x}) := \int_\Gamma \Phi_k(\mathbf{x},\mathbf{y})\phi(\mathbf{y})\,\mathrm{d}s(\mathbf{y}), \quad \mathbf{x} \in \mathbb{R}^d \setminus \Gamma, \tag{7.77}$$

and

$$\mathscr{D}_k\phi(\mathbf{x}) := \int_\Gamma \frac{\partial \Phi_k(\mathbf{x},\mathbf{y})}{\partial \nu(\mathbf{y})}\phi(\mathbf{y})\,ds(\mathbf{y}), \quad \mathbf{x}\in\mathbb{R}^d\setminus\Gamma, \tag{7.78}$$

respectively, where the normal $\nu$ is directed into $\Omega_+$. These layer potentials provide solutions to (7.60) in $\mathbb{R}^d \setminus \Gamma$; moreover, they also automatically satisfy the radiation condition (7.63). In general all the standard BVPs for the Helmholtz equation (7.60) can be formulated as integral equations on $\Gamma$ using these layer potentials.

Specifically, we can use Green's representation theorems, which lead to so-called *direct* BIE formulations (as we shall see in the section entitled "Hybrid Numerical-Asymptotic BEMs for High-Frequency Problems" these lend themselves particularly well to efficient approximation strategies when $k$ is large). Denoting the exterior and interior trace operators, from $\Omega_+$ and $\Omega_-$, respectively, by $\gamma_+$ and $\gamma_-$, and the exterior and interior normal derivative operators by $\partial_\nu^+$ and $\partial_\nu^-$, respectively, we have the following result for interior problems (see [19, Theorem 2.20]).

**Theorem 7.1.** *If $u \in H^1(\Omega_-)\cap C^2(\Omega_-)$ and, for some $k\geq 0$, $\Delta u + k^2 u = 0$ in $\Omega_-$, then*

$$\mathscr{S}_k\partial_\nu^- u(\mathbf{x}) - \mathscr{D}_k\gamma_- u(\mathbf{x}) = \begin{cases} u(\mathbf{x}), & \mathbf{x}\in\Omega_-,\\ 0, & \mathbf{x}\in\Omega_+.\end{cases} \tag{7.79}$$

The following is the corresponding result for exterior problems (see [19, Theorem 2.21]).

**Theorem 7.2.** *If $u \in H^1_{\mathrm{loc}}(\Omega_+)\cap C^2(\Omega_+)$ and, for some $k > 0$, $\Delta u + k^2 u = 0$ in $\Omega_+$ and $u$ satisfies the Sommerfeld radiation condition* (7.63) *in $\Omega_+$, then*

$$-\mathscr{S}_k\partial_\nu^+ u(\mathbf{x}) + \mathscr{D}_k\gamma_+ u(\mathbf{x}) = \begin{cases} u(\mathbf{x}), & \mathbf{x}\in\Omega_+,\\ 0, & \mathbf{x}\in\Omega_-.\end{cases} \tag{7.80}$$

The formulae (7.79) and (7.80) lie at the heart of boundary integral methods. Each expresses the solution throughout the domain in terms of its Dirichlet and Neumann traces on the boundary. Thus for Dirichlet problems, if the Neumann data can be computed, then these formulae immediately give a representation for the solution anywhere in the domain. Likewise, for Neumann or impedance problems, knowledge of the Dirichlet data is sufficient to determine the solution anywhere in the domain.

In order to derive BIEs for (7.60), for which the "unknown" to be computed will be the complementary boundary data required to complete the representation formula for the solution, we need to take Dirichlet and Neumann traces of (7.79) and (7.80). First, we define the acoustic single- and double-layer operators, $S_k : H^{-1/2}(\Gamma)\to H^{1/2}(\Gamma)$ and $D_k : H^{1/2}(\Gamma)\to H^{1/2}(\Gamma)$, by

$$S_k\phi(\mathbf{x}) := \int_\Gamma \Phi_k(\mathbf{x},\mathbf{y})\phi(\mathbf{y})ds(\mathbf{y}), \quad \mathbf{x}\in\Gamma, \tag{7.81}$$

and

$$D_k\phi(\mathbf{x}) := \int_\Gamma \frac{\partial \Phi_k(\mathbf{x},\mathbf{y})}{\partial \nu(\mathbf{y})}\phi(\mathbf{y})ds(\mathbf{y}), \quad \mathbf{x}\in\Gamma, \tag{7.82}$$

respectively, where for Lipschitz $\Gamma$ and $\phi \in H^{1/2}(\Gamma)$, $D_k\phi$ is understood as a Cauchy principal value integral. These appear on taking boundary traces of $\mathscr{S}_k$ and $\mathscr{D}_k$. When we apply the normal derivative operator $\partial_\nu$ to $\mathscr{S}_k$ and $\mathscr{D}_k$, we encounter two additional boundary integral operators, the acoustic adjoint double-layer operator $D'_k : H^{-1/2}(\Gamma) \to H^{-1/2}(\Gamma)$ and the acoustic hypersingular operator $H_k : H^{1/2}(\Gamma) \to H^{-1/2}(\Gamma)$, defined by

$$D'_k\phi(\mathbf{x}) := \int_\Gamma \frac{\partial \Phi_k(\mathbf{x},\mathbf{y})}{\partial \nu(\mathbf{x})}\phi(\mathbf{y})ds(\mathbf{y}) \tag{7.83}$$

and

$$H_k\phi(\mathbf{x}) := \frac{\partial}{\partial \nu(\mathbf{x})}\int_\Gamma \frac{\partial \Phi_k(\mathbf{x},\mathbf{y})}{\partial \nu(\mathbf{y})}\phi(\mathbf{y})ds(\mathbf{y}), \tag{7.84}$$

respectively. For Lipschitz $\Gamma$ and $\phi \in L^2(\Gamma)$, (7.83) makes sense as a Cauchy principal value integral, for almost all $\mathbf{x} \in \Gamma$, while, for $\phi \in H^1(\Gamma)$, $H_k\phi \in L^2(\Gamma)$ is defined in the sense explained in [19, Section 2.3].

The following jump relations are shown in [65], [19, Section 2.3]. As operators on $H^{-1/2}(\Gamma)$,

$$\gamma_\pm \mathcal{S}_k = S_k, \quad \partial_\nu^\pm \mathcal{S}_k = \mp\tfrac{1}{2}I + D'_k, \tag{7.85}$$

where $I$ is the identity operator. Similarly, as operators on $H^{1/2}(\Gamma)$, we have

$$\gamma_\pm \mathcal{D}_k = \pm\tfrac{1}{2}I + D_k, \quad \partial_\nu^\pm \mathcal{D}_k = H_k. \tag{7.86}$$

Applying the Dirichlet and Neumann trace operators to (7.79) and (7.80), and using (7.85) and (7.86), we obtain the Calderón projection:

$$c_\pm u = P_\pm c_\pm u, \tag{7.87}$$

where $c_\pm u = [\gamma_\pm u, \partial_\nu^\pm u]^T$ is the Cauchy data for $u$ on $\Gamma$, and

$$P_\pm = \pm\begin{bmatrix} \gamma_\pm \mathcal{D}_k & -\gamma_\pm \mathcal{S}_k \\ \partial_\nu^\pm \mathcal{D}_k & -\partial_\nu^\pm \mathcal{S}_k \end{bmatrix} = \tfrac{1}{2}I \pm \begin{bmatrix} D_k & -S_k \\ H_k & -D'_k \end{bmatrix}, \tag{7.88}$$

where $I$ is the ($2\times 2$ matrix) identity operator. Explicitly, we can rewrite equation (7.87) as

$$\left(D_k - \tfrac{1}{2}I\right)\gamma_+ u - S_k \partial_\nu^+ u = 0 \tag{7.89}$$

and

$$H_k \gamma_+ u - \left(D'_k + \tfrac{1}{2}I\right)\partial_\nu^+ u = 0 \tag{7.90}$$

for the exterior problem and

$$\left(D_k + \tfrac{1}{2}I\right)\gamma_- u - S_k \partial_\nu^- u = 0 \tag{7.91}$$

and

$$H_k \gamma_- u - \left(D'_k - \tfrac{1}{2}I\right)\partial_\nu^- u = 0 \tag{7.92}$$

for the interior problem. Each of these equations is a linear relationship between the components $\gamma_\pm u$ and $\partial_\nu^\pm u$ of the Cauchy data $c_\pm u$. These key results can be summarized in the following lemma [19, Lemma 2.22].

**Lemma 7.3.** *If $u \in H^1(\Omega_-) \cap C^2(\Omega_-)$ and, for some $k > 0$, $\Delta u + k^2 u = 0$ in $\Omega_-$, then $P_- c_- u = c_- u$. Similarly, if $u \in H^1_{\text{loc}}(\Omega_+) \cap C^2(\Omega_+)$ and, for some $k > 0$, $\Delta u + k^2 u = 0$ in $\Omega_+$ and $u$ satisfies the Sommerfeld radiation condition (7.63) in $\Omega_+$, then $P_+ c_+ u = c_+ u$.*

This lemma is the basis for all the standard direct BIE formulations for interior and exterior acoustic BVPs. For example, if $u$ satisfies the exterior Dirichlet problem (7.66), then it follows immediately from Lemma 7.3 that $\partial_\nu^+ u$ satisfies both

$$S_k \partial_\nu^+ u = \left(D_k - \tfrac{1}{2}I\right) h \tag{7.93}$$

and

$$\left(D_k' + \tfrac{1}{2}I\right) \partial_\nu^+ u = H_k h. \tag{7.94}$$

Similarly, if $u$ satisfies the interior Dirichlet problem (7.64), then it follows from Lemma 7.3 that

$$S_k \partial_\nu^- u = \left(D_k + \tfrac{1}{2}I\right) h \tag{7.95}$$

and

$$\left(D_k' - \tfrac{1}{2}I\right) \partial_\nu^- u = H_k h. \tag{7.96}$$

All these equations are BIEs of the form

$$Av = f, \tag{7.97}$$

where $A$ is a linear boundary integral operator, or a linear combination of such operators and the identity, $v$ is the solution to be determined, and $f$ is given data. Noting that the same operator $A$ can arise from both interior and exterior problems, it is immediately apparent that, although exterior acoustic problems are generically uniquely solvable, the natural BIE formulations of these problems need not be uniquely solvable for all wavenumbers $k$. As a specific instance, we noted above that the homogeneous interior Dirichlet problem has nontrivial solutions at a sequence $k_n$ of positive wavenumbers. If $k = k_n$ and $u$ is such a solution, then $\partial_\nu^- u$ is a nontrivial solution of (7.95) with $h = 0$ (see, e.g., [19, Theorem 2.4]) and so, for $k = k_n$, the BIE (7.93) for the exterior Dirichlet problem (7.66) has infinitely many solutions.

Similar BIE formulations (with the same problems of nonuniqueness) can be derived by utilizing the fact that the layer potentials satisfy (7.60) and (7.63); to satisfy the BVPs, it just remains to take the Dirichlet or Neumann trace of the layer potentials (using the jump relations as above) and then to match with the boundary data. The resulting formulations are known as indirect BIEs; we do not discuss these further here. As discussed above we will focus on direct formulations in which the unknown to be determined is the normal derivative or trace of the solution in the domain; it is possible as we discuss in the section entitled "Hybrid Numerical-Asymptotic BEMs for High-Frequency Problems" to bring high-frequency asymptotics to bear to understand the behavior of these solutions and so design efficient approximation spaces.

In order to derive uniquely solvable BIE formulations, the classical approach (dating back to [17] for the direct and [11, 63, 70] for the indirect formulation) is to solve a linear combination of the two equations arising for each problem from the Calderón projection. Taking a linear combination of (7.93) and (7.94), we obtain

$$A_{k,\eta}' \partial_\nu^+ u = B_{k,\eta} h, \tag{7.98}$$

where $\eta \in \mathbb{C}$ is a parameter that we need to choose and $A_{k,\eta}'$, $B_{k,\eta}$ are the operators

$$A_{k,\eta}' = \tfrac{1}{2}I + D_k' - \mathrm{i}\eta S_k \text{ and } B_{k,\eta} = H_k + \mathrm{i}\eta\left(\tfrac{1}{2}I - D_k\right). \tag{7.99}$$

Both $A'_{k,\eta}$ and $B_{k,\eta}$ are invertible (considered as operators between appropriate pairs of Sobolev spaces) for all $k > 0$ provided $\mathrm{Re}[\eta] \neq 0$; see, e.g., [19, Theorem 2.27].

The corresponding direct formulation for the exterior impedance problem is

$$C_{k,\eta,\beta}\gamma_+ u = A'_{k,\eta} h, \tag{7.100}$$

where

$$C_{k,\eta,\beta}\phi := B_{k,\eta}\phi + \mathrm{i}kA'_{k,\eta}(\beta\phi), \quad \phi \in H^{1/2}(\Gamma), \tag{7.101}$$

is invertible (considered as an operator between an appropriate pair of Sobolev spaces) for all $k > 0$ provided $\mathrm{Re}[\eta] \neq 0$; again see [19, Theorem 2.27].

That the exterior Dirichlet, Neumann, and impedance BVPs can be solved by combined potential direct integral equation formulations follows from, e.g., [19, Corollary 2.28]. Specifically:

**Corollary 7.4.** *Suppose that $k > 0$ and $\eta \in \mathbb{C}$ with $\mathrm{Re}[\eta] \neq 0$. Then both the following statements hold:*

(i) *If $u$ is the unique solution of* (7.66), *then $\partial_\nu^+ u \in H^{-1/2}(\Gamma)$ is the unique solution of* (7.98). *Further, if $h = \gamma_+ u \in H^s(\Gamma)$ with $1/2 < s \leq 1$, then $\partial_\nu^+ u \in H^{s-1}(\Gamma)$.*

(ii) *If $u$ is the unique solution of* (7.67), *then $\gamma_+ u \in H^{1/2}(\Gamma)$ is the unique solution of* (7.100). *Further, if $h = \gamma_+ u \in H^s(\Gamma)$ with $-1/2 < s \leq 0$, then $\gamma_+ u \in H^{s+1}(\Gamma)$.*

Although the combined potential integral equations (7.98) and (7.100) are the most common integral equation formulations for exterior Dirichlet and impedance scattering problems, other formulations are possible. One that is of particular interest for boundary element methods is the so-called *star-combined integral equation*, proposed for the exterior Dirichlet problem in the case when $\Omega_-$ is star-shaped with respect to an appropriately chosen origin in [77].

Specifically, if $u$ satisfies the exterior Dirichlet problem (7.66) with $h \in H^1(\Gamma)$, then for $\eta(\mathbf{x}) := k|\mathbf{x}| + \mathrm{i}(d-1)/2$, $\mathbf{x} \in \Gamma$, we have

$$\mathscr{A}_k \partial_\nu^+ u = (\mathbf{x} \cdot \left[\nu H_k + \nabla_\Gamma D_k - \tfrac{1}{2}\nabla_\Gamma\right] - \mathrm{i}\eta\left(-\tfrac{1}{2}I + D_k\right))h, \tag{7.102}$$

where the star-combined operator $\mathscr{A}_k : L^2(\Gamma) \to L^2(\Gamma)$ is defined by

$$\mathscr{A}_k := \mathbf{x} \cdot \nu\left(\tfrac{1}{2}I + D'_k\right) + \mathbf{x} \cdot \nabla_\Gamma S_k - \mathrm{i}\eta S_k. \tag{7.103}$$

It is shown in [77] that if $\Omega_-$ is star-shaped with respect to the origin, specifically if, for some $c > 0$,

$$\mathbf{x} \cdot \nu \geq c \quad \text{for almost all } \mathbf{x} \in \Gamma, \tag{7.104}$$

then $\mathscr{A}_k : L^2(\Gamma) \to L^2(\Gamma)$ is invertible with $\|\mathscr{A}_k^{-1}\|_{L^2(\Gamma)\leftarrow L^2(\Gamma)} \leq 2/c$. Indeed, $\mathscr{A}_k : L^2(\Gamma) \to L^2(\Gamma)$ is in fact coercive, as defined in Section 7.3.3 below, a much stronger property whose significance for numerical solution is explained in Section 7.3.3.

Turning to the Dirichlet screen problem (7.69), it follows easily from Theorem 7.2 that, where $\Gamma$ has the form (7.68) and $D := \mathbb{R}^d \setminus \overline{\Gamma}$, if $u \in W^1_{\mathrm{loc}}(D) \cap C^2(D)$ and, for some $k > 0$, $\Delta u + k^2 u = 0$ in $D$ and $u$ satisfies the Sommerfeld radiation condition (7.63) in $D$, then

$$-\mathscr{S}_k[\partial_\nu u](\mathbf{x}) + \mathscr{D}_k[u](\mathbf{x}) = u(\mathbf{x}), \quad \mathbf{x} \in D, \tag{7.105}$$

where $[u] := \gamma_+ u - \gamma_- u \in \tilde{H}^{1/2}(\Gamma)$ and $[\partial_\nu u] = \partial_\nu^+ u - \partial_\nu^- u \in \tilde{H}^{-1/2}(\Gamma)$ denote the jumps in $u$ and its normal derivative across the plane $x_d = 0$ containing $\Gamma$. Here the normal is directed in the $x_d$ direction and, for $s \in \mathbb{R}$, the "tilde" space $\tilde{H}^s(\Gamma)$ denotes the set of those $\phi \in H^s(\mathbb{R}^{d-1})$ (here we are identifying $\mathbb{R}^{d-1}$ with the plane containing $\Gamma$) that have support in $\overline{\Gamma}$. It is clear that the jumps in $u$ and its normal derivative across the plane $x_d = 0$, which are zero outside the screen, are in these "tilde" spaces. Thus if $u$ satisfies the Dirichlet screen problem, in which case $[u] = 0$ (see [20, Section 3.3]), it holds that

$$u(\mathbf{x}) = -\mathcal{S}_k[\partial_\nu u](\mathbf{x}), \quad \mathbf{x} \in D,$$

and, taking traces, that

$$S_k[\partial_\nu u](\mathbf{x}) = -h(\mathbf{x}), \quad \mathbf{x} \in \Gamma. \tag{7.106}$$

The operator $S_k$ in this equation, defined on the screen $\Gamma$ by (7.81), is bounded and invertible as an operator $S_k : \tilde{H}^{-1/2}(\Gamma) \to H^{1/2}(\Gamma)$; see [81, 20]. Indeed as noted in [29] (this result uses that $\Gamma$ is planar and is derived by applying the Fourier transform which diagonalizes $S_k$; see [20]), and noting that $H^s(\Gamma)$ is the dual space of $\tilde{H}^{-s}(\Gamma)$ (see [65, 20]), $S_k : \tilde{H}^{-1/2}(\Gamma) \to H^{1/2}(\Gamma)$, like the operator $\mathcal{A}_k$ introduced above, is coercive in the sense of the section entitled "The Galerkin Method" below, moreover with the $k$-dependence of the coercivity constants understood in each case.

We remark that it is often considered surprising that it is possible to write down coercive formulations of BVPs for the Helmholtz equation, whose BVPs are often considered to be "highly indefinite," at least in the high-frequency regime; see the discussion in [67].

If $u$ satisfies the transmission problem (7.70), then both representations (7.79) and (7.80) hold. Applying (7.85) and (7.86) we deduce that a linear operator equation of the form (7.97) holds for the unknown $v = [\gamma_+ u, \partial_\nu^+ u]^T \in H^{1/2}(\Gamma) \times H^{-1/2}(\Gamma)$, with

$$A = \begin{bmatrix} I + D_{k_-} - D_{k_+} & S_{k_+} - S_{k_-} \\ H_{k_-} - H_{k_+} & I + D'_{k_+} - D'_{k_-} \end{bmatrix}, \; f = \begin{bmatrix} -\frac{1}{2}h - D_{k_-}h + S_{k_-}g \\ -\frac{1}{2}g + D'_{k_-}g - H_{k_-}h \end{bmatrix}. \tag{7.107}$$

The operator $A$ is bounded and invertible as an operator on $H^{1/2}(\Gamma) \times H^{-1/2}(\Gamma)$, but also, adapting arguments of [82], as an operator on $H^1(\Gamma) \times L^2(\Gamma)$, and as an operator on $L^2(\Gamma) \times L^2(\Gamma)$ [50].

We conclude this section by stating precisely direct BIE formulations for the sound-soft, impedance, screen, and transmission scattering problems, (7.73), (7.74), (7.75), and (7.76). These follow immediately from the integral equation formulations (7.98) and (7.102) for the sound-soft scattering problem (7.73), and (7.106) for the screen problem (7.75), in each case applying these equations with $u$ replaced by $u^S$ and $h = -u^I|_\Gamma$; from the integral equation (7.100) for the impedance scattering problem, with $u$ replaced by $u^S$ and $h = -(\partial_\nu + ik\beta\gamma)u^I|_\Gamma$; and from (7.97) and (7.107) for the transmission problem (7.76), with $u$ replaced by $u^*$ defined as $u^S$ in $\Omega_+$ and $u$ in $\Omega_-$, and $h = -u^I|_\Gamma$, $g = -\partial_\nu u^I|_\Gamma$. But thanks to the special form of the boundary data $h$ and $g$, we can work with versions of these integral equations with simplified expressions for the inhomogeneous terms. Specifically (see, e.g., [19, Theorem 2.43]), in the case that $u^S$ satisfies the sound-soft scattering problem (7.73) it holds that $\partial_\nu^+ u \in L^2(\Gamma)$ and

$$u(\mathbf{x}) = u^I(\mathbf{x}) - \int_\Gamma \Phi_k(\mathbf{x}, \mathbf{y}) \partial_\nu^+ u(\mathbf{y})\, ds(\mathbf{y}), \quad \mathbf{x} \in \Omega_+; \tag{7.108}$$

similarly, if $u^S$ satisfies the screen scattering problem (7.75), then $[\partial_\nu u] \in \tilde{H}^{-1/2}(\Gamma)$ and

$$u(\mathbf{x}) = u^I(\mathbf{x}) - \int_\Gamma \Phi_k(\mathbf{x}, \mathbf{y}) [\partial_\nu u](\mathbf{y})\, ds(\mathbf{y}), \quad \mathbf{x} \in D. \tag{7.109}$$

In the case that $u^S$ satisfies the impedance scattering problem (7.74) it holds that $\gamma_+ u \in H^1(\Gamma)$ and

$$u(\mathbf{x}) = u^I(\mathbf{x}) + \int_\Gamma \left( \frac{\partial \Phi_k(\mathbf{x},\mathbf{y})}{\partial \nu(\mathbf{y})} + \mathrm{i}k\beta(\mathbf{y})\Phi_k(\mathbf{x},\mathbf{y}) \right) \gamma_+ u(\mathbf{y})\, ds(\mathbf{y}), \quad \mathbf{x} \in \Omega_+. \tag{7.110}$$

Finally, in the case that $u$ satisfies the transmission scattering problem (7.76) it holds that $\gamma_+ u = \gamma_- u \in H^1(\Gamma)$ and $\partial_\nu^+ u = \partial_\nu^- u \in L^2(\Gamma)$, and (e.g., [50])

$$u(\mathbf{x}) = u^I(\mathbf{x}) + \int_\Gamma \left( \frac{\partial \Phi_k(\mathbf{x},\mathbf{y})}{\partial \nu(\mathbf{y})} \gamma_+ u(\mathbf{y}) - \Phi_k(\mathbf{x},\mathbf{y}) \partial_\nu^+ u(\mathbf{y}) \right) ds(\mathbf{y}), \quad \mathbf{x} \in \Omega_+, \tag{7.111}$$

while (7.79) holds in $\Omega_-$.

The direct versions of the combined potential equations are then the integral equation (7.98) for the sound-soft problem (with $u$ replaced by $u^S$ and $h = -u^I|_\Gamma$) and (7.100) for the impedance scattering problem (with $u$ replaced by $u^S$ and $h = -(\partial_\nu + \mathrm{i}k\beta\gamma)u^I|_\Gamma$). Versions of these equations with simplified expressions for the inhomogeneous terms on the right-hand side are as follows (see, e.g., [19, Theorems 2.46, 2.47]). Suppose that $u^S$ satisfies the sound-soft scattering problem (7.73). Then $\partial_\nu^+ u \in L^2(\Gamma)$ satisfies the integral equation

$$A'_{k,\eta} \partial_\nu^+ u = f_{k,\eta} := [\partial_\nu^+ u^I - \mathrm{i}\eta u^I]|_\Gamma, \tag{7.112}$$

and this equation is uniquely solvable for all $k > 0$ if $\mathrm{Re}[\eta] \neq 0$. In the case that $\Omega_-$ is star-shaped with respect to the origin, satisfying (7.104) for some $c > 0$, and $\eta(\mathbf{x}) := k|\mathbf{x}| + \mathrm{i}(d-1)/2$, $\mathbf{x} \in \Gamma$, an alternative BIE formulation is

$$\mathscr{A}_k \partial_\nu^+ u = f_k := [\mathbf{x} \cdot \nabla u^I - \mathrm{i}\eta u^I]|_\Gamma, \tag{7.113}$$

with $\mathscr{A}_k$ defined by (7.103), and this equation is uniquely solvable for all $k > 0$. We will refer to (7.113) as the *star-combined* integral equation.

If $u^S$ satisfies the impedance scattering problem (7.67) and $\eta \in \mathbb{C}$, then $\gamma_+ u \in H^1(\Gamma)$ satisfies the integral equation

$$C_{k,\eta,\beta}\gamma_+ u = -\left[ \frac{\partial u^I}{\partial \nu} - \mathrm{i}\eta u^I \right]\Bigg|_\Gamma, \tag{7.114}$$

with $C_{k,\eta,\beta}$ defined by (7.101). If $\mathrm{Re}[\eta] \neq 0$, then $\gamma_+ u$ is the unique solution in $H^{1/2}(\Gamma)$ of this equation. If $u^S$ satisfies the screen scattering problem (7.75), then from (7.106) it follows that $[\partial_\nu u] \in \tilde{H}^{-1/2}(\Gamma)$ satisfies

$$S_k[\partial_\nu u] = u^I|_\Gamma. \tag{7.115}$$

Finally, if $u$ satisfies the transmission scattering problem (7.76), then (7.97) holds for $v = [\gamma_+ u, \partial_\nu^+ u]^T \in H^1(\Gamma) \times L^2(\Gamma)$ with $A$ given by (7.107) and $f = [u^I|_\Gamma, \partial_\nu u^I|_\Gamma]$ [50].

### 7.3.3 ▪ The Boundary Element Method

In this section we describe methods for the numerical solution of integral equations of the form (7.97), i.e.,

$$Av = f, \tag{7.116}$$

where $A : \mathscr{V} \to \mathscr{V}'$ is a linear boundary integral operator mapping some Hilbert space $\mathscr{V}$ continuously to its dual space $\mathscr{V}'$, or a linear combination of such operators and the identity, $v \in \mathscr{V}$ is the solution to be determined, and $f \in \mathscr{V}'$ is given data. Specifically, for the integral equation (7.112), we have $A = A'_{k,\eta}$, $\mathscr{V} = \mathscr{V}' = L^2(\Gamma)$, $v = \partial_\nu^+ u$, and $f = f_{k,\eta}$; for (7.113), we have $A = \mathscr{A}_k$, $\mathscr{V} = \mathscr{V}' = L^2(\Gamma)$, $v = \partial_\nu^+ u$, and $f = f_k$; for (7.114), we have $A = C_{k,\eta,\beta}$, $\mathscr{V} = H^{1/2}(\Gamma)$, $\mathscr{V}' = H^{-1/2}(\Gamma)$, $v = \gamma_+ u$, and $f = -\left[\frac{\partial u^I}{\partial \nu} - i\eta u^I\right]\Big|_\Gamma$; and for (7.115) we have $A = S_k$, $\mathscr{V} = \tilde{H}^{-1/2}(\Gamma)$, $\mathscr{V}' = H^{1/2}(\Gamma)$, $v = [\partial_\nu u]$, and $f = u_I|_\Gamma$. Finally, for the transmission scattering problem (7.76), $A$ is given by (7.107), we can take $\mathscr{V} = \mathscr{V}' = L^2(\Gamma) \times L^2(\Gamma)$, and $v = \left[\gamma_+ u, \partial_\nu^+ u\right]^T$, $f = \left[u^I|_\Gamma, \partial_\nu u^I|_\Gamma\right]$.

In order that the equation we are solving is well-posed, it is important that the boundary integral operator $A$ is invertible. In order to prove convergence of numerical schemes and deduce error estimates, additional properties of the operator are also required, as we discuss below.

The most commonly used methods for the solution of equations of the form (7.116) arising from scattering problems are the Galerkin, collocation, and Nyström schemes. For an operator equation of the form (7.116) the Galerkin method consists of choosing a finite-dimensional approximating space $\mathscr{V}_N \subset \mathscr{V}$ and then seeking an approximate solution $v_N \in \mathscr{V}_N$ such that

$$\langle A v_N, w_N \rangle = \langle f, w_N \rangle \quad \text{for all } w_N \in \mathscr{V}_N, \tag{7.117}$$

where, for $f \in \mathscr{V}'$, $w \in \mathscr{V}$, $\langle f, w \rangle$ denotes the action of the functional $f$ on $w$. In all of the examples we consider below, this duality pairing is just the $L^2$ inner product. In terms of practical implementation, this requires two integrations (corresponding to the integral operator $A$ and the duality pairing on the left-hand side); to avoid this, the collocation or Nyström method is often preferred. For an operator equation of the form (7.116), with $A : L^\infty(\Gamma) \to L^\infty(\Gamma)$, the collocation method consists of choosing a finite-dimensional approximating space $\mathscr{V}_N$, and a set of $N$ distinct node points $x_1, \ldots, x_N \in \Gamma$, and then seeking an approximate solution $v_N \in \mathscr{V}_N$ such that

$$A v_N(x_i) = f(x_i) \text{ for } i = 1, \ldots, N. \tag{7.118}$$

The Nyström method is even simpler—it consists of just discretizing the integral operator in (7.116) directly, via a quadrature rule, i.e., solving

$$A_N v_N = f, \tag{7.119}$$

where $A_N$, $N = 1, 2, \ldots$, represents a convergent sequence of numerical integration operators (see, e.g., [60, Chapter 12]).

Although the Nyström and collocation methods are both simpler to implement than Galerkin methods, we focus on the Galerkin method here. One part of the rationale for this choice is that a key step (and our major focus below) in designing both Galerkin and collocation methods is designing subspaces $\mathscr{V}_N$ that can approximate the solution accurately, with a relatively low number of degrees of freedom $N$. Everything we say below about designing $\mathscr{V}_N$ for the Galerkin method applies equally to the collocation method, and indeed to other numerical schemes where we select the numerical solution from an approximating subspace. The second part of our rationale is that, for collocation and Galerkin and other related methods, choosing a subspace from which we select the numerical solution is only part of the story. We also have to design our numerical scheme so that the numerical solution selected is "reasonably close" to the best possible approximation from the subspace. It is only for the Galerkin method that any analysis tools have been developed that can guarantee that this is the case for the hybrid numerical-asymptotic schemes we will describe in the section entitled "Hybrid Numerical-Asymptotic BEMs for High-Frequency Problems" below, leading to guaranteed convergence and error bounds, at least for some problems and geometries, that we discuss there. (In fairness we should note that, while it is not known how to carry out the complete error analysis needed, numerical experiments suggest that both the Nyström [12, 48, 14] and collocation [4] hybrid methods can be as effective as Galerkin methods for numerical solution of certain high-frequency scattering problems.) The last part of our rationale is that, solving many of the implementation issues for users, open source Galerkin BEM software is becoming available, for example, the boundary element software package BEM++ that exclusively uses Galerkin methods, which we discuss a little more in the section entitled "BEM with Standard (Piecewise Polynomial) Approximation Spaces" below.

### The Galerkin Method

In this section we present a general framework for the solution of (7.116) by the Galerkin method, and we begin by writing it in weak form as follows: find $v \in \mathcal{V}$ such that

$$\langle Av, w\rangle = \langle f, w\rangle \quad \text{for all } w \in \mathcal{V}. \tag{7.120}$$

The Galerkin method for approximating (7.120) then seeks a solution $v_N \in \mathcal{V}_N \subset \mathcal{V}$, where $\mathcal{V}_N$ is a finite-dimensional subspace, requiring that

$$\langle Av_N, w_N\rangle = \langle f, w_N\rangle \quad \text{for all } w_N \in \mathcal{V}_N. \tag{7.121}$$

The standard analysis of Galerkin methods assumes that the operator $A : \mathcal{V} \to \mathcal{V}'$ is injective and takes the form $A = B + C$, where $C : \mathcal{V} \to \mathcal{V}'$ is compact and $B : \mathcal{V} \to \mathcal{V}'$ is coercive, by which we mean that, for some $\alpha > 0$ (the *coercivity constant*),

$$|\langle Bv, v\rangle| \geq \alpha \|v\|_{\mathcal{V}}^2 \quad \text{for all } v \in \mathcal{V}. \tag{7.122}$$

In this case we say that $A$ is a compactly perturbed coercive operator and the following theorem holds (see, e.g., [80, Theorem 8.11]).

**Theorem 7.5.** *Suppose the bounded linear operator $A : \mathcal{V} \to \mathcal{V}'$ is a compactly perturbed coercive operator and is injective. Suppose moreover that $(\mathcal{V}_N)_{N\in\mathbb{N}}$ is a sequence of approximation spaces converging to $\mathcal{V}$ in the sense that*

$$\inf_{w_N \in \mathcal{V}_N} \|w - w_N\|_{\mathcal{V}} \to 0 \quad \text{as } N \to \infty$$

*for every $w \in \mathcal{V}$. Then there exists $N_0 \in \mathbb{N}$ and $C > 0$ such that, for $N \geq N_0$, (7.121) has exactly one solution $v_N \in \mathcal{V}_N$, which satisfies the quasi-optimal error estimate*

$$\|v - v_N\|_{\mathcal{V}} \leq C \inf_{w_N \in \mathcal{V}_N} \|v - w_N\|_{\mathcal{V}}. \tag{7.123}$$

This theorem is relevant to the combined potential integral equation (7.114) for the impedance scattering problem, for general Lipschitz $\Gamma$, since $C_{k,\eta,\beta} : H^{1/2}(\Gamma) \to H^{-1/2}(\Gamma)$ is a compactly perturbed coercive operator and is also injective for all $k > 0$ if $\mathrm{Re}[\eta] \neq 0$ (see [65, Theorem 7.8] and [19, Theorem 2.27]). Theorem 7.5 also applies to the integral equation (7.112) for the sound-soft scattering problem if $\Gamma$ is $C^1$ and also for the 2D case if $\Gamma$ is Lipschitz and a curvilinear polygon (see [19, Section 2.11] for details). However, it cannot be applied directly to (7.112) for general Lipschitz $\Gamma$, for which case it is an open question whether $A'_{k,\eta}$ is a compactly perturbed coercive operator. Introducing additional operators into the equations it is possible to formulate BIEs for this problem that do satisfy the conditions required by Theorem 7.5, but the downside of this is that this approach requires additional computational effort. For a survey of recent progress in this area we refer to [19, Sections 2.11 and 5.7]; see also [15, 16, 41, 42].

However, even in cases where it can be applied, the classical theory does not tell us how $C$ and $N_0$ depend on $k$. Moreover, suppose that we solve (7.116) using the Galerkin method (7.117) on a family of, for example, $N$-dimensional spaces $\mathscr{V}_N$ of piecewise polynomial functions of fixed degree on a quasi-uniform sequence of meshes on $\Gamma$ with diameter $h \to 0$ (so that $N$ is proportional to $h^{1-d}$). Then, for such standard piecewise polynomial approximation spaces, the best approximation error $\inf_{w_N \in \mathscr{V}_N} \|v - w_N\|_{\mathscr{V}}$ will be highly $k$-dependent. For example, if $v$ is $p+1$ times continuously differentiable on each mesh interval, then standard estimates for piecewise polynomial approximation of degree $p$ suggest that we might be able to bound the difference between $v$ and its best polynomial approximation $p_N$ on each interval by

$$\|v - p_N\|_{\mathscr{V}} \leq C h^{p+1} \|v^{(p+1)}\|_{\mathscr{V}} \tag{7.124}$$

for some constant $C > 0$ (which may depend on $p$). But since $v$ solving (7.116) represents either $\partial_\nu^+ u$ or $\gamma_+ u$, where $u$ solves (7.60), it follows that $v$ will be oscillatory, with the $j$th derivative of $v$ being, in general, of order $k^j$. In this case, (7.124) becomes

$$\|v - p_N\|_{\mathscr{V}} \leq C(hk)^{p+1}, \tag{7.125}$$

with this estimate carrying over naturally to the right-hand side of (7.123), i.e.,

$$\inf_{w_N \in \mathscr{V}_N} \|v - w_N\|_{\mathscr{V}} \leq C(hk)^{p+1}. \tag{7.126}$$

Thus $h$ needs to decrease with $\mathscr{O}(k^{-1})$, and possibly faster, just to keep the error bounded as $k \to \infty$. Hence, standard (piecewise) polynomial BEMs applied directly to approximate the oscillatory solution of scattering problems will have complexity of at least $\mathscr{O}(h^{1-d}) = \mathscr{O}(k^{d-1})$ as $k \to \infty$ (see, e.g., [49] for further details).

In the case that the integral operator $A$ in (7.116) is bounded and coercive, we can follow a slightly different approach. In this case, the Lax–Milgram lemma guarantees that (7.120) has exactly one solution $v \in \mathscr{V}$ for every $w \in \mathscr{V}$, with

$$\|v\|_{\mathscr{V}} \leq \alpha^{-1} \|f\|_{\mathscr{V}'}, \tag{7.127}$$

and existence of the Galerkin solution and quasi optimality is guaranteed by Céa's lemma (see, e.g., [19, Lemma 2.48]).

**Lemma 7.6. Céa's lemma.** *Suppose that the bounded linear operator $A : \mathscr{V} \to \mathscr{V}'$ is coercive (i.e., it satisfies (7.122) for some $\alpha > 0$), and moreover that, for some constant $C > 0$, $|\langle Av, w\rangle| \leq C\|v\|_{\mathscr{V}}\|w\|_{\mathscr{V}}$ for all $v, w \in \mathscr{V}$. Then (7.121) has exactly one solution $v_N \in \mathscr{V}_N$, which satisfies*

$$\|v - v_N\|_{\mathscr{V}} \leq \frac{C}{\alpha} \inf_{w_N \in \mathscr{V}_N} \|v - w_N\|_{\mathscr{V}}. \tag{7.128}$$

This theorem is relevant to the star-combined formulation (7.113) for sound-soft scattering by star-shaped obstacles, to the BIE (7.115) for the screen scattering problem, and also to the standard combined potential formulation (7.112) for a certain range of geometries (see [79] for details). It is sometimes presumed, since the standard domain-based variational formulations of BVPs for the Helmholtz equation are standard examples of indefinite problems where coercivity does not hold, at least for sufficiently large $k$, that the same should hold true for weak formulations arising via integral equation formulations. However, recent results, discussed in [19, Section 5] (see also [77, 10, 9, 79, 20]), show that coercivity holds for these BIEs for a range of geometries, with $\alpha$ bounded away from zero for all sufficiently large $k$, and moreover with the $k$-dependence of $\alpha$ and $C$ in (7.128) explicitly known in many cases.

The advantage of using Lemma 7.6, as opposed to Theorem 7.5, is that if the $k$-dependence of the continuity and coercivity constants is known, then everything in (7.128) is $k$-explicit, as opposed to the unknown $k$-dependence of the constants $C$ and $N_0$ in (7.123). In either case, though, the question of how the best approximation estimate $\inf_{w_N \in \mathscr{V}_N} \|v - w_N\|$ depends on $k$ is crucial.

In the next section we briefly mention recent progress on standard schemes, with piecewise polynomial approximation spaces, where the best approximation estimate may grow with $k$ but other ideas can be applied to improving efficiency and accuracy even at large frequencies. These approaches have the advantage that little need be known a priori about the behavior of the solution. On the other hand, if one can understand the high-frequency asymptotic behavior of the solution, then one can design one's approximation space accordingly in order to efficiently represent the oscillatory solution at high frequencies. This is the idea behind the hybrid numerical-asymptotic approach, which we describe more fully in the section entitled "Hybrid Numerical-Asymptotic BEMs for High-Frequency Problems."

### BEM with Standard (Piecewise Polynomial) Approximation Spaces

As alluded to above, BEMs with standard piecewise polynomial approximation spaces suffer from the restriction that the number of degrees of freedom required to achieve a prescribed level of accuracy will grow rapidly (at least with $O(k)$ for 2D and $O(k^2)$ for 3D problems) as $k$ increases. (Recent progress in the analysis of the standard BEM for the Helmholtz equation, teasing out explicitly the dependence of this error on the wavenumber, is reported in [64, 49].) As BEMs lead to linear systems with dense matrices, this can render the computational cost (in terms of both speed and memory) of standard linear solvers impractical when $k$ is very large.

To ease this problem, much effort has been put into developing preconditioners (see, e.g., [52, 13, 55]), efficient iterative solvers (see, e.g., [2, 26, 43]), fast multipole methods (see, e.g., [30, 38, 31, 25, 69]), and matrix compression techniques (see, e.g., [7, 8]) for Helmholtz and related problems.

These advances have enabled the solution of larger and larger problems, but a common argument put forward against the widespread usage of BEMs is the added investment required by the user to initiate code development incorporating such features, certainly in comparison to more intuitive and less computationally complicated finite difference and particularly finite element methods. Moreover, whereas there exist many widely available computational resources for efficient development of finite element software, until recently comparable resources have been somewhat lacking for boundary element computations. To redress this balance, there has been much recent activity on the development of software packages specifically for BEMs. The key aim of much of this software is to make many of the recent advances outlined above accessible to the more casual user.

We mention in particular the package BEM++, which has recently been developed for the solution of a range of 3D linear elliptic PDEs, including the Helmholtz equation (see [75] for details). BEM++ utilizes $hp$-Galerkin schemes, with low-order approximation spaces (piecewise constant or piecewise linear continuous spaces), or else allows higher-order polynomials on flat triangles. Moreover, BEM++ implements many of the developments listed above within a single framework. For further details we refer to [75]. Many other codes have also been developed in recent years, for example, BETL [56] and HyENA [66]. For more details, we refer to the review of this topic in [75].

### Hybrid Numerical-Asymptotic BEMs for High-Frequency Problems

In this section we describe recent progress in the design, analysis, and implementation of hybrid numerical-asymptotic (HNA) BEMs for BVPs for the Helmholtz equation that model time-harmonic acoustic wave scattering. As alluded to above, the problem (7.60) has solutions that oscillate in space with wavelength $\lambda = 2\pi/k$ (e.g., the plane waves $u(\mathbf{x}) = \exp(ik\mathbf{x}.\hat{\mathbf{a}})$, where $\hat{\mathbf{a}} \in \mathbb{R}^d$ is a unit vector, are solutions). Since the number of oscillations of linear combinations of such solutions will in general grow linearly with $k$ (in each direction), the number of degrees of freedom required to represent this oscillatory solution by conventional (piecewise polynomial) boundary elements must also grow with order $k^{d-1}$. This lack of robustness with respect to increasing values of $k$ (which puts many problems of practical interest beyond the reach of standard algorithms) is the motivation behind the development of HNA algorithms.

The key idea of the HNA approach is to exploit, within the design of the numerical method, detailed information about the oscillations of the solution, based on advances in the understanding of the high-frequency behavior of solutions to the Helmholtz equation. Known highly oscillatory components of $v$ solving (7.116) (with, for example, $v = \partial_\nu^+ u$ or $v = \gamma_+ u$, where $u$ solves (7.60)) are treated exactly in the algorithm, leaving only more slowly varying components to be approximated by piecewise polynomials; i.e., we combine conventional piecewise polynomial approximations with high-frequency asymptotics to build basis functions suitable for representing the oscillatory solutions. This yields methods which require very many fewer degrees of freedom as $k \to \infty$.

More specifically, we approximate the ($k$-dependent) solution $v$ using (in general) an ansatz of the form

$$v(\mathbf{x}) \approx V_0(\mathbf{x},k) + \sum_{m=1}^{M} V_m(\mathbf{x},k)\exp(ik\psi_m(\mathbf{x})), \quad \mathbf{x} \in \Gamma. \tag{7.129}$$

In this representation, $V_0$ is a known (generally oscillatory) function, the phases $\psi_m$ are chosen a priori, and the amplitudes $V_m$, $m = 1,\dots,M$, are approximated numerically. The idea (and in many cases this can be rigorously proven) is that if the phases are carefully chosen, then $V_m(\cdot,k)$, $m = 1,\dots,M$, will be much less oscillatory than $v$ and so can be better approximated by piecewise polynomials than $v$ itself.

The crucial first step then lies in choosing $V_0$ and $\psi_m$, $m = 1,\dots,M$, appropriately, and the exact way in which this is done depends on the geometry and boundary conditions of the problem under consideration. Section 3 of the recent review article [19] provides a historical survey before explaining the key ideas for a number of examples, specifically scattering by smooth convex obstacles (see also [12, 37, 58, 47]), an impedance half plane (see also [24, 61]), sound-soft convex polygons (see also [22, 54]), convex curvilinear polygons (see also [62]), convex impedance polygons (see also [23]), nonconvex polygons (see also [21]), and multiple scattering configurations (see also [48, 39, 3]).

Here, we describe some more recent advances, overlapping only in a fairly minor way with the examples described in [19]. We begin by illustrating the HNA scheme in the context of a problem of scattering by a single sound-soft screen (equivalently, scattering by an aperture in a single sound-hard screen; see [53] for details), then describing how this idea extends to scattering by sound-soft convex polygons, outlining the added difficulties that arise in the case that the obstacle is nonconvex (presenting sharper results in terms of $k$-dependence than those outlined in [19]), and finally considering scattering by penetrable polygons (the transmission problem). We demonstrate that HNA methods have the potential to solve scattering problems accurately in a computation time that is (almost) independent of frequency.

**Scattering by Screens.** To get the main ideas across, we first describe an HNA BEM for a simple 2D geometry, scattering by a single planar sound-soft screen. This problem, indeed the more general problem of scattering by an arbitrary collinear array of such screens, has been treated by HNA BEM methods with a complete numerical analysis in [53].

To be precise then, we consider the 2D problem of scattering of the time-harmonic incident plane wave (7.72) by a sound-soft screen

$$\Gamma := \{(x_1, 0) \in \mathbb{R}^2 : 0 < x_1 < L\}, \tag{7.130}$$

where $L$ is the length of the screen and $\mathbf{x} = (x_1, x_2) \in \mathbb{R}^2$, i.e., we solve the BVP (7.75) with $d = 2$. An example of our scattering configuration, for $k = 5$ and for $k = 20$, is shown in Figure 7.8. The increased oscillations for $k = 20$ can be clearly seen. The solution of this BVP satisfies (7.109), and hence our problem reduces to solving the BIE (7.115).

**Figure 7.8.** Re[$u$], *scattering by a screen of length* $L = 2\pi$, $k = 5$ *(left) and* $k = 20$ *(right).*

The HNA method for solving (7.115) uses an approximation space that is specially adapted to the high-frequency asymptotic behavior of the solution $[\partial_\nu u]$ on $\Gamma$, which we now consider. Representing a point $\mathbf{x} \in \Gamma$ parametrically by $\mathbf{x}(s) := (s, 0)$, where $s \in (0, L)$, the following theorem is proved in [53]. (This is derived directly from (7.75) using an elementary representation for the solution in the half plane above the screen in terms of a Dirichlet half plane Green's function. For details see [53] and cf. [22, Theorem 3.2 and Corollary 3.4] and [54, Section 3].)

**Theorem 7.7.** *Suppose that* $k \geq k_0 > 0$. *Then*

$$[\partial_\nu u](\mathbf{x}(s)) = \Psi(\mathbf{x}(s)) + v^+(s)\mathrm{e}^{iks} + v^-(L-s)\mathrm{e}^{-iks}, \qquad s \in (0, L), \tag{7.131}$$

*where* $\Psi := 2\partial u^I / \partial \nu$, *and the functions* $v^\pm(s)$ *are analytic in the right half plane* $\mathrm{Re}[s] > 0$, *where they satisfy the bound*

$$|v^\pm(s)| \leq C_1 \mathcal{M} k\, |ks|^{-\frac{1}{2}},$$

*where*

$$\mathcal{M} := \sup_{\mathbf{x}\in D} |u(\mathbf{x})| \le C(1+k),$$

*and the constants $C, C_1 > 0$ depend only on $k_0$ and $L$.*

The representation (7.131) is of the form (7.129), with $V_0(\mathbf{x}(s),k) = \Psi(\mathbf{x}(s))$, $M = 2$, $V_1(\mathbf{x}(s),k) = v^+(s)$, $V_2(\mathbf{x}(s),k) = v^-(s)$, $\psi_1(\mathbf{x}(s)) = s$, and $\psi_2(\mathbf{x}(s)) = -s$. Using this representation we can design an appropriate approximation space $V_{N,k}$ to represent

$$\varphi(s) := \frac{1}{k}([\partial_\nu u](\mathbf{x}(s)) - \Psi(\mathbf{x}(s))), \quad s \in (0,L). \tag{7.132}$$

Here $N$ denotes the total number of degrees of freedom in the method, and the subscript $k$ in $V_{N,k}$ serves as a reminder that the hybrid approximation space depends explicitly on the wavenumber $k$. The function $\varphi$, which we seek to approximate, can be thought of as the scaled difference between $[\partial_\nu u]$ and its "physical optics" approximation $\Psi$ (which represents the direct contribution of the incident and reflected waves), with the $1/k$ scaling ensuring that $\varphi$ is nondimensional, and reflecting that $[\partial_\nu u] = O(k)$ as $k \to \infty$. The second and third terms on the right-hand side of (7.131) represent the diffracted rays emanating from the ends of the screen at $s = 0$ and at $s = L$, respectively. As alluded to earlier, instead of approximating $\varphi$ directly by conventional piecewise polynomials we instead use the representation (7.131) with $v^+(s)$ and $v^-(L-s)$ replaced by piecewise polynomials supported on overlapping geometric meshes, graded toward the singularities at $s = 0$ and $s = L$, respectively. We proceed by describing our mesh, which is illustrated in Figure 7.9.

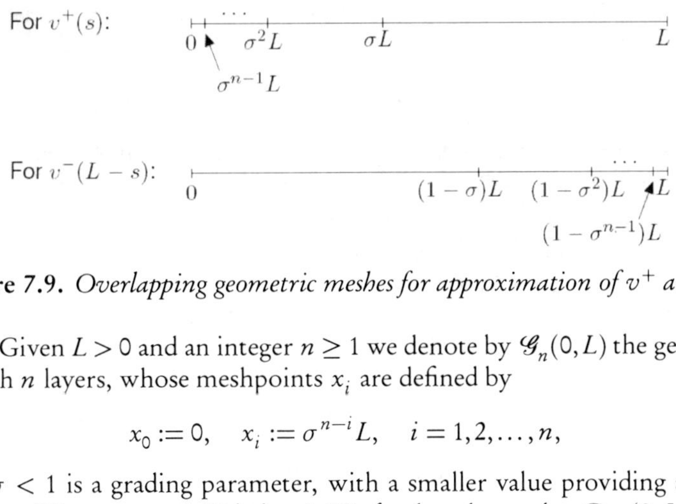


**Figure 7.9.** *Overlapping geometric meshes for approximation of $v^+$ and $v^-$.*

**Definition.** Given $L > 0$ and an integer $n \ge 1$ we denote by $\mathcal{G}_n(0,L)$ the geometric mesh on $[0,L]$ with $n$ layers, whose meshpoints $x_i$ are defined by

$$x_0 := 0, \quad x_i := \sigma^{n-i}L, \quad i = 1,2,\dots,n,$$

where $0 < \sigma < 1$ is a grading parameter, with a smaller value providing a more severe mesh grading. (We use $\sigma = 0.15$ below.) We further denote by $\mathcal{P}_{p,n}(0,L)$ the space of piecewise polynomials of order $p$ on the geometric mesh $\mathcal{G}_n(0,L)$, i.e.,

$$\mathcal{P}_{p,n}(0,L) := \left\{\rho : [0,L] \to \mathbb{C} : \rho|_{(x_{i-1},x_i)} \text{ is a polynomial of degree} \le p\right\}.$$

We are now in a position to define the approximation space $V_{N,k}$. Let

$$V^+_{p,n} := \left\{\rho(s)e^{iks} : \rho \in \mathcal{P}_{p,n}(0,L)\right\},$$
$$V^-_{p,n} := \left\{\rho(L-s)e^{-iks} : \rho \in \mathcal{P}_{p,n}(0,L)\right\},$$

with which we approximate, respectively, the terms $v^+(s)e^{iks}$ and $v^-(L-s)e^{-iks}$ in the representation (7.131). We choose values for $p$ and $n$ and then define

$$V_{N,k} := \text{span}\left\{V_{p,n}^+ \cup V_{p,n}^-\right\}, \tag{7.133}$$

which has a total number of degrees of freedom $N = 2n(p+1)$. The following best approximation result is shown in [53, Theorem 5.1].

**Theorem 7.8.** *Let $n$ and $p$ satisfy $n \geq cp$ for some constant $c > 0$ and suppose that $k \geq k_0 > 0$. Then, there exist constants $C, \tau > 0$, dependent only on $k_0$, $L$, and $c$, such that*

$$\inf_{w_N \in V_{N,k}} \|\varphi - w_N\|_{\tilde{H}^{-\frac{1}{2}}(\Gamma)} \leq Cke^{-p\tau}.$$

Identifying $\Gamma$ with $(0,L)$, here $\tilde{H}^{-1/2}(\Gamma) = \tilde{H}^{-1/2}(0,L) \subset H^{-1/2}(\mathbb{R})$, $\tilde{H}^{-1/2}(0,L)$ is just the subspace of those $\psi \in H^{-1/2}(\mathbb{R})$ that have support in $[0,L]$. And then $\|\cdot\|_{\tilde{H}^{-1/2}(\Gamma)}$ is just the standard norm on the Sobolev space $H^{-1/2}(\mathbb{R})$ (see, e.g., [65]).[8]

Having designed an appropriate approximation space $V_{N,k}$, we use a Galerkin method to select an element so as to efficiently approximate $\varphi$. That is, we seek $\varphi_N \in V_{N,k}$ such that

$$\langle S_k\varphi_N, w_N\rangle_\Gamma = \frac{1}{k}\left\langle u^I - S_k\Psi, w_N\right\rangle_\Gamma \quad \text{for all } w_N \in V_{N,k}, \tag{7.134}$$

where the duality pairings in (7.134) are simply $L^2(\Gamma)$ inner products. The following error estimate follows from [53, Theorem 6.1].

**Theorem 7.9.** *If the assumptions of Theorem 7.8 hold, then there exist constants $C, \tau > 0$, dependent only on $k_0$, $L$, and $c$, such that*

$$\|\varphi - \varphi_N\|_{\tilde{H}^{-1/2}(\Gamma)} \leq Ck^{3/2}e^{-p\tau}. \tag{7.135}$$

Note that if $n$ is chosen proportional to $p$, precisely so that $c_2p \geq n \geq c_1p$, for some constants $c_2 > c_1 > 0$, then the total number of degrees of freedom $N = O(p^2)$, and Theorem 7.9 implies that we can achieve any required accuracy with $N$ growing like $\log^2 k$ as $k \to \infty$, rather than like $k$ as for a standard BEM.

We now present numerical results for the solution of (7.134). The screen is of length $L = 2\pi$, and hence our scatterer is $k$ wavelengths long (recall that $\lambda = 2\pi/k$). The angle of incidence is $\pi/6$ measured anticlockwise from the downward vertical, as in Figure 7.8. In our experiments we choose $n = 2(p+1)$, so that the total number of degrees of freedom is $N = 2n(p+1) = 4(p+1)^2$. Since the total number of degrees of freedom depends only on $p$, we adjust our notation by defining $\psi_p(s) := \varphi_N(s)$, and we take the "reference" solution to be $\psi_7$. In Figure 7.10 we plot $|\psi_7| \approx |\varphi|$ for $k = 20$ and for $k = 10240$. The singularities at the edge of the screen can be clearly seen, as can the increased oscillations for larger $k$ (the apparently shaded area is an artifact of the rapidly oscillating solution).

In Figure 7.11 we plot on a logarithmic scale the relative $L^1$ errors

$$\frac{\left\|\psi_7 - \psi_p\right\|_{L^1(\Gamma)}}{\|\psi_7 + \Psi/k\|_{L^1(\Gamma)}}, \tag{7.136}$$

[8] We note that the bounds in [53] are in fact in terms of a natural wavenumber-dependent Sobolev norm $\|\cdot\|_{\tilde{H}_k^{-1/2}(\Gamma)}$, but it is an easy calculation, this feeding into the results we state here, that $\|\cdot\|_{\tilde{H}_k^{-1/2}(\Gamma)} \leq \max(1, k^{1/2})\|\cdot\|_{\tilde{H}^{-1/2}(\Gamma)}$.

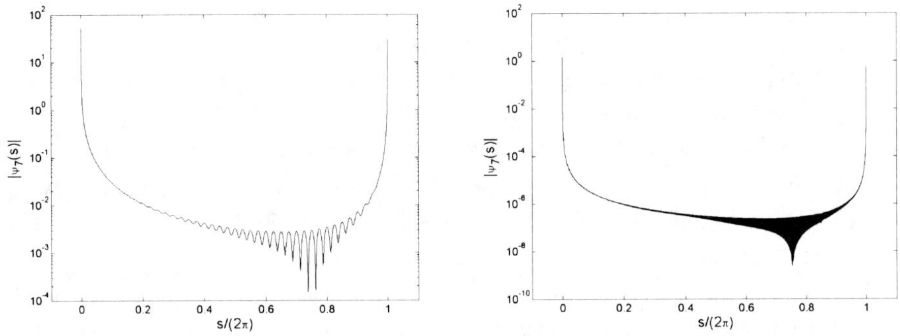

**Figure 7.10.** *The boundary solution* $|\psi_7| \approx |\varphi|$, *as given by* (7.132), *for* $k = 20$ *(left) and* $k = 10240$ *(right), scattering by a screen.*

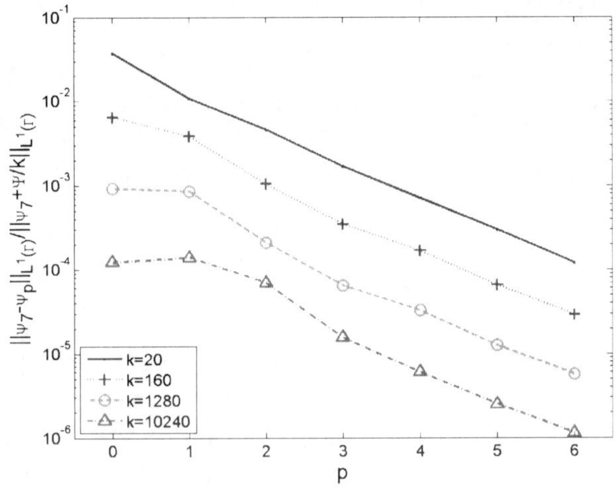

**Figure 7.11.** *Relative errors in our approximation to* $[\partial_\nu u]$, *scattering by a screen.*

against $p$ for a range of $k$. This is the relative error in $L^1(\Gamma)$-norm in our approximation, $\psi_p + \Psi/k$, to the dimensionless quantity $\frac{1}{k}[\partial_\nu u]$ (recall (7.132)). (We calculate $L^1$-rather than $\tilde{H}^{-1/2}(\Gamma)$ norms for simplicity of computation.) The $L^1$-norms are calculated using a high-order Gaussian quadrature routine on a mesh graded toward the endpoint singularities. The linear plots demonstrate exponential decay as the polynomial degree, $p$, increases, as predicted in Theorem 7.9. More significantly, the relative error decreases as $k$ increases for fixed $p$. This behavior is better than might be expected from Theorem 7.9, and in particular suggests that the error estimate in Theorem 7.9 is not sharp in its $k$-dependence. Further results, including calculations of errors measured in an $||\cdot||_{\tilde{H}^{-1/2}(\Gamma)}$-norm, computations for an array of collinear screens, and computations of the solution in the domain and the far field pattern, can be found in [53].

**Scattering by Convex Polygons.** The ideas outlined above for the screen problem can also be applied to the case of scattering by polygons. First, we consider the 2D problem of scattering of the time-harmonic incident plane wave (7.72) by a sound-soft polygon with boundary $\Gamma$, i.e., problem (7.73). The solution $u$ then has the representation (7.108), where $\partial_\nu^+ u$ satisfies (7.112) and, if $\Omega_-$ is star-shaped, (7.113).

We denote the number of sides of the polygon by $n_s$ and the corners (labeled in order counterclockwise) by $P_j$, $j = 1,\dots,n_s$. We set $P_{n_s+1} := P_1$, and then for $j = 1,\dots,n_s$ we denote the side of the polygon connecting the corners $P_j$ and $P_{j+1}$ by $\Gamma_j$, with length $L_j$. We denote the exterior angle at the corner $P_j$ by $\omega_j$, $j = 1,\dots,n_s$. We say that a side $\Gamma_j$ is illuminated by the incident plane wave given by (7.72) if $\hat{\mathbf{a}}\cdot\nu < 0$ on $\Gamma_j$ and is in shadow if $\hat{\mathbf{a}}\cdot\nu \geq 0$ on $\Gamma_j$.

As for the screen problem, the HNA method for solving (7.112) or (7.113) uses an approximation space that is specially adapted to the high-frequency asymptotic behavior of $\partial_\nu^+ u$ on $\Gamma$. The nature of this behavior is highly dependent on the geometry of the polygon: in particular, we consider convex and nonconvex polygons separately, considering first the case that the polygon is convex and treating the nonconvex case in the next subsection.

Denoting the distance of $\mathbf{x} \in \Gamma_j$ from $P_j$ by $s$, the following theorem follows from [54, Theorems 3.2 and 4.3].

**Theorem 7.10.** *Let $k \geq k_0 > 0$. Then on any side $\Gamma_j$*

$$\partial_\nu^+ u\,(\mathbf{x}(s)) = \Psi\,(\mathbf{x}(s)) + v_j^+(s)\mathrm{e}^{iks} + v_j^-\left(L_j - s\right)\mathrm{e}^{-iks}, \qquad \mathbf{x}(s) \in \Gamma_j, \tag{7.137}$$

*where*

$$\Psi := \begin{cases} 2\partial u^I/\partial\nu & \text{if } \Gamma_j \text{ is illuminated}, \\ 0 & \text{if } \Gamma_j \text{ is in shadow} \end{cases} \tag{7.138}$$

*and the functions $v_j^\pm(s)$ are analytic in the right half plane* $\mathrm{Re}[s] > 0$, *with*

$$|v_j^\pm(s)| \leq C \begin{cases} k^{3/2}\log^{1/2}(2+k)|ks|^{-\delta_j^\pm}, & 0 < |s| \leq 1/k, \\ k^{3/2}\log^{1/2}(2+k)|ks|^{-1/2}, & |s| > 1/k, \end{cases} \tag{7.139}$$

*where $\delta_j^\pm \in (0,1/2)$ are given by $\delta_j^+ := 1 - \pi/\omega_j$ and $\delta_j^- := 1 - \pi/\omega_{j+1}$, and the constant $C > 0$ depends only on $k_0$ and $\Gamma$.*

Using the representation (7.137) on each side of the polygon leads to a representation of the form (7.129) with $V_0(\mathbf{x}(s),k) = \Psi(\mathbf{x}(s))$, $M = 2n_s$, $V_{2j-1}(\mathbf{x}(s),k) = v_j^+(s)$ if $\mathbf{x}(s) \in \Gamma_j$ and zero otherwise, $V_{2j}(\mathbf{x}(s),k) = v_j^-(s)$ if $\mathbf{x}(s) \in \Gamma_j$ and zero otherwise, $\psi_{2j-1}(\mathbf{x}(s)) = \mathbf{x}(s)\cdot\hat{d}_j$, and $\psi_{2j}(\mathbf{x}(s)) = -\mathbf{x}(s)\cdot\hat{d}_j$, $j = 1,\dots,n_s$, where $\hat{d}_j$ is a unit vector parallel to $\Gamma_j$. Our approximation space $\tilde{V}_{N,k}$ for

$$\varphi(s) := \frac{1}{k}\left(\partial_\nu^+ u\,(\mathbf{x}(s)) - \Psi\,(\mathbf{x}(s))\right), \quad \mathbf{x}(s) \in \Gamma, \tag{7.140}$$

is then identical on each side of the polygon to that for the screen (7.133), in effect replacing $v_j^+(s)$ and $v_j^-(L_j - s)$ by piecewise polynomials supported on overlapping geometric meshes, graded toward the singularities at $s = 0$ and $s = L_j$, respectively. As for the screen, the function $\varphi$, which we seek to approximate, can be thought of as the scaled difference between $\partial_\nu^+ u$ and its physical optics approximation $\Psi$, which again represents the direct contribution of the incident and reflected waves (when they are present). The second and third terms in (7.137) represent the diffracted rays emanating from the corners $P_j$ and $P_{j+1}$, respectively.

The following best approximation estimate follows from [54, Theorems 4.3 and 5.5].

**Theorem 7.11.** *If $c, k_0 > 0$ and $n \geq cp$, $k \geq k_0$, then, for some $C, \tau > 0$, depending only on $c$, $k_0$, and $\Gamma$,*

$$\inf_{w_N \in \tilde{V}_{N,k}} \|\partial_\nu^+ u - w_N\|_{L^2(\Gamma)} \leq C k^{1/2+\alpha} \log^{1/2}(2+k) e^{-p\tau}, \tag{7.141}$$

*where $\alpha := 1 - \min_{m=1,\ldots,n_s}(1-\pi/\omega_m) \in (1/2, 1)$.*

Having designed an appropriate approximation space $\tilde{V}_{N,k}$ we use the Galerkin method to select an element to approximate $\varphi$. Since convex polygons are star-shaped, in this case we can use the integral equation formulation (7.113); i.e., we seek $\varphi_N \in \tilde{V}_{N,k}$ such that

$$\langle \mathscr{A}_k \varphi_N, w_N \rangle_\Gamma = \frac{1}{k} \langle f_k - \Psi, w_N \rangle_\Gamma \quad \text{for all } w_N \in \tilde{V}_{N,k}. \tag{7.142}$$

Thanks to the coercivity of the integral operator $\mathscr{A}_k$, we have the following error estimate (cf. [54, Corollary 6.2]).

**Theorem 7.12.** *If the assumptions of Theorem 7.11 hold, then there exist constants $C, \tau > 0$, dependent only on $k_0$, $c$, and $\Gamma$, such that*

$$\|\varphi - \varphi_N\|_{L^2(\Gamma)} \leq C k^\alpha \log^{1/2}(2+k) e^{-p\tau}. \tag{7.143}$$

To compute the solution in the domain, we rearrange (7.140) to get

$$\partial_\nu^+ u(\mathbf{x}(s)) = k\varphi(s) + \Psi(\mathbf{x}(s)) \approx k\varphi_N(s) + \Psi(\mathbf{x}(s)), \quad \mathbf{x}(s) \in \Gamma, \tag{7.144}$$

and then we insert this approximation to $\partial_\nu^+ u$ into the representation formula (7.108) to get an approximation to $u$, which we denote by $u_N$. We then have the following error estimate (cf. [54, Theorem 6.3], [21, Corollary 64]).

**Theorem 7.13.** *If the assumptions of Theorem 7.11 hold, then there exist constants $C, \tau > 0$, dependent only on $k_0$, $c$, and $\Gamma$, such that*

$$\frac{\|u - u_N\|_{L^\infty(\Omega_+)}}{\|u\|_{L^\infty(\Omega_+)}} \leq C k \log(2+k) e^{-p\tau}. \tag{7.145}$$

Similarly, we can derive an approximation to the far field pattern of the scattered field, given explicitly for $\hat{\mathbf{x}} = \mathbf{x}/|\mathbf{x}|$ by

$$F(\hat{\mathbf{x}}) = -\int_\Gamma e^{-ik\hat{\mathbf{x}}\cdot\mathbf{y}} \partial_\nu^+ u(\mathbf{y})\, ds(\mathbf{y}), \quad \hat{\mathbf{x}} \in \mathbb{S}^1, \tag{7.146}$$

where $\mathbb{S}^1$ denotes the unit circle. Efficient computation of the far field patten is of interest in many applications; see, e.g., [27]. To compute an approximation $F_N$ to $F$, we again just insert the approximation (7.144) into the integral (7.146). We then have the following estimate (cf. [54, Theorem 6.4], [21, Corollary 64]).

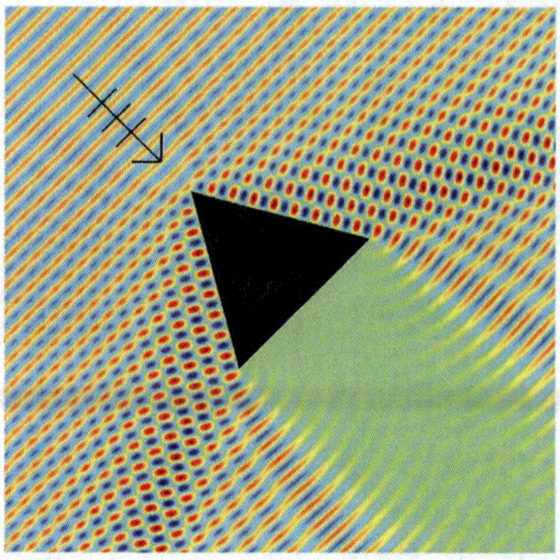

**Figure 7.12.** *Total field, scattering by a triangle.*

**Theorem 7.14.** *If the assumptions of Theorem* 7.11 *hold, then there exist constants* $C, \tau > 0$, *dependent only on* $k_0$, $c$, *and* $\Gamma$, *such that*

$$||F - F_N||_{L^\infty(\mathbb{S}^1)} \leq C k^{1+\alpha} \log^{1/2}(2+k) e^{-p\tau}. \tag{7.147}$$

Note that the estimates above for the solution in the domain and the far field pattern follow from results in [21] and are actually a little sharper than those in [54].

The algebraically $k$-dependent prefactors in the error estimates of Theorems 7.12, 7.13, and 7.14 can be absorbed into the exponentially decaying factors by allowing $p$ to grow modestly ($O(\log^2 k)$) with increasing $k$. In practice, numerical results [54, 19] suggest that this is pessimistic and that in many cases a fixed accuracy of approximation can be achieved without any requirement for the number of degrees of freedom to increase with $k$.

To illustrate the approach described above, we present numerical results for the problem of scattering by a sound-soft equilateral triangle, of side length $2\pi$, so that the number of wavelengths per side is equal to $k$. The total field for $k = 10$ is plotted in Figure 7.12.

In our computations we choose $n = 2(p+1)$, as for the screen results above, so that the total number of degrees of freedom is $N = 6n(p+1) = 12(p+1)^2$. Since the total number of degrees of freedom depends only on $p$, we again adjust our notation by defining $\psi_p(s) := \varphi_N(s)$. In Figure 7.13 we plot on a logarithmic scale the relative $L^2$ errors

$$\frac{\left\|\psi_6 - \psi_p\right\|_{L^2(\Gamma)}}{\left\|\frac{1}{k}\partial_\nu^+ u\right\|_{L^2(\Gamma)}},$$

against $p$ for a range of values of $k$, this quantity being an estimate of the relative error in our approximation (7.144) to $\partial_\nu^+ u$. (Again we take the "reference" solution, our approximation to the true solution $\varphi$, to be $\psi_6$.) This example is identical to one that appears in [54], except that here we show results for much higher values of $k$ (in [54] the largest value of $k$ tested was $k = 5120$), and here we plot relative errors in our approximation to $\partial_\nu^+ u$, rather than the relative errors in the approximation $\psi_p$ to $\varphi$ computed in [54]. The $L^2$-norms are calculated using a high-order Gaussian quadrature routine on a mesh graded toward the endpoint singularities; see [54] for details. For fixed $k$, as $p$ increases the error decays exponentially, as predicted by Theorem 7.12. For fixed $p$, the error seems

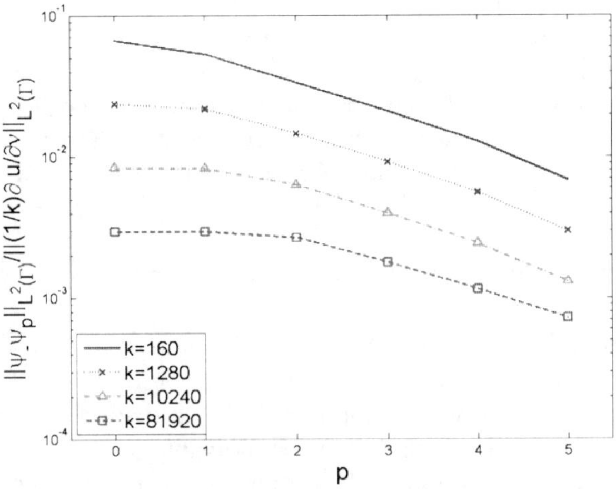


**Figure 7.13.** *Relative $L^2$ errors in $\partial_\nu^+ u$.*

Table 7.1: Relative $L^2$ errors for scattering by a triangle, fixed $p = 3$ (and hence $N = 192$), various $k$.

| $k$ | $\frac{N}{L/\lambda}$ | $\|\psi_7 - \psi_4\|_{L^2(\Gamma)} / \|\frac{1}{k}\partial_\nu^+ u\|_{L^2(\Gamma)}$ | COND | cpt(s) |
|---|---|---|---|---|
| 5 | $1.28\times 10^{+1}$ | $7.91\times 10^{-2}$ | $5.36\times 10^{+1}$ | $4.68\times 10^{+2}$ |
| 10 | $6.40\times 10^{+0}$ | $6.18\times 10^{-2}$ | $2.38\times 10^{+1}$ | $3.84\times 10^{+2}$ |
| 20 | $3.20\times 10^{+0}$ | $4.76\times 10^{-2}$ | $2.97\times 10^{+1}$ | $3.40\times 10^{+2}$ |
| 40 | $1.60\times 10^{+0}$ | $3.64\times 10^{-2}$ | $3.85\times 10^{+1}$ | $3.72\times 10^{+2}$ |
| 80 | $8.00\times 10^{-1}$ | $2.77\times 10^{-2}$ | $5.08\times 10^{+1}$ | $3.55\times 10^{+2}$ |
| 160 | $4.00\times 10^{-1}$ | $2.10\times 10^{-2}$ | $6.76\times 10^{+1}$ | $4.31\times 10^{+2}$ |
| 320 | $2.00\times 10^{-1}$ | $1.60\times 10^{-2}$ | $9.00\times 10^{+1}$ | $3.57\times 10^{+2}$ |
| 640 | $1.00\times 10^{-1}$ | $1.21\times 10^{-2}$ | $1.20\times 10^{+2}$ | $4.93\times 10^{+2}$ |
| 1280 | $5.00\times 10^{-2}$ | $9.18\times 10^{-3}$ | $1.60\times 10^{+2}$ | $5.33\times 10^{+2}$ |
| 2560 | $2.50\times 10^{-2}$ | $6.96\times 10^{-3}$ | $2.13\times 10^{+2}$ | $5.23\times 10^{+2}$ |
| 5120 | $1.25\times 10^{-2}$ | $5.27\times 10^{-3}$ | $2.81\times 10^{+2}$ | $5.44\times 10^{+2}$ |
| 10240 | $6.25\times 10^{-3}$ | $3.99\times 10^{-3}$ | $3.67\times 10^{+2}$ | $6.40\times 10^{+2}$ |
| 20480 | $3.13\times 10^{-3}$ | $3.03\times 10^{-3}$ | $4.75\times 10^{+2}$ | $6.12\times 10^{+2}$ |
| 40960 | $1.56\times 10^{-3}$ | $2.29\times 10^{-3}$ | $6.04\times 10^{+2}$ | $6.34\times 10^{+2}$ |
| 81920 | $7.81\times 10^{-4}$ | $1.79\times 10^{-3}$ | $7.48\times 10^{+2}$ | $7.83\times 10^{+2}$ |

to decrease as $k$ increases. To investigate this further, in Table 7.1 we show the relative $L^2$ errors (computed exactly as in Figure 7.13) as computed with fixed $p = 3$ (and hence $N = 192$) for a wider range of values of $k$. The relative error decreases as $k$ increases, for a fixed number of degrees of freedom, suggesting that the $k$-dependence of our error estimate in Theorem 7.12 is pessimistic.

We also show $N/(L/\lambda)$, the average number of degrees of freedom per wavelength, the condition number (COND) of the linear system that we solve (details of our implementation can be found in [54]), and the computing time (cpt) in seconds to set up and solve our linear system. All computations were carried out in MATLAB, using a standard desktop PC. We surmise that it might be possible to reduce these computing times with some effort to optimize the code. The key point of these last two columns though is that both the condition number and the computing time grow only very slowly as $k$ increases for fixed $p$. Implementation of our scheme includes the need to evaluate oscillatory integrals; details of the approach we use for that can be found in [19, Section 4]. Whereas for standard BEMs standard practice suggests that 10 degrees of freedom per wavelength are required for "engineering accuracy," the results in Table 7.1 suggest that $\partial_\nu^+ u$ can be computed with much lower than 1% relative error with a fixed number of degrees of freedom per wavelength (indeed, over 1000 wavelengths per degree of freedom for the case $k = 81920$) with low condition numbers for the linear system and a total computing time of the order of 10 minutes or so (and increasing only slowly as $k$ increases).

**Scattering by Nonconvex Polygons.** For nonconvex polygons, we encounter behavior that does not occur for convex polygons, as a result of which the leading-order asymptotic behavior on $\Gamma$ is more complicated. In particular, in this case we may see partial illumination of a side of the nonconvex polygon (whereas for a convex polygon a side is either completely illuminated or completely in shadow) and/or rereflections (where a wave that has been reflected from one side of the polygon may be incident on another side of the polygon), as illustrated in Figure 7.14.

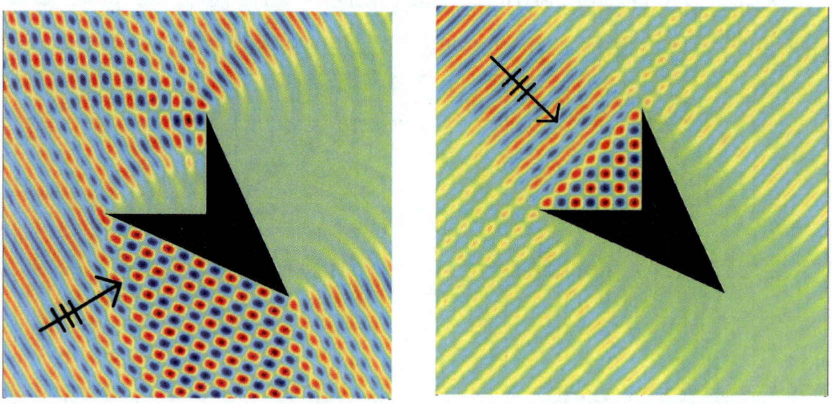

**Figure 7.14.** *Partial illumination (left) and rereflections (right).*

We restrict our attention to a particular class of nonconvex polygons that satisfy the following assumptions. (A description of how the approach described below can be extended to polygons that do not satisfy these assumptions can be found in [21, Section 8].)

**Assumption 1.** *Each exterior angle $\omega_j$, $j = 1, \ldots, n_s$, is either a right angle or greater than $\pi$.*

**Assumption 2.** *At each right angle, the obstacle lies within the dashed lines shown in Figure 7.15.*

Polygons satisfying these criteria may or may not be star-shaped. For each side $\Gamma_j$, $j = 1, \ldots, n_s$, if either $\omega_j$ or $\omega_{j+1}$ is a right angle, then we define that side to be a "nonconvex" side; otherwise we say it is a "convex" side, as illustrated for a particular non-star-shaped example in Figure 7.16. On convex sides, $\partial_\nu^+ u$ behaves exactly as in the convex case, and the approximation results above hold. However, on nonconvex sides we need to consider the possibilities of partial illumination and/or rereflections. To illustrate our approach,

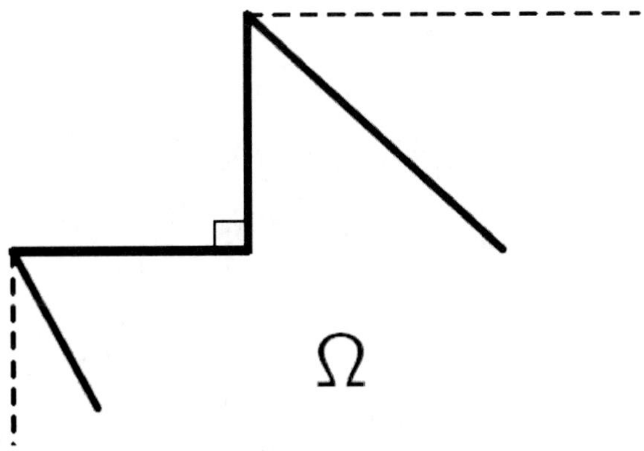


**Figure 7.15.** *Assumption 2 on geometry of nonconvex polygon is that it lies entirely within the semi-infinite dashed lines.*

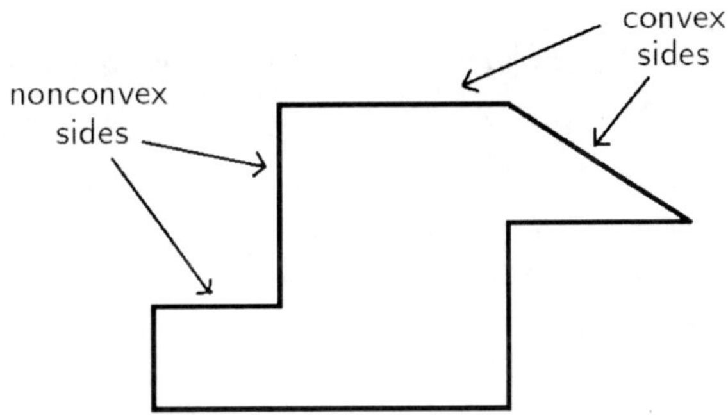


**Figure 7.16.** *Convex and nonconvex sides for a non-star-shaped example.*

we consider the behavior at a point $\mathbf{x}(s)$ on a nonconvex side $\Gamma_j$, distance $s$ from $P_j$ and $r$ from $P_{j-1}$, as illustrated in Figure 7.17. Then $\Gamma_j$ will be fully illuminated if $\pi \le \alpha < 3\pi/2$ (where $\alpha$ is the incident angle shown in Figure 7.17), $\Gamma_j$ will be partially illuminated for some values of $\alpha$ in the range $\pi/2 < \alpha < \pi$ (e.g., in the case that $L_j = L_{j-1}$, $\Gamma_j$ will be partially illuminated for $3\pi/4 < \alpha < \pi$), and $\Gamma_j$ will be in shadow otherwise. There will be reflections from $\Gamma_{j-1}$ onto $\Gamma_j$ if $\pi < \alpha < 3\pi/2$. Whatever the value of $\alpha$, there will be diffraction from $P_{j-1}$ and $P_{j+1}$ (either directly from the incident wave or from waves that have traveled around $\Gamma$).

For $\mathbf{x}(s) \in \Gamma_j$, where $s$ denotes distance from $P_j$ as in Figure 7.17, the following theorem follows from [21, Theorem 36].

**Theorem 7.15.** *Let $k \ge k_0 > 0$. Then, for $\mathbf{x}(s) \in \Gamma_j$,*

$$\partial_\nu^+ u(\mathbf{x}(s)) = \Psi^d(\mathbf{x}(s)) + v_j^+(L_j + s)e^{iks} + v_j^-\left(L_j - s\right)e^{-iks} + \tilde{v}_j(s)e^{ikr}, \tag{7.148}$$

*where*

$$\Psi^d := \begin{cases} 2\partial u^d/\partial \nu & \text{if } \frac{\pi}{2} \le \alpha \le \frac{3\pi}{2}, \\ 0 & \text{otherwise,} \end{cases} \tag{7.149}$$

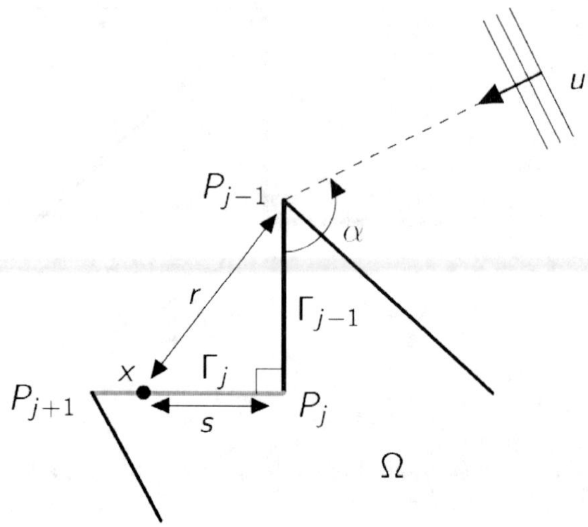


**Figure 7.17.** *Geometry of a nonconvex side* $\Gamma_j$.

*the functions* $v_j^{\pm}(t)$ *have the same properties as those for the convex sides and in particular are analytic in the right half plane* $\mathrm{Re}[t] > 0$ *and satisfy the bounds* (7.139), *and the function* $\tilde{v}_j$ *is analytic in a complex* $k$*-independent neighborhood* $D_\epsilon$ *of the side* $\Gamma_j$ *with*

$$\left|\tilde{v}_j(t)\right| \leq Ck\log^{1/2}(2+k), \quad t \in D_\epsilon, \tag{7.150}$$

*where the constant* $C > 0$ *depends only on* $k_0$ *and* $\Gamma$.

Here $u^d$ is the known solution of a canonical diffraction problem, namely that of scattering by a semi-infinite "knife edge," precisely scattering of $u^I$ by the semi-infinite sound-soft screen starting at $P_{j-1}$ and extending vertically down through $P_j$; for details, see [21, Lemma 35].

Using the representation (7.148) on each nonconvex side of the polygon (and (7.137) on convex sides) again leads to a representation of the form (7.129). Our approximation space $\hat{V}_{N,k}$ for

$$\varphi(s) := \frac{1}{k}\left(\partial_\nu^+ u(\mathbf{x}(s)) - \Psi^d(\mathbf{x}(s))\right), \quad \mathbf{x}(s) \in \Gamma, \tag{7.151}$$

is then identical on each convex side of the polygon to that for the convex polygon and the screen (7.133), again replacing $v_j^+(s)$ and $v_j^-(L_j - s)$ by piecewise polynomials supported on overlapping geometric meshes, graded toward the singularities at $s = 0$ and $s = L_j$, respectively. However, on nonconvex sides it is slightly different. In this case, we note first that in the representation (7.148) we have $v_j^+(L_j + s)$ rather than $v_j^+(s)$ (compare with (7.137)); hence $v_j^+$ is not singular on $\Gamma_j$ when it is nonconvex, and we approximate it by a polynomial (of order $p$) supported on the whole side $\Gamma_j$. Second, we notice from the analyticity of $\tilde{v}_j$ and the bound (7.150) that it is also sufficient to approximate $\tilde{v}_j$ by a polynomial (of order $p$) supported on the whole side $\Gamma_j$.

The following best approximation estimate follows from [21, Theorem 56].

**Theorem 7.16.** *If $c, k_0 > 0$ and $n \geq cp$, $k \geq k_0$, then, for some $C, \tau > 0$, depending only on $c$, $k_0$, and $\Gamma$,*

$$\inf_{w_N \in \hat{V}_{N,k}} \|\partial_\nu^+ u - w_N\|_{L^2(\Gamma)} \leq C k^{1/2+\alpha} \log^{1/2}(2+k) e^{-p\tau}, \tag{7.152}$$

*where $\alpha := 1 - \min_{m=1,\ldots,n_s,\, \omega_m \neq \pi/2}(1-\pi/\omega_m) \in (1/2, 1)$.*

Comparing Theorems 7.11 and 7.16 we see that the best approximation estimate is identical for convex and nonconvex polygons, again implying that we can achieve any required accuracy with $N$ growing like $\log^2 k$ as $k \to \infty$, rather than like $k$ as for a standard BEM. Note also that the result (7.152) is sharper (in terms of $k$-dependence) than that in [19, Theorem 3.13].

Having designed an appropriate approximation space $\hat{V}_{N,k}$ we again use the Galerkin method to select an element so as to effectively approximate $\varphi$. For star-shaped polygons satisfying Assumptions 1–2, we can again use the integral equation formulation (7.113), i.e., we seek $\varphi_N \in \hat{V}_{N,k}$ such that

$$\langle \mathscr{A}_k \varphi_N, w_N \rangle_\Gamma = \frac{1}{k} \langle f_k - \Psi^d, w_N \rangle_\Gamma \quad \text{for all } w_N \in \hat{V}_{N,k}, \tag{7.153}$$

in which case, due to the coercivity of the integral operator $\mathscr{A}_k$, we have the following error estimates (cf. [21, Theorems 61–63]).

**Theorem 7.17.** *For star-shaped polygons (satisfying also Assumptions 1–2), if the assumptions of Theorem 7.16 hold, then there exist constants $C, \tau > 0$, dependent only on $k_0$, $c$, and $\Gamma$, such that*

$$\|\varphi - \varphi_N\|_{L^2(\Gamma)} \leq C k^\alpha \log^{1/2}(2+k) e^{-p\tau},$$

$$\frac{\|u - u_N\|_{L^\infty(\Omega_+)}}{\|u\|_{L^\infty(\Omega_+)}} \leq C k \log(2+k) e^{-p\tau},$$

$$\|F - F_N\|_{L^\infty(\mathbb{S}^1)} \leq C k^{1+\alpha} \log^{1/2}(2+k) e^{-p\tau}.$$

Our approximations $u_N$ and $F_N$ to the solution in the domain and the far field pattern are constructed exactly as for the convex case, as outlined above. As for the convex polygonal case, our conclusion is again that $N$ proportional to $p^2$, growing like $\log^2 k$ as $k$ increases, provably maintains accuracy.

We present numerical results for the two cases shown in Figure 7.14. These examples have also been studied in [21], but our results here are new, as detailed below. The nonconvex sides of the scatterer have length $2\pi$ and the convex sides have length $4\pi$, so the total length of the boundary is $12\pi$, which is $6k$ wavelengths (recalling that the wavelength is $\lambda = 2\pi/k$). In our computations we again choose $n = 2(p+1)$, so that for this example the total number of degrees of freedom is $N = 12p^2 + 28p + 16$. Since $N$ depends only on $p$, we again adjust our notation by defining $\psi_p(s) := \varphi_N(s)$, and we take our "reference" solution to be $\psi_7$. In Figure 7.18 we plot on a logarithmic scale the relative $L^2$ errors

$$\frac{\|\psi_7 - \psi_p\|_{L^2(\Gamma)}}{\left\|\frac{1}{k}\partial_\nu^+ u\right\|_{L^2(\Gamma)}},$$

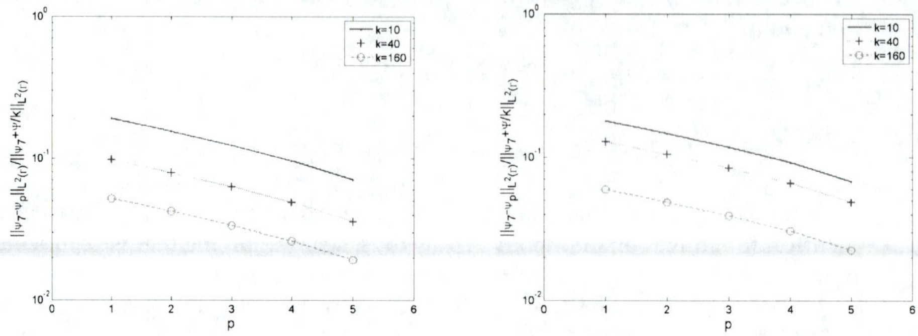


**Figure 7.18.** *Relative $L^2$ errors in $\partial_\nu^+ u$, scattering by a nonconvex polygon, with partial illumination (left) and rereflections (right), as in Figure 7.14.*

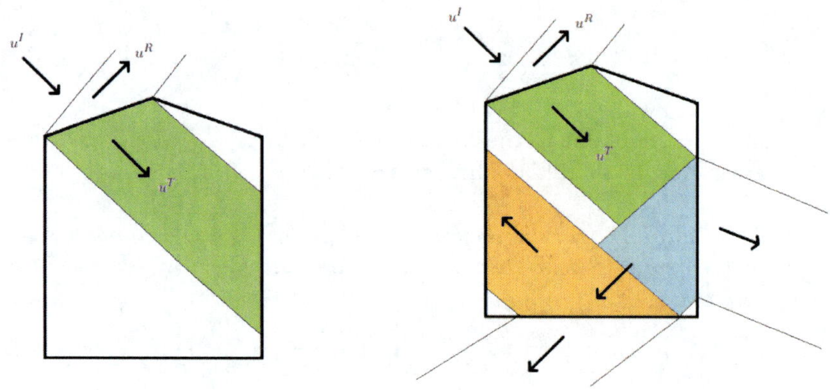


**Figure 7.19.** *Illustration, for the transmission scattering problem, of incident ($u^I$), reflected ($u^R$), and primary transmitted ($u^T$) fields (left), with multiple internal rereflections (right).*

against $p$ for a range of values of $k$. As for the convex polygon, we scale the relative error by the quantity that we are actually trying to approximate, namely $(1/k)\partial_\nu^+ u = \varphi + \Psi^d/k \approx \psi_7 + \Psi^d/k$ (7.151). This is in contrast to results in [21], where the relative errors were scaled by $||\psi_7||_{L^2(\Gamma)}$, i.e., by the "reference" solution to the BIE, which is only a component of the quantity we are seeking to approximate (compare Figure 7.18 with [21, Figure 7 and Table 1]). As for the convex polygon, the $L^2$-norms are again calculated using a high-order Gaussian quadrature routine on a mesh graded toward the endpoint singularities. For fixed $k$, as $p$ increases, the error decays exponentially, as predicted by Theorem 7.17. For fixed $p$, the error seems to decrease as $k$ increases.

**Transmission Scattering Problems.** Finally we consider the transmission scattering problem (7.76), for which the behavior is more complicated still, incorporating as it does multiple internal rereflections. To illustrate this, consider the scenario illustrated in Figure 7.19, in which we show an incident wave striking a penetrable polygonal scatterer. On the left of Figure 7.19 we show what happens as the incident wave $u^I$ strikes one side of the boundary $\Gamma$. Following Snell's law (see, e.g., [50, Appendix A]), this gives rise to a "reflected" wave $u^R$ (traveling from $\Gamma$ into $\Omega_+$) and a "transmitted" wave $u^T$ that

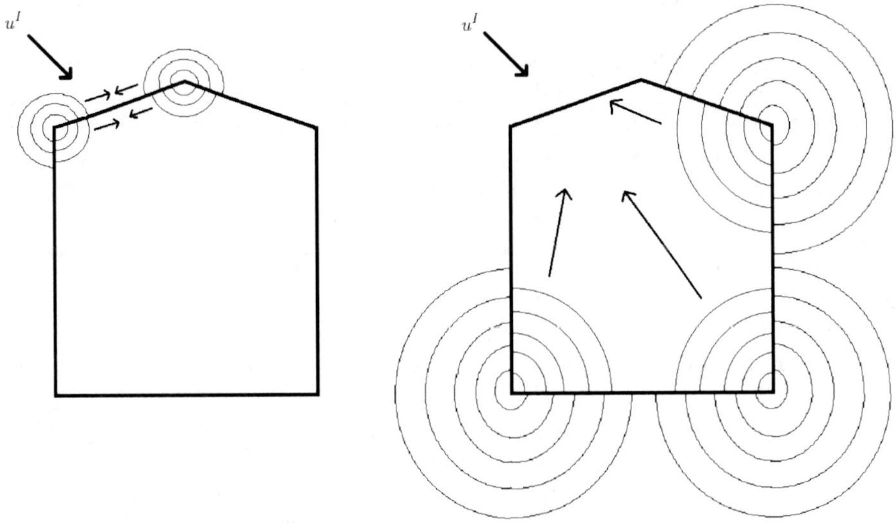


**Figure 7.20.** *Illustration of waves "diffracted" by the corners of a particular side of* $\Gamma$ *(left) and by other corners of the polygon (right).*

passes into $\Omega_-$. For the case that the scatterer is impenetrable (as in all the other problems considered earlier in this section), only the reflected wave would be present here. As the transmitted wave passes through $\Omega_-$ it may decay (if $\mathrm{Im}\, k_- > 0$), but if $k_-$ is constant, then its direction does not change. As this transmitted wave strikes $\Gamma$ it again gives rise to another transmitted wave (that now passes through $\Gamma$ into $\Omega_+$) and another reflected wave (which passes through $\Omega_-$ again, in a new direction). This reflected wave again may lose energy as it passes through $\Omega_-$, but again it will strike $\Gamma$, leading to further reflected and transmitted waves, as shown on the right of Figure 7.19. These multiple internally reflected waves are not present for impenetrable scatterers and make the task of designing a hybrid approximation space (based on the ansatz (7.129)) extremely challenging.

We restrict attention here to the case of convex polygonal scatterers and outline briefly the approach presented in [50]. For nonconvex scatterers or for scatterers with curved surfaces the problem would be significantly harder, and we do not discuss such generalizations. Utilizing Snell's law, we can write down explicit formulae for each term in the (in general infinite) series of "rereflected" waves (the first three of which are illustrated on the right of Figure 7.19). We refer to [50] for details, but note that these formulae rely on a complete understanding of what happens when a plane wave passes from one (possibly absorbing) homogeneous medium to another, and further that there appear to be some misconceptions in the literature regarding the solution to that canonical problem, which are addressed fully in [50, Appendix A]. Framing this in the context of (7.129), this corresponds to expressing $V_0$ as an infinite series of these "rereflected" waves, which must be truncated in any numerical algorithm.

The part of the total field that is not represented by this series corresponds primarily to the "diffracted" waves emanating from the corners of the polygon (though other wave components, e.g., lateral waves, may also be present; again we refer to the discussion in [50] for details). These may originate from the incident field (e.g., as illustrated on the left of Figure 7.20), or else they may originate from the waves that travel through the interior of the polygon striking corners that are on the "shadow" side of the polygon (using the same definition as for impenetrable scatterers above). Either way these "diffracted" waves travel through $\Omega_+$ with speed determined by $\mathrm{Re}[k_+]$ and through $\Omega_-$ with speed determined by $\mathrm{Re}[k_-]$ (and decay determined by $\mathrm{Im}\, k_-$). These differing wave speeds in $\Omega_-$ and $\Omega_+$ of course imply differing wavelengths, as illustrated in Figure 7.20. A full consideration of the wave behavior of the solution would also need to take into account reflection of these waves from each side of $\Gamma$. We do not consider such rereflections of diffracted waves here.

We turn now to the integral equation formulation for the solution of the transmission scattering problem (7.76). We need to solve (7.97) for $v = \left[\gamma_+ u, \partial_\nu^+ u\right]^T \in H^1(\Gamma) \times L^2(\Gamma)$, with $A$ given by (7.107) and $f = \left[u^I|_\Gamma, \partial_\nu u^I|_\Gamma\right]$. The HNA ansatz for $v$ then needs to incorporate the behavior shown in Figures 7.19 and 7.20. On any given side $\Gamma_j$ (of length $L_j$) of the polygon, with $s$ representing distance from $P_{j-1}$, we represent the solution as

$$v(s) = v_0(s) + v_+^+(s)\mathrm{e}^{\mathrm{i}k_+ s} + v_+^-(L_j - s)\mathrm{e}^{-\mathrm{i}k_+ s} + v_-^+(s)\mathrm{e}^{\mathrm{i}k_- s} + v_-^-(L_j - s)\mathrm{e}^{-\mathrm{i}k_- s} + \sum_{m=1}^{n_s-2} v_m(s)\mathrm{e}^{\mathrm{i}k_- r_m(s)}, \quad (7.154)$$

where $v_0$ represents a (known) truncated series of rereflected plane waves (referred to as "beams" in [50]), i.e., an approximation to the behavior shown in Figure 7.19; $v_\pm^\pm(\cdot)\mathrm{e}^{\pm \mathrm{i}k_\pm s}$ represents diffraction along $\Gamma_j$ from adjacent corners, i.e., the behavior as shown on the left of Figure 7.20, where the phase functions $\mathrm{e}^{\pm \mathrm{i}k_\pm s}$ capture the oscillations of these diffracted waves, but the amplitudes $v_\pm^\pm$ are unknown, and are approximated by piecewise polynomials on overlapping graded meshes (due to singularities at the corners) as for the sound-soft convex polygonal scattering problem described above; finally $v_m(s)\mathrm{e}^{\mathrm{i}k_- r_m(s)}$ represents diffraction on $\Gamma_j$ emanating from nonadjacent corners, with $r_m(s)$ denoting the distance from the $m$th corner to the point $x(s)$ parametrized by $s$ on $\Gamma_j$, i.e., the behavior as shown on the right of Figure 7.20, where the phase functions $\mathrm{e}^{\mathrm{i}k_- r_m(s)}$ capture the oscillations of these diffracted waves, but the amplitudes $v_m$ are unknown, and are each approximated by piecewise polynomials (of order $p$) on a mesh on $\Gamma_j$, with meshpoints lined up with potential discontinuities in $v_m$ arising from our sharp "cutting off" of the beams, as shown in Figure 7.19 (for full details, see [50, Section 3.2.3]).

This scheme is rather more complicated than the corresponding ones for impenetrable scatterers, but numerical results in [50, Section 4] suggest that the best approximation achieved by this HNA approximation space has a relative error that decreases as $k$ increases and further that represents a significant improvement over using $v_0$ (essentially the geometrical optics solution) on its own.

Here, we compare results from [50, Table 1] with those presented in Table 7.1 above for the problem of sound-soft scattering by an impenetrable convex polygon. We consider the same geometry as described above for the case of scattering by a sound-soft convex polygon, i.e., an equilateral triangle of side length $2\pi$, but now we allow the wave to pass through the triangle, as illustrated in Figure 7.21. In the results below we take the refractive index of the media to be 1.31, so that for any given exterior wavenumber $k_+$, the interior wavenumber is $k_- = k_+(1.31 + \xi\mathrm{i})$, where $\xi$ determines the level of absorption. We fix $p = 4$, in which case the total number of degrees of freedom required by the approximation space outlined above is 193 (compared to 192 for $p = 3$ for the sound-soft convex polygon example, presented above). In Figure 7.22 we plot the relative error in best approximation to both $u$ and $\partial_\nu u$ on $\Gamma$, as computed with 193 degrees of freedom, for $\xi = 0$ (zero absorption), for $\xi = 0.025$, and for $\xi = 0.05$. The best approximation

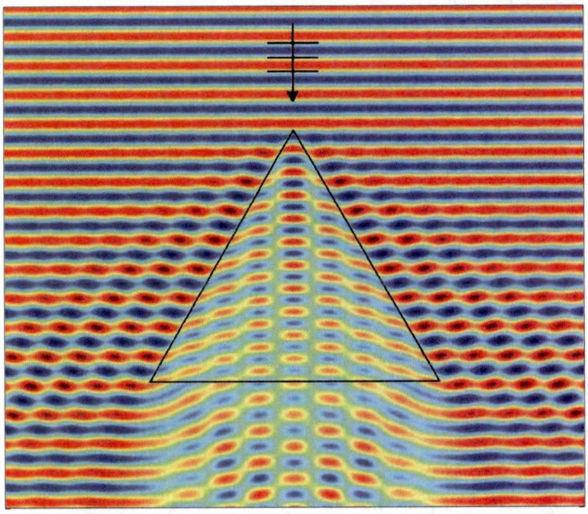

**Figure 7.21.** *Real part of the total field for scattering of a plane wave by a penetrable equilateral triangle.*

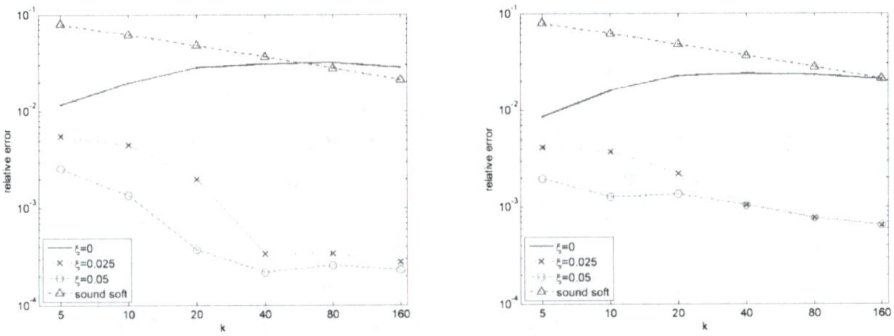


**Figure 7.22.** *Relative $L^2(\Gamma)$ best approximation errors in $u$ (left) and $\partial_\nu u$ (right) for scattering by a penetrable triangle of varying absorption, and a comparison with the HNA BEM relative errors for scattering by a sound soft polygon.*

is computed via matching with a known "reference" solution computed using standard BEM on a very fine mesh—see [50] for details. For comparison, on each plot we also show the relative error in the HNA BEM solution to the problem of scattering by the sound-soft convex polygon, as computed with 192 degrees of freedom (i.e., some of the results from Table 7.1). The comparison here is not completely fair—for the sound-soft problem we have computed the solution by solving the BIE using an HNA approximation space, whereas for the transmission problem we have merely fitted the best approximation from an HNA approximation space, and there is no guarantee that solving the BIE numerically would achieve this best approximation. Also, in solving the sound-soft problem, the relative error is in our approximation to $\partial_\nu u$, and we compare that result here with the best approximation to both $u$ and $\partial_\nu u$ for the transmission problem. However, having said all that, we note that, except for the most challenging case of zero absorption, the error in best approximation from the HNA approximation space for the transmission problem is comparable to or better than that for the HNA BEM solution for the much simpler problem of scattering by a sound-soft convex polygon.

Although in this case there is no rigorous theory, unlike for the simpler cases of the screen and the impenetrable polygon, the results in Figure 7.22 (and see [50] for a much wider range of examples) demonstrate that the HNA BEM has the potential to be successful even for this significantly more complicated scenario. Further results in [50], also for the calculation of the solution in the domain and the far field pattern, show that good results can be achieved for a range of scatterers with different absorptions; as absorption reduces, so the influence of diffraction from nonadjacent corners increases, and we surmise that, in this case, it may be necessary to add additional terms to the ansatz (7.154) in order to achieve higher levels of accuracy. We note though that results in [50] suggest that the ansatz outlined above is sufficient to achieve 1% relative error in the far field pattern for any absorption and frequency.

**Other Boundary Conditions and 3D Problems.** We have focused mostly in this section on sound-soft scattering problems. There is no difficulty in extending the algorithms and much of the analysis to sound-hard or impedance scattering problems. In particular, the HNA approach has been very successfully applied to the problem of scattering by convex polygons with impedance boundary conditions (see [23] for details), solving (7.114) with $\eta = 0$. We do not include specific details of that case here, but note that the approximation space is very similar to that for scattering by sound-soft convex polygons, as detailed above, and also that a summary of the approach in that case and further numerical results can be found in [19].

Much more challenging is extension to three dimensions, because of the greater complexity of the high-frequency asymptotics of the solution, in particular the much larger number of possible contributing ray paths and associated oscillatory phases. But at least for significant classes of 3D problems it seems likely that this methodology will be effective, leading to substantially reduced computational cost compared to standard BEMs.

As a starting point, we reported in [19, Section 7.6] initial numerical experiments by Hewett for scattering in three dimensions by a sound-soft square screen (the 3D version of the problem tackled in the section above entitled "Scattering by Screens"), exploring whether the high-frequency ansatz (7.129) is able to approximate the exact solution with a small number of degrees of freedom, experimenting with different choices for the phases $\psi_m$ and the number $M$ of oscillatory phases included. In work in progress this ansatz has been implemented in a Galerkin scheme analogous to that in the section above entitled "Scattering by Screens" [51]. These results suggest that the best approximation results for certain 3D problems are broadly achieved by HNA BEM in practice, and that inclusion of appropriate oscillatory basis functions in the approximation space as outlined here can lead to high accuracy at reasonably high frequencies with a relatively small number of degrees of freedom, certainly compared to standard BEM, even in the 3D case.

### 7.3.4 ▪ Links to the Unified Transform Method

As noted in the introduction, this paper appears in a collection of articles in significant part focused on the so-called unified transform or Fokas transform introduced by Fokas in 1997 [44] and developed further by Fokas and collaborators since then (see, e.g., [32] and the other papers in this collection). This method is on the one hand an analytical transform technique which can be employed to solve linear BVPs with constant coefficients in canonical domains, for which it serves as a generalization of classical transform and separation of variables techniques (for a high-frequency application, to acoustic scattering by a circular domain, see [46]). But this method also, when applied in general domains, has some aspects in common with BIE and BEMs, as discussed by Spence [76] in this book.

The investigation of the unified transform method as a numerical scheme for the Helmholtz equation is arguably in its infancy. The focus to date has been on numerical experiments for particular geometries for the 2D interior Dirichlet problem (7.64), and general methodologies and associated theoretical numerical analysis results are limited to date. Spence [76, Section 6.1.10] in this book reviews these developments in detail, proves some new theoretical convergence results for one particular proposed implementation, and makes connections with BIEs and BEMs and other numerical schemes. We add to this discussion in the next section, in particular pointing out that one particular implementation (see Theorem 7.19 below), approximating the unknown Neumann data from a space of traces of plane waves, computes precisely the best approximation from that space. We also make additional connections to related methods: the *least squares method* and the *method of fundamental solutions*; see Remarks 48 and 49.

The unified transform method, as articulated in the next section and in [78, 76], does not apply to exterior problems for the Helmholtz equation, at any rate to exterior problems set in the exterior of a bounded set $\Omega_-$, such as the exterior Dirichlet problem (7.66). The issue is that plane wave (and generalized plane wave) solutions of the Helmholtz equation, which are fundamental to the method (see the next section) do not satisfy the standard Sommerfeld radiation condition (7.63) (for more discussion see the section entitled "Diffraction Gratings and the Unified Transform Method" below, and note that [45] *does* achieve an implementation for a particular exterior problem for the *modified* Helmholtz equation, i.e., (7.60) with $k$ pure imaginary). But the unified transform method can be applied to so-called *rough surface scattering problems*, where the scatterer takes the form

$$\Omega_- := \{\mathbf{x} = (\tilde{\mathbf{x}}, x_d) \in \mathbb{R}^d : x_d < f(\tilde{\mathbf{x}})\}, \tag{7.155}$$

for some bounded, Lipschitz continuous function $f : \mathbb{R}^{d-1} \to \mathbb{R}$ so that $\Gamma$ is the graph of $f$ and the BVP to be solved is posed in the perturbed half-space $\Omega_+ := \mathbb{R}^d \setminus \overline{\Omega_-}$. Generalized plane waves (as defined in the next section) that propagate upward or decay in the vertical direction satisfy the appropriate radiation conditions in this case: in two dimensions these are the so-called *Rayleigh expansion radiation conditions*, (7.163) below, in the case when $f$ is periodic, and the *upward propagating radiation condition* [18] more generally.

Not only can the unified transform method be applied to these rough surface scattering problems, *it already has been applied in these cases*, developed independently in papers by DeSanto and co-authors from 1981 onward [33, 35, 36, 34, 5]. In the section entitled "Diffraction Gratings and the Unified Transform Method" below, we recall this method for the simplest of these problems, the 2D sound-soft scattering problem (7.73) in the particular case when the scatterer $\Omega_-$ is a one-dimensional *diffraction grating* by which we will mean that $d = 2$ and $f : \mathbb{R} \to \mathbb{R}$ is periodic (this case considered in particular in [35, 5]). We point out that the so-called *spectral-coordinate (SC)* and *spectral-spectral (SS)* methods proposed in [35] correspond to two implementations of the unified transform methods with different choices of approximation space. We note that the $SS^*$ method proposed in [5], a variant of the SS method, corresponds precisely to the method for the interior Dirichlet problem (7.64) analyzed in Theorem 7.19 below, and we prove a new result (Theorem 7.21) characterizing and proving convergence of this method, sharpening [5, Lemma 4.1]. We also discuss the conditioning of the linear systems that arise from these methods.

### The Unified Transform Method for the Interior Dirichlet Problem

At the heart of the unified transform method is the so-called *global relation*. For linear elliptic PDEs with constant coefficients this global relation follows from the divergence theorem. In particular, as described in [78, 76] for the Helmholtz equation (7.60) (and the Laplace and modified Helmholtz equations), applying Green's second theorem to a solution $u \in H^1(D)$ of (7.60) and a function $v \in \mathscr{R} := \{v \in H^1(D) \cap C^2(D) : \Delta u + k^2 u = 0 \text{ in } D\}$ in a bounded 2D Lipschitz domain $D$ gives that

$$\int_\Gamma \partial_\nu u \overline{\gamma v} \, ds = \int_\Gamma \gamma u \overline{\partial_\nu v} \, ds \quad \forall v \in \mathscr{R}. \tag{7.156}$$

We remark that the lower part of (7.79), i.e., (7.79) with $\mathbf{x} \in \Omega_+$, is precisely a particular instance of (7.156).

For (7.60) with $k > 0$, the *global relation* [78, 76] is (7.156) restricted to the subset $\mathscr{P} \subset \mathscr{R}$ of separable solutions of (7.60) in Cartesian coordinates. In two dimensions $\mathscr{P}$ is the one-parameter family $\mathscr{P} = \{v(\cdot, \theta) : \theta \in \mathbb{C}\}$, where $v(\mathbf{x}, \theta) := \exp(ik(\cos\theta \, x_1 + \sin\theta \, x_2))$. For $\theta \in \mathbb{R}$, $v(\cdot, \theta)$ is a plane wave traveling in the direction $\hat{\mathbf{a}} = (\cos\theta, \sin\theta)$. For complex $\theta$ we term $v(\cdot, \theta)$ a *generalized* plane wave (often alternatively an *evanescent* or *inhomogeneous* plane wave). The following lemma notes linear independence and density properties of $\mathscr{P}$.

**Lemma 7.18.** *Suppose that $-k^2$ is not a Dirichlet eigenvalue of the Laplacian in D. Then*

(i) *for $N \in \mathbb{N}$, $\gamma v(\cdot, \theta_1), \ldots, \gamma v(\cdot, \theta_N)$ are linearly independent if $\theta_1, \ldots, \theta_N \in \mathbb{C}$ are distinct;*

(ii) *The linear span of $\{\gamma v(\cdot, \theta) : 0 \leq \theta < 2\pi\}$ is dense in $L^2(\Gamma)$.*

***Proof.*** (i) This is a development of standard arguments, e.g., [1], which demonstrate the linear independence of ordinary plane waves. Suppose that $c_1, \ldots, c_N \in \mathbb{C}$ and $v := \sum_{n=1}^N c_n v(\cdot, \theta_n) = 0$ on $\Gamma$. Then $v = 0$ in $D$ and, since $v \in C^2(\mathbb{R}^2)$ and satisfies (7.60) in $\mathbb{R}^2$, it follows by analyticity of $v$ [28, page 72] that $v = 0$ in $\mathbb{R}^2$.

For $n = 1, \ldots, N$ set $\alpha_n := k \cos\theta_n$ and $\beta_n := k \sin\theta_n$. Suppose, without loss of generality, that $N = 2M$ is even, that $N \geq 4$, and that $\alpha_1 = 1$, $\alpha_2 = -1$, and $\alpha_{2m-1} = \alpha_{2m}$, in which case $\beta_{2m-1} = -\beta_{2m} \neq 0$, for $m = 2, \ldots, M$. Since $v((t, 0)) = 0$ for $t \in \mathbb{R}$, it holds that

$$\sum_{m=1}^{M+1} d_m e^{i\eta_m t} = 0, \quad t \in \mathbb{R}, \tag{7.157}$$

where $d_m := c_m$, $\eta_m := \alpha_m$, for $m = 1, 2$, and $\eta_{m+1} := \alpha_{2m-1}$, $d_{m+1} := c_{2m-1} + c_{2m}$, for $m = 2, \ldots, M$. Since the $\theta_n$, $n = 1, \ldots, N$, are distinct, so are the $\eta_m$, $m = 1, \ldots, M+1$. Let $n \in \{1, \ldots, M+1\}$ be such that $\operatorname{Im} \eta_n \leq \operatorname{Im} \eta_m$ for $m = 1, \ldots, M+1$. Then, multiplying (7.157) by $\exp(-i\eta_n)$ and integrating it follows that

$$\sum_{m=1}^{M} \frac{d_m}{A} \int_A^{2A} e^{i(\eta_m - \eta_n)t} \, dt = 0, \quad A > 0.$$

Taking the limit $A \to \infty$ we see that $d_n = 0$. Repeating this argument we deduce that $d_m = 0$, $m = 1, \ldots, M+1$, so that $c_1 = c_2 = 0$ and $c_{2m-1} = -c_{2m}$, $m = 2, \ldots, M$.

To conclude we note also that $\partial v(x)/\partial x_2 = 0$ for $x = (t, 0)$ and $t \in \mathbb{R}$. This implies that

$$\sum_{m=3}^{M+1} e_m e^{i\eta_m t} = 0, \quad t \in \mathbb{R},$$

where $e_m = i\beta_{2m}(c_{2m} - c_{2m-1})$. Arguing as for (7.157) we deduce that $e_m = 0$, for $m = 2, \ldots, M+1$, so that $c_m = 0$, for $m = 1, \ldots, 2M$.

(ii) This is an easy consequence of a standard result on the Herglotz integral operator [27, Theorem 5.24]. □

Spence [76] reviews the implementations to date of the unified transform method for the (2D) interior Dirichlet problem (7.64). These implementations impose (7.156) (with $\gamma u = h$) for $v \in \mathscr{P}_N$, where $\mathscr{P}_N$ is an $N$-dimensional *test space* of generalized plane waves: explicitly, for some distinct $\theta_{j,N} \in \mathbb{C}$, $j = 1, \ldots, N$, $\mathscr{P}_N$ is the space spanned by $\{v(\cdot, \theta_{j,N}) : 1 \leq j \leq N\}$. An approximation $\phi_M$ to $\partial_\nu u$, which is an element of $Q_M \subset H^{-1/2}(\Gamma)$, the $M$-dimensional *trial space* with $M \geq N$, is obtained by requiring that

$$\int_\Gamma \phi_M \overline{\gamma v_N} \mathrm{d}s = \int_\Gamma h \overline{\partial_\nu v_N} \mathrm{d}s \quad \forall v_N \in \mathscr{P}_N, \tag{7.158}$$

this equation overdetermined if $M > N$, in which case it is imposed, e.g., in a least squares sense. Spence [76] tabulates the implementations to date, which vary in the choice of approximation space $Q_M$ and in the choice of generalized plane waves, i.e., in the choice of $\theta_{j,N} \in \mathbb{C}$, $j = 1, \ldots, N$.

Assume now that $-k^2$ is not a Dirichlet eigenvalue of the Laplacian, in other words (see the discussion below (7.64)) that the interior Dirichlet problem (7.64) has a unique solution for all $h \in H^{1/2}(\Gamma)$. In that case there is a well-defined operator $P_{\mathrm{DtN}} : H^{1/2}(\Gamma) \to H^{-1/2}(\Gamma)$, the *Dirichlet-to-Neumann map*, that takes the Dirichlet data $h \in H^{1/2}(\Gamma)$ to $\partial_\nu u \in H^{-1/2}(\Gamma)$, where $u$ is the solution of the BVP (7.64). From (7.91) we see that, explicitly,

$$P_{\mathrm{DtN}} = S_k^{-1}(D_k + \tfrac{1}{2}I). \tag{7.159}$$

With this notation the global relation (7.156) for the interior Dirichlet problem can be written equivalently as

$$\int_\Gamma \phi \overline{\psi} \mathrm{d}s = \int_\Gamma h \overline{P_{\mathrm{DtN}} \psi} \mathrm{d}s \quad \forall \psi \in H^{1/2}(\Gamma), \tag{7.160}$$

where $\phi = \partial_\nu u \in H^{-1/2}(\Gamma)$.

The implementations of the unified transform method reviewed in [76] can be viewed as Petrov–Galerkin methods for the variational equation (7.160), *Petrov*–Galerkin since the trial and test spaces differ, reflecting that the trial and test spaces, $H^{-1/2}(\Gamma)$ and $H^{1/2}(\Gamma)$, respectively, differ in (7.160) at the continuous level. But in the case that the Dirichlet data is sufficiently smooth, precisely when $h \in H^1(\Gamma)$ in which case, by Corollary 7.4, $\phi \in L^2(\Gamma)$, we can think of (7.160) alternatively as a variational problem on $L^2(\Gamma) \times L^2(\Gamma)$. (Note that, from (7.159) and [19, Theorem 2.25], $P_{\mathrm{DtN}}$ extends to a bounded mapping from $L^2(\Gamma)$ to $H^{-1}(\Gamma)$, this the dual space of $H^1(\Gamma)$ with respect to the $L^2(\Gamma)$ inner product, so that, when $h \in H^1(\Gamma)$, the right-hand side of (7.160) defines a continuous antilinear functional on $L^2(\Gamma)$.) Indeed, in this setting, (7.160) is the simplest possible variational problem: (7.160) corresponds to (7.120) with $\mathscr{V} = \mathscr{V}' = L^2(\Gamma)$ and $A = I$, the identity operator! Of course this problem is trivially continuous and coercive, with continuity and coercivity constants, $C$ and $\alpha$ in Lemma 7.6, equal to one. Thus it is attractive to choose $M = N$ and $Q_M := \mathscr{P}_N^\Gamma$, where $\mathscr{P}_N^\Gamma := \gamma(\mathscr{P}_N) = \{\gamma v : v \in \mathscr{P}\}$ is a space of restrictions of plane waves to $\Gamma$, for in that case (7.158) is a standard Galerkin method and, by Céa's lemma (Lemma 7.6) or standard properties of orthogonal projection in Hilbert spaces, e.g., [71, Theorem 12.14], one has the following result.

**Theorem 7.19.** *Suppose that $-k^2$ is not a Dirichlet eigenvalue, so that (7.64) is uniquely solvable, and $h \in H^1(\Gamma)$ so that $\phi := \partial_\nu u \in L^2(\Gamma)$. Suppose also that $M = N \in \mathbb{N}$ and $Q_M := \mathscr{P}_N^\Gamma := \gamma(\mathscr{P}_N)$, where $\mathscr{P}_N$ is some $N$-dimensional subspace of $\mathscr{P}$. Then (7.158) has a unique solution which is $\phi_N = P_N\phi$, where $P_N : L^2(\Gamma) \to \mathscr{P}_N^\Gamma$ is orthogonal projection, so that $\phi_N$ is the best $L^2(\Gamma)$ approximation to $\phi$ in $\mathscr{P}_N^\Gamma$.*

With the choices made in this theorem it holds that

$$\phi_N = \sum_{n=1}^{N} c_n \gamma v(\cdot,\theta_{n,N})$$

for some complex coefficients $c_n$, and (7.158) is equivalent to the linear system

$$\sum_{n=1}^{N} a_{mn} c_n = \int_\Gamma h\, \overline{\partial_\nu v(\cdot,\theta_{m,N})} \mathrm{d}s, \quad m = 1,\dots,N, \tag{7.161}$$

where $a_{mn} = \int_\Gamma \gamma v(\cdot,\theta_{n,N}) \overline{\gamma v(\cdot,\theta_{m,N})} \mathrm{d}s$. It is easy to see that the matrix $[a_{mn}]$ is Hermitian and positive semidefinite, indeed positive definite in view of Lemma 7.18(i): see, e.g., the discussion in [5, Section 5].

**Remark 48.** *The same matrix $[a_{mn}]$ and a similar right-hand side arise when solving the interior Dirichlet problem by a least squares method (see e.g., [1]). Seek an approximation $u_N$ to the solution $u$ of (7.64) in the form*

$$u_N = \sum_{n=1}^{N} c_n' v(\cdot,\theta_{n,N}),$$

*choosing the complex coefficients $c_n'$ to minimize $\|h - \gamma u_N\|_{L^2(\Gamma)}$. Then the coefficients $c_n'$ satisfy*

$$\sum_{n=1}^{N} a_{mn} c_n' = \int_\Gamma h \overline{\gamma v(\cdot,\theta_{m,N})} \mathrm{d}s, \quad m = 1,\dots,N.$$

**Remark 49.** *The focus above is on the* unified method *as a numerical method, i.e., on (7.158). But we note that the same analysis and results apply (and in two and three dimensions) if we replace $\mathscr{P}_N$ in (7.158) by any $N$-dimensional subspace of $\mathscr{R}$. Not least, Theorem 7.19 holds (in two and three dimensions) with $\mathscr{P}_N$ any $N$-dimensional subspace of $\mathscr{R}$, and (7.161) holds with $v(\cdot,\theta_{j,N})$, $j = 1,\dots,N$, replaced by any choice of basis for this amended $\mathscr{P}_N$.*

*One possible choice of $\mathscr{P}_N \subset \mathscr{R}$, this choice suggested by the lower part of (7.79), is to take $\mathscr{P}_N$ to be the space spanned by $\{\Phi(\cdot,\mathbf{x}_n) : 1 \le n \le N\}$, where $\mathbf{x}_1,\dots,\mathbf{x}_N$ are distinct points in $\mathbb{R}^d \setminus \overline{D}$. With this choice (7.158) is a variant on the so-called* method of fundamental solutions*; see, e.g.,* [1].

### Diffraction Gratings and the Unified Transform Method

The previous subsection described the unified transform method as a numerical method for interior problems, in particular the interior Dirichlet problem (7.64), but our focus in this section is acoustic *scattering*, in which we are solving exterior problems. As noted above, we cannot see how the unified transform method, as currently formulated, can be applied to any of the exterior or scattering problems that we have stated in Section 7.3.2, where we need to solve the Helmholtz equation in the exterior of a bounded set $\Omega_-$, whose boundary is denoted by $\Gamma$. In particular, the global relation (7.156) does not hold for any generalized plane wave $v$ in this case.

But, as outlined above, the global relation *does* hold for some generalized plane waves in the rough surface scattering case, where the scatterer is the nonlocally perturbed half plane (in two dimensions) or half-space (in three dimensions), given by (7.155). For this geometry, in both two and three dimensions, versions of the global relation and the unified transform method have been developed independently (and with different terminology) by DeSanto and co-workers [33, 35, 36, 34, 5].

In this section we describe (and elaborate on) this work for the simplest case, for which the numerical analysis is most complete [33, 35, 5], namely the 2D sound-soft scattering problem and associated Dirichlet problem in the case when $f:\mathbb{R}\to\mathbb{R}$ in (7.155) is periodic, with some period $L$. For this scattering problem, in the case when the incident field $u^I$ is the plane wave (7.72) for some $\hat{\mathbf{a}}=(\sin\theta^I,-\cos\theta^I)$, where $\theta^I\in(-\pi/2,\pi/2)$ is the angle of incidence, it is natural to look for a solution $u$ to the scattering problem which is *quasi-periodic* with period $L$ and phase shift $\mu=k\sin\theta^I$, meaning that

$$u((x_1+L,x_2))=\exp(\mathrm{i}\mu L)u(\mathbf{x})\quad \text{for } \mathbf{x}\in\Omega_+. \tag{7.162}$$

Let $\Omega_+^L:=\{\mathbf{x}=(x_1,x_2)\in\Omega_+:0<x_1<L\}$, $\Gamma^L:=\{\mathbf{x}\in\Gamma:0<x_1<L\}$, and, for $\mu\in\mathbb{R}$, let $C_\mu^2(\Omega_+^L)$ denote those $u\in C^2(\Omega_+)$ that satisfy (7.162). The standard radiation condition imposed on the scattering field $u^S:=u-u^I$ for this problem can be obtained by applying separation of variables in $\Omega_+^L$ to the Helmholtz equation (7.60) under the constraint that $u\in C_\mu^2(\Omega_+)$. This leads to an expression for $u^S$ as a countable linear combination of generalized plane waves. Discarding those generalized plane waves which are growing exponentially away from $\Gamma$, or are plane waves propagating toward $\Gamma$, leads to the representation (the *Rayleigh expansion radiation condition* (RERC)) that

$$u^S(\mathbf{x})=\sum_{n\in\mathbb{Z}}c_n\exp(\mathrm{i}k[\alpha_n x_1+\beta_n x_2])\quad \text{for } x_2>f_+:=\max(f), \tag{7.163}$$

for some complex coefficients $c_n$, where

$$\alpha_n:=\mu/k+2\pi n/(kL)\ \text{and}\ \beta_n:=\begin{cases}\sqrt{1-\alpha_n^2}, & |\alpha_n|\le 1,\\ \mathrm{i}\sqrt{\alpha_n^2-1}, & |\alpha_n|>1.\end{cases} \tag{7.164}$$

The standard Dirichlet problem in this case is then

$$\begin{aligned}&\text{Given } h\in H_\mu^{1/2}(\Gamma^L),\ \text{find } u\in C_\mu^2(\Omega_+^L)\cap H_{\mathrm{loc}}^1(\Omega_+^L)\\ &\text{such that (7.60) holds in } \Omega_+,\ \gamma u=h \text{ on } \Gamma^L,\\ &\text{and } u \text{ satisfies the RERC (7.163).}\end{aligned} \tag{7.165}$$

Here, for $\mu\in\mathbb{R}$, $H_\mu^{1/2}(\Gamma^L)$ is the closure in $H^{1/2}(\Gamma^L)$ of those $\phi\in C^\infty(\Gamma)$ that satisfy (7.162) for $\mathbf{x}\in\Gamma$. That (7.165) is uniquely solvable is shown in [40].

For $\mu\in\mathbb{R}$, let

$$\mathscr{R}_\mu:=\{v\in C_\mu^2(\Omega_+^L)\cap H_{\mathrm{loc}}^1(\Omega_+^L): v \text{ satisfies (7.60) and (7.163)}\},$$

and note that $u \in \mathscr{R}_\mu$ if $u$ is a solution of (7.165). The numerical schemes in [35] derive from the observation that if $u$ satisfies (7.165), then (where $\nu$ is the unit normal directed into $\Omega_+$)

$$\int_{\Gamma^L} \partial_\nu u \gamma v \,\mathrm{d}s = \int_{\Gamma^L} h \partial_\nu v \,\mathrm{d}s \quad \text{for all } v \in \mathscr{R}_{-\mu}, \tag{7.166}$$

this identity (7.166) derived by applying Green's second theorem to $u$ and $v$ in $\{\mathbf{x} \in \Omega_+^L : x_2 < H\}$ for some $H > f_+$. The identity (7.166) holds, in particular, for those generalized plane waves $v(\cdot,\theta)$ that are in the set $\mathscr{P}_\mu := \{v(\cdot,\theta_n) : n \in \mathbb{Z}\}$, where $\theta_n \in \mathbb{C}$ is defined by

$$(\cos\theta_n, \sin\theta_n) = (-\alpha_n, \beta_n),$$

with $\alpha_n$ and $\beta_n$ given by (7.164). These are the generalized plane waves that are elements of $\mathscr{R}_{-\mu}$.

Thus a version of the global relation holds that

$$\int_{\Gamma^L} \partial_\nu u \, \overline{\gamma v} \,\mathrm{d}s = \int_{\Gamma^L} h \, \overline{\partial_\nu v} \,\mathrm{d}s, \text{ for all } v \in \mathscr{P}_\mu^*, \tag{7.167}$$

where

$$\mathscr{P}_\mu^* := \{v(\cdot,\theta) : (\cos\theta, \sin\theta) = (\alpha_n, -\overline{\beta_n}) \text{ and } n \in \mathbb{Z}\} \subset \mathscr{P}.$$

Explicitly, (7.167) is the sequence of equations

$$\int_{\Gamma^L} \partial_\nu u \, v(\cdot,\theta_n) \,\mathrm{d}s = \int_{\Gamma^L} h \, \partial_\nu v(\cdot,\theta_n) \,\mathrm{d}s, \quad n \in \mathbb{Z}. \tag{7.168}$$

Equation (7.167), which is equivalent to (7.168), uniquely determines $\partial_\nu u$ because of the following density result (cf. Lemma 7.18).

**Lemma 7.20.** (i) *(see* [5, Corollary 3.2]*) The functions* $\{\gamma v(\cdot,\theta_n) : n \in \mathbb{Z}\} \subset L^2(\Gamma^L)$ *are linearly independent.*

(ii) *(see* [5, Lemma 3.1]*) The linear span of* $\mathscr{P}_\mu^*$*, which is the linear span of* $\{v(\cdot,\theta_n) : n \in \mathbb{Z}\}$*, is dense in* $L^2(\Gamma^L)$.

In the sound-soft scattering problem, the incident field $u^I$ is the plane wave (7.72), with $\hat{\mathbf{a}} = (\sin\theta^I, -\cos\theta^I) = (\alpha_0, -\beta_0)$, and the scattered field $u^S$ is the solution of (7.165) with $h = -u^I|_{\Gamma^L}$. So (7.168) holds with $u$ replaced by $u^S$ and $h = -u^I|_{\Gamma^L}$. But also, applying Green's second theorem to $u^I$ and $v(\cdot,\theta_n)$ in $\{\mathbf{x} \in \Omega_+^L : x_2 < H\}$ for some $H > f_+$, we deduce that

$$\int_{\Gamma^L} \partial_\nu u^I \, v(\cdot,\theta_n) \,\mathrm{d}s = \int_{\Gamma^L} \gamma u^I \, \partial_\nu v(\cdot,\theta_n) \,\mathrm{d}s - 2ikL\delta_{0,n}, \quad n \in \mathbb{Z}, \tag{7.169}$$

where $\delta_{m,n}$ is the Kronecker delta. Adding (7.168) (with $u$ replaced by $u^S$ and $h = -\gamma u^I$) and (7.169), we see that the total field $u = u^I + u^S$ satisfies

$$\int_{\Gamma^L} \partial_\nu u \, \gamma v(\cdot,\theta_n) \,\mathrm{d}s = -2ikL\delta_{0,n}, \quad n \in \mathbb{Z}, \tag{7.170}$$

this equation dating back to [33] (and see [35, Section 4] and [5]). It follows, e.g., from Rellich identities, that $\partial_\nu u \in L^2(\Gamma^L)$ [40].

DeSanto et al. [35] propose discretizations of (7.170) in which we seek an approximation to $\partial_\nu u$ in $Q_N$, some $N$-dimensional subspace of $L^2(\Gamma^L)$, and we impose (7.170) for $n = n_1,\ldots,n_N$ for distinct integers $n_j$; in other words we impose (7.170) for $v(\cdot,\theta_n) \in \mathscr{P}_N$, the $N$-dimensional space spanned by $\{v(\cdot,\theta_{n_m}) : 1 \le m \le N\}$. If $\{\chi_1,\ldots,\chi_N\}$ is a basis for $Q_N$, and $\psi_n := \overline{\gamma v(\cdot,\theta_n)}$, then the numerical method is to approximate

$$\partial_\nu u \approx \phi_N = \sum_{m=1}^{N} c_m \chi_m, \tag{7.171}$$

where the coefficients $c_m$ satisfy the linear system

$$\sum_{m=1}^{N} \int_{\Gamma^L} \chi_m \overline{\psi_{n_j}} \, \mathrm{d}s \, c_m = -2\mathrm{i}kL\delta_{0,n_j}, \quad j = 1,\ldots,N. \tag{7.172}$$

DeSanto et al. [35] propose two different choices of space $Q_N$ and basis functions $\chi_m$, the SC method, using a "pulse" basis (piecewise constant basis functions), in which $\chi_m$ is the characteristic function of $\{\mathbf{x} \in \Gamma^L : (m-1)h < x_1 < mh\}$, where $h = L/N$, and the SS method, in which $\chi_m := \gamma v(\cdot,\theta_{n_m} + \pi)$ for $m = 1,\ldots,N$. Numerical experiments with the SC and SS methods are carried out in [35]. The methods are extended to transmission problems in [36] and the SC method to 3D sound-soft and sound-hard scattering problems for doubly periodic, diffraction gratings surfaces in [34].

In [5] a variant of the SS method is proposed, the SS* method, characterized by the choice $\chi_m = \psi_{n_m} = \overline{\gamma v(\cdot,\theta_{n_m})}$, so that $Q_N = \mathscr{P}_N^\Gamma$, the space spanned by $\{\overline{\gamma v(\cdot,\theta_{n_m})} : 1 \le m \le N\}$. An attraction of this method, as observed in [5], is that, as with (7.161), this leads to a coefficient matrix $A_N = [a_{jm}]$, in this case with $a_{jm} = \int_{\Gamma^L} \psi_{n_m} \overline{\psi_{n_j}} \, \mathrm{d}s$, that is Hermitian and positive definite.

Analogously to (7.160), (7.170) can be viewed as a variational formulation problem on $L^2(\Gamma^L) \times L^2(\Gamma^L)$: (7.170) corresponds to (7.120) with $\mathscr{V} = \mathscr{V}' = L^2(\Gamma^L)$ and $A = I$, the identity operator. Like (7.160) this formulation is trivially continuous and coercive, with continuity and coercivity constants, $C$ and $\alpha$ in Lemma 7.6, equal to one. The choice $Q_N = \mathscr{P}_N^\Gamma$ is a Galerkin method for (7.170), and by Céa's lemma (Lemma 7.6) one has the following result (cf. Theorem 7.19).

**Theorem 7.21.** *Suppose that $Q_N = \mathscr{P}_N^\Gamma$. Then $\phi_N$, given by* (7.171) *and* (7.172), *is the best $L^2(\Gamma^L)$ approximation to $\partial_\nu u$ in $\mathscr{P}_N^\Gamma$.*

We note that this result improves on [5, Lemma 4.1], where it is shown, under the assumptions of this theorem, that

$$\|\partial_\nu u - \phi_N\|_{L^2(\Gamma^L)} \le 2 \inf_{\psi \in \mathscr{P}_N^\Gamma} \|\partial_\nu u - \psi\|_{L^2(\Gamma^L)}.$$

Combining Theorem 7.21 with Lemma 7.20(ii) we obtain the following corollary, guaranteeing convergence of the SS* method.

**Corollary 7.22 (see [5, Corollary 4.1]).** *Suppose that the sequence of subspaces $\mathscr{P}_1^\Gamma$, $\mathscr{P}_2^\Gamma,\ldots$, is chosen so that, for every $n \in \mathbb{Z}$, $\psi_n \in \mathscr{P}_N^\Gamma$, for all sufficiently large $N$, and that $Q_N = \mathscr{P}_N^\Gamma$, for all $N \in \mathbb{N}$. Then*

$$\|\partial_\nu u - \phi_N\|_{L^2(\Gamma^L)} \to 0 \quad \text{as } N \to \infty.$$

**Remark 50.** *Numerical experiments in* [5] *compare the performance of the SC, SS, and SS* methods—to emphasize these can all be viewed as numerical implementations of the unified transform method for this sound-soft diffraction grating problem—for the case when the surface profile $f$ is sinusoidal, and for various sizes of $kL$ and $ka$, where $a$ is the surface amplitude. Perhaps predictably, given Theorem 7.21, the SS* method is most reliable, but, for some geometries and some angles of incidence, all three methods perform very well, producing highly accurate results with around one degree of freedom per wavelength (see* [5, Section 6] *for more details).*

**Remark 51.** *A potential difficulty with all of the SC, SS, and SS* methods is that the linear systems that arise can be very ill-conditioned. For the SS* method rigorous upper and lower bounds for* $\mathrm{cond}(A_N)$, *the condition number of the system matrix $A_N$, are computed in* [5], *and the effects of this ill-conditioning on the computed solution are estimated. Regarding the behavior of* $\varkappa_N := \mathrm{cond}(A_N)$, *the main results are that $\varkappa_N$ remains low as long as $\mathscr{P}_N^\Gamma$ contains only plane waves, i.e., $\overline{\gamma v(\cdot,\theta_n)}$ with $\theta_n \in \mathbb{R}$, but necessarily eventually grows exponentially as $N \to \infty$.*

**Remark 52.** *The system matrix in the SS* method is the transpose of the matrix to be solved when the same scattering problem is solved by a least squares method: see* [5, Section 5] *for more detail, and cf.* Remark 48.

# Bibliography


[1] C. J. S. Alves and S. S. Vaitchev, Numerical comparison of two meshfree methods for acoustic wave scattering, Eng. Anal. Boundary Elements **29** (2005), 371–382.

[2] S. Amini and N. D. Maines, Preconditioned Krylov subspace methods for boundary element solution of the Helmholtz equation, Internat. J. Numer. Methods Engrg. **41** (1998), 875–898.

[3] A. Anand, Y. Boubendir, F. Ecevit, and F. Reitich, Analysis of multiple scattering iterations for high-frequency scattering problems. II: The three-dimensional scalar case, Numer. Math. **114** (2010), 373–427.

[4] S. Arden, S. N. Chandler-Wilde, and S. Langdon, A collocation method for high-frequency scattering by convex polygons, J. Comput. Appl. Math., **204** (2007), 334–343.

[5] T. Arens, S. N. Chandler-Wilde, and J. A. DeSanto, On integral equation and least squares methods for scattering by diffraction gratings, Commun. Comput. Phys. **1** (2006), 1010–1042.

[6] K. E. Atkinson, The Numerical Solution of Integral Equations of the Second Kind, Cambridge Monogr. Appl. Comput. Math., Cambridge University Press, 1997.

[7] L. Banjai and W. Hackbusch, Hierarchical matrix techniques for low and high frequency Helmholtz equation, IMA J. Numer. Anal. **28** (2008), 46–79.

[8] M. Bebendorf, Hierarchical Matrices: A Means to Efficiently Solve Elliptic Boundary Value Problems, Lecture Notes in Comput. Sci. Eng. 63, Springer, Berlin, 2008.

[9] T. Betcke, J. Phillips, and E. A. Spence, Spectral decompositions and nonnormality of boundary integral operators in acoustic scattering, IMA J. Numer. Anal. **34** (2014), 700–731.

[10] T. Betcke and E. A. Spence, Numerical estimation of coercivity constants for boundary integral operators in acoustic scattering, SIAM J. Numer. Anal. **49** (2011), 1572–1601.

[11] H. Brakhage and P. Werner, Über das Dirichletsche Aussenraumproblem für die Helmholtzsche Schwingungsgleichung, Ark. Math. **16** (1965), 325–329.

[12] O. P. Bruno, C. A. Geuzaine, J. A. Monro, Jr., and F. Reitich, Prescribed error tolerances within fixed computational times for scattering problems of arbitrarily high frequency: The convex case, Philos. Trans. R. Soc. Lond. Ser. A **362** (2004), 629–645.

[13] O. P. Bruno and S. K. Lintner, A high-order integral solver for scalar problems of diffraction by screens and apertures in three-dimensional space, J. Comput. Phys. **252** (2013), 250–274.

[14] O. P. Bruno and F. Reitich, High Order Methods for High-Frequency Scattering Applications, in Modeling and Computations in Electromagnetics, H. Ammari, ed., Lecture Notes in Comput. Sci. Eng. 59, Springer, Berlin, 2007, 129–164.

[15] A. Buffa and R. Hiptmair, A coercive combined field integral equation for electromagnetic scattering, SIAM J. Numer. Anal. **42** (2005), 621–640.

[16] A. Buffa and S. Sauter, On the acoustic single layer potential: stabilization and Fourier analysis, SIAM J. Sci. Comput. **28** (2006), 1974–1999.

[17] A. J. Burton and G. F. Miller, The application of integral equation methods to the numerical solution of some exterior boundary-value problems, Proc. R. Soc. Lond. Ser. A Math. Phys. Sci. **323** (1971), 201–210.

[18] S. N. Chandler-Wilde, Boundary value problems for the Helmholtz equation in a half-plane, in Proceedings of the Third International Conference on Mathematical and Numerical Aspects of Wave Propagation, G. Cohen, ed., SIAM, 1995, 188–197.

[19] S. N. Chandler-Wilde, I. G. Graham, S. Langdon, and E. A. Spence, Numerical-asymptotic boundary integral methods in high-frequency acoustic scattering, Acta Numer. **21** (2012), 89–305.

[20] S. N. Chandler-Wilde and D. P. Hewett, Acoustic Scattering by Fractal Screens: Mathematical Formulations and Wavenumber-Explicit Continuity and Coercivity Estimates, preprint MPS-2013-17, University of Reading (2013).

[21] S. N. Chandler-Wilde, D. P. Hewett, S. Langdon, and A. Twigger, A high frequency boundary element method for scattering by a class of nonconvex obstacles, Numer. Math., ISSN 0029-599X doi:10.1007/s00211-014-0648-7 (in press).

[22] S. N. Chandler-Wilde and S. Langdon, A Galerkin boundary element method for high frequency scattering by convex polygons, SIAM J. Numer. Anal. **45** (2007), 610–640.

[23] S. N. Chandler-Wilde, S. Langdon, and M. Mokgolele, A high frequency boundary element method for scattering by convex polygons with impedance boundary conditions, Commun. Comput. Phys. **11** (2012), 575–593.

[24] S. N. Chandler-Wilde, S. Langdon, and L. Ritter, A high-wavenumber boundary-element method for an acoustic scattering problem, Philos. Trans. R. Soc. London Ser. A Math. Phys. Eng. Sci. **362** (2004), 647–671.

[25] H. Cheng, W. Y. Crutchfield, Z. Gimbutas, L. F. Greengard, J. F. Ethridge, J. Huang, V. Rokhlin, N. Yarvin, and J. Zhao, A wideband fast multipole method for the Helmholtz equation in three dimensions, J. Comput. Phys. **216** (2006), 300–325.

[26] S. H. Christiansen and J. C. Nédélec, Preconditioners for the numerical solution of boundary integral equations from electromagnetism, C.R. Acad. Sci. I Math **331** (2000), 733–738.

[27] D. Colton and R. Kress, Inverse Acoustic and Electromagnetic Scattering Theory, Springer-Verlag, Berlin, 1992.

[28] D. L. Colton and R. Kress, Integral Equation Methods in Scattering Theory, John Wiley, New York, 1983.

[29] M. Costabel, Time-dependent problems with the boundary integral equation method, in Encyclopedia of Computational Mechanics, Wiley (2004).

[30] E. Darrigrand, Coupling of fast multipole method and microlocal discretization for the 3-D Helmholtz equation, J. Comput. Phys. **181** (2002), 126–154.

[31] E. Darve and P. Havé, A fast multipole method for Maxwell equations stable at all frequencies, Philos. Trans. R. Soc. Lond. Ser. A Math. Phys. Eng. Sci. **362** (2004), 603–628.

[32] B. Deconinck, T. Trogdon, and V. Vasan, The method of Fokas for solving linear partial differential equations, SIAM Rev. **56** (2014), 159–186.

[33] J. A. DeSanto, Scattering from a perfectly reflecting arbitrary periodic surface: An exact theory, Radio Sci. **16** (1981), 1315–1326.

[34] J. A. DeSanto, G. Erdmann, W. Hereman, B. Krause, M. Misra, and E. Swim, Theoretical and computational aspects of scattering from periodic surfaces: two-dimensional perfectly reflecting surfaces using the spectral-coordinate method, Waves Random Media **11** (2001), 455–487.

[35] J. A. DeSanto, G. Erdmann, W. Hereman, and M. Misra, Theoretical and computational aspects of scattering from rough surfaces: One-dimensional perfectly reflecting surfaces, Waves Random Media **8** (1998), 385–414.

[36] J. A. DeSanto, G. Erdmann, W. Hereman, and M. Misra, Theoretical and computational aspects of scattering from rough surfaces: One-dimensional transmission interfaces, Waves Random Media **11** (2001), 425–453.

[37] V. Dominguez, I. G. Graham, and V. P. Smyshlyaev, A hybrid numerical-asymptotic boundary integral method for high-frequency acoustic scattering, Numer. Math. **106** (2007), 471–510.

[38] K. C. Donepudi, J. M. Jin, and W. C. Chew, A grid-robust higher-order multilevel fast multipole algorithm for analysis of 3-d scatterers, Electromagnetics **23** (2003), 315–330.

[39] F. Ecevit and F. Reitich, Analysis of multiple scattering iterations for high-frequency scattering problems. I: The two-dimensional case, Numer. Math. **114** (2009), 271–354.

[40] J. Elschner and M. Yamamoto, An inverse problem in periodic diffractive optics: Reconstruction of Lipschitz grating profiles, Appl. Anal. **81** (2001), 1307–1328.

[41] S. Engleder and O. Steinbach, Modified boundary integral formulations for the Helmholtz equation, J. Math. Anal. Appl. **331** (2007), 396–407.

[42] S. Engleder and O. Steinbach, Stabilized boundary element methods for exterior Helmholtz problems Numer. Math. **110** (2008), 145–160.

[43] O. G. Ernst and M. J. Gander, Why It Is Difficult to Solve Helmholtz Problems with Classical Iterative Methods, in Numerical Analysis of Multiscale Problems, I. G. Graham, T. Y. Hou, O. Lakkis, and R. Scheichl, eds., Lecture Notes in Comput. Sci. Eng. 83, Springer, Berlin, 2012, 325–363.

[44] A. S. Fokas, A unified transform method for solving linear and certain nonlinear pdes, Proc. R. Soc. Lond. Ser. A **453** (1997), 1411–1443.

[45] A. S. Fokas and J. Lenells, The Unified Transform for the Modified Helmholtz Equation in the Exterior of a Square, in Unified Transform for Boundary Value Problems: Applications and Advances, A. S. Fokas and B. Pelloni, eds., SIAM, 2015.

[46] A. S. Fokas and E. A. Spence, Synthesis, as opposed to separation, of variables, SIAM Rev. **54** (2012), 291–324.

[47] M. Ganesh and S. C. Hawkins, A fully discrete Galerkin method for high frequency exterior acoustic scattering in three dimensions, J. Comput. Phys. **230** (2011), 104–125.

[48] C. Geuzaine, O. Bruno, and F. Reitich, On the O(1) solution of multiple-scattering problems, IEEE Trans. Magn. **41** (2005), 1488–1491.

[49] I. G. Graham, M. Löhndorf, J. M. Melenk, and E. A. Spence, When is the error in the $h$-BEM for solving the Helmholtz equation bounded independently of $k$?, BIT. ISSN 0006-3835 doi:10.1007/s10543-014-0501-5 (in press).

[50] S. P. Groth, D. P. Hewett, and S. Langdon, Hybrid numerical-asymptotic approximation for high frequency scattering by penetrable convex polygons, IMA J. Appl. Math., doi:10.1093/imamat/hxt040 (in press).

[51] J. A. Hargreaves, D. P. Hewett, Y. W. Lam, and S. Langdon, A high frequency boundary element method for scattering by three-dimensional screens (in preparation).

[52] P. J. Harris and K. Chen, On efficient preconditioners for iterative solution of a Galerkin boundary element equation for the three-dimensional exterior Helmholtz problem, J. Comput. Appl. Math. **156** (2003), 303–318.

[53] D. P. Hewett, S. Langdon, and S. N. Chandler-Wilde, A frequency-independent boundary element method for scattering by two-dimensional screens and apertures, IMA J. Numer. Anal., doi:10.1093/imanum/dru043 (in press).

[54] D. P. Hewett, S. Langdon, and J. M. Melenk, A high frequency $hp$ boundary element method for scattering by convex polygons, SIAM J. Numer. Anal. **51** (2013), 629–653.

[55] R. Hiptmair, C. Jerez-Hanckes, and C. Urzua, Optimal operator preconditioning for boundary elements on open curves, Tech. report 2013-48, Seminar for Applied Mathematics, ETH Zürich, 2013.

[56] R. Hiptmair and L. Kielhorn, BETL: A Generic Boundary Element Template Library, Tech. report 2012-36, Seminar for Applied Mathematics, ETH Zürich, Switzerland, 2012.

[57] G. C. Hsaio and W. L. Wendland, Boundary element methods: Foundation and error analysis, in Fundamentals, E. Stein, R. de Borst, and T. J. R. Hughes, eds., Encyclopedia of Computational Mechanics 1, John Wiley, New York (2004), 339–373.

[58] D. Huybrechs and S. Vandewalle, A sparse discretization for integral equation formulations of high frequency scattering problems, SIAM J. Sci. Comput. **29** (2007), 2305–2328.

[59] F. Ihlenburg, Finite Element Analysis of Acoustic Scattering, Appl. Math. Sci. **132**, Springer-Verlag, Berlin (1998).

[60] R. Kress, Linear Integral Equations, 2nd ed., Springer-Verlag, New York (1999).

[61] S. Langdon and S. N. Chandler-Wilde, A wavenumber independent boundary element method for an acoustic scattering problem, SIAM J. Numer. Anal. **43** (2006), 2450–2477.

[62] S. Langdon, M. Mokgolele, and S. N. Chandler-Wilde, High frequency scattering by convex curvilinear polygons, J. Comput. Appl. Math. **234** (2010), 2020–2026.

[63] R. Leis, Zur dirichletschen randwertaufgabe des aussenraumes der schwingungsgleichung, Math. Z. 90 (1965), 205–211.

[64] M. Löhndorf and J. M. Melenk, Wavenumber-explicit hp-BEM for high frequency scattering, SIAM J. Numer. Anal. **49** (2011), 2340–2363.

[65] W. McLean, Strongly Elliptic Systems and Boundary Integral Equations, Cambridge University Press, 2000.

[66] M. Messner, M. Messner, P. Urthaler, and F. Rammerstorfer, Hyperbolic and Elliptic Numerical Analysis (HyENA), 2010.

[67] A. Moiola and E. A. Spence, Is the Helmholtz equation really sign-indefinite?, SIAM Rev. **56** (2014), 274–312.

[68] J.-C. Nédélec, Acoustic and electromagnetic equations: integral representations for harmonic problems, Springer-Verlag New York, 2001.

[69] G. Of, O. Steinbach, and W. L. Wendland, The fast multipole method for the symmetric boundary integral formulation, IMA J. Numer. Anal. **26** (2006), 272–296.

[70] O. I. Panich, On the question of the solvability of exterior boundary-value problems for the wave equation and for a system of Maxwell's equations (in Russian), Uspekhi Mat. Nauk **20** (1965), 221–226.

[71] W. Rudin, Functional Analysis, 2nd ed., McGraw–Hill, New York, 1991.

[72] S. A. Sauter and C. Schwab, Boundary Element Methods, Springer Ser. in Comput. Math. 39, Springer-Verlag, Berlin, 2011.

[73] I. H. Sloan, Error analysis of boundary integral methods, Acta Numer. **1** (1992), 287–339.

[74] I. H. Sloan, Boundary element methods, in Theory and Numerics of ODEs and PDEs, M. Ainsworth, J. Levesley, W. A. Light, and M. Marletta, eds., Adv. Numer. Anal. 4, Oxford Science Publications, 1995, 143–180.

[75] W. Śmigaj, S. Arridge, T. Betcke, J. Phillips, and M. Schweiger, Solving boundary integral problems with BEM++, ACM Trans. Math. Software, to appear.

[76] E. A. Spence, "When All Else Fails, Integrate By Parts": New and Old Variational Formulations for Linear Elliptic PDEs, in Unified Transform for Boundary Value Problems: Applications and Advances, A. S. Fokas and B. Pelloni, eds., SIAM, 2015.

[77] E. A. Spence, S. N. Chandler-Wilde, I. G. Graham, and V. P. Smyshlyaev, A new frequency-uniform coercive boundary integral equation for acoustic scattering, Comm. Pure Appl. Math. **64** (2011), 1384–1415.

[78] E. A. Spence and A. S. Fokas, A new transform method I: domain-dependent fundamental solutions and integral representations, Proc. R. Soc. A **466** (2010), 2259–2281.

[79] E. A. Spence, I. V. Kamotski, and V. P. Smyshlyaev, Coercivity of combined boundary integral equations in high-frequency scattering, Comm. Pure Appl. Math., doi:10.1002/cpa.21543 (in press).

[80] O. Steinbach, Numerical Approximation Methods for Elliptic Boundary Value Problems: Finite and Boundary Elements, Springer, New York, 2008.

[81] E. P. Stephan, Boundary integral equations for screen problems in $\mathbb{R}^3$, Integral Equations Operator Theory **10** (1987), 236–257.

[82] R. H. Torres and G. V. Welland, The Helmholtz equation and transmission problems with Lipschitz interfaces, Indiana Univ. Math. J. **42** (1993), 1457–1486.

# 7.4 ▪ Novel Multitrace Boundary Integral Equations for Transmission Boundary Value Problems

*X. Claeys, R. Hiptmair, C. Jerez-Hanckes, and S. Pintarelli*

## 7.4.1 ▪ Introduction

This is the story of a marriage between BEMs and domain decomposition. In fact, viewed from the angle of BEMs, this relationship may be labeled a forced marriage, because, as the reader will certainly remember, BEMs can only cope with linear boundary value problems with constants coefficients. Piecewise constant coefficients are still within their scope, but in this case, the computational domain has to be decomposed into *subdomains*, on which the coefficients are constant. Subsequently, boundary integral equations have to be devised for the resulting transmission problems. They feature traces on the interfaces between subdomains as unknowns. Then, for the sake of discretization, these interfaces are triangulated and the degrees of freedom of the boundary element method will be located on the union of the interfaces, the so-called *skeleton*. All this very much resembles what is done in domain decomposition methods, even in the context of finite elements.

The most important representatives of these skeleton-based BEM approaches rely on what we have dubbed the (direct) *classical single-trace boundary integral formulation* (STF). It had long been known for simple transmission problems comprising two subdomains [24] and was presented in full generality in [79] in the setting of strongly elliptic second-order scalar problems. More exotic, indirect variants have been developed, for instance, in [68, 45, 69]. Second-kind STFs are known, too, and covered in [17, 13, 31, 78, 81] for scalar transmission problems and in [58, section 23] for electromagnetics. The classical STF is particularly popular in computational electromagnetics, where it is known as the Poggio–Miller–Chew–Harrington–Wu–Tsai (PMCHWT) integral equations [67, 10, 80, 33]. Their numerical analysis for two subdomains was first accomplished in [9] and later extended in [7].

As unknowns the direct STF features a full set of Cauchy data, that is, pairs of Dirichlet and Neumann traces, on each interface between adjacent subdomains. In variational form, its associated bilinear form is built from local subdomain contributions, very much in the spirit of domain decomposition. This dispenses with global interactions in the assembly of the discrete boundary integral operators. The benefit of localization may be so big that it can be worthwhile to pursue "genuine domain decomposition" by introducing extra artificial interfaces inside regions with constant coefficients.

However, the classical STF renounces the spirit of domain decomposition in the choice of unknowns, which establish the coupling between the subdomains in strong form. As a consequence, the variational STF employs a function space that fails to be a simple product space of subdomain contributions. At first glance this occurs without a penalty, since the STF turns out to be unconditionally well posed; see [79] and [16, Section 3.2]. A brief review of the derivation and analysis of STF will be given in Section 7.4.3 of this article.

A drawback of the strong coupling imposed through the function space has surfaced recently: it compounds the difficulties of designing preconditioners. This matters, because modern boundary element applications are inconceivable without the use of local low-rank matrix compression implemented in techniques like fast multipole methods [30, 25], $\mathcal{H}$-matrix compression [32], or adaptive cross approximation [5, Chapter 3]. Compressed matrices allow only the use of iterative solvers, whose speed of convergence will deteriorate for ill-conditioned linear systems. Yet, standard low-order boundary element Galerkin discretization of the classical STF, which amounts to a *first-kind boundary integral equation*, will invariably produce ill-conditioned linear systems on fine triangulations. Thus, effective preconditioning becomes crucial.

Many preconditioning strategies have been suggested for discretized first-kind boundary integral equations. Among them are geometric multilevel subspace correction methods (two-grid or multigrid) [71, 1, 75, 36, 43, 61, 52], as well as attempts to bring algebraic multigrid to bear on BEM [60, 48, 49]. However, no idea has revolutionized preconditioning for BEM as much as an approach known as *Calderón preconditioning*, invented by Steinbach and co-workers in [72, 57] and later extended by Buffa and Christiansen in [12, 8]. It has had and continues to have a massive impact, for instance, in BEM-based simulations in computational electromagnetism; witness the flurry of papers that has been devoted to its use [77, 4, 76, 2, 3, 21]. We point out that Calderón preconditioning fits the more general policy of *operator preconditioning* introduced in [38, 55]; see also [16, Section 4].

Awareness of the gist of operator preconditioning as presented in [38] is key to understanding why it encounters problems for the STF. Thus, we briefly recall the main result of [38].

**Theorem 7.23 (Theorem 2.1 of [38]).** *Let $X$, $Y$ be Hilbert spaces, and let $X_h := \mathrm{span}\{\varphi_i\}_{i=0}^N \subset X$ and $Y_h := \mathrm{span}\{\phi_j\}_{j=0}^M \subset Y$ be finite-dimensional subspaces with bases $\{\varphi_i\}_{i=0}^N$ and $\{\phi_j\}_{j=0}^M$. Further, let $\mathbf{a} \in L(X \times X, \mathbb{R})$ and $\mathsf{b} \in L(Y \times Y, \mathbb{R})$ be continuous bilinear forms (with norms $\|\mathbf{a}\|$ and $\|\mathsf{b}\|$, respectively), each satisfying discrete inf-sup conditions with constants $c_A, c_B > 0$ on $X_h$ and $Y_h$, respectively. If there is a continuous bilinear form $\mathsf{d} \in L(X \times Y, \mathbb{R})$ that also satisfies a discrete inf-sup condition on $X_h \times Y_h$ with constant $c_T > 0$, then the associated Galerkin matrices*

$$\mathbf{A}_h := \left(\mathbf{a}(\varphi_i, \varphi_j)\right)_{i,j=1}^N, \quad \mathbf{B}_h := \left(\mathsf{b}(\phi_i, \phi_j)\right)_{i,j=1}^M, \quad \mathbf{D}_h := \left(\mathsf{d}(\varphi_i, \phi_j)\right)_{i,j=1}^{N,M}$$

*satisfy*

$$\varkappa(\mathbf{D}_h^{-1}\mathbf{B}_h\mathbf{D}_h^{-T}\mathbf{A}_h) \le \frac{\|\mathbf{a}\|\,\|\mathsf{b}\|\,\|\mathsf{d}\|^2}{c_A c_B c_T^2}, \tag{7.173}$$

*where $\varkappa$ designates the spectral condition number.*

This theorem clearly reveals what it takes to build a viable "operator preconditioner" $\mathbf{M}_h := \mathbf{D}_h^{-1}\mathbf{B}_h\mathbf{D}_h^{-T}$ when given a stable Galerkin discretization of a variational problem involving the bilinear form $\mathbf{a}$. We have to find

- a space $Y$, equipped with a bilinear form $\mathsf{b}$,
- a suitable Galerkin trial space $Y_h \subset Y$ that yields a stable discretization of $\mathsf{b}$,
- and a "pairing bilinear form" $\mathsf{d}$ that induces a stable discrete duality pairing between $X_h$ and $Y_h$ (independent of the choice of discretization parameters)
- and that gives rise to a square Galerkin matrix $\mathbf{D}_h$, for which linear systems can be solved with little computational effort.

In the context of boundary element methods the last item usually means that the Galerkin matrix $\mathbf{D}_h$ is *sparse*. Thus, its formal inverse in the definition of the preconditioner $\mathbf{M}_h$ can be evaluated by means of direct Gaussian elimination. In general, sparsity of $\mathbf{D}_h$ can be achieved only if $\mathsf{d}$ is local, for instance, a simple $L^2$-inner product.

Unfortunately, for situations with more than two subdomains, so far no space $Y$ has been found that is in duality with the variational space $X$ of the STF with respect to a local pairing bilinear form d. It goes without saying that suitable boundary element spaces $Y_h$ also remain elusive. This seems to be a fundamental obstacle to the application of Calderón preconditioning to the STF, as explained in [16, Section 4.5].

$L^2$-dual pairs of trace spaces are well known for boundaries of individual subdomains. Thus, if a stable variational boundary integral equation can be posed on their product space, local operator preconditioning becomes a straightforward option. This insight was the main motivation behind the development of *multitrace formulations* (MTFs). Truly in the spirit of domain decomposition, they rely on simple products of local trace spaces and impose coupling between different subdomains weakly. They owe their name to the use of two pairs of unknown Cauchy traces on every interface. It is this new class of boundary integral formulations that this article is devoted to.

MTFs come in two different flavors: we distinguish between *global* and *local* MTFs. The former can be deduced from the STF through a vanishing gap limit and they will be treated in Section 7.4.4. The latter employ local transmission conditions and they are discussed in Section 7.4.5, with numerical results reported in Section 7.4.6. We emphasize that this article is largely meant to be a review. Occasionally, rigorous proofs and technical details are skipped. Those can be found in the original publications:

- [18], as concerns the global MTF for acoustic scattering,
- [14], where the global MTF for electromagnetic scattering was introduced,
- [39], which proposed the local MTF,
- [40], where the local MTF is extended to more general transmission conditions,
- [15], where a global MTF for scattering problems with homogeneous Dirichlet boundary conditions is derived with focus on avoiding spurious resonances by means of combined field integral equations.

Parallel to the developments in numerical analysis, local multitrace BIEs have recently been devised for large-scale parallel simulations in computational electromagnetism by Lee, Peng, and collaborators [64, 65, 66]. This underscores their relevance for computational engineering.

In parts, this manuscript runs parallel to the survey article [16]. What is new is the treatment of essential boundary conditions, because, apart from [15], earlier work has always been concerned with (scattering) transmission problems posed on the entire space $\mathbb{R}^d$, $d = 2, 3$. Also new are the numerical investigations of the local MTFs in three dimensions reported in Section 7.4.6.

Admittedly, MTFs are by no means the only attempt to harness ideas from domain decomposition for boundary element methods. Prominently, this was also pursued by Steinbach and co-workers with the boundary element tearing and interconnecting method [62, 47, 51, 50, 53]. This is a specimen of the class of BEM-based domain decomposition methods that involve approximate realizations of Dirichlet-to-Neumann maps by means of boundary elements [53, 73]. Other approaches marrying boundary elements and domain decomposition employ Lagrangian multipliers; cf. [46].

A word of warning: throughout we address domain decomposition in the volume. It must not be mixed up with domain decomposition on surfaces meant to break apart the linear systems arising from BEMs. Only recently various kinds of such schemes have been proposed; see [11, 37, 35, 20, 63].

Table 7.2: List of Notation.

| | |
|---|---|
| $\Omega_j, \Omega_Z$ | Lipschitz domains, "subdomains," see (7.174) |
| $\mathscr{L}_j$ | second-order diffusion operator associated with $\Omega_j$, see (7.176a), Page 231 |
| $\mathsf{T}_{D,i}, \mathsf{T}_{N,i}$ | Dirichlet and Neumann trace operators onto $\partial\Omega_i$ |
| $\langle \cdot, \cdot \rangle_{\partial\Omega}$ | $L^2$ duality pairing between $H^{\frac{1}{2}}(\partial\Omega)$ and $H^{-\frac{1}{2}}(\partial\Omega)$ |
| $\mathbb{T}_i$ | Cauchy trace operator on $\partial\Omega_i$ |
| $\mathscr{T}(\partial\Omega_i)$ | Cauchy trace space on $\partial\Omega_i$, see Page 232 |
| $\langle\!\langle \cdot, \cdot \rangle\!\rangle_{\partial\Omega}$ | duality pairing in Cauchy trace space, see (7.183), Page 232 |
| $\mathbb{G}_i$ | potential associated with $\Omega_i$, see (7.184), Page 232 |
| $\mathbb{P}_i$ | Calderón projector onto $\mathbb{T}_i$, see (7.186), Page 233 |
| $\mathbb{A}_i$ | Calderón operator belonging to $\Omega_i$, see (7.187), Page 233 |
| $\mathscr{MT}(\Sigma)$ | multitrace space, see (7.188), Page 233 |
| $\mathbb{L}_i$ | localization operator $\mathscr{MT}(\Sigma) \top \mathscr{T}(\partial\Omega_i)$, see (7.189), Page 233 |
| $\langle\!\langle \cdot, \cdot \rangle\!\rangle_{\Sigma}$ | self-duality pairing for $\mathscr{MT}(\Sigma)$, see (7.190), Page 233 |
| $\mathscr{ST}(\Sigma), \mathscr{ST}_0(\Sigma)$ | single-trace spaces, see (7.191), Page 234, (7.193), Page 234 |
| $\widehat{\mathscr{MT}}(\Sigma), \widehat{\mathscr{MT}}_0(\Sigma)$ | clipped multitrace spaces, see (7.208), (7.209), Page 238 |
| $\mathbb{C}_{i \to j}$ | remote coupling operators, see (7.212), Page 239 |
| $\mathscr{G}_i$ | triangular surface mesh on $\partial\Omega_i$ |
| $\mathscr{T}_h(\partial\Omega_i)$ | boundary element Cauchy trace space on $\partial\Omega_i$, see (7.214), Page 241 |
| $\mathbb{S}_{ij}$ | trace restriction operator, see (7.220), Page 242 |
| $\mathbb{X}_{i \to j}$ | trace transfer operator, see (7.220), Page 242 |
| $\widetilde{\mathscr{T}}_{\text{pw}}(\partial\Omega_i)$ | piecewise Cauchy trace space, see (7.224), Page 243 |
| $\mathscr{MT}_0(\Sigma), \widetilde{\mathscr{MT}}_0(\Sigma)$ | multitrace spaces with local regularity, see (7.228a), Page 244 |
| $\widetilde{\mathscr{MT}}_0^{\sharp}(\Sigma)$ | multitrace space with locally regular Neumann components, see (7.232), Page 245 |

## 7.4.2 ▪ Exterior Transmission BVPs

As a model problem we study a scalar second-order elliptic transmission BVP in the exterior $\mathbb{R}^d \setminus \overline{\Omega}_Z$ of a bounded Lipschitz domain $\Omega_Z \subset \mathbb{R}^d$, $d = 2, 3$. We restrict ourselves to spatially varying diffusion coefficients $\mu = \mu(\boldsymbol{x})$, which are *piecewise constant* with respect to a partition of $\mathbb{R}^d \setminus \overline{\Omega}_Z$,

$$\mathbb{R}^d \setminus \Omega_Z = \overline{\Omega}_0 \cup \overline{\Omega}_1 \cup \ldots \cup \overline{\Omega}_N, \quad N \in \mathbb{N}, \tag{7.174}$$

into $N+1$ subdomains $\Omega_i$, of which only $\Omega_0$ is unbounded; see Figure 7.23 for an illustration. The two subdomains $\Omega_i$ and $\Omega_j$, $i \neq j$, are separated by the interface $\Gamma_{ij}$. The piece of boundary separating $\Omega_i$ and $\Omega_Z$ is denoted by $\Gamma_{iZ}$. The union of all boundaries forms the skeleton $\Sigma$: $\Sigma = \bigcup_i \partial\Omega_i$. If $\Omega_Z \neq \emptyset$ or $N > 1$, junction points will usually occur and then the skeleton may not be orientable nor be a manifold.

Given diffusion coefficients $\mu_i > 0$, $i = 0, \ldots, N$, and Dirichlet data $g \in H^{\frac{1}{2}}(\partial\Omega_Z)$, our model transmission problem seeks $U \in H^1_{\text{loc}}(\mathbb{R}^d \setminus \overline{\Omega}_Z)$ that satisfies the Dirichlet boundary conditions

$$U = g \quad \text{on } \partial\Omega_Z, \tag{7.175}$$

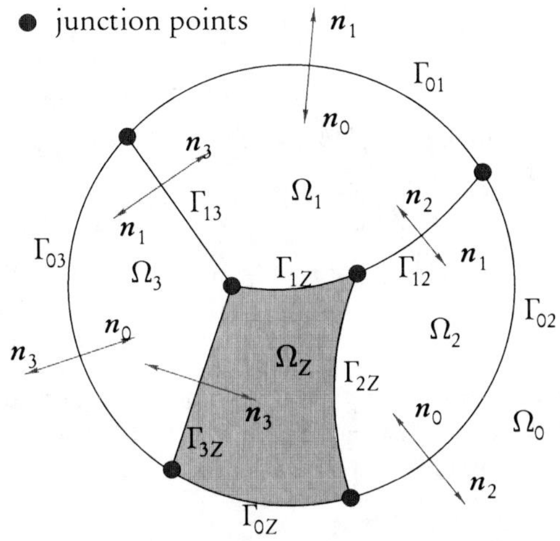


**Figure 7.23.** *Geometric situation for the second-order scalar elliptic exterior transmission model problem. The $n_j$'s stand for the* exterior *unit normal vector fields on the subdomain boundaries $\partial\Omega_j$.*

which complies with suitable decay conditions at infinity (depending on the space dimension $d$ (see [56, Chapter 8]), and whose restrictions $U_i := U|_{\Omega_i} \in H^1_{\mathrm{loc}}(\Omega_i)$, $i = 0, \ldots, N$, fulfill

$$\mathscr{L}_i U_i := -\operatorname{div}(\mu_i \,\mathbf{grad}\, U_i) = 0 \quad \text{in } \Omega_i\,, \tag{7.176a}$$

$$U_i|_{\Gamma_{ij}} - U_j\Big|_{\Gamma_{ij}} = 0\,, \quad \mu_i \frac{\partial U_i}{\partial \boldsymbol{n}_i}\bigg|_{\Gamma_{ij}} + \mu_j \frac{\partial U_j}{\partial \boldsymbol{n}_j}\bigg|_{\Gamma_{ij}} = 0 \quad \text{on } \Gamma_{ij}\,. \tag{7.176b}$$

Equivalently, $U$ can be characterized as the solution of a variational problem posed on a Sobolev space with weighted $H^1(\mathbb{R}^d \setminus \overline{\Omega}_Z)$-norm; cf. [70, Section 2.9.2.4.] and [26, Chapter XI, Part B]. Existence and uniqueness of solutions of this variational problem can be established by standard techniques; see [70, Section 2.10.2.2] or [56, Chapter 8].

**Remark 53 (More general boundary conditions on $\partial\Omega_Z$).** *We confine ourselves to Dirichlet boundary conditions on $\partial\Omega_Z$ for the sake of simplicity. Neumann boundary conditions can be treated alike and BIEs can even accommodate a mixture of both following the ideas of* [70, Section 3.5.2].

**Remark 54.** *As remarked in the introduction, multitrace BIEs can be derived for many more transmission problems beside the simple model problem* (7.176). *In fact, in* [16] *and* [18] *multitrace BIEs were first proposed and analyzed for acoustic scattering. Recall that the* acoustic transmission scattering problem *involves the local partial differential equations*

$$-\operatorname{div}(\mu_i \,\mathbf{grad}\, U_i) - \varkappa_i^2 U_i = 0 \quad \textit{in } \Omega_i \tag{7.177}$$

*with locally constant wavenumbers $\varkappa_i > 0$ and Sommerfeld radiation conditions at infinity* [19, Chapter 2], [59, Chapter 2]. *The transmission conditions* (7.176b) *and boundary conditions* (7.175) *apply unchanged. An extension to transmission problems for time-harmonic* *electromagnetic waves is pursued in* [14]. *Those read as*

$$\mathbf{U}\times\mathbf{n}_Z=\mathbf{g} \quad on\ \partial\Omega_Z\,, \tag{7.178}$$

$$\mathbf{curl}(\mu_i\,\mathbf{curl}\,\mathbf{U}_i)-\varkappa_i^2\mathbf{U}_i=0 \quad in\ \Omega_i\,, \quad i=0,\ldots,N\,, \tag{7.179}$$

$$\left.\begin{aligned} \boldsymbol{n}_i\times(\mathbf{U}_i|_{\Gamma_{ij}}\times\boldsymbol{n}_i)-\boldsymbol{n}_j\times(\mathbf{U}_j\big|_{\Gamma_{ij}}\times\boldsymbol{n}_j)=0\,,\\ \mu_i\,\mathbf{curl}\,\mathbf{U}_i|_{\Gamma_{ij}}\times\boldsymbol{n}_i+\mu_j\,\mathbf{curl}\,\mathbf{U}_j\big|_{\Gamma_{ij}}\times\boldsymbol{n}_j=0 \end{aligned}\right\} \quad on\ \Gamma_{ij}\,, \tag{7.180}$$

$$+\quad \textit{Silver-Müller radiation conditions at } \infty \textit{ for } \mathbf{U}-\mathbf{U}_{\mathrm{inc}}.$$

*A detailed and comprehensive presentation of multitrace BIEs for the scattering transmission problems* (7.177) *and* (7.178) *is given in* [16]. *In all these works essential boundary conditions are not taken into account.*

### 7.4.3 ▪ Single-Trace BIEs

#### Calderón Projectors

Two trace operators are naturally associated with the second-order scalar differential operator $\mathscr{L}_i\,U_i := -\mathrm{div}(\mu_i\,\mathbf{grad}\,U_i)$. These are the Dirichlet trace $\mathsf{T}_{D,i}$, and Neumann trace $\mathsf{T}_{N,i}$, defined for smooth functions $V$ on $\overline{\Omega}_i$ through

$$\mathsf{T}_{D,i}\,U := U|_{\partial\Omega_i}\,,\quad \mathsf{T}_{N,i}\,U := \mu_i\,\mathbf{grad}\,U\cdot\boldsymbol{n}_i|_{\partial\Omega_i}\,. \tag{7.181}$$

They can be extended to continuous and surjective operators [70, Sections 2.6, 2.7],[9]

$$\mathsf{T}_{D,i}:H^1(\Omega_i)\to H^{\frac12}(\partial\Omega_i),\quad \mathsf{T}_{N,i}:H(\Delta,\Omega_i)\to H^{-\frac12}(\partial\Omega_i). \tag{7.182}$$

Their ranges are known as *trace spaces* and form the Gelfand triple $H^{\frac12}(\partial\Omega_i)\subset L^2(\partial\Omega_i)\subset H^{-\frac12}(\partial\Omega_i)$ with a duality $H^{\frac12}(\partial\Omega_i)'\cong H^{-\frac12}(\partial\Omega_i)$ effected by the $L^2(\partial\Omega_i)$ inner product. We write $\langle\cdot,\cdot\rangle_{\partial\Omega_i}$ for the associated duality pairing, which agrees with the $L^2(\partial\Omega_i)$-inner product for sufficiently regular functions. Trace spaces and operators may be combined into *Cauchy trace spaces* and *Cauchy trace operators*:

$$\mathscr{T}(\partial\Omega_i):=H^{\frac12}(\partial\Omega_i)\times H^{-\frac12}(\partial\Omega_i),\quad \mathbb{T}_i:\left\{\begin{array}{ccc} H(\Delta,\Omega_i) & \to & \mathscr{T}(\partial\Omega_i),\\ U & \mapsto & (\mathsf{T}_{D,i}\,U,\mathsf{T}_{N,i}\,U). \end{array}\right.$$

The ranges of the $\mathbb{T}_i$ are dense in $\mathscr{T}(\partial\Omega_i)$ [22, Lemma 3.5]. Obviously, the Cauchy trace spaces are in self-duality with respect to the skew-symmetric pairing[10]

$$\langle\!\langle \mathfrak{u},\mathfrak{v}\rangle\!\rangle_{\partial\Omega_i} := \langle u,\varphi\rangle_{\partial\Omega_i}-\langle v,\nu\rangle_{\partial\Omega_i}\,,\quad \mathfrak{u}:=\begin{pmatrix}u\\ \nu\end{pmatrix},\ \mathfrak{v}:=\begin{pmatrix}v\\ \varphi\end{pmatrix}\in\mathscr{T}(\partial\Omega)\,. \tag{7.183}$$

Beside trace operators, potential *representation formulae* are the linchpin of the derivation of BIEs. Here, the term "potential" is reserved for mappings that take functions on boundaries to functions in domains. These functions usually provide solutions of the homogeneous PDE. For our concrete operators $\mathscr{L}_i := -\mathrm{div}(\mu_i\,\mathbf{grad}\,\cdot)$ the key potential is formally given by

$$\mathbb{G}_i(\mathfrak{u})(\boldsymbol{x}) := \langle\!\langle \mathbb{T}_i\,\Phi(\boldsymbol{x}-\cdot),\mathfrak{u}\rangle\!\rangle_{\partial\Omega_i}\,,\quad \boldsymbol{x}\in\mathbb{R}^d\setminus\partial\Omega_i,\ \mathfrak{u}\in\mathscr{T}(\partial\Omega_i)\,, \tag{7.184}$$

[9] As usual, $H(\Delta,\Omega):=\{U\in H^1(\Omega):\ \Delta U\in L^2(\Omega)\}$.

[10] Fraktur font is used to designate functions in the Cauchy trace space, whereas Roman typeface is reserved for Dirichlet traces and Greek symbols for Neumann traces.

where $\Phi$ is the fundamental solution for $\mathscr{L}_i$; see [70, Equation (3.3)]. A comprehensive discussion of potentials can be found in [70, Section 3.1]. Then, *every* $U \in H_{\mathrm{loc}}(\Delta,\Omega_i)$ that satisfies $\mathscr{L}_i\, U = 0$ and, for $i = 0$, the appropriate decay conditions at $\infty$, has the representation (in the sense of distributions)

$$U = \mathbb{G}_i(\mathbb{T}_i\, U); \tag{7.185}$$

cf. [70, Section 3.11] and [56, Chapter 6]. Applying the Cauchy trace to the potential yields the *Calderón projector* (see [16, Section 2.3], [70, Section 3.6], and [44, Section 5.6]):

$$\mathbb{P}_i := \mathbb{T}_i\, \mathbb{G}_i : \mathscr{T}(\partial\Omega_i) \to \mathscr{T}(\partial\Omega_i)\,. \tag{7.186}$$

This operator turns out to be a projector: $\mathbb{P}_i^2 = \mathbb{P}_i$. The Calderón projectors owe their importance to the following fundamental theorem [16, Theorem 2.6], [70, Propositon 3.6.2(ii)].

**Theorem 7.24.** *$U_i \in H(\Delta,\Omega_i)$ solves $\mathscr{L}_i\, U_i = 0$ in $\Omega_i$ (and satisfies appropriate decay conditions at $\infty$ for $i = 0$) if and only if $(\mathbb{P}_i - \mathsf{Id})\,\mathbb{T}_i\, U_i = 0$.*

We remark that the Calderón projector $\mathbb{P}_i$ implicitly contains the customary four different boundary integral operators associated with second-order scalar PDEs [70, Equation (3.122)], because the *Calderón operator*

$$\mathbb{A}_i := \mathbb{P}_i - \tfrac{1}{2}\mathsf{Id} \quad \text{can be written as} \quad \mathbb{A}_i = \begin{pmatrix} -\mathsf{K}_i & \mathsf{V}_i \\ \mathsf{W}_i & \mathsf{K}_i' \end{pmatrix}, \tag{7.187}$$

where we have adopted the notations from [70, Section 3.1]: $\mathsf{K}_i$, $\mathsf{V}_i$, $\mathsf{W}_i$, and $\mathsf{K}_i'$ stand for the double-layer, single-layer, hypersingular, and adjoint double-layer boundary integral operators on $\partial\Omega_i$, respectively.

### Skeleton Trace Spaces

Boundary integral equations arising from the transmission problem (7.176) invariably involve unknown functions in trace spaces on the skeleton $\Sigma$. The largest and simplest such space is the (skeleton) *multitrace space*:

$$\mathscr{M}\mathscr{T}(\Sigma) := \mathscr{T}(\partial\Omega_0) \times \mathscr{T}(\partial\Omega_1) \times \cdots \times \mathscr{T}(\partial\Omega_N)\,, \tag{7.188}$$

which comprises completely decoupled local traces. It owes its name to the fact that on each interface $\Gamma_{ij}$ a function $\underline{\mathfrak{u}} \in \mathscr{M}\mathscr{T}(\Sigma)$[11] comprises two pairs of Dirichlet and Neumann data, each stemming from a subdomain on either side. The simple component projections $\mathbb{L}_i : \mathscr{M}\mathscr{T}(\Sigma) \to \mathscr{T}(\partial\Omega_i)$ isolate the contribution of individual subdomains:

$$\mathbb{L}_i\, \underline{\mathfrak{u}} := \begin{pmatrix} u_i \\ \nu_i \end{pmatrix} \text{ for } \underline{\mathfrak{u}} := \left( \begin{pmatrix} u_0 \\ \nu_0 \end{pmatrix}, \ldots, \begin{pmatrix} u_N \\ \nu_N \end{pmatrix} \right) \in \mathscr{M}\mathscr{T}(\Sigma)\,. \tag{7.189}$$

Properties of its local components carry over to the multitrace space, for instance, self-duality with respect to the symplectic $L^2$-type bilinear pairing (cf. (7.183)),

$$\langle\!\langle \underline{\mathfrak{u}}, \underline{\mathfrak{v}} \rangle\!\rangle_\Sigma := \sum_{i=0}^{N} \langle\!\langle \mathbb{L}_i\, \underline{\mathfrak{u}}, \mathbb{L}_i\, \underline{\mathfrak{v}} \rangle\!\rangle_{\partial\Omega_i}\,, \quad \underline{\mathfrak{u}}, \underline{\mathfrak{v}} \in \mathscr{M}\mathscr{T}(\Sigma)\,. \tag{7.190}$$

[11] Functions in a multitrace space are tagged by an underline, e.g., $\underline{\mathfrak{u}}$, $\underline{\mathfrak{v}}$.

**Corollary 7.25.** *The pairing* (7.190) *induces an isomorphism* $\mathcal{MT}(\Sigma) \cong \mathcal{MT}(\Sigma)'$.

Generically, multitrace functions on different subdomains are utterly disconnected. If we impose the continuity stipulated by the transmission conditions (7.176b), we arrive at the so-called *single-trace spaces*

$$\mathcal{ST}(\Sigma) := \left\{ \left( \begin{pmatrix} u_i \\ v_i \end{pmatrix} \right)_{i=0}^{N} \in \mathcal{MT}(\Sigma) : \begin{array}{l} \exists V \in H^1(\mathbb{R}^d) : u_i = \mathsf{T}_{D,i} V \\ \exists \mathbf{W} \in \boldsymbol{H}(\mathrm{div}, \mathbb{R}^d) : v_i = \mathsf{T}_{n,i} \mathbf{W} \end{array} \right\}. \tag{7.191}$$

Here, $\mathsf{T}_{n,i} \mathbf{W}$ is the normal component trace $(\mathbf{W} \cdot \boldsymbol{n}_i)|_{\partial\Omega_i}$ on $\partial\Omega_i$ for a vector field $\mathbf{W}$ in $\Omega_i$. For traces in $L^2(\partial\Omega_i)$, the formal definition (7.191) implies

$$\left( \begin{pmatrix} u_i \\ v_i \end{pmatrix} \right)_{i=0}^{N} \in \mathcal{ST}(\Sigma) \quad \Rightarrow \quad \begin{array}{c} u_i|_{\Gamma_{ij}} = u_j\big|_{\Gamma_{ij}} \text{ and } v_i|_{\Gamma_{ij}} = -v_j\big|_{\Gamma_{ij}}, \\ 0 \le i, j \le N; \end{array} \tag{7.192}$$

that is, the $H^{\frac{1}{2}}$-components are continuous across interfaces, and so are the $H^{-\frac{1}{2}}$-components, the latter up to a change of sign. The sign change reflects the opposite relative orientation of an interface with respect to the two adjacent subdomains; see Figure 7.23. The definition (7.191) may be modified slightly in order to impose zero Dirichlet boundary conditions on $\partial\Omega_Z$.

$$\mathcal{ST}_0(\Sigma) := \left\{ \begin{pmatrix} u_i \\ v_i \end{pmatrix}_{i=0}^{N} : \begin{array}{l} \exists U \in H^1_{\partial\Omega_Z}(\mathbb{R}^d \setminus \Omega_Z) : u_i = \mathsf{T}_{D,i} U \\ \exists \mathbf{V} \in \boldsymbol{H}(\mathrm{div}, \mathbb{R}^d) : v_i = \mathsf{T}_{n,i} \mathbf{V} \end{array} \right\}. \tag{7.193}$$

Here, we wrote $H^1_{\partial\Omega_Z}(\mathbb{R}^d \setminus \Omega_Z)$ for the space of functions in $H^1(\mathbb{R}^d \setminus \Omega_Z)$ that vanish on $\partial\Omega_Z$.

Obviously, the restriction of single traces to $\partial\Omega_Z$ is well defined by taking the point trace $\mathsf{T}_{D,Z} V$ and normal component trace $\mathsf{T}_{n,Z} \mathbf{W}$ onto $\partial\Omega_Z$, respectively, of their extensions $V$ and $\mathbf{W}$ according to (7.191). For this restriction operation we introduce the localization onto $\partial\Omega_Z$

$$\mathbb{L}_Z = \begin{pmatrix} \mathsf{L}_{D,Z} \\ \mathsf{L}_{N,Z} \end{pmatrix} : \mathcal{ST}(\Sigma) \to \mathcal{T}(\partial\Omega_Z) := H^{\frac{1}{2}}(\partial\Omega_Z) \times H^{-\frac{1}{2}}(\partial\Omega_Z). \tag{7.194}$$

Thanks to (7.192) the pairing of single trace functions leads to massive cancellations, because for sufficiently smooth multitrace functions we can rewrite (7.190) as a sum of contributions of interfaces:

$$\langle\!\langle \underline{\mathfrak{u}}, \underline{\mathfrak{v}} \rangle\!\rangle_\Sigma = \sum_{\substack{i,j=0 \\ i<j}}^{N} \int_{\Gamma_{ij}} u_i \psi_i - v_i v_i + u_j \psi_j - v_j v_j \, \mathrm{d}S + \sum_{i=0}^{N} \int_{\Gamma_{iZ}} u_i \psi_i - v_i v_i \, \mathrm{d}S, \tag{7.195}$$

where $\underline{\mathfrak{u}} = \left(\binom{u_0}{v_0}, \dots, \binom{u_N}{v_N}\right)$, $\underline{\mathfrak{v}} = \left(\binom{v_0}{\psi_0}, \dots, \binom{v_N}{\psi_N}\right)$. Remember that $\Gamma_{iZ}$ are the interfaces separating subdomains and $\Omega_Z$. The identity (7.195) holds since each interior interface is visited twice in the evaluation of the pairing. Now, combine (7.195) with the insight from (7.192) and conclude for $\underline{\mathfrak{u}}, \underline{\mathfrak{v}} \in \mathcal{S}\mathcal{T}(\Sigma)$ that

$$\langle\langle \underline{\mathfrak{u}}, \underline{\mathfrak{v}} \rangle\rangle_\Sigma = \sum_{i,j} \int_{\Gamma_{ij}} \underbrace{(u_i - u_j)}_{=0}\psi_i - \underbrace{(v_i - v_j)}_{=0}\nu_i \, dS + \sum_{i=0}^{N} \int_{\Gamma_{iZ}} u_i \psi_i - v_i \nu_i \, dS .$$

We immediately get

$$\langle\langle \underline{\mathfrak{u}}, \underline{\mathfrak{v}} \rangle\rangle_\Sigma = -\langle\langle \mathbb{L}_Z \underline{\mathfrak{u}}, \mathbb{L}_Z \underline{\mathfrak{v}} \rangle\rangle_{\partial\Omega_Z} \quad \forall \underline{\mathfrak{u}}, \underline{\mathfrak{v}} \in \mathcal{S}\mathcal{T}(\Sigma) \tag{7.196}$$

and note that the minus sign is due to the opposite orientation of normals on $\partial\Omega_Z$.

**Remark 55.** *A fundamental result is the characterization of single-trace spaces as complete Lagrangian subspaces w.r.t. the pairing $\langle\langle \cdot, \cdot \rangle\rangle_\Sigma$ of multitrace spaces as given in* [18, Proposition 2.1] *and* [16, Theorem 2.23]. *It can be adapted to the homogeneous Dirichlet boundary on $\partial\Omega_Z$:*

$$\mathcal{S}\mathcal{T}_0(\Sigma) = \{\underline{\mathfrak{u}} \in \mathcal{M}\mathcal{T}(\Sigma) : \langle\langle \underline{\mathfrak{u}}, \underline{\mathfrak{v}} \rangle\rangle_\Sigma = 0 \; \forall \underline{\mathfrak{v}} \in \mathcal{S}\mathcal{T}_0(\Sigma)\} .$$

*The intuition behind this characterization is clear from* (7.195)*: Varying $\underline{\mathfrak{v}} \in \mathcal{S}\mathcal{T}_0(\Sigma)$ enforces* (7.192) *and vanishing Dirichlet components on the boundaries $\Gamma_{iZ}$.*

### First-Kind BIEs

Assume that $U \in H^1_{\text{loc}}(\mathbb{R}^d)$ solves the transmission problem (7.176). Then Theorem 7.24 permits us to conclude that

$$(\mathbb{P}_j - \mathsf{Id})\mathbb{T}_j U = 0 \quad \text{in } \mathcal{T}(\partial\Omega_j), \quad j = 0, \ldots, N . \tag{7.197}$$

Using (7.187), this implies that for all $\underline{\mathfrak{v}} \in \mathcal{S}\mathcal{T}_0(\Sigma)$

$$\sum_{j=0}^{N} \langle\langle (\mathbb{A}_j - \tfrac{1}{2}\mathsf{Id})\mathbb{T}_j U, \mathbb{L}_j \underline{\mathfrak{v}} \rangle\rangle_{\partial\Omega_j} = 0 . \tag{7.198}$$

Next, recall (7.196), that the Dirichlet components of $\underline{\mathfrak{v}}$ vanish on $\partial\Omega_Z$, $\mathsf{L}_{D,Z}\underline{\mathfrak{v}} = 0$, and that $U$ satisfies the Dirichlet boundary conditions (7.175), that is, $\mathsf{T}_{D,Z} U = g$ on $\partial\Omega_Z$. Thus, we find ($\mathbb{T}_Z U$ is the Cauchy trace of $U$ on $\partial\Omega_Z$)

$$\sum_{j=0}^{N} \langle\langle -\tfrac{1}{2}\mathbb{T}_j U, \mathbb{L}_j \underline{\mathfrak{v}} \rangle\rangle_{\partial\Omega_j} = -\tfrac{1}{2}\langle\langle \mathbb{T}_Z U, \mathbb{L}_Z \underline{\mathfrak{v}} \rangle\rangle_{\partial\Omega_Z} = -\tfrac{1}{2}\langle g, \mathsf{L}_{N,Z}\underline{\mathfrak{v}} \rangle_{\partial\Omega_Z},$$

and (7.198) can be converted into the equation

$$\sum_{j=0}^{N} \langle\langle \mathbb{A}_j \mathbb{T}_j U, \mathbb{L}_j \underline{\mathfrak{v}} \rangle\rangle_{\partial\Omega_j} = -\tfrac{1}{2}\langle g, \mathsf{L}_{N,Z}\underline{\mathfrak{v}} \rangle_{\partial\Omega_Z} . \tag{7.199}$$

Eventually, we have derived a variational equation satisfied by $\mathbb{T}_\Sigma U := (\mathbb{T}_i U)_{i=0}^N \in \mathcal{S}\mathcal{T}(\Sigma)$. In order to balance trial and test spaces, we employ the customary offset function technique. We rely on a function $G \in H^1_{\text{loc}}(\mathbb{R}^d)$, whose point trace on $\partial\Omega_Z$ agrees with the Dirichlet data $g$: $\mathsf{T}_{D,Z} G = g$. Thus, we can define the skeleton extension of the Dirichlet boundary values, $\underline{\mathfrak{g}} := \binom{G|_\Sigma}{0} \in \mathcal{S}\mathcal{T}(\Sigma)$. Introducing the skeleton Cauchy trace of $U$ minus this extension as unknown $\underline{\mathfrak{u}}$, that is, $\underline{\mathfrak{u}} = \mathbb{T}_\Sigma U - \underline{\mathfrak{g}}$ yields the classical STF: seek $\underline{\mathfrak{u}} \in \mathscr{S}\mathscr{T}_0(\Sigma)$ such that

$$\boxed{\sum_{j=0}^{N} \langle\!\langle \mathbb{A}_j \mathbb{L}_j \underline{\mathfrak{u}}, \mathbb{L}_j \underline{\mathfrak{v}} \rangle\!\rangle_{\partial\Omega_j} = -\tfrac{1}{2}\langle g, \mathsf{L}_{N,Z}\underline{\mathfrak{v}}\rangle_{\partial\Omega_Z} - \sum_{j=0}^{N} \langle\!\langle \mathbb{A}_j \mathbb{L}_j \underline{\mathfrak{g}}, \mathbb{L}_j \underline{\mathfrak{v}} \rangle\!\rangle_{\partial\Omega_j},} \tag{7.200}$$

for all $\underline{\mathfrak{v}} \in \mathscr{S}\mathscr{T}_0(\Sigma)$. This is a first-kind BIE in weak form. We write $\mathsf{m}_{\mathrm{STF}}$ for its associated bilinear form $\mathsf{m}_{\mathrm{STF}} : \mathscr{S}\mathscr{T}_0(\Sigma) \times \mathscr{S}\mathscr{T}_0(\Sigma) \to \mathbb{R}$.

### Existence and Uniqueness of Solutions

The bilinear form $\mathsf{m}_{\mathrm{STF}}$ from (7.200) is a sum of local contributions associated with the integral operators $\mathbb{A}_j$, $j = 0, \dots, N$. This permits us to appeal to well-established results asserting the ellipticity of single layer and hypersingular boundary integral operators. The proofs can be found in [56, Corollary 3.13, Theorem 8.18], [70, Theorem 3.5.3], and [74, Section 6.6.1].

**Theorem 7.26.** *Let $D \subset \mathbb{R}^d$ be a Lipschitz domain, $\partial D$ connected and bounded, and write $\mathsf{V}$ and $\mathsf{W}$ for the single-layer boundary integral operator and hypersingular boundary integral operator on $\partial D$, respectively, associated with $-\Delta$. If $d = 2$, assume that $\partial D$ has a diameter smaller than 1. Then there are constants $c_V > 0$ and $c_W > 0$ such that*

$$\langle \mathsf{V}\varphi, \varphi\rangle_{\partial D} \geq c_V \|\varphi\|^2_{H^{-\frac{1}{2}}(\partial d)} \qquad \forall \varphi \in H^{-\frac{1}{2}}(\partial D), \tag{7.201}$$

$$\langle \mathsf{W}v, v\rangle_{\partial D} \geq c_W \|v\|^2_{H^{-\frac{1}{2}}(\partial d)} \qquad \forall v \in H^{\frac{1}{2}}(\partial D)/\mathbb{R}. \tag{7.202}$$

*If $D$ is not bounded, then (7.202) holds on the entire space $H^{\frac{1}{2}}(\partial D)$.*

To apply this theorem recall the symmetry of the double-layer boundary integral operators for $-\Delta$: $\langle \mathsf{K}v, \varphi\rangle_{\partial D} = \langle v, \mathsf{K}'\varphi\rangle_{\partial D}$ for all $v \in H^{\frac{1}{2}}(\partial D)$ and $\varphi \in H^{-\frac{1}{2}}(\partial D)$. Thus, we obtain $D = \Omega_j$

$$\begin{aligned}\left\langle\!\!\left\langle \mathbb{A}_j \begin{pmatrix} v \\ \varphi \end{pmatrix}, \begin{pmatrix} -v \\ \varphi \end{pmatrix} \right\rangle\!\!\right\rangle_{\partial\Omega_j} &= \left\langle\!\!\left\langle \begin{pmatrix} -\mathsf{K}_j v + \mathsf{V}_j \varphi \\ \mathsf{W}_j v + \mathsf{K}_j' \varphi \end{pmatrix}, \begin{pmatrix} -v \\ \varphi \end{pmatrix} \right\rangle\!\!\right\rangle_{\partial\Omega_j} \\ &= \langle \mathsf{V}_j \varphi, \varphi\rangle_{\partial\Omega_j} + \langle \mathsf{W}_j v, v\rangle_{\partial\Omega_j},\end{aligned} \tag{7.203}$$

and from Theorem 7.26 we immediately conclude that there are constants $c_j > 0$, $j = 1, \dots, N$, such that

$$\langle\!\langle \mathbb{A}_j \mathfrak{v}, \Xi_j \mathfrak{v}\rangle\!\rangle_{\partial\Omega_j} \geq c_j \inf_{\alpha\in\mathbb{R}} \|\mathfrak{v} - \left(\begin{smallmatrix}\alpha\\0\end{smallmatrix}\right)\|^2_{\mathscr{T}(\partial\Omega_j)} \quad \forall \mathfrak{v} \in \mathscr{T}(\partial\Omega_j), \tag{7.204}$$

where the simple isometric isomorphism $\Xi_j : \mathscr{T}(\partial\Omega_j) \to \mathscr{T}(\partial\Omega_j)$ is defined as $\Xi_j\left(\begin{smallmatrix}v\\\varphi\end{smallmatrix}\right) := \left(\begin{smallmatrix}-v\\\varphi\end{smallmatrix}\right)$. Again the quotient norm is redundant for $j = 0$:

$$\exists c_0 > 0: \quad \langle\!\langle \mathbb{A}_0 \mathfrak{v}, \Xi_0 \mathfrak{v}\rangle\!\rangle_{\partial\Omega_0} \geq c_0 \|\mathfrak{v}\|^2_{\mathscr{T}(\partial\Omega_0)} \quad \forall \mathfrak{v} \in \mathscr{T}(\partial\Omega_0). \tag{7.205}$$

In the interest of concise notation we merge the local sign-change isomorphisms $\Xi_j$ into the operator $\Xi$ on $\mathscr{M}\mathscr{T}(\Sigma)$ that amounts to componentwise application of the $\Xi_j$'s.

**Theorem 7.27 ($\Xi$-ellipticity of STF bilinear form).** *The STF bilinear form* $\mathsf{m}_{\mathrm{STF}}$ *satisfies the* $\Xi$*-ellipticity estimate*

$$\mathsf{m}_{\mathrm{STF}}(\underline{\mathfrak{v}},\Xi\underline{\mathfrak{v}}) \geq c\,\|\mathfrak{v}\|^2_{\mathcal{M}\mathcal{T}(\Sigma)} \qquad \forall \mathfrak{v} \in \mathcal{S}\mathcal{T}_0(\Sigma)$$

*for some constant* $c > 0$.

***Proof.*** Invoking the definition of $\mathsf{m}_{\mathrm{STF}}$ from (7.200) together with (7.204) and (7.205) confirms that $\mathsf{m}_{\mathrm{STF}}$ is positive semidefinite with an at most finite-dimensional kernel

$$\mathsf{m}_{\mathrm{STF}}(\underline{\mathfrak{v}},\Xi\underline{\mathfrak{v}}) \geq c_0\,\|\mathfrak{v}\|^2_{\mathcal{T}(\partial\Omega_0)} + \sum_{j=1}^{N} c_j \inf_{\alpha\in\mathbb{R}} \left\|\mathfrak{v} - \binom{\alpha}{0}\right\|^2_{\mathcal{T}(\partial\Omega_j)} .$$

It remains to show that the kernel can only be trivial. Let us consider the variational problem with vanishing right-hand side: $\mathsf{m}_{\mathrm{STF}}(\mathfrak{u},\mathfrak{v}) = 0$ for all $\mathfrak{v} \in \mathcal{S}\mathcal{T}_0(\Sigma)$. Choosing $\underline{\mathfrak{v}} = \underline{\mathfrak{u}}$ we infer from (7.204) that $\mathbb{L}_j\,\mathfrak{u} = \binom{c_j}{0}$ for some $c_j \in \mathbb{R}$, that is, the Neumann components of $\mathfrak{u}$ vanish and its Dirichlet components are constant on the subdomain boundaries. In fact, by (7.205), on $\partial\Omega_0$ those have to vanish, too. Hence, they have to be zero on all subdomains abutting $\partial\Omega_0$, and a simple induction arguments confirms $\mathfrak{u} = 0$. □

**Corollary 7.28.** *The variational problem* (7.200) *has a unique solution for any* $g \in H^{\frac{1}{2}}(\partial\Omega_Z)$.

***Proof.*** The continuity of the bilinear form $\mathsf{m}_{\mathrm{STF}}$ from (7.200) and of the right-hand side linear form on $\mathcal{S}\mathcal{T}_0(\Sigma)$ is obvious. Moreover, the ellipticity result of Theorem 7.27 implies an inf-sup condition for $\mathsf{m}_{\mathrm{STF}}$. □

### Obstruction to Operator Preconditioning

According to Theorem 7.27 the single-trace variational BIE (7.200) is stable. Moreover, its boundary element discretization is straightforward, once a triangulation of the skeleton is given.

However, (7.200) is posed on the single-trace space $\mathcal{S}\mathcal{T}_0(\Sigma)$ and there is no known trace space that is dual to $\mathcal{S}\mathcal{T}_0(\Sigma)$ with respect to an $L^2$-type duality pairing. Of course, uniformly stable discrete $L^2$-dualities are even more elusive for conventional boundary element subspaces of $\mathcal{S}\mathcal{T}_0(\Sigma)$. Thus, as remarked in the introduction, operator preconditioning for (7.200) is *not possible*. This is further elaborated in [16, Section 4.5].

## 7.4.4 ▪ Global Multitrace BIEs

### Preface

In this and the next section we present multitrace BIEs for the transmission problem (7.176). They have in common that they give rise to variational problems posed on products of simple trace spaces on subdomain boundaries.

To understand the benefit of a product space framework in terms of operator preconditioning, recall Theorem 7.23 and its straightforward generalization to a setting where the bilinear form $\mathsf{a}$ is defined on a product of Hilbert spaces[12] $X := X_1 \times \ldots X_n$, $n \in \mathbb{N}$. We also need Hilbert spaces $Y_1,\ldots,Y_n$ equipped with continuous bilinear forms

[12]The notation introduced in the context of Theorem 7.23 is used tacitly.

$\mathsf{b}_j \in L(Y_j \times Y_j, \mathbb{R})$. Then, given finite-dimensional subspaces $X_{j,h} \subset X_j$, $Y_{j,h} \subset Y_j$, assuming inf-sup conditions for a on $X_h := X_{1,h} \times \cdots \times X_{n,h}$ and for the $\mathsf{b}_j$'s on $Y_{j,h}$, and with continuous and stable discrete pairings $\mathsf{d}_j : X_{j,h} \times Y_{j,h} \to \mathbb{R}$ at our disposal, we find that the matrix

$$\mathbf{M}_h^{\times} := \sum_{j=1}^{n} \mathbf{D}_j^{-1} \mathbf{B}_j \mathbf{D}_j^{-T} \tag{7.206}$$

provides operator preconditioning for the Galerkin matrix $\mathbf{A}_h$ of a on $X_h$. Here, the matrices $\mathbf{B}_j$ and $\mathbf{D}_j$ are the Galerkin matrices spawned by $\mathsf{b}_j$ on $Y_{j,h} \times Y_{j,h}$ and $\mathsf{d}_j$ on $X_{j,h} \times Y_{j,h}$.

### Heuristic Gap Construction

Consider the special situation that none of the subdomains $\Omega_1, \ldots, \Omega_N$ and $\Omega_Z$ touch as in Figure 7.24 (right). In this case the skeleton can be partitioned according to

$$\Sigma = \partial\Omega_0 = \partial\Omega_1 \cup \cdots \cup \partial\Omega_N \cup \partial\Omega_Z \,. \tag{7.207}$$

We arrive at this arrangement of "separated subdomains" when introducing a small gap between all bounded subdomains, which is illustrated in Figure 7.24.

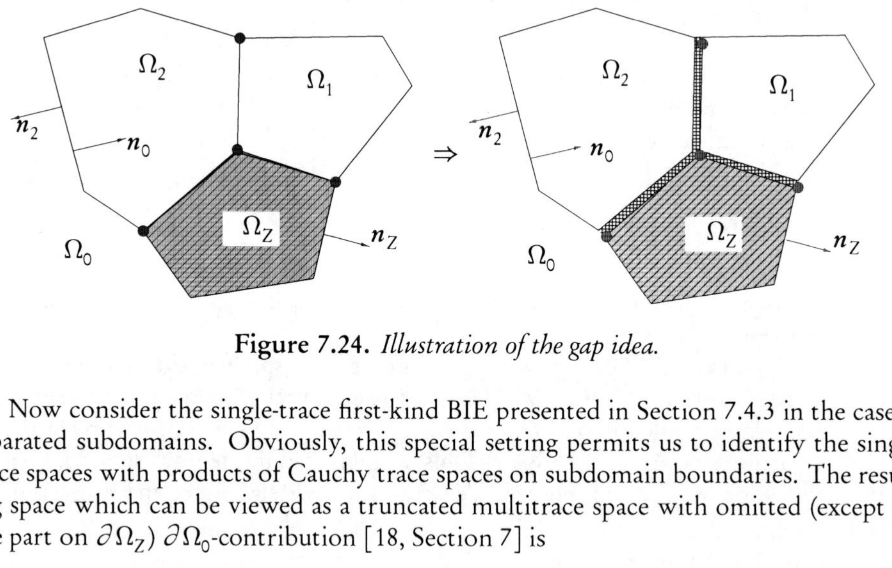


**Figure 7.24.** *Illustration of the gap idea.*

Now consider the single-trace first-kind BIE presented in Section 7.4.3 in the case of separated subdomains. Obviously, this special setting permits us to identify the single-trace spaces with products of Cauchy trace spaces on subdomain boundaries. The resulting space which can be viewed as a truncated multitrace space with omitted (except for the part on $\partial\Omega_Z$) $\partial\Omega_0$-contribution [18, Section 7] is

$$\mathcal{ST}(\Sigma) \cong \widehat{\mathcal{MT}}(\Sigma) := \mathcal{T}(\partial\Omega_1) \times \cdots \times \mathcal{T}(\partial\Omega_N) \times \mathcal{T}(\partial\Omega_Z), \tag{7.208}$$

$$\mathcal{ST}_0(\Sigma) \cong \widehat{\mathcal{MT}}_0(\Sigma) := \mathcal{T}(\partial\Omega_1) \times \cdots \times \mathcal{T}(\partial\Omega_N) \times H^{-\frac{1}{2}}(\partial\Omega_Z). \tag{7.209}$$

The related isomorphisms are $\mathbb{I} : \widehat{\mathcal{MT}}(\Sigma) \to \mathcal{ST}(\Sigma)$ and $\mathbb{I}_0 : \widehat{\mathcal{MT}}_0(\Sigma) \to \mathcal{ST}(\Sigma)$. Specifically, with $\mathfrak{v}_j = \binom{v_j}{\psi_j} \in \mathcal{T}(\partial\Omega_j)$, $j = 1, \ldots, N, Z$, we have

$$\mathbb{I}(\mathfrak{v}_1, \ldots, \mathfrak{v}_N, \mathfrak{v}_Z) := \left( \begin{pmatrix} v_1 \vee \cdots \vee v_N \vee v_Z \\ -(\psi_1 \vee \cdots \vee \psi_N \vee \psi_Z) \end{pmatrix}, \mathfrak{v}_1, \ldots, \mathfrak{v}_N \right), \tag{7.210}$$

where $\vee$ designates the joining of functions on subdomain boundaries to form a function on $\partial\Omega_0$. This is made possible by the partitioning (7.207). The minus sign reflects the opposite orientations of the normals $\boldsymbol{n}_0$ and $\boldsymbol{n}_j$.

We observe that for separated subdomains the STF gives rise to a variational problem on a genuine product Hilbert space of subdomain trace spaces. As explained above in the section entitled "Preface," this facilitates the construction of operator preconditioners of the form (7.206). Algorithmic details are postponed to the section entitled "Global MTF: Operator Preconditioning in Three Dimensions."

Next, we take a closer look at the STF variational formulation (7.200) in the situation of separated subdomains, with the aim of recasting it into a problem posed on $\widehat{\mathcal{MT}}_0(\Sigma)$. Since

$$\mathbb{L}_j \, \mathbb{I}(\mathfrak{v}_1,\dots,\mathfrak{v}_N,\mathfrak{v}_Z) = \mathfrak{v}_j\,, \quad j=1,\dots,N \text{ or } j=Z\,, \tag{7.211}$$

it is only the term in (7.200) contributed by $\partial\Omega_0$ that needs closer scrutiny. We write $\widehat{\underline{\mathfrak{u}}} = (\mathfrak{u}_1,\dots,\mathfrak{u}_N,\mathfrak{u}_Z) \in \widehat{\mathcal{MT}}(\Sigma)$, $\widehat{\underline{\mathfrak{v}}} = (\mathfrak{v}_1,\dots,\mathfrak{v}_N,\mathfrak{v}_Z) \in \widehat{\mathcal{MT}}(\Sigma)$, appeal to the lengthy manipulations from [16, Equation (5.8)] or [18, Section 8], and end up with

$$\langle\!\langle \mathbb{A}_0 \mathbb{L}_0 \mathbb{I}\widehat{\underline{\mathfrak{u}}}, \mathbb{L}_0 \mathbb{I}\widehat{\underline{\mathfrak{v}}} \rangle\!\rangle_{\partial\Omega_0} = \sum_{j=1,Z}^{N} \Big( \langle\!\langle \mathbb{A}_j^{\mu_0} \mathfrak{u}_j, \mathfrak{v}_j \rangle\!\rangle_{\partial\Omega_j} + \sum_{\substack{i=1,Z\\ i\neq j}}^{N} \langle\!\langle \mathbb{T}_j^{\mu_0} \mathbb{G}_i^{\mu_0}(\mathfrak{u}_i), \mathfrak{v}_j \rangle\!\rangle_{\partial\Omega_j} \Big).$$

Here, $Z$ in the range of summation indices means that $\partial\Omega_Z$ is also covered by the sum. Moreover, a superscript $\mu_0$ indicates that the Calderón operators $\mathbb{A}_j^{\mu_0}$, potentials $\mathbb{G}_j^{\mu_0}$, or Cauchy trace operators $\mathbb{T}_j^{\mu_0}$ are defined using the diffusion coefficient $\mu_0$ but still live on $\partial\Omega_j$, $j=1,\dots,N,Z$.

For the sake of brevity, we introduce the "remote coupling operators" for $i \neq j$,

$$\mathbb{C}_{i\to j} : \mathcal{T}(\partial\Omega_i) \to \mathcal{T}(\partial\Omega_j), \quad \mathbb{C}_{i\to j}(\mathfrak{v}_i) := \mathbb{T}_j^{\mu_0}\big(\mathbb{G}_i^{\mu_0}(\mathfrak{v}_i)\big)\,. \tag{7.212}$$

Observe that these operators take the trace of a potential at another boundary. We also note that for separated subdomains the offset function $\underline{\mathfrak{g}}$ from (7.200) may be supported on $\partial\Omega_Z$ alone, that is, we can choose $\underline{\mathfrak{g}} = \mathbb{I}\big(0,\dots,0,\binom{g}{0}\big)$. In particular, this means that $\mathbb{L}_j\,\underline{\mathfrak{g}} = 0$ for all $j=1,\dots,N$. Then, using the identity stated above, we can rewrite (7.200) as the following variational problem posed on $\widehat{\mathcal{MT}}_0(\Sigma)$: seek $\widehat{\underline{\mathfrak{u}}} = (\mathfrak{u}_1,\dots,\mathfrak{u}_N,\nu_Z) \in \widehat{\mathcal{MT}}_0(\Sigma)$ such that

$$\boxed{\begin{aligned} &\langle \mathsf{V}_Z \nu_Z, \psi_Z \rangle_{\partial\Omega} + \sum_{j=1}^{N} \langle\!\langle (\mathbb{A}_j^{\mu_0} + \mathbb{A}_j)\mathfrak{u}_j, \mathfrak{v}_j \rangle\!\rangle_{\partial\Omega_j} \\ &\qquad + \sum_{i=1}^{N} \langle\!\langle \mathbb{C}_{i\to Z}\, \mathfrak{u}_i, \tbinom{0}{\psi_Z} \rangle\!\rangle_{\partial\Omega_Z} + \sum_{j=1}^{N} \sum_{\substack{i=1\\ i\neq j}}^{N} \langle\!\langle \mathbb{C}_{i\to j}\, \mathfrak{u}_i, \mathfrak{v}_j \rangle\!\rangle_{\partial\Omega_j} \\ &\qquad\qquad + \sum_{j=1}^{N} \langle\!\langle \mathbb{C}_{Z\to j} \tbinom{0}{\nu_Z}, \mathfrak{v}_j \rangle\!\rangle_{\partial\Omega_j} \\ &\qquad\qquad\qquad = -\tfrac{1}{2} \langle (\mathsf{Id} - 2\mathsf{K}_Z^{\mu_0}) g, \psi_Z \rangle_{\partial\Omega_Z} - \sum_{j=1}^{N} \langle\!\langle \mathbb{C}_{Z\to j} \tbinom{g}{0}, \mathfrak{v}_j \rangle\!\rangle_{\partial\Omega_j} \end{aligned}} \tag{7.213}$$

for all $\widehat{\underline{\mathfrak{v}}} = (\mathfrak{v}_1,\dots,\mathfrak{v}_N,\psi_Z) \in \widehat{\mathcal{MT}}_0(\Sigma)$.

So far, this bulky variational BIE was derived for separated subdomains, depicted on the right in Figure 7.24. The key insight gleaned in [18, Section 5, "Gap idea"] and also discussed in [16, Section 5.2, "Gap idea"] is *that* (7.213) *remains well defined even without a gap* between the $\Omega_j$ and $\Omega_Z$, $j = 1,\ldots,N$, as shown on the left in Figure 7.24. Indeed, reading the trace $\mathbb{T}_j^{\mu_0}$ in (7.212) as an *exterior* Cauchy trace onto $\partial\Omega_j$ with respect to $\Omega_j$, the remote coupling operators remain well-defined continuous mappings. Hence, (7.213) defines the variational form of the *global multitrace BIE formulation* for the transmission problem (7.176) even in the general setting outlined in Section 7.4.2.

**Remark 56.** *As hinted at by their name, in* (7.213) *the "remote coupling operators"* $\mathbb{C}_{i\to j}$ *from* (7.212) *establish a variational coupling between subdomains, even if they do not have a common interface. These long-range interactions made us choose the attribute "global" for this kind of MTF.*

## Global MTF: Existence and Uniqueness of Solutions

In light of the "gap idea" it is not surprising that the variational problem (7.213) inherits quite a few of the remarkable properties of the original STF (7.200). The first result mirrors the "$\Xi$-ellipticity" of $\mathsf{m}_{\mathrm{STF}}$ asserted in Theorem 7.27.

**Theorem 7.29.** *The global multitrace bilinear form defined by the left-hand side of* (7.213) *is* continuous *and* $\Xi$-elliptic *on* $\widehat{\mathcal{MT}}_0(\Sigma)$.

***Proof.*** Continuity of the remote coupling operators $\mathbb{C}_{i\to j}$ is immediate from the mapping properties of potentials and traces. This implies continuity of the bilinear form.

As regards $\Xi$-ellipticity on $\widehat{\mathcal{MT}}_0(\Sigma)$, we merely sketch the idea of the proof. First note that the first line of (7.213) enjoys $\widehat{\mathcal{MT}}_0(\Sigma)$-ellipticity on a subspace, defined by imposing vanishing averages of the Dirichlet components on $\partial\Omega_z,\ldots,\partial\Omega_N,\partial\Omega_Z$. The remaining constant Dirichlet components are taken care of by the remote coupling terms in lines 2 and 3 of (7.213); see the proof of [18, Proposition 10.3] for details. □

The next result confirms that the unique solution of (7.213) yields the Cauchy traces of the unique solution of the transmission problem (7.176).

**Theorem 7.30.** *If* $\widehat{\mathfrak{u}} = (\mathfrak{u}_1,\ldots,\mathfrak{u}_N,\nu_Z) \in \widehat{\mathcal{MT}}_0(\Sigma)$ *solves* (7.213), *then* $\mathfrak{u}_j = \mathbb{T}_j U - \binom{\mathsf{T}_{D,j} G}{0}$ *and* $\nu_Z = \mathsf{T}_{N,Z} U$, *where $U$ is the unique solution of* (7.176).

***Proof.*** We refer to [18, Section 9], in particular the proof of [18, Theorem 9.1], which carries over to (7.213) with only slight modifications. □

## Global MTF: Operator Preconditioning in Three Dimensions

We restrict ourselves to $d = 3$ and the simplest boundary element Galerkin discretization of (7.213). To define trial and test spaces we *independently* equip each boundary $\partial\Omega_j$, $j = 1,\ldots,N$, and $\partial\Omega_Z$ with a conforming triangular surface mesh $\mathcal{G}_j$ [70, Section 4.1.2]. On these meshes we define the spaces $S^{-1}(\mathcal{G}_j) \subset H^{-\frac{1}{2}}(\partial\Omega_j)$ of (discontinuous) $\mathcal{G}_j$-piecewise constant functions [70, Example 4.1.16] and $S^0(\mathcal{G}_j) \subset H^{\frac{1}{2}}(\partial\Omega_j)$ of $\mathcal{G}_j$-piecewise linear and *continuous* functions [70, Example 4.1.37]. Their product provides discrete local Cauchy trace spaces

$$\mathcal{T}_h(\partial\Omega_j) := S^0(\mathcal{G}_j)\times S^{-1}(\mathcal{G}_j)\subset\mathcal{T}(\partial\Omega_j),\quad j=1,\dots,N, \tag{7.214}$$

which are the building blocks for the discrete counterpart of $\widehat{\mathcal{MT}}_0(\Sigma)$ (as defined in (7.209)):

$$\widehat{\mathcal{MT}}_{0,h}(\{\mathcal{G}_j\}) := \mathcal{T}_h(\partial\Omega_1)\times\cdots\times\mathcal{T}_h(\partial\Omega_N)\times S^{-1}(\mathcal{G}_Z)\subset\widehat{\mathcal{MT}}_0(\Sigma). \tag{7.215}$$

This completely defines the Galerkin BEM for (7.213). If standard localized nodal basis functions are used for $S^{-1}(\mathcal{G}_j)$ and $S^0(\mathcal{G}_j)$, then for shape-regular and quasi-uniform families of surface meshes we will encounter a growth of the spectral condition numbers of the resulting linear systems of equations like $O(h^{-2})$ as the mesh width $h\to 0$ [70, Section 4.5].

Operator preconditioning will be carried out in the product space framework outlined in the section entitled "Preface." Retaining the notation from there we have $n=N+1$, $X_j := \mathcal{T}(\partial\Omega_j)$, $j=1,\dots,N$, $X_n := H^{-\frac{1}{2}}(\partial\Omega_Z)$, and the bilinear form $\mathsf{a}$ is that of (7.213), for which Theorem 7.29 guarantees uniform stability. From (7.209) we see that the role of $X_{j,h}$ is played by $\mathcal{T}_h(\partial\Omega_j)$, $j=1,\dots,N$, and that of $X_{n,h}$ by $S^{-1}(\mathcal{G}_Z)$.

As suggested by $L^2$-duality, we pick

$$Y_j := X_j,\quad j=1,\dots,N,\quad\text{and}\quad Y_n := H^{\frac{1}{2}}(\partial\Omega_Z), \tag{7.216}$$

because we want to employ the local $L^2$-type pairings on $\partial\Omega_j$ and $\partial\Omega_Z$, respectively, as pairing bilinear forms $\mathsf{d}_j$, that is,

$$\mathsf{d}_j := \langle\!\langle\cdot,\cdot\rangle\!\rangle_{\partial\Omega_j},\quad j=1,\dots,N,\quad \mathsf{d}_n := \langle\cdot,\cdot\rangle_{\partial\Omega_Z}. \tag{7.217}$$

Next, the Calderón operators from (7.187) supply the bilinear forms $\mathsf{b}_j$, $j=1,\dots,N$,

$$\mathsf{b}_j(\mathfrak{u}_j,\mathfrak{v}_j) := \langle\!\langle \mathbb{A}_j\,\mathfrak{u}_j,\mathfrak{v}_j\rangle\!\rangle_{\partial\Omega_j},\quad \mathfrak{u}_j,\mathfrak{v}_j\in\mathcal{T}(\partial\Omega_j), \tag{7.218}$$

whereas the hypersingular boundary integral operator on $\partial\Omega_Z$ gives us $\mathsf{b}_n$: $\mathsf{b}_n(u_Z,v_Z) := \langle \mathsf{W}_Z u_Z, v_Z\rangle_{\partial\Omega_Z}$, $u_Z,v_Z\in H^{\frac{1}{2}}(\partial\Omega_Z)$.

The choice of the boundary element spaces $Y_{j,h}\subset Y_j$ poses a challenge, because just using lowest-order boundary elements on the same meshes $\mathcal{G}_j$ fails to deliver uniformly stable discretizations of the pairing bilinear forms $\mathsf{d}_j$ even on sequences of shape-regular and quasi-uniform meshes. Instead we resort to the breakthrough idea from [72] and [8] and use boundary element spaces on *dual meshes* $\widehat{\mathcal{G}}_j$ defined via barycentric subdivisions of $\mathcal{G}_j$ as displayed in Figure 7.25 [73, Section 2.2].

The spaces $S^{-1}(\widehat{\mathcal{G}}_j)\subset H^{-\frac{1}{2}}(\partial\Omega_j)$ comprise functions that are piecewise constant on dual cells [8, Section 2, Figure 3]. Further, $S^0(\widehat{\mathcal{G}}_j)$ is spanned by continuous functions that are piecewise linear on the barycentric refinement of $\mathcal{G}_j$ and whose values at nodes of $\mathcal{G}_j$ and midpoints of edges of $\mathcal{G}_j$ are determined by the average of their values in adjacent barycenters [8, Section 2, Figure 1]. Then the theoretical developments of [73, Section 2] confirm that the pairs of spaces $S^{-1}(\mathcal{G}_j)\times S^0(\widehat{\mathcal{G}}_j)$ and $S^{-1}(\widehat{\mathcal{G}}_j)\times S^0(\mathcal{G}_j)$ provide stable

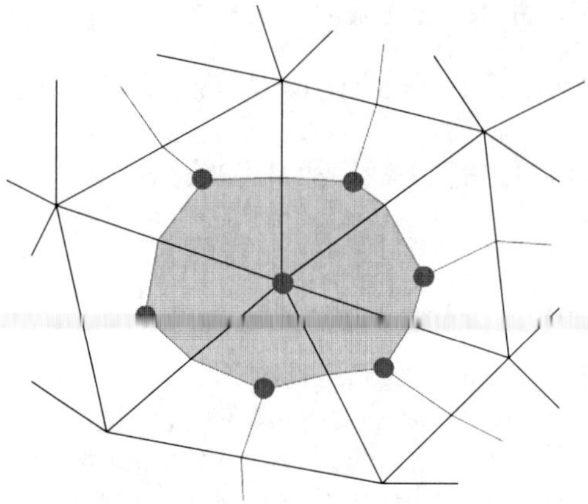

**Figure 7.25.** *Barycentric dual mesh for a triangular primal mesh (black lines): the shaded region marks a dual cell associated with a primal node (center disk), the finer lines represent dual edges, the peripheral disks dual nodes.*

Galerkin discretizations of the duality pairing $\langle\cdot,\cdot\rangle_{\partial\Omega_j} : H^{-\frac{1}{2}}(\partial\Omega_j)\times H^{\frac{1}{2}}(\partial\Omega_j)\to\mathbb{R}$. This hinges on certain assumptions on the geometry of $\mathcal{G}_j$, which are satisfied for shape-regular and quasi-uniform families and even for a wide range of locally refined meshes. Appealing to this theory, the next theorem is a consequence of Theorem 7.23.

**Theorem 7.31.** *Operator preconditioning of the Galerkin boundary element discretization of* (7.213) *as outlined in the section entitled "Preface" leads to uniformly bounded spectral condition numbers of the preconditioned linear systems in the case of shape-regular and quasi-uniform families of triangular surface meshes.*

## 7.4.5 ▪ Local Multitrace BIEs

### Localized Transmission Conditions

In the global multitrace formulation introduced in Section 7.4.4 the transmission conditions are implicitly contained in the variational formulation. Conversely, the local multitrace approach takes into account the transmission conditions in their local form (7.176b), which can be expressed as

$$\begin{pmatrix}\mathsf{T}_{D,i}\\ \mathsf{T}_{N,i}\end{pmatrix}U_i=\begin{pmatrix}\mathsf{Id} & 0\\ 0 & -\mathsf{Id}\end{pmatrix}\begin{pmatrix}\mathsf{T}_{D,j}\\ \mathsf{T}_{N,j}\end{pmatrix}U_j \quad \text{on } \Gamma_{ij}\,, \tag{7.219}$$

for which we embrace the compact notation

$$\mathbb{S}_{ij}\,\mathbb{T}_i\,U_i=\mathbb{X}_{j\to i}\,\mathbb{T}_j\,U_j\,. \tag{7.220}$$

Here $\mathbb{S}_{ji}:\mathcal{T}(\partial\Omega_j)\to\mathcal{T}(\Gamma_{ij})$ restricts traces on $\partial\Omega_j$ to the interface $\Gamma_{ij}\subset\partial\Omega_j$. We used the natural notation $\mathcal{T}(\Gamma_{ij}):=H^{\frac{1}{2}}(\Gamma_{ij})\times H^{-\frac{1}{2}}(\Gamma_{ij})$. The action of the operator $\mathbb{X}_{j\to i}:\mathcal{T}(\partial\Omega_j)\to\mathcal{T}(\Gamma_{ij})$ is immediate from (7.219): $\mathbb{X}_{j\to i}\binom{v}{\psi}:=\mathbb{S}_{ji}\binom{v}{-\psi}$; in addition to restricting traces, $\mathbb{X}_{j\to i}$ flips the sign of the Neumann component, thus adjusting it to the orientation of the other subdomain boundary $\partial\Omega_i$.

Since we aim for BIEs in weak form, we need to cast (7.220) into a variational equation. Formally, this can be accomplished by pairing with test functions in the dual space of $\mathcal{T}(\Gamma_{ij})$ (with respect to the pivot space $L^2(\Gamma_{ij})$). However, be aware that, in contrast to $\mathcal{T}(\partial\Omega_j)$, the $L^2(\Gamma_{ij})$ inner product does *not* induce a self-duality of $\mathcal{T}(\Gamma_{ij})$. Rather, the dual space is [56, Chapter 3], [70, Section 2.4.2]

$$\mathcal{T}(\Gamma_{ij})' = (H^{\frac{1}{2}}(\Gamma_{ij}))' \times (H^{-\frac{1}{2}}(\Gamma_{ij}))' \cong \widetilde{H}^{-\frac{1}{2}}(\Gamma_{ij}) \times \widetilde{H}^{\frac{1}{2}}(\Gamma_{ij}) . \tag{7.221}$$

Here $\widetilde{H}^{\frac{1}{2}}(\Gamma_{ij})$ and $\widetilde{H}^{-\frac{1}{2}}(\Gamma_{ij})$ designate distributions on $\Gamma_{ij}$, whose extensions by zero to $\partial\Omega_i$ belong to $H^{\frac{1}{2}}(\partial\Omega_i)$ and $H^{-\frac{1}{2}}(\partial\Omega_i)$, respectively. Thus, $\widetilde{H}^{\frac{1}{2}}(\Gamma_{ij})$ and $\widetilde{H}^{-\frac{1}{2}}(\Gamma_{ij})$ can be identified with subspaces of $H^{\frac{1}{2}}(\Gamma_{ij})$ and $H^{-\frac{1}{2}}(\Gamma_{ij})$, respectively, which are actually *dense*. Yet, the norms of $\widetilde{H}^{\frac{1}{2}}(\Gamma_{ij})$ and $\widetilde{H}^{-\frac{1}{2}}(\Gamma_{ij})$ are *strictly stronger* than those of $H^{\frac{1}{2}}(\Gamma_{ij})$ and $H^{-\frac{1}{2}}(\Gamma_{ij})$.

We conclude that an equivalent weak form of (7.220) is

$$\langle\!\langle \mathbb{S}_{ij}\,\mathbb{T}_i\,U_i - \mathbb{X}_{j\to i}\,\mathbb{T}_j\,U_j, \mathfrak{v}_{ij} \rangle\!\rangle_{\Gamma_{ij}} = 0 \quad \forall\, \mathfrak{v}_{ij} \in \widetilde{\mathcal{T}}(\Gamma_{ij}) , \tag{7.222}$$

where $\widetilde{\mathcal{T}}(\Gamma_{ij}) := \widetilde{H}^{\frac{1}{2}}(\Gamma_{ij}) \times \widetilde{H}^{-\frac{1}{2}}(\Gamma_{ij})$. Here, the *continuous* pairing $\langle\!\langle \cdot,\cdot \rangle\!\rangle_{\Gamma_{ij}} : \mathcal{T}(\Gamma_{ij}) \times \widetilde{\mathcal{T}}(\Gamma_{ij}) \to \mathbb{R}$ of Cauchy traces on $\Gamma_{ij}$ is defined in analogy to (7.183) based on $L^2(\Gamma_{ij})$-inner products.

### Local MTF: Variational Formulation

We now combine Theorem 7.24 in the form of (7.197) and (7.222) into a set of variational equations that *every* solution $U \in H^1_{\mathrm{loc}}(\mathbb{R}^d)$ of (7.176) will satisfy:

$$\text{(7.197)} \quad \Rightarrow \quad \langle\!\langle (\mathsf{Id} - \mathbb{P}_j)\mathbb{T}_j\,U, \mathfrak{v}_j \rangle\!\rangle_{\partial\Omega_j} = 0 \quad \forall\, \mathfrak{v}_j \in \mathcal{T}(\partial\Omega_j) , \tag{7.223a}$$

$$\text{(7.222)} \quad \Rightarrow \quad \langle\!\langle \mathbb{S}_{ji}\,\mathbb{T}_j\,U - \mathbb{X}_{i\to j}\,\mathbb{T}_i\,U, \mathbb{S}_{ji}\,\widetilde{\mathfrak{v}}_j \rangle\!\rangle_{\Gamma_{ij}} = 0 \quad \forall\, \widetilde{\mathfrak{v}}_j \in \widetilde{\mathcal{T}}_{\mathrm{pw}}(\partial\Omega_j) . \tag{7.223b}$$

Note that the duality (7.221), which underlies (7.222), enforces the use of the following special test space in the second equation:

$$\widetilde{\mathcal{T}}_{\mathrm{pw}}(\partial\Omega_j) := \{\widetilde{\mathfrak{v}} \in \mathcal{T}(\partial\Omega_j) : \mathbb{S}_{ji}\,\widetilde{\mathfrak{v}} \in \widetilde{\mathcal{T}}(\Gamma_{ij}),\ i \in \mathcal{N}_j\} \subset \mathcal{T}(\partial\Omega_j) . \tag{7.224}$$

Here and below, we write $\mathcal{N}_j \subset \{0,\dots,N\}$ for the set of indices of subdomains that have an interface in common with $\partial\Omega_j$. Obviously, the restriction operators map continuously $\mathbb{S}_{ji} : \widetilde{\mathcal{T}}_{\mathrm{pw}}(\partial\Omega_j) \to \widetilde{\mathcal{T}}(\Gamma_{ij})$.

By adding (7.223a) and several instances of (7.223b), from (7.223) we conclude that a solution $U \in H^1_{\mathrm{loc}}(\mathbb{R}^d)$ of (7.176) also fulfills (cf. [40, Section 3.2])

$$\langle\!\langle (\mathsf{Id} - \mathbb{P}_j)\mathbb{T}_j\,U, \widetilde{\mathfrak{v}}_j \rangle\!\rangle_{\partial\Omega_j} + \sum_{i\in\mathcal{N}_j} \sigma_{ij} \langle\!\langle \mathbb{S}_{ji}\,\mathbb{T}_j\,U - \mathbb{X}_{i\to j}\,\mathbb{T}_i\,U, \mathbb{S}_{ji}\,\widetilde{\mathfrak{v}}_j \rangle\!\rangle_{\Gamma_{ij}} = 0 \tag{7.225}$$

for all $\widetilde{\mathfrak{v}}_j \in \widetilde{\mathcal{T}}_{\mathrm{pw}}(\partial\Omega_j)$, $j = 0,\dots,N$. The $\sigma_{ij} \in \mathbb{R}$ are arbitrary nonzero combination coefficients.

Again, we employ an offset function technique to deal with the nonhomogeneous Dirichlet boundary conditions (7.175). As in the section above entitled “First-Kind BIEs” we introduce a function $G \in H^1_{\mathrm{loc}}(\mathbb{R}^d)$, whose Dirichlet trace on $\partial\Omega_Z$ agrees with the given data $g \in H^{\frac{1}{2}}(\partial\Omega_Z)$: $\mathsf{T}_{D,Z}\, G = g$. Then $\bar{U} := U - G \in \{V \in H^1_{\mathrm{loc}}(\mathbb{R}^d):\ V|_{\partial\Omega_Z} = 0\}$ solves

$$\begin{aligned}
&\langle\!\langle (\mathsf{Id} - \mathbb{P}_j)\mathbb{T}_j\,\bar{U}, \widetilde{\mathfrak{v}}_j \rangle\!\rangle_{\partial\Omega_j} + \sum_{i\in\mathcal{N}_j} \sigma_{ij} \langle\!\langle \mathbb{S}_{ji}\,\mathbb{T}_j\,\bar{U} - \mathbb{X}_{i\to j}\,\mathbb{T}_i\,\bar{U}, \mathbb{S}_{ji}\,\widetilde{\mathfrak{v}}_j \rangle\!\rangle_{\Gamma_{ij}} \\
&\quad = \underbrace{-\langle\!\langle (\mathsf{Id} - \mathbb{P}_j)\mathbb{T}_j\, G, \widetilde{\mathfrak{v}}_j \rangle\!\rangle_{\partial\Omega_j} - \sum_{i\in\mathcal{N}_j} \sigma_{ij} \langle\!\langle \mathbb{S}_{ji}\,\mathbb{T}_j\, G - \mathbb{X}_{i\to j}\,\mathbb{T}_i\, G, \mathbb{S}_{ji}\,\widetilde{\mathfrak{v}}_j \rangle\!\rangle_{\Gamma_{ij}}}_{=:\Psi_j(\widetilde{\mathfrak{v}}_j)}
\end{aligned} \tag{7.226}$$

for all $\widetilde{\mathfrak{v}}_j \in \widetilde{\mathcal{T}}_{\mathrm{pw}}(\partial\Omega_j)$, $j = 0, \ldots, N$.

In order to arrive at a variational BIE for the unknown Cauchy traces of $\bar{U}$, in the next key step we introduce all local subdomain traces $\mathfrak{u}_j := \binom{u_j}{\mu_j} := \mathbb{T}_j\,\bar{U} \in \mathcal{T}(\partial\Omega_j)$, $j = 0, \ldots, N$, as unknowns. By construction, they satisfy $u_j|_{\Gamma_{jZ}} = 0$. In addition we use $\mathbb{A}_j = \mathbb{P}_j - \frac{1}{2}\mathsf{Id}$ from (7.187) and obtain a *first-kind BIE*: seek $\underline{\mathfrak{u}} = (\mathfrak{u}_0, \ldots, \mathfrak{u}_N) \in \mathcal{MT}_0(\Sigma)$ such that

$$\langle\!\langle (\mathbb{A}_j - \tfrac{1}{2}\mathsf{Id})\mathfrak{u}_j, \widetilde{\mathfrak{v}}_j \rangle\!\rangle_{\partial\Omega_j} \quad + \quad \sum_{i\in\mathcal{N}_j} \sigma_{ij} \langle\!\langle \mathbb{S}_{ji}\,\mathfrak{u}_j - \mathbb{X}_{i\to j}\,\mathfrak{u}_i, \mathbb{S}_{ji}\,\widetilde{\mathfrak{v}}_j \rangle\!\rangle_{\Gamma_{ij}} \quad = \quad \Psi_j(\widetilde{\mathfrak{v}}_j) \tag{7.227}$$

for all $\underline{\mathfrak{v}} := (\widetilde{\mathfrak{v}}_0, \ldots, \widetilde{\mathfrak{v}}_N) \in \widetilde{\mathcal{MT}}_0(\Sigma)$, $j = 0, \ldots, N$. This variational problem is posed on multitrace spaces that respect homogeneous Dirichlet boundary conditions on $\partial\Omega_Z$:

$$\mathcal{MT}_0(\Sigma) := \left\{ \left( \begin{pmatrix} v_j \\ \nu_j \end{pmatrix} \right)_{j=0}^{N} \in \mathcal{MT}(\Sigma):\ v_j\Big|_{\Gamma_{jZ}} = 0 \right\}, \tag{7.228a}$$

$$\widetilde{\mathcal{MT}}_0(\Sigma) := \left( \widetilde{\mathcal{T}}_{\mathrm{pw}}(\partial\Omega_0) \times \cdots \times \widetilde{\mathcal{T}}_{\mathrm{pw}}(\partial\Omega_N) \right) \cap \mathcal{MT}_0(\Sigma). \tag{7.228b}$$

Again, we point out that the trial functions must allow extension by zero from each interface $\Gamma_{ij}$ to the associated subdomain boundaries $\partial\Omega_i$ and $\partial\Omega_j$.

There is a "magic" choice for the parameters $\sigma_{ij}$, because, by density arguments,

$$\sum_{i\in\mathcal{N}_j} \langle\!\langle \mathbb{S}_{ji}\,\mathfrak{u}_j, \mathbb{S}_{ji}\,\widetilde{\mathfrak{v}}_j \rangle\!\rangle_{\Gamma_{ij}} = \langle\!\langle \mathfrak{u}_j, \widetilde{\mathfrak{v}}_j \rangle\!\rangle_{\partial\Omega} \quad \forall \mathfrak{u}_j \in \mathcal{T}(\partial\Omega_j),\ \widetilde{\mathfrak{v}}_j \in \widetilde{\mathcal{T}}_{\mathrm{pw}}(\partial\Omega_j).$$

Hence, if we set $\sigma_{ij} = \frac{1}{2}$ in (7.227), we can benefit from cancellation and convert (7.227) into the following: seek $\underline{\mathfrak{u}} = (\mathfrak{u}_0, \ldots, \mathfrak{u}_N) \in \mathcal{MT}_0(\Sigma)$ such that

$$\boxed{\langle\!\langle \mathbb{A}_j\,\mathfrak{u}_j, \widetilde{\mathfrak{v}}_j \rangle\!\rangle_{\partial\Omega_j} - \frac{1}{2} \sum_{i\in\mathcal{N}_j} \langle\!\langle \mathbb{X}_{i\to j}\,\mathfrak{u}_i, \mathbb{S}_{ji}\,\widetilde{\mathfrak{v}}_j \rangle\!\rangle_{\Gamma_{ij}} = \Psi_j(\widetilde{\mathfrak{v}}_j)} \tag{7.229}$$

for all $\underline{\mathfrak{v}} := (\widetilde{\mathfrak{v}}_0, \ldots, \widetilde{\mathfrak{v}}_N) \in \widetilde{\mathcal{MT}}_0(\Sigma)$. This is the "classical" MTF for the transmission problem (7.176) as proposed in [39] and discussed in [16, Section 6] and [40, Section 3.3]. It is a variational problem posed on $\mathcal{MT}_0(\Sigma) \times \widetilde{\mathcal{MT}}_0(\Sigma)$, whose underlying bilinear form we are going to denote by $\mathsf{m}_{\mathrm{LMF}}$:

$$\mathsf{m}_{\mathrm{LMF}}(\underline{\mathfrak{u}}, \underline{\mathfrak{v}}) := \sum_{j=0}^{N} \langle\!\langle \mathbb{A}_j\,\mathfrak{u}_j, \widetilde{\mathfrak{v}}_j \rangle\!\rangle_{\partial\Omega_j} - \sum_{j=0}^{N} \sum_{i\in\mathcal{N}_j} \langle\!\langle \mathbb{X}_{i\to j}\,\mathfrak{u}_i, \mathbb{S}_{ji}\,\widetilde{\mathfrak{v}}_j \rangle\!\rangle_{\Gamma_{ij}}, \tag{7.230}$$

$\underline{\mathfrak{u}} = (\mathfrak{u}_0, \ldots, \mathfrak{u}_N) \in \mathcal{MT}_0(\Sigma)$, $\underline{\mathfrak{v}} = (\widetilde{\mathfrak{v}}_0, \ldots, \widetilde{\mathfrak{v}}_N) \in \widetilde{\mathcal{MT}}_0(\Sigma)$.

In light of (7.183) and (7.187) and the sign flip effected by $\mathbb{X}_{i\to j}$, the compact notation (7.229) can be unravelled into an explicit variational problem for Dirichlet and Neumann traces: seek $\underline{\mathrm{u}} = \left(\binom{u_0}{\nu_0},\ldots,\binom{u_N}{\nu_N}\right) \in \mathcal{MT}_0(\Sigma)$ such that

$$\begin{array}{rcrcrcl} \left\langle \mathsf{V}_j \nu_j, \widetilde{\varphi}_j \right\rangle_{\partial\Omega_j} & - & \left\langle \mathsf{K}_j u_j, \widetilde{\varphi}_j \right\rangle_{\partial\Omega_j} & - & \frac{1}{2}\sum_{i\in\mathcal{N}_j} \left\langle u_i|_{\Gamma_{ij}}, \widetilde{\varphi}_j\Big|_{\Gamma_{ij}} \right\rangle_{\Gamma_{ij}} & = & \cdots, \\ -\left\langle \mathsf{K}'_j \nu_j, \widetilde{v}_j \right\rangle_{\partial\Omega_j} & - & \left\langle \mathsf{W}_j u_j, \widetilde{v}_j \right\rangle_{\partial\Omega_j} & - & \frac{1}{2}\sum_{i\in\mathcal{N}_j} \left\langle \nu_i|_{\Gamma_{ij}}, \widetilde{v}_j\Big|_{\Gamma_{ij}} \right\rangle_{\Gamma_{ij}} & = & \cdots, \end{array} \tag{7.231}$$

for all $\binom{\widetilde{v}_j}{\widetilde{\varphi}_j} \in \mathcal{T}_{\mathrm{pw}}(\partial\Omega_j)$, $j = 0,\ldots,N$, which satisfy $\widetilde{v}_j\,|_{\Gamma_{jZ}} = 0$. Please refer to (7.187), Page 233, for the definition of the boundary integral operators. The right-hand-side functionals $\Psi_j$ have been suppressed for the sake of brevity.

**Remark 57.** *As explained in* [39, Section 3.2.4], *the MTF bilinear form* $\mathsf{m}_{\mathrm{LMF}}$ *from* (7.230) *will remain well defined, when* both arguments *belong to*

$$\widetilde{\mathcal{MT}}{}_0^{\sharp}(\Sigma) := \left\{ \left(\binom{v_j}{\nu_j}\right)_{j=0}^{N} \in \mathcal{MT}(\Sigma): \begin{array}{l} v_j\Big|_{\Gamma_{jZ}} = 0, \\ \nu_j \in \widetilde{H}_{\mathrm{pw}}^{-\frac{1}{2}}(\partial\Omega_j) \end{array} \right\}, \tag{7.232}$$

*where* $\widetilde{H}_{\mathrm{pw}}^{-\frac{1}{2}}(\partial\Omega_j)$ *comprises functions in* $H^{-\frac{1}{2}}(\partial\Omega_j)$, *whose restrictions to any interface* $\Gamma_{ij}$ *belong to* $\widetilde{H}^{-\frac{1}{2}}(\Gamma_{ij})$: $\mathsf{m}_{\mathrm{LMF}} \in L(\widetilde{\mathcal{MT}}{}_0^{\sharp}(\Sigma) \times \widetilde{\mathcal{MT}}{}_0^{\sharp}(\Sigma), \mathbb{R})$. *This paves the way to an alternative formulation of the local MTF with equal trial and test spaces* [39, Section 3.2].

**Remark 58.** *From* [39, Section 3.2.5] *recall that the variational formulation* (7.200) *of the STF can formally be obtained by restricting* (7.229) *to test and trial functions in* $\mathcal{ST}_0(\Sigma)$.

## Local MTF: Existence and Uniqueness of Solutions

The choice of the particular broken trace spaces was made to ensure the continuity of the MTF bilinear form.

**Lemma 7.32.** *The MTF bilinear form* $\mathsf{m}_{\mathrm{LMF}} : \mathcal{MT}_0(\Sigma) \times \widetilde{\mathcal{MT}_0}(\Sigma) \to \mathbb{R}$ *from* (7.230) *is continuous.*

***Proof.*** Thanks to the continuity properties of the boundary integral operators the first term in (7.230) is even continuous on $\mathcal{MT}(\Sigma) \times \mathcal{MT}(\Sigma)$.

The continuity of $\mathbb{S}_{ji} : \mathcal{T}(\partial\Omega_j) \to \mathcal{T}(\Gamma_{ij})$ and $\mathbb{S}_{ji} : \widetilde{\mathcal{T}}_{\mathrm{pw}}(\partial\Omega_j) \to \widetilde{\mathcal{T}}(\Gamma_{ij})$ is clear. Besides, the $L^2(\Gamma_{ij})$-inner product can be extended to continuous pairings on $H^{\frac{1}{2}}(\Gamma_{ij}) \times \widetilde{H}^{-\frac{1}{2}}(\Gamma_{ij})$ and $\widetilde{H}^{\frac{1}{2}}(\Gamma_{ij}) \times H^{-\frac{1}{2}}(\Gamma_{ij})$. This ensures the continuity of the second part of (7.230). □

The following lemma reveals a particular "block skew-symmetric" structure of (7.229). It makes use of the isometric local "sign change isomorphisms" $\Xi_j : \mathcal{T}(\partial\Omega_j) \to \mathcal{T}(\partial\Omega_j)$, $\Xi_j\binom{v}{\varphi} := \binom{-v}{\varphi}$, introduced in the section entitled "Existence and Uniqueness of Solutions," p. 236.

**Lemma 7.33 (see [39, Section 2.2.3, Lemma 1]).** *For all* $\mathfrak{u}_i \in \mathcal{T}(\partial\Omega_i)$ *and* $\widetilde{\mathfrak{v}} \in \widetilde{\mathcal{T}}_{\mathrm{pw}}(\partial\Omega_j)$ *it holds that*

$$\langle\!\langle \mathbb{X}_{i\to j}\,\mathfrak{u}_i, \mathbb{S}_{ij}\,\Xi_j\widetilde{\mathfrak{v}}_j \rangle\!\rangle_{\Gamma_{ij}} = -\langle\!\langle \mathbb{X}_{j\to i}\,\widetilde{\mathfrak{v}}_j, \mathbb{S}_{ji}\,\Xi_i\mathfrak{u}_i \rangle\!\rangle_{\Gamma_{ij}} .$$

*Proof.* The identity follows from straightforward computations using the definitions of the operators and the pairings. We write $\mathfrak{u}_i = \binom{u}{\nu}$, $\widetilde{\mathfrak{v}}_j = \binom{\widetilde{v}}{\widetilde{\psi}}$ and find

$$\langle\!\langle \mathbb{X}_{i\to j}\,\mathfrak{u}_i\big|_{\Gamma_{ij}}, \mathbb{S}_{ij}\,\Xi_i\widetilde{\mathfrak{v}}_j\big|_{\Gamma_{ij}} \rangle\!\rangle_{\Gamma_{ij}} = \langle\!\langle \mathbb{S}_{ij}\begin{pmatrix} u \\ -\nu \end{pmatrix}, \mathbb{S}_{ji}\begin{pmatrix} -\widetilde{v} \\ \widetilde{\psi} \end{pmatrix} \rangle\!\rangle_{\Gamma_{ij}} = \left\langle u, \widetilde{\psi}\right\rangle_{\Gamma_{ij}} - \langle \widetilde{v}, \nu\rangle_{\Gamma_{ij}} ,$$

$$\langle\!\langle \mathbb{X}_{j\to i}\,\widetilde{\mathfrak{v}}_j\big|_{\Gamma_{ij}}, \mathbb{S}_{ji}\,\Xi_i\mathfrak{u}_i\big|_{\Gamma_{ij}} \rangle\!\rangle_{\Gamma_{ij}} = \langle\!\langle \mathbb{S}_{ji}\begin{pmatrix} \widetilde{v} \\ -\widetilde{\psi} \end{pmatrix}, \mathbb{S}_{ij}\begin{pmatrix} -u \\ \nu \end{pmatrix} \rangle\!\rangle_{\Gamma_{ij}} = \langle \widetilde{v}, \nu\rangle_{\Gamma_{ij}} - \left\langle u, \widetilde{\psi}\right\rangle_{\Gamma_{ij}} . \quad \square$$

Very much in analogy to Theorem 7.29 for the global MTF, the relationship from Lemma 7.33 guarantees ellipticity of the bilinear form of (7.229) up to a simple local change of sign.

**Theorem 7.34 ("Ξ-ellipticity" of bilinear form for MTF).** *There is a constant* $c > 0$ *such that*

$$\mathsf{m}_{\mathrm{LMF}}(\underline{\mathfrak{v}}, \Xi\underline{\mathfrak{v}}) \geq c\,\|\underline{\mathfrak{v}}\|^2_{\mathcal{MT}(\Sigma)} \quad \forall \underline{\mathfrak{v}} \in \widetilde{\mathcal{MT}}_0(\Sigma) .$$

*Proof.* Throughout the proof we write $\underline{\mathfrak{v}} = (\widetilde{\mathfrak{v}}_0, \dots, \widetilde{\mathfrak{v}}_N) = \left(\binom{\widetilde{v}_0}{\widetilde{\psi}_0}, \dots, \binom{\widetilde{v}_N}{\widetilde{\psi}_N}\right) \in \widetilde{\mathcal{MT}}_0(\Sigma)$.

❶ As a consequence of Lemma 7.33 all the "off-diagonal" coupling terms cancel and we end up with (cf. (7.203))

$$\mathsf{m}_{\mathrm{LMF}}(\underline{\mathfrak{v}}, \Xi\underline{\mathfrak{v}}) = \sum_{j=0}^{N} \langle\!\langle \mathbb{A}_j\,\mathfrak{v}, \Xi_j\,\mathfrak{v} \rangle\!\rangle_{\partial\Omega} = \sum_{j=0}^{N} \left\langle \mathsf{V}_j\widetilde{\psi}_j, \widetilde{\psi}_j\right\rangle_{\partial\Omega_j} + \left\langle \mathsf{W}_j\widetilde{v}_j, \widetilde{v}_j\right\rangle_{\partial\Omega_j} . \tag{7.233}$$

Then (7.204) immediately shows that $\mathsf{m}_{\mathrm{LMF}}(\underline{\mathfrak{v}}, \Xi\underline{\mathfrak{v}}) \geq 0$ for all $\underline{\mathfrak{v}} \in \widetilde{\mathcal{MT}}_0(\Sigma)$, because $\mathcal{T}_{\mathrm{pw}}(\partial\Omega_j) \subset \mathcal{T}(\partial\Omega_j)$. In addition, $\underline{\mathfrak{v}} \mapsto \mathsf{m}_{\mathrm{LMF}}(\underline{\mathfrak{v}}, \Xi\underline{\mathfrak{v}})$ inherits $\mathcal{MT}(\Sigma)$-coercivity (up to finite-dimensional perturbations) from the Calderón operators; recall (7.204) and (7.205).

❷ Next, single out a $\underline{\mathfrak{v}} \in \widetilde{\mathcal{MT}}_0(\Sigma)$ such that $\mathsf{m}_{\mathrm{LMF}}(\underline{\mathfrak{v}}, \Xi\underline{\mathfrak{v}}) = 0$. Then (7.233) and (7.204) imply that $\widetilde{\psi}_j = 0$ for all $j = 0, \dots, N$ and that $\widetilde{v}_j$ is *constant* on $\partial\Omega_j$, $j = 0, \dots, N$. So, $\underline{\mathfrak{v}}$ belongs to a space of finite dimension and it remains to show the injectivity of $\mathsf{m}_{\mathrm{LMF}}$ on this space.

To begin, we conclude from (7.205) that $\widetilde{v}_0 = 0$. Moreover, if the subdomain $\Omega_j$ has a common interface with $\Omega_Z$, then the definition (7.228b) makes $\widetilde{v}_j$ vanish on $\partial\Omega_j \cap \partial\Omega_Z$ and, since it is constant on $\partial\Omega_j$, it has to vanish on the entire boundary $\partial\Omega_j$.

Now, assume that $\mathsf{m}_{\mathrm{LMF}}(\underline{\mathfrak{v}}, \underline{\mathfrak{w}}) = 0$ for all $\underline{\mathfrak{w}} \in \widetilde{\mathcal{MT}}_0(\Sigma)$. In particular, we can choose $\underline{\mathfrak{w}}$ such that its Neumann component is equal to 1 on the boundary of a single subdomain $\Omega_k$ and zero everywhere else. If $\Omega_k$ is adjacent to a subdomain $\Omega_l$, where $\widetilde{\mathfrak{v}}_l = 0$ is already known, the coupling terms enforce that $\widetilde{v}_k|_{\Gamma_{lk}} = 0$ and $\widetilde{\mathfrak{v}}_k$ has to be zero, too. Thus, we can work our way through all subdomains, because $\mathbb{R}^d \setminus \Omega_Z$ is supposed to be connected. This finally establishes $\underline{\mathfrak{v}} = 0$. □

Unfortunately, this theorem does not settle the issue of existence and uniqueness of solutions of (7.229), because we encounter a mismatch of spaces as observed in [39, Section 3.2.8]: $\mathsf{m}_{\mathrm{LMF}}$ is $\mathcal{MT}_0(\Sigma)$-elliptic but continuous only on $\mathcal{MT}_0(\Sigma) \times \widetilde{\mathcal{MT}_0}(\Sigma)$, so that we cannot instantly conclude an inf-sup condition from $\Xi$-ellipticity. We have to rely on a more sophisticated result known as "Lion's projection lemma" [54, Chapter III, Theorem 1.1]; see also [23, Section 2] and [28, Section 2].

**Lemma 7.35 (see [39, Lemma 9]).** *Let $H$ be a Hilbert space and $V$ be a subspace of $H$ (not necessarily closed in $H$). Moreover, let $\mathsf{b} : H \times V \to \mathbb{R}$ be a bilinear form satisfying the following properties:*

1. *For every $\varphi \in V$, the linear form $u \mapsto \mathsf{b}(u, \varphi)$ is continuous on $H$.*

2. *There exists $c > 0$ such that*

$$|\mathsf{b}(\varphi, \varphi)| \geq c \, \|\varphi\|_H^2 \quad \forall \varphi \in V\,. \tag{7.234}$$

*Then for each continuous linear form $l \in H'$, there exists $u_0 \in H$ such that*

$$\mathsf{b}(u_0, \varphi) = \langle l, \varphi \rangle_H \quad \forall \varphi \in V \quad \text{and} \quad \|u_0\|_H \leq \frac{1}{c} \|l\|_{H'}\,. \tag{7.235}$$

Clearly, Theorem 7.34 suggests that we apply this lemma with $\mathsf{b} := \mathsf{m}_{\mathrm{LMF}}$, $H := \mathcal{MT}_0(\Sigma)$, and $V := \widetilde{\mathcal{MT}_0}(\Sigma)$. We immediately conclude existence of solutions of the MTF variational problem (7.229).

Yet, as pointed out in [54, Chapter III, Remark 31], despite (7.234), Lemma 7.35 does not ensure uniqueness of solutions. To obtain it, we resort to considerations that directly exploit the boundary integral equations.

**Theorem 7.36.** *Solutions of (7.229) are unique.*

*Proof.* As in the proof of [39, Theorem 9], we show that (7.229) with $\Psi_j = 0$ involves $\underline{\mathfrak{u}} = 0$. Hence, let $\underline{\mathfrak{u}} \in \mathcal{MT}_0(\Sigma)$ satisfy

$$\mathsf{m}_{\mathrm{LMF}}(\underline{\mathfrak{u}}, \underline{\mathfrak{v}}) = 0 \quad \forall \mathfrak{v} \in \widetilde{\mathcal{MT}_0}(\Sigma)\,. \tag{7.236}$$

❶ We set $U_j := \mathbb{G}_j(\mathfrak{u}_j)$, $j = 0, \ldots, N$, and from (7.186) we get

$$\mathbb{T}_j\, U_j := (\mathbb{A}_j + \tfrac{1}{2}\mathsf{Id})\mathfrak{u}_j\,. \tag{7.237}$$

On the other hand, from (7.236) and (7.229) we infer

$$\langle\!\langle (\mathbb{A}_j + \tfrac{1}{2}\mathsf{Id})\mathfrak{u}_j, \widetilde{\mathfrak{v}}_j \rangle\!\rangle_{\partial\Omega_j} = \tfrac{1}{2} \langle\!\langle \mathfrak{u}_j, \widetilde{\mathfrak{v}}_j \rangle\!\rangle_{\partial\Omega_j} + \tfrac{1}{2} \sum_{i \in \mathcal{N}_j} \langle\!\langle \mathbb{X}_{i\to j}\, \mathfrak{u}_i, \mathbb{S}_{ji}\, \widetilde{\mathfrak{v}}_j \rangle\!\rangle_{\Gamma_{ij}} \tag{7.238}$$

for all $\widetilde{\mathfrak{v}} \in \mathcal{T}_{\mathrm{pw}}(\partial\Omega_j)$.

Now, we single out an interface $\Gamma_{ij}$, choose arbitrary functions $\widetilde{v} \in \widetilde{H}^{\frac{1}{2}}(\Gamma_{ij})$ and $\widetilde{\psi} \in \widetilde{H}^{-\frac{1}{2}}(\Gamma_{ij})$, and obtain the components of the test functions $\widetilde{\mathfrak{v}}_j \in \widetilde{\mathcal{T}}_{\mathrm{pw}}(\partial\Omega_j)$ and $\widetilde{\mathfrak{v}}_i \in \widetilde{\mathcal{T}}_{\mathrm{pw}}(\partial\Omega_i)$ by extending $\widetilde{v}$ and $\widetilde{\psi}$ by zero onto $\partial\Omega_j$ and $\partial\Omega_i$, respectively. For $\widetilde{\mathfrak{v}}_i$

we also change the sign of the Neumann component. For these special test functions (7.238) together with (7.237) yields

$$\begin{aligned}
\langle\!\langle \mathbb{T}_j\, U_j, \mathbb{S}_{ji}\, \mathfrak{v}_j \rangle\!\rangle_{\partial\Omega} &= \tfrac{1}{2}\langle\!\langle \mathfrak{u}_j, \widetilde{\mathfrak{v}}_j \rangle\!\rangle_{\partial\Omega_j} + \tfrac{1}{2}\langle\!\langle \mathbb{X}_{i\to j}\, \mathfrak{u}_i, \mathbb{S}_{ji}\, \widetilde{\mathfrak{v}}_j \rangle\!\rangle_{\Gamma_{ij}}, \\
\langle\!\langle \mathbb{X}_{i\to j}\, \mathbb{T}_i\, U_i, \mathbb{S}_{ji}\, \mathfrak{v}_j \rangle\!\rangle_{\partial\Omega} &= -\langle\!\langle \mathbb{T}_i\, U_i, \mathbb{S}_{ij}\, \mathfrak{v}_i \rangle\!\rangle_{\partial\Omega} \\
&= -\tfrac{1}{2}\langle\!\langle \mathfrak{u}_i, \widetilde{\mathfrak{v}}_i \rangle\!\rangle_{\partial\Omega_i} - \tfrac{1}{2}\langle\!\langle \mathbb{X}_{j\to i}\, \mathfrak{u}_j, \mathbb{S}_{ij}\, \widetilde{\mathfrak{v}}_i \rangle\!\rangle_{\Gamma_{ij}}.
\end{aligned}$$

Owing to the local support of the test functions, a closer inspection reveals that

$$\langle\!\langle \mathfrak{u}_j, \widetilde{\mathfrak{v}}_j \rangle\!\rangle_{\partial\Omega_j} = -\langle\!\langle \mathbb{X}_{j\to i}\, \mathfrak{u}_j, \mathbb{S}_{ij}\, \widetilde{\mathfrak{v}}_i \rangle\!\rangle_{\Gamma_{ij}} \quad \text{and} \quad \langle\!\langle \mathbb{X}_{i\to j}\, \mathfrak{u}_i, \mathbb{S}_{ji}\, \widetilde{\mathfrak{v}}_j \rangle\!\rangle_{\Gamma_{ij}} = -\langle\!\langle \mathfrak{u}_i, \widetilde{\mathfrak{v}}_i \rangle\!\rangle_{\partial\Omega_i},$$

which means that

$$\langle\!\langle \mathbb{T}_j\, U_j, \mathbb{S}_{ji}\, \mathfrak{v}_j \rangle\!\rangle_{\partial\Omega} = \langle\!\langle \mathbb{X}_{i\to j}\, \mathbb{T}_i\, U_i, \mathbb{S}_{ji}\, \mathfrak{v}_j \rangle\!\rangle_{\partial\Omega}. \tag{7.239}$$

As a consequence, we find that

$$\mathbb{T}_j\, U_j = \mathbb{X}_{i\to j}\, \mathbb{T}_i\, U_i \quad \text{on } \Gamma_{ij}. \tag{7.240}$$

In words, the function $U \in L^2_{\text{loc}}(\mathbb{R}^d)$ that is obtained by patching together the $U_j$'s, satisfies the transmission conditions (7.176b). Moreover, the representation formula guarantees $\mathscr{L}_j\, U_j = 0$. Finally, using (7.236) and (7.238) with test functions supported on $\Gamma_{jZ} \subset \partial\Omega_Z$, confirms that $U|_{\partial\Omega_Z} = 0$. Summing up, $U$ solves (7.176) with zero Dirichlet data on $\partial\Omega_Z$. Uniqueness of solutions of (7.176) tells us that $U = 0$.

❷ Having established $U_j = 0$, combining (7.237) and (7.238) yields

$$\langle\!\langle \mathfrak{u}_j, \widetilde{\mathfrak{v}}_j \rangle\!\rangle_{\partial\Omega_j} = -\sum_{i\in\mathcal{N}_j} \langle\!\langle \mathbb{X}_{i\to j}\, \mathfrak{u}_i, \mathbb{S}_{ji}\, \widetilde{\mathfrak{v}}_j \rangle\!\rangle_{\Gamma_{ij}} \quad \forall \widetilde{\mathfrak{v}} \in \mathscr{T}_{\text{pw}}(\partial\Omega_j), \tag{7.241}$$

which means $\mathfrak{u}_j = -\mathbb{X}_{i\to j}\, \mathfrak{u}_i$ on $\Gamma_{ij}$. In words, the Cauchy data $\mathfrak{u}_j$ satisfy *sign-flipped transmission conditions* across interfaces $\Gamma_{ij}$.

❸ Taking the Cauchy trace $\mathbb{T}_j$ of $U_j = \mathbb{G}_j(\mathfrak{u}_j) = 0$ reveals that $\mathbb{P}_j\, \mathfrak{u}_j = 0$. Hence, using the fact that interior and exterior Calderón projectors add up to zero [39, Section 2.3.3] and Theorem 7.24 on the complement domain $\Omega_j^c := \mathbb{R}^d \setminus \Omega_j$, we find that $\mathfrak{u}_j = \mathbb{T}_j^c\, V_j$ for a function $V_j \in H_{\text{loc}}(\Delta, \Omega_j^c)$ that satisfies $\mathscr{L}_j\, V_j = 0$ in $\Omega_j^c$ and appropriate decay conditions at $\infty$ for $j \neq 0$. Here, $\mathbb{T}_j^c$ is the Cauchy trace operator on $\Omega_j^c$.

We adapt an idea from the proof of [39, Theorem 9]: For a "sign vector"

$$\boldsymbol{\sigma} := (\sigma_1, \ldots, \sigma_N) \in \{-1, +1\}^N,$$

we define the *multivalued function*

$$V^{\boldsymbol{\sigma}} := \sigma_j\, V_j \quad \text{on } \Omega_j^c, \qquad V^{\boldsymbol{\sigma}} := V_0 \quad \text{on } \Omega_0^c. \tag{7.242}$$

Case (i). Assume that there is a sign vector $\boldsymbol{\sigma}$ such that $\sigma_i = -\sigma_j$ for every interface $\Gamma_{ij}$; we are dealing with a bipartite connected graph of complement subdomains (whose edges correspond to nonempty interfaces $\Gamma_{ij}$). Then, by virtue of (7.241), $V^{\boldsymbol{\sigma}}$ is a *multivalued* solution of a transmission problem of the type (7.176) on $\bigcup_{j=0}^N \Omega_j^c$, which features the right decay conditions at $\infty$ and has zero Dirichlet boundary conditions on $\partial\Omega_Z$. Hence, $V^{\boldsymbol{\sigma}} = 0$, which implies $V_j = 0$, and, immediately, $\mathfrak{u}_j = 0$.

Case (ii). Assume that the graph of complement subdomains is not bipartite. Regard two complement domains $\Omega_j^c$ and $\Omega_i^c$ as "linked" if they share an interface and if $\sigma_i = -\sigma_j$. We denote by $V_\sharp^{\sigma}$ the restriction of $V^{\sigma}$ to the union of complement domains, for which there is a chain of links to $\Omega_0^c$. Again thanks to (7.241), this multivalued function satisfies transmission conditions (7.176b) between linked complement domains.

Next, we appeal to the *strong unique continuation principle* for solutions of transmission problems of type (7.176) [34, Section 3.4.1]. It confirms that any two functions $V_\sharp^{\sigma}$ agree on complement domains, on which they are both defined, because they are the same on $\Omega_0^c$. Now, if the graph is not bipartite, for some $k \in \{1,\dots,N\}$ we can find two sign vectors $\boldsymbol{\sigma}, \boldsymbol{\sigma}'$ such that $\sigma_k = -\sigma_k'$ and such that for both $\boldsymbol{\sigma}$ and $\boldsymbol{\sigma}'$ there is a chain of links from $\Omega_k^c$ to $\Omega_0^c$. We conclude that both $V^{\boldsymbol{\sigma}}|_{\Omega_k^c} = V^{\boldsymbol{\sigma}'}|_{\Omega_k^c}$ (from unique continuation) and $V^{\boldsymbol{\sigma}}|_{\Omega_k^c} = -V^{\boldsymbol{\sigma}'}|_{\Omega_k^c}$ (by definition), which implies $V_k = 0$. By unique continuation we then conclude that $V_j = 0$ for all $j = 0,\dots,N$, and also $\mathfrak{u}_j = 0$. □

**Remark 59.** *In Remark* 57 *we pointed out that the variational formulation* (7.229) *of the classical MTF may be lifted to trace spaces $\widetilde{\mathcal{MT}}_0^{\sharp}(\Sigma)$, for which additional regularity is imposed on the Neumann components only.*

*Now assume Dirichlet data $g \in H^1(\partial\Omega_Z)$. Then elliptic regularity results* [56, Chapter 4] *ensure that all local Neumann traces $\mathsf{T}_{N,j}\, U$ and $\mathsf{T}_{N,Z}\, U$ of the solution $U$ of the transmission problem* (7.176) *belong to $L^2(\partial\Omega_j)$, $j = 1,\dots,N,Z$.*

*As a consequence the unique solution of* (7.229) *is contained in the space $\widetilde{\mathcal{MT}}_0^{\sharp}(\Sigma)$ from* (7.232), *Page* 245. *Thus, the solution of* (7.229) *will be preserved when we switch to the trial space $\widetilde{\mathcal{MT}}_0^{\sharp}(\Sigma)$. The Neumann components of Cauchy traces in that space possess more regularity compared to those of $\mathcal{MT}_0(\Sigma)$. This enables us to use test functions with less regular Dirichlet components than stipulated by $\widetilde{\mathcal{MT}}_0(\Sigma)$. More precisely, testing with functions merely belonging to $\widetilde{\mathcal{MT}}_0^{\sharp}(\Sigma)$ becomes possible, which yields a variant of* (7.229) *with the same trial and test space;* cf. [39, Section 3.2.1].

## Local MTF: Operator Preconditioning in Three Dimensions

The developments are largely parallel to that of the section entitled "Global MTF: Operator Preconditioning in Three Dimensions" and we reuse the notation introduced there mostly without further mention. In the context of Galerkin boundary element discretization it is advisable to adopt the perspective of Remarks 57 and 59 and lift the MTF variational problem (7.229) into the space $\widetilde{\mathcal{MT}}_0^{\sharp}(\Sigma)$. Then, since piecewise polynomial approximation invariably provides functions in $\widetilde{H}_{\mathrm{pw}}^{-\frac{1}{2}}(\partial\Omega_j)$, we deal with an $\widetilde{\mathcal{MT}}_0^{\sharp}(\Sigma)$-conforming boundary element Galerkin approximation.

In order to take into account the Dirichlet boundary conditions, we rely on the boundary element spaces ($j = 0,\dots,N$)

$$S_Z^0(\mathcal{G}_j) := \{v_h \in S^0(\mathcal{G}_j) : \; v_h|_{\partial\Omega_Z} = 0\}, \tag{7.243}$$

$$\mathcal{T}_{h,Z}(\partial\Omega_j) := S_Z^0(\mathcal{G}_j) \times S^{-1}(\mathcal{G}_j) \subset H^{\frac{1}{2}}(\partial\Omega_j) \times \widetilde{H}_{\mathrm{pw}}^{-\frac{1}{2}}(\partial\Omega_j) \tag{7.244}$$

to build the trial and test space for (7.229):

$$\mathcal{MT}_{0,h}(\{\mathcal{G}_j\}) := \mathcal{T}_{h,Z}(\partial\Omega_0) \times \cdots \times \mathcal{T}_{h,Z}(\partial\Omega_N) \subset \widetilde{\mathcal{MT}}_0^{\sharp}(\Sigma) . \tag{7.245}$$

Existence, uniqueness, and convergence of Galerkin solutions in the absence of the impenetrable object $\Omega_Z$ have been established in [39, Section 4].

From (7.245) we identify the spaces $\mathcal{T}_{h,Z}(\partial\Omega_j)$ as the $X_j$'s in the product space setting for operator preconditioning (see the preface to Section 7.4.4 on Page 237). As in the case of the global MTF discussed in the section entitled "Global MTF: Operator Preconditioning in Three Dimensions," the bilinear forms $\mathsf{b}_j$ are chosen according to (7.218), and the $Y_j$ are boundary element spaces on dual meshes $\widehat{\mathcal{G}}_j$, $j = 0, \ldots, N$, arising from barycentric refinement:

$$Y_j := S^0(\widehat{\mathcal{G}}_j) \times S_0^{-1}(\widehat{\mathcal{G}}_j) \subset H^{\frac{1}{2}}(\partial\Omega_j) \times H^{-\frac{1}{2}}(\partial\Omega_j) . \tag{7.246}$$

Note that the $\widehat{\mathcal{G}}_j$-piecewise constant functions in the spaces $S_0^{-1}(\mathcal{G}_j)$ have to vanish on dual cells associated with nodes of $\mathcal{G}_j$ that are located on $\partial\Omega_Z$. This ensures equal dimensions of $S_Z^0(\mathcal{G}_j)$ and $S_0^{-1}(\mathcal{G}_j)$ and follows the policy of [8, Section 4.2]. The local pairing bilinear forms $\mathsf{d}_j$ are again chosen as duality pairings $\langle\!\langle \cdot,\cdot \rangle\!\rangle_{\partial\Omega_j}$.

The analysis of operator preconditioning is slightly more difficult than in the section entitled "Global MTF: Operator Preconditioning in Three Dimensions," because the local MTF bilinear form $\mathsf{m}_{\mathrm{LMF}}$ enjoys ellipticity in $\mathcal{MT}_0(\Sigma)$ (recall Theorem 7.34) but is continuous only on $\widetilde{\mathcal{MT}}_0^{\sharp}(\Sigma)$ as defined in (7.232). The norm of the latter space,

$$\|\underline{\mathfrak{v}}\|^2_{\widetilde{\mathcal{MT}}_0^{\sharp}(\Sigma)} := \sum_{j=0}^{N}\Big( \|v_j\|^2_{H^{-\frac{1}{2}}(\partial\Omega_j)} + \sum_{i\in\mathcal{N}_j} \|v_j|_{\Gamma_{ij}}\|^2_{\tilde{H}^{-\frac{1}{2}}(\Gamma_{ij})}\Big), \tag{7.247}$$

is stronger than that of the former. This mismatch compounded the theoretical difficulties encountered in the section entitled "Local MTF: Existence and Uniqueness of Solutions" and thwarts the straightforward application of Theorem 7.23.

To cope with this situation, we have to make another assumption concerning the meshes: we assume that each interface $\Gamma_{ij}$ is resolved by cells of the meshes $\mathcal{G}_j$ and $\mathcal{G}_i$. The same should apply to $\mathcal{G}_j$ and the boundary parts $\Gamma_{jZ}$.

**Theorem 7.37.** *In the setting detailed above let us consider families of surface meshes generated by* regular refinement *of coarse initial meshes. Then the spectral condition numbers of the preconditioned linear systems grow moderately like* $O(L^{\frac{3}{2}}) = O(|\log h|^{\frac{3}{2}})$ *in the level* $L$ *of refinement as* $L \to \infty$ *(*$h \to 0$*).*

***Proof.*** We observe that the inf-sup constants and norms entering the bound in the estimate (7.173) of Theorem 7.23 refer to the discrete setting. In it all norms are equivalent and we can resort to the $\mathcal{MT}(\Sigma)$-norm throughout.

On boundary element spaces we have to use *inverse inequalities* to relate the norms of $\widetilde{\mathcal{MT}}_0^{\sharp}(\Sigma)$ and $\mathcal{MT}(\Sigma)$. In particular, we appeal to the estimate of [41, Theorem 2.2]

$$\|v_h\|_{\tilde{H}^{-\frac{1}{2}}(\Gamma_{ij})} \leq C L^{\frac{3}{2}} \|v_h\|_{H^{-\frac{1}{2}}(\Gamma_{ij})} \quad \forall v_h \in S^{-1}(\mathcal{G}_j) \tag{7.248}$$

when $\mathscr{G}_j$ is on level $L$ of the refinement hierarchy ($C > 0$ is constant that depends only on the geometry of $\Gamma_{ij}$ and the coarsest mesh). This implies, with another constant $C > 0$ independent of the level $L$ of refinement,

$$\|\underline{\mathfrak{v}}_h\|_{\widetilde{\mathcal{MT}}^{\sharp}_0(\Sigma)} \leq C L^{\frac{3}{2}} \|\underline{\mathfrak{v}}_h\|_{\mathcal{MT}(\Sigma)} \quad \forall \underline{\mathfrak{v}}_h \in \mathcal{MT}_{0,h}(\{\mathscr{G}_j\}). \tag{7.249}$$

As a consequence, the norm of the bilinear form a on $\mathcal{MT}_{0,h}(\{\mathscr{G}_j\})$, when measured in the weaker $\mathcal{MT}(\Sigma)$-norm, increases mildly like $O(L^{\frac{3}{2}})$ when we keep on refining the meshes and send $L \to \infty$. Since all other norms and inf-sup constants do not depend on $L$ (see [8] for results on the pairing bilinear forms $\mathsf{d}_j$), this accounts for the assertion of the theorem. □

### 7.4.6 ▪ Numerical Studies

In this section, we report numerical tests of the performance of operator preconditioning for the *local MTF* implementing the algorithms described in the section entitled "Local MTF: Operator Preconditioning in Three Dimensions." All computations were carried out with the C++ boundary element template library [42].[13] The matrices arising from Galerkin discretization were subject to local low-rank compression using the AHMED library[14] [5]. However, the compression parameters were chosen so that its impact can safely be ignored.

Due to its saddle point structure, the operator preconditioner based on (7.218) fails to be positive definite. Thus, the conjugate gradient method is not an option and the preconditioned generalized minimal residual method (GMRES) without restart is used as an iterative solver. The iterations are stopped as soon as a relative decrease of the Euclidean norm of the residual vector by a factor of $10^{-7}$ was achieved. The initial guess is zero throughout.

Quasi-uniform and shape-regular sequences (finite, of course) of triangular surface meshes with flat triangles are used for all experiments. They were produced by the mesh generator Gmsh[15] [29] and all of them are compatible with the interfaces.

In all experiments the behavior of the GMRES iterative solver without preconditioner and with operator preconditioning on meshes of different resolution is recorded. Total iteration counts versus global mesh widths are tabulated.

#### Experiment I: Two Half-Spheres

The first experiment uses $\Omega_Z = \{x \in \mathbb{R}^3 : \|x\| < 1, x_3 > 0\}$, $\Omega_1 = \{x \in \mathbb{R}^3 : \|x\| < 1, x_3 < 0\}$ and Dirichlet data $g(x) := \|x\|^{-1}|_{\partial\Omega_Z}$. The geometric situation is displayed in Figure 7.26. Measured data on GMRES convergence are given in Figure 7.27 and Table 7.3.

#### Experiment II: Glued Boxes

The geometry of the second experiment is $\Omega_Z = \{x \in \mathbb{R}^3 : -\frac{1}{2} \leq x_1, x_3 \leq \frac{1}{2}, -1 \leq x_2 \leq 0\}$, $\Omega_1 = \{x \in \mathbb{R}^3 : -\frac{1}{2} \leq x_1, x_3 \leq \frac{1}{2}, 0 \leq x_2 \leq 1\}$, with Dirichlet data $g(x) := \|x\|^{-1}\big|_{\partial\Omega_Z}$; see Figure 7.28. GMRES behavior is documented in Figure 7.29 and Table 7.4.

[13] http://www.sam.math.ethz.ch/betl.

[14] http://bebendorf.ins.uni-bonn.de/AHMED.html.

[15] http://geuz.org/gmsh.

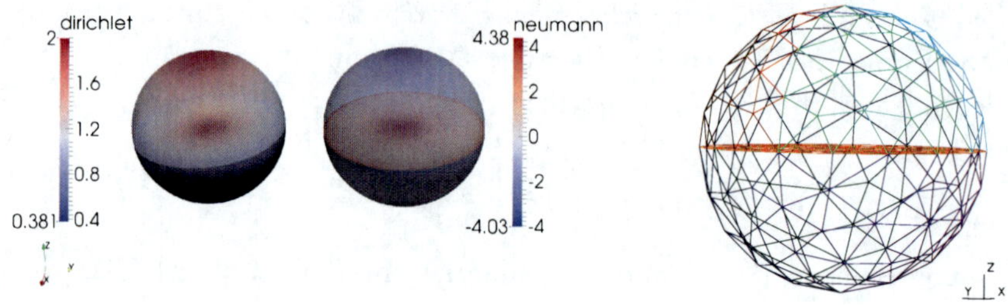


**Figure 7.26.** *Experiment* I *(Two half-spheres): Dirichlet and Neumann traces of the solution (left) for* $\mu_0 = 10, \mu_1 = 1$, 23,480 *elements, and coarse mesh with* 296 *flat triangles (right).*

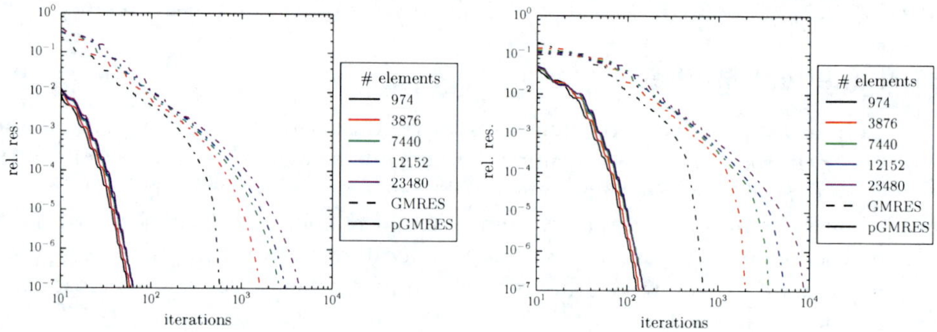


(a). $\mu_0 = 1.5$, $\mu_1 = 1$ (b). $\mu_0 = 10$, $\mu_1 = 1$

**Figure 7.27.** *Experiment* I: *(Preconditioned) GMRES iteration history; decay of relative Euclidean norm of the residual. Different diffusion coefficients were used on different subdomains.*

Table 7.3: Experiment I: Iteration counts for (preconditioned) GMRES for different choices of diffusion coefficients. $N$ denotes number of elements.

| | | $\mu_0 = 1.5$, $\mu_1 = 1$ | | $\mu_0 = 10$, $\mu_1 = 1$ | |
|---|---|---|---|---|---|
| $N$ | Mesh width | GMRES | pGMRES | GMRES | pGMRES |
| 974 | $1{,}25 \cdot 10^{-1}$ | 581 | 56 | 688 | 128 |
| 3876 | $6{,}18 \cdot 10^{-2}$ | 1617 | 59 | 1999 | 139 |
| 7440 | $4{,}50 \cdot 10^{-2}$ | 2668 | 65 | 3574 | 152 |
| 12152 | $3{,}52 \cdot 10^{-2}$ | 3098 | 64 | 5366 | 153 |
| 23480 | $2{,}52 \cdot 10^{-2}$ | 4426 | 64 | 8728 | 155 |

### Experiment III: Cube Split into Eight Smaller Cubes

The subdomains $\Omega_i$, $i = 1,\dots,8$, are the equal cubes of edge length $\frac{1}{2}$ created by splitting the unit cube. $\partial\Omega_z$ is centered at the origin, Dirichlet data are $g(x) := \|x\|^{-1}|_{\partial\Omega_z}$, and the same diffusion coefficient $\mu_i = 1$ was used on all subdomains; see Figure 7.30. Information about the convergence of GMRES is provided in Figure 7.30 and Table 7.5.

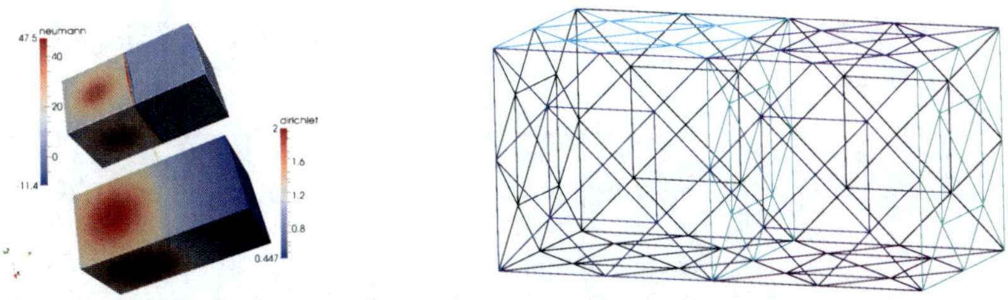


**Figure 7.28.** *Experiment* II*: Left: Dirichlet (front) and Neumann traces (rear) of the solution. Right: coarsest mesh with* 176 *triangles.*

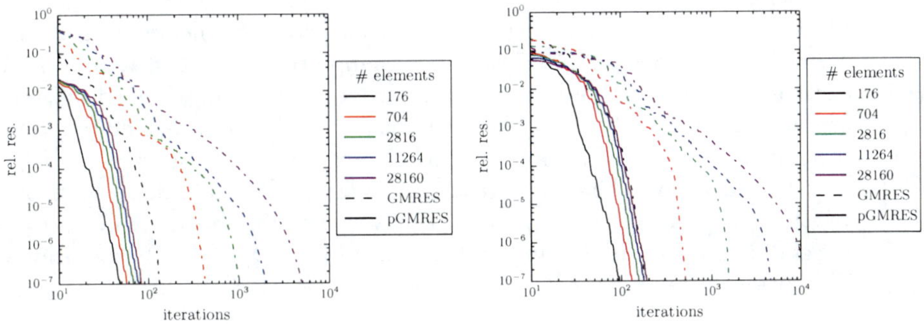


(a). $\mu_0 = 1.5$, $\mu_1 = 1$

(b). $\mu_0 = 10$, $\mu_1 = 1$

**Figure 7.29.** *Experiment* II*: (Preconditioned) GMRES iteration history. Diffusion coefficients attain different values on different subdomains.*

Table 7.4: Experiment II: Iteration counts for (preconditioned) GMRES, $N$.

| | | $\mu_0 = 1.5$, $\mu_1 = 1$ | | $\mu_0 = 10$, $\mu_1 = 1$ | |
|---|---|---|---|---|---|
| $N$ | Mesh width | GMRES | pGMRES | GMRES | pGMRES |
| 176 | $2{,}64 \cdot 10^{-1}$ | 133 | 49 | 179 | 95 |
| 704 | $1{,}32 \cdot 10^{-1}$ | 422 | 57 | 501 | 130 |
| 2816 | $6{,}59 \cdot 10^{-2}$ | 1017 | 68 | 1564 | 156 |
| 11264 | $3{,}30 \cdot 10^{-2}$ | 1982 | 78 | 4552 | 176 |
| 28160 | $2{,}14 \cdot 10^{-2}$ | 4891 | 83 | 10961 | 189 |

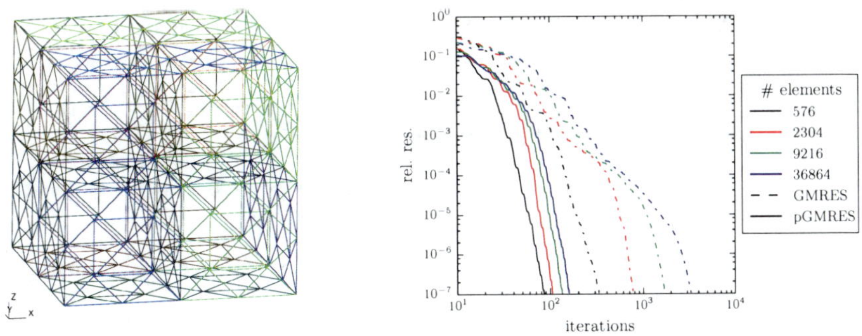


**Figure 7.30.** *Experiment* III*: Coarsest mesh (left) and behavior of relative residual during (preconditioned) GMRES iterations (right).*

Table 7.5: Experiment III: Iteration counts for (preconditioned) GMRES vs. number of triangles.

| $N$ | 576 | 2304 | 9216 | 36864 |
|---|---|---|---|---|
| Mesh width | $1{,}32 \cdot 10^{-1}$ | $6{,}59 \cdot 10^{-2}$ | $3{,}30 \cdot 10^{-2}$ | $1{,}65 \cdot 10^{-2}$ |
| GMRES | 339 | 765 | 1728 | 3304 |
| pGMRES | 87 | 106 | 135 | 159 |

**Summary of observations.** Obviously, in all numerical tests operator preconditioning substantially accelerates the convergence of GMRES. A moderate dependence of the number of iterations on the mesh width seems to persist, at least in Experiments II and III. We point out that matching the empiric data with the theoretical predictions of Theorem 7.37 is problematic because (i) preasymptotic behavior may prevail in numerical experiments, (ii) logarithmic factors are hard to tell from measured data, and (iii) convergence rates of GMRES seem to be governed not by the spectral condition number but by the numerical range of the nonsymmetric preconditioned system matrix [6, 27].

The data collected hint at a strong dependence of the iteration counts on the relative variation of the diffusion coefficients between the subdomains. Operator preconditioning as described in the section entitled "Local MTF: Operator Preconditioning in Three Dimensions" does not seem to be robust with respect to the size of jumps of the coefficients.

## Bibliography

[1] M. Ainsworth and W. McLean, Multilevel diagonal scaling preconditioners for boundary element equations on locally refined meshes, Numer. Math. **93** (2003), 387–413.

[2] F. P. Andriulli, K. Cools, H. Bagci, F. Olyslager, A. Buffa, S. Christiansen, and E. Michielssen, A multiplicative Calderon preconditioner for the electric field integral equation, IEEE Trans. Antennas Propagation **56** (2008), 2398–2412.

[3] F. P. Andriulli, K. Cools, and E Michielssen, A Calderon multiplicative preconditioner for PEC structures containing junctions, in Proceedings of the IEEE Antennas and Propagation Society International Symposium, (2011); https://biblio.ugent.be/publication/2029048.

[4] H. Bagci, F. P. Andriulli, K. Cools, F. Olyslager, and E. Michielssen, A Calderón Multiplicative Preconditioner for the Combined Field Integral Equation, IEEE Trans. Antennas Propagation, **57** (2009), 3387–3392.

[5] M. Bebendorf, Hierarchical Matrices: A Means to Efficiently Solve Elliptic Boundary Value Problems, Lecture Notes in Comput. Sci. Engrg. 63, Springer-Verlag, New York, 2008.

[6] B. Beckermann, S. A. Goreinov, and E. E. Tyrtyshnikov, Some remarks on the Elman estimate for GMRES, SIAM J. Matrix Anal. Appl. **27** (2006), 772–778.

[7] A. Buffa, Remarks on the discretization of some noncoercive operator with applications to heterogeneous Maxwell equations, SIAM J. Numer. Anal. **43** (2005), 1–18.

[8] A. Buffa and S.H. Christiansen, A dual finite element complex on the barycentric refinement, Math. Comp. **76** (2007), 1743–1769.

[9] A. Buffa and R. Hiptmair, Galerkin Boundary Element Methods for Electromagnetic Scattering, in Topics in Computational Wave Propagation. Direct and Inverse Problems (M. Ainsworth, P. Davis, D. Duncan, P. Martin and B. Rynne, eds.), Lecture Notes in Comput. Sci. Engrg. 31, Springer, Berlin, 2003, 83–124.

[10] Y. Chang and R. F. Harrington, A surface formulation or characteristic modes of material bodies, IEEE Trans. Antennas Propagation **25** (1977), 789–795.

[11] F. Chouly and N. Heuer, A Nitsche-based domain decomposition method for hypersingular integral equations, Numer. Math. **121** (2012), 705–729.

[12] S. H. Christiansen and J.-C. Nédélec, Des préconditionneurs pour la résolution numérique des équations intégrales de frontiére de l'acoustique, C.R. Acad. Sci. Paris Ser. I Math **330** (2000), 617–622.

[13] X. Claeys, A Single Trace Integral Formulation of the Second-Kind for Multiple Sub-Domain Scattering, Report 2011-14, Seminar for Applied Mathematics, ETH Zürich, Zürich, Switzerland, 2011.

[14] X. Claeys and R. Hiptmair, Electromagnetic scattering at composite objects: A novel multi-trace boundary integral formulation, Math. Model. Numer. Anal. **46** (2012), 1421–1445.

[15] X. Claeys and R. Hiptmair, Integral Equations for Acoustic Scattering by Partially Impenetrable Composite Objects, Integral Eqns. and Operator Theory, to appear.

[16] X. Claeys, R. Hiptmair, and C. Jerez-Hanckes, Multi-trace boundary integral equations, Direct and Inverse Problems in Wave Propagation and Applications (I.G. Graham, U. Langer, J.M. Melenk, and M. Sini, eds.), Radon Ser. Comput. Appl. Math. 14, De Gruyter, Berlin, 2013, 51–100.

[17] X. Claeys, R. Hiptmair, and E. Spindler, A Second-Kind Galerkin Boundary Element Method for Scattering at Composite Objects, BIT, 2014, doi:10.007/510543-014-0496-Y.

[18] X. Claeys and R. Hiptmair, Multi-Trace Boundary Integral Formulation for Acoustic Scattering by Composite Structures, Commun. Pure Appl. Math. **66** (2013), 1163–1201.

[19] D. Colton and R. Kress, Inverse Acoustic and Electromagnetic Scattering Theory, 2nd ed., Appl. Math. Sci. 93, Springer, Heidelberg, 1998.

[20] K. Cools, Mortar Boundary Elements for the EFIE Applied to the Analysis of Scattering by PEC Junctions, in Proceedings of the 2012 Asia-Pacific Symposium on Electromagnetic Compatibility (APEMC), 2012, 165–168.

[21] K. Cools, F. P. Andriulli, and F. Olyslager, A Calderón Preconditioned PMCHWT Equation, in Proceedings of the International Conference on Electromagnetics in Advanced Applications, Torino, Italy, 2009, pp. 521–524.

[22] M. Costabel, Boundary integral operators on Lipschitz domains: Elementary results, SIAM J. Math. Anal. **19** (1988), 613–626.

[23] M. Costabel, Boundary integral operators for the heat equation, Integral Equations Oper. Theory **13** (1990), 498–552.

[24] M. Costabel and E. Stephan, A direct boundary integral equation method for transmission problems, J. Math. Anal. Appl. **106** (1985), 367–413.

[25] E. Darve and P. Have, Efficient fast multipole method for low-frequency scattering, J. Comput. Phys. **197** (2004), 341–363.

[26] R. Dautray and J.-L. Lions, Mathematical Analysis and Numerical Methods for Science and Technology, Springer, Berlin, 1990.

[27] S. C. Eisenstat, H. C. Elman, and M. H. Schultz, Variational iterative methods for nonsymmetric systems of linear equations, SIAM J. Numer. Anal. **20** (1983), 345–357.

[28] F. Frühauf, B. Gebauer, and O. Scherzer, Detecting interfaces in a parabolic-elliptic problem from surface measurements, SIAM J. Numer. Anal. **45** (2007), 810–836.

[29] C. Geuzaine and J.-François Remacle, Gmsh: A 3-D finite element mesh generator with built-in pre- and post-processing facilities, Internat. J. Numer. Methods Engrg. **79** (2009), 1309–1331.

[30] L. Greengard, J.-F. Huang, V. Rokhlin, and S. Wandzura, Accelerating fast multipole methods for the Helmholtz equation at low frequencies, IEEE Comput. Sci. Engrg. **5** (1998), 32–38.

[31] L. Greengard and J.-Y. Lee, Stable and accurate integral equation methods for scattering problems with multiple material interfaces in two dimensions, J. Comput. Phys. **231** (2012), 2389–2395.

[32] W. Hackbusch, Hierarchische Matrizen. Algorithmen und Analysis, Springer, Heidelberg, 2009.

[33] R.F. Harrington, Boundary integral formulations for homogeneous material bodies, J. Electromagnetic Waves Appl. **3** (1989), 1–15.

[34] C. Hazard and M. Lenoir, On the solution of time-harmonic scattering problems for Maxwell's equations, SIAM J. Math. Anal. **27** (1996), 1597–1630.

[35] M. Healey, The Mortar Boundary Element Method, PhD thesis, School of Information Systems, Computing and Mathematics, Brunel University, London, 2010.

[36] N. Heuer and E. P. Stephan, Iterative substructuring for hypersingular integral equations in $\mathbb{R}^3$, SIAM J. Sci. Comput. **20** (1999), 739–749.

[37] N. Heuer and S. Meddahi, Discontinuous Galerkin hp-BEM with quasi-uniform meshes, Numer. Math. **125** (2013), 679–703.

[38] R. Hiptmair, Operator preconditioning, Comput. Math. Appl. **52** (2006), 699–706.

[39] R. Hiptmair and C. Jerez-Hanckes, Multiple traces boundary integral formulation for Helmholtz transmission problems, Adv. Appl. Math. **37** (2012), 39–91.

[40] R. Hiptmair, C. Jerez-Hanckes, J.-F. Lee, and Z. Peng, Domain Decomposition for Boundary Integral Equations via Local Multi-Trace Formulations, in Domain Decomposition Methods, in Science and Engineering, Lecture Notes in Computational Science and Engineering 98, 2014.

[41] R. Hiptmair, C. Jerez-Hanckes, and S.-P. Mao, Extension By Zero In Discrete Trace Spaces: Inverse Estimates, Report 2012-33, Seminar for Applied Mathematics, ETH Zürich, Switzerland, 2012, Math. Comp., to appear.

[42] R. Hiptmair and L. Kielhorn, BETL: A generic boundary element template library, Report 2012-36, Seminar for Applied Mathematics, ETH Zürich, Switzerland, 2012.

[43] R. Hiptmair and S.-P. Mao, Stable Multilevel Splittings of Boundary Edge Element Spaces, BIT **52** (2012), 661–685.

[44] G. C. Hsiao and W. L. Wendland, Boundary Integral Equations, Appl. Math. Sci. 164, Springer-Verlag, Berlin, 2008.

[45] R. E. Kleinman and P. A. Martin, On single integral equations for the transmission problem of acoustics, SIAM J. Appl. Math. **48** (1988), 307–325.

[46] A. R. Laliena, M.-L. Rapún and F.-J. Sayas, Symmetric boundary integral formulations for Helmholtz transmission problems, Appl. Numer. Math. **59** (2009), 2814–2823.

[47] U. Langer and C. Pechstein, All-floating coupled data-sparse boundary and interface-concentrated finite element tearing and interconnecting methods, Comput. Vis. Sci. **11** (2008), 307–317.

[48] U. Langer and D. Pusch, Data-sparse algebraic multigrid methods for large scale boundary element equations, Appl. Numer. Math. **54** (2005), 406–424.

[49] U. Langer, D. Pusch and S. Reitzinger, Efficient preconditioners for boundary element matrices based on grey-box algebraic multigrid methods, Internat. J. Numer. Methods Engrg. **58** (2003), 1937–1953.

[50] U. Langer and O. Steinbach, Boundary Element Tearing and Interconnecting Methods, Computing **71** (2003), 205–228.

[51] U. Langer and C. Pechstein, Coupled finite and boundary element tearing and interconnecting solvers for nonlinear potential problems, ZAMM Z. Angew. Math. Mech. **86** (2006), 915–931.

[52] U. Langer and D. Pusch, Convergence analysis of geometrical multigrid methods for solving data-sparse boundary element equations, Comput. Vis. Sci. **11** (2008), 181–189.

[53] U. Langer and O. Steinbach, Coupled finite and boundary element domain decomposition methods, Boundary Element Analysis, Lecture Notes in Appl. Comput. Mech. 29, Springer, Berlin, 2007, pp. 61–95.

[54] J.-L. Lions, Équations différentielles opérationnelles et problèmes aux limites, Grundlehren Math. Wiss. 111, Springer-Verlag, Berlin, 1961.

[55] K.-A. Mardal and R. Winther, Preconditioning discretizations of systems of partial differential equations, Numer. Linear Algebra Appl. **18** (2011), 1–40.

[56] W. McLean, Strongly Elliptic Systems and Boundary Integral Equations, Cambridge University Press, Cambridge, UK, 2000.

[57] W. McLean and O. Steinbach, Boundary element preconditioners for a hypersingular integral equations on an interval, Adv. Comput. Math. **11** (1999), 271–286.

[58] C. Müller, Foundations of the Mathematical Theory of Electromagnetic Waves, Grundlehren Math. Wiss. 155, Springer, Berlin, 1969.

[59] J.-C. Nédélec, Acoustic and Electromagnetic Equations: Integral Representations for Harmonic Problems, Appl. Math. Sci. 44, Springer-Verlag, Berlin, 2001.

[60] G. Of, An efficient algebraic multigrid preconditioner for a fast multipole boundary element method, Computing **82** (2008), 139–155.

[61] P. Oswald, Multilevel norms for $H^{-1/2}$, Computing **61** (1998), 235–255.

[62] C. Pechstein, Boundary element tearing and interconnecting methods in unbounded domains, Appl. Numer. Math. **59** (2009), 2824–2842.

[63] Z. Peng, K.-H. Lim and J.-F. Lee, A discontinuous Galerkin surface integral equation method for electromagnetic wave scattering from non-penetrable eargets, IEEE Trans. Antennas Propagation **61** (2013), 3617–3628.

[64] Z. Peng, K.-H. Lim and J.-F. Lee, Computations of Electromagnetic Wave Scattering From Penetrable Composite Targets Using a Surface Integral Equation Method With Multiple Traces, IEEE Trans. Antennas Propagation **61** (2013), 256–270.

[65] Z. Peng, K.-H. Lim and J.-F. Lee, Non-conformal Domain Decompositions for Solving Large Multiscale Electromagnetic Scattering Problems, Proc. IEEE **101** (2013), 298–319.

[66] Z. Peng, X.-C. Wang and J.-F. Lee, Integral Equation Based Domain Decomposition Method for Solving Electromagnetic Wave Scattering From Non-Penetrable Objects, IEEE Trans. Antennas Propagation **59** (2011), 3328–3338.

[67] A.J. Poggio and E.K. Miller, Integral Equation Solution of Three-Dimensional Scattering Problems, in Computer Techniques for Electromagnetics (R. Mittra, ed.), Pergamon, New York, 1973, 159–263.

[68] M.-L. Rapún and F.-J. Sayas, Mixed boundary integral methods for Helmholtz transmission problems, J. Comput. Appl. Math. **214** (2008), 238–258.

[69] M.-L. Rapún and F.-J. Sayas, Indirect Methods with Brakhage-Werner Potentials for Helmholtz Transmission Problems, Numer. Math. Adv. Appl., Springer, Berlin, 2006, 1146–1154.

[70] S. Sauter and C. Schwab, Boundary Element Methods, Springer Ser. Comput. Math. 39, Springer, Heidelberg, 2010.

[71] G. Schmidlin, C. Lage, and C. Schwab, Rapid solution of first kind boundary integral equations in $\mathbb{R}^3$, Eng. Anal. Bound. Elem. **27** (2003), 469–490.

[72] O. Steinbach and W. Wendland, The construction of some efficient preconditioners in the boundary element method, Adv. Comput. Math. **9** (1998), 191–216.

[73] O. Steinbach, Stability Estimates for Hybrid Coupled Domain Decomposition Methods, Lecture Notes in Math. 1809, Springer-Verlag, Berlin, 2003.

[74] O. Steinbach, Numerical Approximation Methods for Elliptic Boundary Value Problems, Springer, New York, 2008.

[75] E. Stephan, M. Maischak, and F. Leydecker, Some Schwarz methods for integral equations on surfaces- $h$- and $p$-versions, Comput. Vis. Sci. **8** (2005), 211–216.

[76] M. B. Stephanson and J.-F. Lee, Preconditioned electric field integral equation using Calderon identities and dual loop/star basis functions, IEEE Trans. Antennas Propagation, **57** (2009), 1274–1279.

[77] F. Valdes, F. P. Andriulli, K. Cools, and E. Michielssen, High-order div- and quasi curl-conforming basis functions for Calderón multiplicative preconditioning of the EFIE, IEEE Trans. Antennas Propagation **59** (2011), 1321–1337.

[78] S. Venakides, M.A. Haider and V. Papanicolaou, Boundary integral calculations of two-dimensional electromagnetic scattering by photonic crystal Fabry–Perot structures, SIAM J. Appl. Math. **60** (2000), 1686–1706.

[79] T. von Petersdorff, Boundary integral equations for mixed Dirichlet, Neumann and transmission problems, Math. Methods Appl. Sci. **11** (1989), 185–213.

[80] T.-K. Wu and L.-L. Tsai, Scattering from arbitrarily-shaped lossy dielectric bodies of revolution, Radio Sci. **12** (1977), 709–718.

[81] P. Yla-Oijala and M. Taskinen, Well-conditioned Muller formulation for electromagnetic scattering by dielectric objects, IEEE Trans. Antennas Propagation, **53** (2005), 3316–3323.

# Chapter 8

# Numerical Methods for Evolution Problems

*T. Trogdon*

## 8.1 ▪ A Unified Numerical Approach for the Nonlinear Schrödinger Equations

### 8.1.1 ▪ Introduction

A new method for analyzing boundary-value problems for both linear and integrable nonlinear partial differential equations (PDEs) was introduced by Fokas in [1] and extended by several authors; see [2]. This method is usually referred to as the unified transform method (UTM) or the Fokas transform method. For integrable nonlinear PDEs the UTM provides a generalization of the inverse scattering transform for analyzing initial-boundary value as opposed to initial value problems. For linear PDEs, the UTM provides a powerful method for obtaining analytical solutions (see the reviews [4, 5]) and also has given rise to new numerical techniques for solving both evolution [6] as well as elliptic PDES (see [7, 8] and the articles by Fokas and Lenells and Fokas, Iserles, and Smitheman in this volume).

For linear PDEs the original formulation of the UTM is based on the observation of Fokas and Gelfand [9] that linear PDEs admit a Lax pair formulation. The first goal of this paper is to show that this formulation is also useful for the numerical integration of the initial value problem of linear evolution PDEs. For integrable nonlinear PDEs the UTM gives rise to a matrix Riemann–Hilbert problem (RHP). The second goal of this paper is to show that now there exists an efficient numerical technique for the numerical solution of such RHPs. The contour on which the RHP is posed is more complicated for boundary value problems. Thus, for economy of presentation, we will analyze only the case of the simple contours of the initial value problem; the more complicated contours of the boundary value problem will be analyzed in the future.

The basic claim is the following. The UTM provides a unique and robust framework for computation. It is understood in the case of linear PDEs with periodic boundary conditions that the explicit Fourier series solution is an effective method of computation.

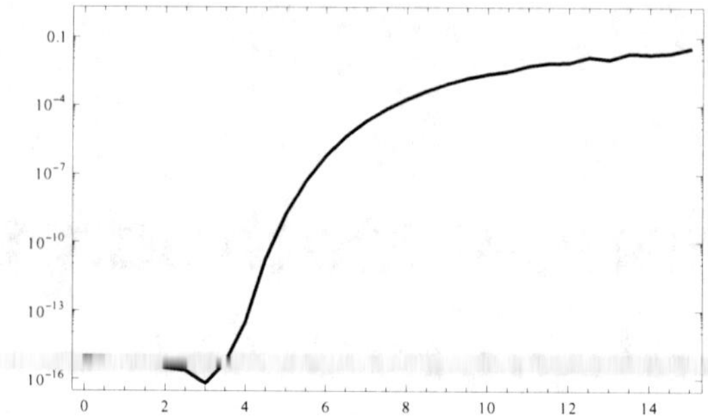


**Figure 8.1.** *Let $q_{FS}(x,t)$ be the solution* (8.2) *with $L = 50$ and $q_{CI}(x,t)$ be the solution of* (8.3) *computed with the method in Section* 8.1.2. *We use $q_0(x) = e^{-x^2}$ as an initial condition for both solutions. We plot $|q_{FS}(-10,t) - q_{CI}(-10,t)|$ versus $t$. It is clear that the solution of the periodic problem diverges from the solution of the problem on the line as $t$ increases.*

For example, consider solving

$$\begin{aligned} -iq_t + q_{xx} &= 0, \\ q(x,0) = q_0(x) &\in C^\infty(\mathbb{R}), \quad L > 0, \\ q_0(x+2L) &= q_0(x). \end{aligned} \tag{8.1}$$

A solution is found with classical separation of variables

$$q(x,t) = \sum_{n=-\infty}^{\infty} \hat{q}_n e^{i\frac{\pi n}{L}x + i\frac{\pi^2 n^2}{L^2}t}, \quad \hat{q}_n = \frac{1}{2L}\int_{-L}^{L} e^{-i\frac{\pi n}{L}x} q_0(x)dx. \tag{8.2}$$

In practice, depending on the smoothness of $q_0(x)$, the coefficients $\hat{q}_n$ are approximated well with the discrete Fourier transform applied to $q_0(x)$ sampled on a uniform grid

$$\{-L, -L + 2L/N, -L + 4L/N, \ldots, L - 2L/N\}.$$

Once approximations of $\hat{q}_n$ are known, (8.2) provides an effective formula for computation for all $x$ and $t$.

The solution of the Cauchy problem on the line with decaying initial data

$$\begin{aligned} -iq_t + q_{xx} &= 0, \\ q(x,0) = q_0(x) &\in \mathscr{S}(\mathbb{R}) \end{aligned} \tag{8.3}$$

is more difficult to compute. Here $\mathscr{S}(\mathbb{R})$ is the Schwartz class of rapidly decaying functions. One could approximate (8.3) with (8.1) with $L$ sufficiently large. This works well for small $t$. Indeed, (8.3) is a dispersive equation and oscillatory radiation is seen to move rapidly in both the positive and negative $x$ directions. Once the radiation reaches $L$ or $-L$, the quality of the periodic approximation is reduced. We demonstrate this error in Figure 8.1. The solution of (8.3) in this figure is computed with the method described in Section 8.1.2.

The method in Section 8.1.2 derives its effectiveness from contour integration. This is also the fundamental idea that drives the UTM: contour integration is used to solve linear PDEs. The connections between computation and the UTM run deeper. We have seen that the Lax pair for linear PDEs is used to derive the transform pair and the explicit solution. This involves the forward and inverse transforms. For (8.3) the forward transform is just the Fourier transform

$$q(x) \mapsto \int_{-\infty}^{\infty} e^{-ikx} q(x) dx, \tag{8.4}$$

and the inverse transform at time $t$ is given by

$$\hat{q}(k) \mapsto \frac{1}{2\pi} \int_{-\infty}^{\infty} e^{ikx+ik^2 t} \hat{q}(k) dk.$$

Numerical techniques for differential equations applied to the Lax pair provide an effective method for computing (8.4). Therefore, the effective numerical solution of (8.3) directly shadows the UTM: accurate computation utilizes both ordinary differential equations and contour integrals.

The solution of integrable nonlinear PDEs on the line proceeds in an analogous way. In this case, the forward transform is known as direct scattering and the inverse transform as inverse scattering. We see in Section 8.1.3 that the direct scattering transform is effectively computed via the Lax pair providing a close parallel with Section 8.1.2. Furthermore, the inverse scattering transform is often expressed in terms of a solution of a matrix RHP. A method for the numerical solution of this RHP is used to effectively compute the inverse scattering transform. In this sense, RHPs are the appropriate generalization of contour integrals for nonlinear problems.

In this chapter we concentrate on solving PDEs on the line. Boundary value problems are the next logical step after the material presented here. We begin by discussing numerical methods for contour integrals, ordinary differential equations, and RHPs. We try to be as detailed as possible but in some cases implementational details are left to references. We combine quadrature methods with methods for ordinary differential equations and solve (8.3) for all $x$ and $t$ accurately. Next, we combine a numerical method for RHPs with methods for differential equations to solve integrable nonlinear PDEs accurately for all $x$ and $t$. This demonstrates that the solution formula for the problem provides an effective method of computation.

### 8.1.2 ▪ Chebyshev Spectral Methods

Before solving PDEs with the methods described above we pause to discuss some necessary Chebyshev numerical analysis techniques. We give a brief overview of everything from computing a Chebyshev series expansion of a given function to the discretization of singular integral equations for the solution of RHPs. More detailed references for the for the discrete cosine transform, the quadrature, and the solution of differential equations are [10, 12, 13]. A detailed development of the method for the solution of RHPs that we sketch here can be found in [14]. The numerical methods that follow rely on numerous well-understood mathematical properties of the Chebyshev polynomials.

The Chebyshev polynomials of the first kind, denoted $T_n(x)$, are orthogonal polynomials on $[-1,1]$ that satisfy [11]

$$\int_{-1}^{1} T_n(x) T_m(x) \frac{dx}{\sqrt{1-x^2}} = \begin{cases} 0 & \text{if } n \neq m, \\ \pi & \text{if } n = m = 0, \\ \pi/2 & \text{if } n = m \neq 0. \end{cases}$$

**Remark 60.** *Additional important properties also follow:*

- $\int_{-1}^{x} T_n(s)ds = \frac{1}{2}\left(\frac{T_{n+1}(x)}{n+1} - \frac{T_{n-1}(x)}{n-1}\right) - \frac{(-1)^n}{n^2-1}$ *for* $n > 1$.
- $\int_{-1}^{x} T_0(s)ds = T_0(x) + T_1(x)$.
- $\int_{-1}^{x} T_1(s)ds = T_2(x)/4 - T_0(x)/4$.
- $\int_{-1}^{1} T_n(s)ds = \frac{1+(-1)^n}{1-n^2}$ *for* $n \neq 1$.

### Discrete Cosine Transform

The discrete cosine transform (DCT) is considered a *fast transform* because it requires $\mathcal{O}(N \log N)$ operations to process $N$ data points. This is in stark contrast to the inverse of a triangular linear system ($\mathcal{O}(N^2)$ operations) or the inversion of a dense linear system ($\mathcal{O}(N^3)$ operations).

We motivate the DCT though Fourier series. Given $f(x) \in C^\infty([-1,1])$ we consider

$$g(\theta) = f(\cos(\theta)). \tag{8.5}$$

It follows that $g$ is smooth, even, and periodic on $[-\pi, \pi]$. It has a nicely convergent cosine series:

$$g(\theta) = \sum_{n=0}^{\infty} \hat{g}_n \cos(n\theta).$$

Inverting the transformation

$$g(\arccos x) = f(x) = \sum_{n=0}^{\infty} \hat{g}_n \cos(n \arccos x).$$

But $\cos(n \arccos x)$ is a polynomial! Indeed, $T_n(x) = \cos(n \arccos x)$ is a definition of the $n$th-order Chebyshev polynomial of the first kind. To find a Chebyshev expansion of $f$ it suffices to find a cosine series expansion for $g$. This is precisely the utility of the DCT.

Rather than build the DCT from the ground up, we assume the reader has some knowledge of the discrete Fourier transform (DFT). Reference [10] provides an especially comprehensive overview. In summary, given data $\{g_j\}_{j=0}^{2N}$ the DFT returns coefficients $\{\hat{g}_j\}_{j=-N}^{N}$ so that

$$g_N(\theta) = \sum_{j=-N}^{N} \hat{g}_j e^{ij\theta}$$

satisfies

$$g_N(\theta_j) = g_j, \quad \theta_j = -\pi + 2\pi \frac{j}{2N+1}, \quad j = 0, \ldots, 2N.$$

In other words, $g_N$ interpolates the data $\{g_j\}_{j=0}^{2N}$ at the points $\{\theta_j\}_{j=0}^{2N}$. In the case that $\{g_j\}_{j=0}^{2N}$ comes from an even function sampled at $\{\theta_j\}_{j=0}^{2N}$ it follows that $g_N(x)$ is even. Therefore $\hat{g}_j = \hat{g}_{-j}$ and

$$g_N(x) = \sum_{j=0}^{N} \gamma_j \cos(jx), \quad \gamma_0 = \hat{g}_0, \quad \gamma_j = 2\hat{g}_j, \quad j > 0.$$

*ALGORITHM 8.1.* **Simplified Chebyshev DCT.**
We present an algorithm for approximating the Chebyshev coefficients of a function $f(x) \in C^\infty([-1,1])$.

1. Fix $N > 0$.
2. Define $g_j = f(\theta_j)$, $j = 0,\ldots,2N$.
3. Apply the DFT to $\{g_j\}_{j=0}^{2N}$ to obtain the approximate Fourier coefficients $\{\hat{g}_j\}_{j=-N}^{N}$.
4. The approximate Chebyshev coefficients of $f$ are given by $\{\gamma_j\}_{j=0}^{N}$ with $\gamma_0 = \hat{g}_0$, $\gamma_j = 2\hat{g}_j$, $j > 0$.

**Remark 61.** *Algorithm* 8.1 *is called a simplified DCT in for two reasons. First, in practice with most DFT algorithms, one should use a power of* 2 *for the number of points. Second, when the DFT is used to compute the DCT one has to sample at double the number of necessary points. Many stock DCT methods remove this complexity.*

### Clenshaw–Curtis Quadrature

The Chebyshev expansion discussed in the previous section is used for accurate quadrature. For $f \in C^\infty([-1,1])$ we apply Algorithm 8.1 and find

$$f(x) \approx \sum_{j=0}^{N} \gamma_j T_j(x).$$

Integrating both sides assuming at least $L^1([-1,1])$ convergence of the approximation we find

$$\int_{-1}^{1} f(x)dx \approx \sum_{j=0}^{\lfloor N/2 \rfloor} \gamma_{2j} \frac{2}{1-4j^2}.$$

This method is called the *Clenshaw–Curtis quadrature*.

A simple change of variables produces a quadrature method for piecewise linear contours. Let $\Gamma = \Gamma_1 \cup \cdots \cup \Gamma_\ell \subset \mathbb{C}$ be an oriented contour. For simplicity, assume $\Gamma_k = \{a_k x + b_k : x \in [-1,1]\}$ for $a_k, b_k \in \mathbb{C}$. It suffices to discuss quadrature on a single contour $\Gamma_k$. Let $f_k \in C^\infty(\Gamma_k)$. It follows that $F_k(x) = f_k(a_k x + b_k)$ is an infinitely smooth function defined on $[-1,1]$ and

$$\int_{\Gamma_k} f_k(z)dz = a_k \int_{-1}^{1} F_k(x)dx.$$

Since the Clenshaw–Curtis quadrature may be applied to $F(x)$ we arrive at the formula

$$\int_{\Gamma_k} f_k(z)dz \approx a_k \sum_{j=0}^{\lfloor N/2 \rfloor} \gamma_{2j}^k \frac{2}{1-4j^2},$$

where $\{\gamma_j^k\}_{j=0}^{N}$ are obtained with Algorithm 8.1 applied to $F_k(x)$. For $f \in C^\infty(\Gamma)$ we let $f_k(z) = f|_{\Gamma_k}(z)$ be the restriction of $f$ and obtain the approximation

$$\int_{\Gamma} f(z)dz \approx \sum_{k=1}^{\ell} \sum_{j=0}^{\lfloor N/2 \rfloor} a_k \gamma_{2j}^k \frac{2}{1-4j^2}.$$

**Remark 62.** *For the results that we present here, contours can be assumed to be piecewise affine without loss of generality. Analyticity allows the deformation of general contours to piecewise affine contours. Furthermore, we expect all functions to be analytic in a neighborhood of the integration contour implying rapid (spectral) convergence of the Clenshaw–Curtis quadrature; see Section* 8.1.2.

### Solving Ordinary Differential Equations

Some of the techniques discussed below require the numerical solution of ordinary differential equations (ODEs). While many techniques exist, we keep with the theme of using Chebyshev polynomials. Before progressing, we briefly discuss two different discretization methods. One may express an unknown function in terms of a Chebyshev series with unknown coefficients. Once this expression for the unknown function is substituted into the differential equation, an appropriate inner product is used to obtain a linear system for the unknown coefficients. When this type of method is used we say we are working in *coefficient space*, i.e., expressing operations in terms of their action on coefficients. This produces a true spectral method.

Taking a different approach, one may express an approximation of the unknown function in terms of its (unknown) values at the Chebyshev points

$$\boldsymbol{x}_N = \{\cos\theta_j\}_{j=0}^N.$$

From the Chebyshev DCT, we know how to construct the unique interpolant. We may substitute this interpolant into the differential equation and enforce that the equation holds at the Chebyshev points. Thus we set up a linear system for the unknown function values at the Chebyshev points. This type of method is said to act in *value space* and the method is a type of pseudospectral method. Often, pseudospectral methods are easier to implement and for this reason we concentrate almost solely on operators in value space.

Consider the ODE

$$\mu_x = A(x)\mu + f(x), \quad \mu(-L) = \mu_0 \in \mathbb{C}^{n\times 1}, \quad A:\mathbb{R}\to\mathbb{C}^{n\times n}.$$

For our purposes it is convenient to convert this to an integral equation by integrating from $-L$ to $x > -L$,

$$\mu(x) - \int_{-L}^{x} A(s)\mu(s)ds = \mu_0 + \int_{-L}^{x} f(s)ds. \tag{8.6}$$

We describe a method to obtain a finite-dimensional matrix approximation of the operator $\mu(x) \mapsto \int_{-L}^{x}\mu(s)ds$. First, we construct the *DCT matrix*. The Chebyshev DCT described above is a linear map from function values at $\boldsymbol{x}_N$ to Chebyshev coefficients. Abstractly, we call this map $\mathscr{F}_N$. Therefore its matrix representation is given by

$$\boldsymbol{F}_N = \begin{bmatrix} \mathscr{F}_N(\boldsymbol{e}_1) & \cdots & \mathscr{F}_N(\boldsymbol{e}_{N+1}) \end{bmatrix},$$

where $\boldsymbol{e}_j$ is the basis (column) vector of length $N+1$ with zeros in every component with the exception of a one in the $j$th component.

The inverse DCT can also be constructed. Each step in Algorithm 8.1 is performed in reverse, replacing the DFT with the inverse DFT. Therefore

$$\boldsymbol{F}_N^{-1} = \begin{bmatrix} \mathscr{F}_N^{-1}(\boldsymbol{e}_1) & \cdots & \mathscr{F}_N^{-1}(\boldsymbol{e}_{N+1}) \end{bmatrix}.$$

The matrix $\boldsymbol{F}_N$ is a map to coefficient space and its inverse is a map to value space.

Let $\gamma = (\gamma_0, \ldots, \gamma_N)^T$ be a vector of Chebyshev coefficients and let

$$f_N(x) = \sum_{j=0}^{N} \gamma_j T_n(x). \tag{8.7}$$

We find

$$\int_{-1}^{x} f_N(s)ds = \int_{-1}^{x} \sum_{j=0}^{N} \gamma_j T_j(s)ds = \gamma_0 T_1(x) + \gamma_1 \frac{T_2(x)}{4}$$
$$+ \sum_{j=2}^{N} \gamma_j \frac{1}{2} \frac{T_{j+1}(x)}{j+1} - \sum_{j=2}^{N} \gamma_j \frac{1}{2} \frac{T_{j-1}(x)}{j-1} + C.$$

The constant $C$ is chosen so that the integral vanishes when $x = -1$:

$$C = -\sum_{j=2}^{N} \gamma_j \left( \frac{1}{2} \frac{T_{j+1}(-1)}{j+1} - \frac{1}{2} \frac{T_{j-1}(-1)}{j-1} \right) - \gamma_0 T_1(-1) - \gamma_1 \frac{T_2(-1)}{4}.$$

This expression simplifies using $T_j(-1) = (-1)^j$. An approximation of the coefficients $\eta = (\eta_0, \ldots, \eta_N)$ of this antiderivative of $f_N$ is given by

$$\eta = \boldsymbol{I}_N \gamma,$$

$$\boldsymbol{I}_N = \begin{bmatrix} 1 & -1/4 & 1/6-1/2 & 1/4-1/8 & 1/10-1/6 & \cdots & \cdots & 0 \\ 1 & 0 & -1/2 & & & & \cdots & 0 \\ 0 & 1/4 & 0 & -1/4 & 0 & & \cdots & 0 \\ 0 & 0 & 1/6 & 0 & -1/6 & 0 & \cdots & 0 \\ & & & \ddots & & \ddots & & \vdots \\ & & & & \ddots & & \ddots & \vdots \\ & & & & & 1/2(N-1) & 0 & -1/2(N-1) \\ & & & & & & 1/2N & 0 \end{bmatrix}.$$

This representation follows from Remark 60. Note that this is an approximation because we have neglected the contribution from $T_{N+1}(x)$ which appears after integration of the right-hand side of (8.7).

For $\mu$ defined on $[-1,1]$, $\boldsymbol{F}_N^{-1}\boldsymbol{I}_N\boldsymbol{F}_N\mu(\boldsymbol{x}_N)$ produces an approximation of $\int_{-1}^{x} \mu(s)ds$ evaluated at the points $\boldsymbol{x}_N$. Define $\boldsymbol{x}_{N,-L} = L/2(\boldsymbol{x}_N - 1)$, the shifted Chebyshev points that discretize the interval $[-L,0]$. If $\mu$ is defined on $[-L,0]$, then $L/2\boldsymbol{F}_N^{-1}\boldsymbol{I}_N\boldsymbol{F}_N\mu(\boldsymbol{x}_{N,-L})$ is an approximation of $\int_{-L}^{x} \mu(s)ds$ at the points $\boldsymbol{x}_{N,-L}$. The factor of $L/2$ arises from a straightforward change of variables. In the case that (8.6) is scalar ($n = 1$) we have the discretization

$$\left( I_{N+1} + \frac{L}{2}\boldsymbol{F}_N^{-1}\boldsymbol{I}_N\boldsymbol{F}_N \operatorname{diag}(A(\boldsymbol{x}_{N,-L})) \right) \mu_N = \mu_0 + \frac{L}{2}\boldsymbol{F}_N^{-1}\boldsymbol{I}_N\boldsymbol{F}_N f(\boldsymbol{x}_{N,-L}), \tag{8.8}$$
$$\mu_0 = (\mu_0, \mu_0, \ldots, \mu_0)^T,$$

where $\mu_N$ is an approximation of $\mu$ at $\boldsymbol{x}_{N,-L}$ and $I_N$ is the $N \times N$ identity matrix.

This is easily extended to the vector-valued case for $\mu : \mathbb{R} \to \mathbb{C}^{n\times 1}$. Define

$$\boldsymbol{I}_{N,-L} = \overbrace{\frac{L}{2}\boldsymbol{F}_N^{-1}\boldsymbol{I}_N\boldsymbol{F}_N \oplus \cdots \oplus \frac{L}{2}\boldsymbol{F}_N^{-1}\boldsymbol{I}_N\boldsymbol{F}_N}^{n-1 \text{ times}}. \tag{8.9}$$

Since $\mu$ is vector valued we define a vector of unknowns

$$\mu = (\mu_1(\boldsymbol{x}_{N,-L}), \mu_2(\boldsymbol{x}_{N,-L}), \ldots, \mu_n(\boldsymbol{x}_{N,-L}))^T.$$

For ease of notation, let $\nu = A(\boldsymbol{x}_{N,-L})\mu$, where $A(\boldsymbol{x}_{N,-L})$ is defined by

$$(\nu_j, \nu_{j+(N+1)}, \ldots, \nu_{j+n(N+1)})^T = A(\boldsymbol{x}_{N,-L,j})\mu(\boldsymbol{x}_{N,-L,j}).$$

The subscript $j$ refers to components. This is the appropriate generalization of $\operatorname{diag}(A(\boldsymbol{x}_{N,-L}))$ to the vector case. Therefore a discretization of (8.6) with $f = 0$ is given by

$$(I_{n(N+1)} + \boldsymbol{I}_{N,-L} A(x_{N,-L}))\mu = \mu_0,$$

where the components of $\mu_0$ are defined by the equality

$$(\mu_{0,j}, \mu_{0,j+1}, \ldots, \mu_{0,j+N+1})^T = \mu_0.$$

In what follows, we need a discretization of the operator $\mu(x) \mapsto \int_x^L \mu(s)ds$. It is clear that $\int_x^L = \int_{-L}^L - \int_{-L}^x$ so that we only have to express $\int_{-L}^L$ in matrix form. To map from coefficients $\gamma = (\gamma_j{}_{j=0}^N)$ to the definite integral over $[-1,1]$ we use

$$\begin{bmatrix} \int_{-1}^1 \sum_j \gamma_j T_j(x)dx \\ 0 \\ \vdots \end{bmatrix} = \begin{bmatrix} 2 & 0 & -2/3 & 0 & -2/15 & \cdots & ((-1)^N - 1)/(N^2 - 1) \\ 0 & 0 & 0 & 0 & 0 & 0 & 0 \\ \vdots & \vdots & \vdots & \vdots & \vdots & \vdots & \vdots \end{bmatrix} \gamma.$$

We use $\boldsymbol{I}_{N,0}$ to denote this matrix, after applying $n-1$ direct sums as in (8.9). It follows that $\boldsymbol{I}_{N,0} - \boldsymbol{I}_N$ is a discretization of the desired operator. Define another set of shifted Chebyshev points $\boldsymbol{x}_{N,L} = L/2(\boldsymbol{x}_N + 1)$. From the ideas used above, it is clear that

$$\boldsymbol{I}_{N,L}\mu(\boldsymbol{x}_{N,L}) = \frac{L}{2}\boldsymbol{F}_N^{-1}(\boldsymbol{I}_{N,0} - \boldsymbol{I}_N)\boldsymbol{F}_N \mu(\boldsymbol{x}_{N,L})$$

is an approximation of $\int_x^L \mu(s)ds$ at the points $\boldsymbol{x}_{N,L}$. In this way we may solve boundary value problems when boundary conditions are imposed on the right side of the domain.

### A Numerical Example

Consider the boundary value problem

$$\mu_x + ik\mu = e^{-x^2},$$
$$\mu(-10) = 0.$$

After integrating from $-10$ to $x$ we find the integral equation

$$\mu(x) + ik \int_{-10}^x \mu(s)ds = \int_{-10}^x e^{-s^2} ds.$$

This is discretized with the linear system

$$(I_{N+1} + i5k\boldsymbol{F}_N^{-1}\mathcal{I}_N \boldsymbol{F}_N)\mu_N = 5\boldsymbol{F}_N^{-1}\mathcal{I}_N \boldsymbol{F}_N e^{-\boldsymbol{x}_{N,-10}^2}.$$

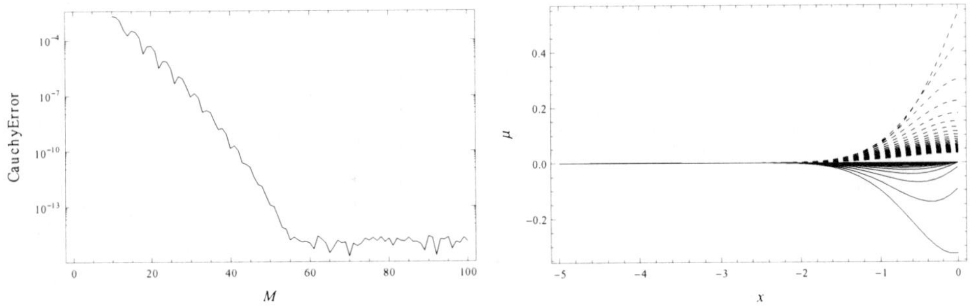


**Figure 8.2.** (a) *A plot of the Cauchy error* $|\mu_{M,-1} - \mu_{200,-1}|$ *as* $M$ *varies.* (b) *A plot of* $\mu_N$ *as* $k$ *varies from 2 to 30 (solid: real part; dashed: imaginary part). The amplitude of the solution is a decreasing function of* $|k|$.

To test the convergence of the method we look at the Cauchy error. Following programming syntax for indexing arrays, we let $\mu_{N,-1}$ be the last entry in the array. For our purposes, the Cauchy error is defined as $|\mu_{M,-1} - \mu_{N,-1}|$. We demonstrate the convergence of the Cauchy error in Figure 8.2(a). We also look at solutions for varying $k$ in Figure 8.2(b).

### Computing Cauchy Integrals

The next topic we discuss is computing integrals of the form

$$\mathscr{C}_\Gamma f(z) = \frac{1}{2\pi} \int_\Gamma \frac{f(s)}{s-z} dz, \quad z \in \mathbb{C} \setminus \Gamma,$$

where $\Gamma$ is a piecewise affine oriented contour. We also compute the boundary values of this integral:

$$\mathscr{C}_\Gamma^+ f(s) = \lim_{\substack{z \to s \\ z \text{ left of } \Gamma}} \mathscr{C}_\Gamma f(z),$$

$$\mathscr{C}_\Gamma^- f(s) = \lim_{\substack{z \to s \\ z \text{ right of } \Gamma}} \mathscr{C}_\Gamma f(z).$$

The method we describe can be found in [14] and is implemented in [19]. By its very nature, computing Cauchy integrals is more involved. Singular limits of analytic functions are computed and this is the reason for the technical nature of what follows.

The operators $\mathscr{C}_\Gamma^\pm$ have many important properties. See [30] for an overview. When $\Gamma$ is piecewise Lipschitz continuous with transverse self-intersections[16] $\mathscr{C}_\Gamma^\pm$ are bounded linear operators on $L^2(\Gamma)$. Furthermore, the *Plemelj lemma* holds: $\mathscr{C}_\Gamma^+ - \mathscr{C}_\Gamma^- = I$.

Another relevant property is the action of these operators on function spaces. Define $H^k(\Gamma)$ to be the space of functions such that the first $k$ weak derivatives on $\Gamma \setminus \gamma_0$ are square integrable, where $\gamma_0$ is the set of self-intersections of $\Gamma$. With this in mind we make a couple definitions.

[16]This condition can be relaxed. The operators $\mathscr{C}_\Gamma^\pm$ are bounded on $L^2(\Gamma)$ if and only if $\Gamma$ is a Carleson curve [15]. All curves we consider here are Carleson curves.

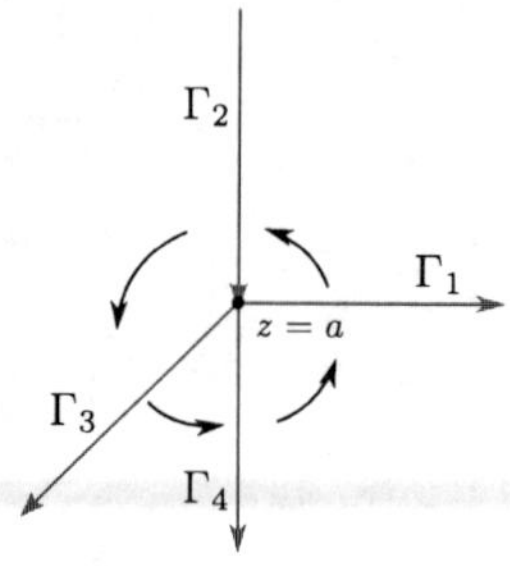


**Figure 8.3.** *A counterclockwise ordering of subcomponents of* $\Gamma$ *that contain* $z = a \in \gamma_0$.

**Definition 8.1.** *Assume that* $a \in \gamma_0$ *and let* $\Gamma_1, \ldots, \Gamma_\ell$ *be a counterclockwise ordering of subcomponents of* $\Gamma$ *(see Figure 8.3) which contain* $z = a$ *as an endpoint. For* $f \in H^k(\Gamma)$, *define*

$$f_i^{(j)} = \begin{cases} -\lim_{z\to a} \left(\frac{d}{dz}\right)^j f|_{\Gamma_i}(z) & \text{if } \Gamma_i \text{ is oriented away from } a, \\ \lim_{z\to a} \left(\frac{d}{dz}\right)^j f|_{\Gamma_i}(z) & \text{if } \Gamma_i \text{ is oriented toward } a. \end{cases} \tag{8.10}$$

*We say that* $f$ *satisfies the* $(k-1)$*th-order zero-sum condition if for all* $a \in \gamma_0$

$$\sum_{i=1}^{\ell} f_i^{(j)} = 0 \quad \text{for } j = 0, \ldots, k-1. \tag{8.11}$$

**Definition 8.2.** *Define*

$$H_z^k(\Gamma) = \{f \in H^k(\Gamma) : f \text{ satisfies the } (k-1)\text{th-order zero-sum condition}\}.$$

**Theorem 8.3 (see [30]).** *The operators* $\mathscr{C}_\Gamma^\pm$ *are bounded on* $H_z^k(\Gamma)$ *provided* $\Gamma$ *is piecewise smooth with transverse self-intersections.*

The important consequence of this theorem for what follows is that if $u \in H_z^k(\Gamma)$, then $\mathscr{C}_\Gamma^\pm u$ is a bounded function on $\Gamma$. This is used below.

Define $\mathbb{T}$ to be the unit circle with typical counterclockwise orientation and $\mathbb{I} = [-1,1]$ with left-to-right orientation. The numerical method relies on a matrix representation of the action of $\mathscr{C}_\Gamma^\pm$ on the Laurent monomials, $L_k(z) = z^k$, in the case $\Gamma = \mathbb{T}$ and the Chebyshev polynomials of the first kind in the case $\Gamma = \mathbb{I}$. The following representations are derived in [16, 14]. For $\Gamma = \mathbb{T}$ the action is given by

$$\mathscr{C}_\mathbb{T}^+ L_k(s) = \begin{cases} L^k(s) & \text{if } k \geq 0, \\ 0 & \text{otherwise}, \end{cases}$$

$$\mathscr{C}_\mathbb{T}^- L_k(s) = \begin{cases} -L^k(s) & \text{if } k < 0, \\ 0 & \text{otherwise}. \end{cases}$$

If $\Gamma = \mathbb{I}$, the expressions are much more complicated. The idea used is similar to computing the DCT. First, $\mathbb{T}$ is mapped to $\mathbb{I}$ using the map $z \mapsto T(z) = \frac{1}{2}(z + z^{-1}) = \cos(\arg z)$. This allows us to map functions on $\mathbb{I}$ to functions on $\mathbb{T}$. Furthermore, if we map a polynomial on $\mathbb{I}$ we obtain a Laurent polynomial on $\mathbb{T}$ and we know how the Cauchy operator acts there. Using the correct inverses of the map $T$ with additional modifications we obtain $\mathscr{C}_{\mathbb{I}}^{\pm}$ [14]. The inverses needed are

$$T_+^{-1}(x) = x - \sqrt{x-1}\sqrt{1+x}, \quad T_-^{-1}(x) = x + \sqrt{x-1}\sqrt{1+x},$$
$$T_\downarrow^{-1}(x) = x - i\sqrt{1-x}\sqrt{1+x}, \quad T_\uparrow^{-1}(x) = x + i\sqrt{1-x}\sqrt{1+x}.$$

Define the auxiliary functions

$$\mu_m(z) = \sum_{j=1}^{\lfloor \frac{m+1}{2} \rfloor} \frac{z^{2j-1}}{2j-1}, \quad \psi_0(z) = \frac{2}{i\pi} \operatorname{arctanh} z,$$
$$\psi_m(z) = z^m \left( \psi_0(z) - \frac{2}{i\pi} \begin{cases} \mu_{-m-1}(z) & \text{if } m < 0 \\ \mu_m(1/z) & \text{if } m > 0 \end{cases} \right).$$

Then (see [14])

$$\mathscr{C}_{\mathbb{I}} T_k(x) = -\frac{1}{2} \left( \psi_k(T_+^{-1})) + \psi_{-k}(T_+^{-1}(x)) \right), \tag{8.12}$$

and taking limits,

$$\begin{aligned}
\mathscr{C}_{\mathbb{I}}^{+} T_k(x) &= -\frac{1}{2} \left( \psi_k(T_\downarrow^{-1}(x)) + \psi_{-k}(T_\downarrow^{-1}(x)) \right) \\
&= -\frac{2}{i\pi} T_k(x) \operatorname{arctanh} T_\downarrow^{-1}(x) + \frac{1}{i\pi} \sum_{j=1}^{\lfloor \frac{k+1}{2} \rfloor} \frac{T_{k-2j+1}(x)}{2j-1} \begin{cases} 1 & \text{if } k-2j+1=0, \\ 2 & \text{otherwise,} \end{cases} \\
\mathscr{C}_{\mathbb{I}}^{-} T_k(x) &= -\frac{1}{2} \left( \psi_k(T_\uparrow^{-1}(x)) + \psi_{-k}(T_\uparrow^{-1}(x)) \right) \\
&= -\frac{2}{i\pi} T_k(x) \operatorname{arctanh} T_\uparrow^{-1}(x) + \frac{1}{i\pi} \sum_{j=1}^{\lfloor \frac{k+1}{2} \rfloor} \frac{T_{k-2j+1}(x)}{2j-1} \begin{cases} 1 & \text{if } k-2j+1=0, \\ 2 & \text{otherwise.} \end{cases}
\end{aligned} \tag{8.13}$$

**Remark 63.** *In practice $\psi_m$ is computed stably using the ${}_2F_1$ hypergeometric function. See* [14] *for details.*

**Remark 64.** *Notice that if $f(\pm 1) \neq 0$, then we expect an unbounded logarithmic singularity of $\mathscr{C}_{\mathbb{I}} f$ at the endpoints.*

We need to deal with more complicated contours. Assume $\Gamma$ is a contour for which there exists a affine mapping $M : \mathbb{I} \to \Gamma$. Assume $f$ is smooth. $\mathscr{C}_\Gamma^{\pm} u$ is the unique locally integrable function, analytic off $\Gamma$, which satisfies $F^+(k) - F^-(k) = f(k)$ for $k \in \Gamma$ with $F(\infty) = 0$. Then we have $F^+(M(z)) - F^-(M(z)) = u(M(z))$ for $z \in \mathbb{I}$ and $F(M(\infty)) = 0$. From the Plemelj lemma

$$\begin{aligned}
F(M(z)) &= \mathscr{C}_{\mathbb{I}}(u \circ M)(z), \\
\Rightarrow F(k) &= \mathscr{C}_{\mathbb{I}}(u \circ M)(M^{-1}(k)).
\end{aligned}$$

The last task is to deal with contours that have corners and self-intersections. The details for this are left to reference [14]. The end result is the following. If $\Gamma = \Gamma_1 \cup \cdots \cup \Gamma_\ell$

where each $\Gamma_j$ is an affine mapping of the unit interval $\Gamma_j = M_j((-1,1))$, $M_j(x) = a_j x + b_j$ and $f_n : \Gamma \to \mathbb{C}$ such that

$$f_n(z) = \sum_{j=0}^{n} \gamma_{j,n} T_j(M_j^{-1}(z)) \quad \text{for} \quad z \in \Gamma_j, \tag{8.14}$$

then $\mathscr{C}_\Gamma f_n$ and $\mathscr{C}_\Gamma^\pm f_n$ can be computed exactly on each segment $\Gamma_j$ provided $f_n \in H_z^1(\Gamma)$. Furthermore, for $f \in H_z^1(\Gamma)$ the Chebyshev DCT applied to $f \circ M_j$ produces an approximation of the form (8.14). Because $f \in H_z^1(\Gamma)$ and the endpoints of the segments $\Gamma_j$ are interpolation points, this approximation $f_n$ will also lie in $H_z^1(\Gamma)$ and $\mathscr{C}_\Gamma f$ provides an accurate approximation of $\mathscr{C}_\Gamma f_n$.

### Solving RHPs

RHPs are a fundamental object of study in what follows. They are boundary value problems in the complex plane. In this section we give a brief overview of the relevant aspects of RHPs. We also discuss a method for the numerical solution of RHPs. The numerical method presented is derived from the formulae for the Cauchy integral in Section 8.1.2.

### Riemann–Hilbert Problems

An RHP is the task of finding a (hopefully unique) function $\Phi(z)$ that satisfies the jump condition

$$\lim_{z\to s,\ z\in\Omega_+} \Phi(z) = \left(\lim_{z\to s,\ z\in\Omega_-} \Phi(z)\right) G(s) + f(s), \quad s \in \Gamma,$$

for some definite functions $f, G$ defined on $\Gamma$. Here $\Omega_+$ ($\Omega_-$) is the region lying just to the left (right) of the contour $\Gamma$. We assume the limit is taken in a nontangential manner. In what follows we use the notation

$$\Phi^+(z) = \lim_{\substack{z\to s\\ z\in\Omega_+}} \Phi(z), \quad \Phi^-(z) = \lim_{\substack{z\to s\\ z\in\Omega_-}} \Phi(z).$$

The jump condition

$$\Phi^+(s) = \Phi^-(s)G(s) + f(s) \tag{8.15}$$

will be used in the case that $\Phi$ is matrix or vector valued. In this case $G$ is matrix valued and $f$ is a matrix or vector valued. We concentrate almost exclusively on the case where $\Phi$ is a $2 \times 2$ matrix. A normalization condition is also imposed: $\Phi(\infty) = I$.

In most cases, RHPs are equivalent to singular integral equations. To see this one must yield to the theory of Hardy spaces [17, 30]. Provided that $\Phi^\pm(s) - I$ is square integrable it follows that

$$\Phi(z) = I + \mathscr{C}_\Gamma u(z)$$

for some square-integrable function $u$. The Plemelj lemma shows that $\Phi^+(s) - \Phi^-(s) = u(s)$ and we can rewrite the jump condition (8.15) in terms of $u$:

$$\begin{aligned}
\Phi^+(s) &= \Phi^-(s)G(s) + f(s),\\
\Phi^+(s) - \Phi^-(s) &= \Phi^-(s)(G(s) - I) + f(s),\\
u(s) - \mathscr{C}_\Gamma^- u(s)(G(s) - I) &= G(s) - I + f(s).
\end{aligned} \tag{8.16}$$

The final equation represents a singular integral equation for the solution of the RHP. For notational purposes we use

$$\mathscr{C}[G;\Gamma]u = u - \mathscr{C}_\Gamma^- u \cdot (G - I)$$

to denote the relevant singular integral operator.

Additional conditions on $G$ must be imposed for this operator to be well behaved.

**Definition 8.4.** *Assume $a \in \gamma_0$, the set of self-intersections of $\Gamma$. Let $\Gamma_1, \ldots, \Gamma_\ell$ be a counterclockwise ordering of subcomponents of $\Gamma$ which contain $z = a$ as an endpoint. For $G \in H^k(\Gamma)$ we define $\hat{G}_i$ by $G|_{\Gamma_i}$ if $\Gamma_i$ is oriented outward and by $(G|_{\Gamma_i})^{-1}$ otherwise. We say that $G$ satisfies the $(k-1)$th-order product condition if, using the $(k-1)$th-order Taylor expansion of each $\hat{G}_i$, we have*

$$\prod_{i=1}^{\ell} \hat{G}_i = I + \mathcal{O}\left(|z-a|^{k-\frac{1}{2}}\right) \ \forall a \in \gamma_0. \tag{8.17}$$

The product condition is necessary to eliminate branch-like singularities as all the jumps which emanate from an intersection point are traversed. In what follows we always assume $G$ satisfies the first-order product condition. In practice, due to analyticity, it satisfies the $k$th-order product condition for every $k$.

### A Numerical Method

From the method described in Section 8.1.2 we are able to compute $\mathscr{C}_\Gamma^- u$ accurately at each point in $M_i(\boldsymbol{x}_N)$ for $i = 1, \ldots, \ell$ if the values of $u$ are known at these same points. Recall that $M_i$ is the affine transformation such that $M_i([-1,1]) = \Gamma_i$. Therefore we expect to be able to compute $\mathscr{C}[G;\Gamma]$ accurately at these points. The finite-dimensional approximation is best described in terms of projections. Let $\mathscr{P}_N$ be the projection that maps a function to its unique piecewise-mapped Chebyshev interpolant, i.e., for $s \in \Gamma_i$

$$\mathscr{P}_N f(s) = \sum_{j=0}^{N} \alpha_j^i T_j(M_i^{-1}(s)),$$

such that $\mathscr{P}_N f(M_i(\boldsymbol{x}_N)) = f(M_i(\boldsymbol{x}_N))$. It is clear that $\mathscr{P}_N \mathscr{P}_N = \mathscr{P}_N$. Let $X_N = \operatorname{ran}(\mathscr{P}_N)$, be the set of mapped Chebyshev polynomials of degree at most $N$. The operator equation we solve is given by

$$\mathscr{P}_N \mathscr{C}[G;\Gamma]u_N = \mathscr{P}_N(G-I), \quad u_N \in X_N. \tag{8.18}$$

See [14] for a comprehensive overview of the method along with implementation details for constructing matrices to solve this system.

There is a subtlety present. We can apply $\mathscr{P}_N$ to a function defined on $\Gamma$ provided that it takes finite values and has well-defined limits at every point in $M_i(\boldsymbol{x}_N)$. Recall that all intersection points are present in at least one of these sets. Thus for $\mathscr{C}_\Gamma^- u$ to satisfy this well-defined/boundedness condition we require that $u \in H_z^1(\Gamma)$ (so that $\mathscr{C}_\Gamma^\pm u \in H_z^1(\Gamma)$). It is not evident a priori that any solution (8.18) satisfies this condition. This question is settled in [14].

**Theorem 8.5 (see [14]).** *If the linear system* (8.18) *is nonsingular and $G$ satisfies the zeroth-order product condition, then the calculated $u_N(x)$ satisfies the zero-sum condition.*

In conclusion, the linear system being nonsingular implies that the numerically constructed solution of the RHP $I+\mathscr{C}_\Gamma u_N$ satisfies the correct jumps at the collocation points $M_i(x_N)$, $i=1,\ldots,\ell$.

### Error Analysis

In some of what follows, we'll be interested in the uniform convergence of a numerical method to the solution of a given PDE. That is, we want the computational cost to compute the solution at a point $(x,t)$, to within a given tolerance, to be bounded for all $(x,t)$. We present error analysis for the Clenshaw–Curtis quadrature and the numerical solution of RHPs. The bulk of the computational results that follow can be read without this section.

### Clenshaw–Curtis Error Analysis

A good reference for the material in this section is [18]. We begin with a theorem for the behavior of the Chebyshev coefficients of a smooth function. Define a weighted seminorm

$$||u||_C = \int_{-1}^{1} \frac{|u'(s)|}{\sqrt{1-s^2}} ds.$$

**Theorem 8.6 (see [18]).** *Assume $u^{(m)}$, $m=0,\ldots,k$, are absolutely continuous and $||u^{(k)}||_C<\infty$ for $k\geq 1$. Define $i(u)=\int u(x)dx$ and $i_N(u)$ to be the approximation of $i(u)$ obtained with the Clenshaw–Curtis quadrature. Then there exists $N_k>0$ such that for $N>N_k$*

$$|i(u)-i_N(u)| \leq \frac{32||u^{(k)}||_C}{15\pi k(2N+1-k)^k}.$$

The important aspect of this theorem for the work here is the following almost trivial corollary.

**Corollary 8.7.** *Let $u(x;t)$ be so that $\partial_x^m u(x;t)$, $m=0,\ldots,k$, are absolutely continuous for each fixed $t$ and satisfy $\sup_{x\in[-1,1]}|\partial_x^{m+1}u(x;t)|\leq M$ for all $t$. Then $i_N(u(\cdot;t))$ converges to $i(u(\cdot;t))$ uniformly in $t$. More precisely, there exists $N_k>0$ such that for $N>N_k$*

$$\sup_t |i(u(\cdot;t))-i_N(u(\cdot;t))| \leq \frac{32M}{15k(2N+1-k)^k}.$$

If we bound derivatives of the integrand uniformly in a parameter, Clenshaw–Curtis converges uniformly with respect to that parameter. This type of behavior is referred to as *uniform approximation*.

### Riemann–Hilbert Error Analysis

We present a brief overview of the framework [20] for the study of the error of the numerical method for RHPs presented in Section 8.1.2. Define $X_{N,z}=X_N\cap H_z^1(\Gamma)$ and $Y_N=\mathscr{P}_N\mathscr{C}[G;\Gamma]X_{N,z}$. Both spaces are equipped with the $L^2(\Gamma)$ inner product. We make an assumption that can be found in a different form in [20].

**Assumption 8.1.** *We assume that if $G-I \in H^1(\Gamma)$, $G$ satisfies the zeroth-order product condition and $\mathscr{C}[G;\Gamma]$ is invertible on $L^2(\Gamma)$, then*

$$\dim X_{N,z} = \dim Y_N$$

*and*

$$\|(\mathscr{P}_N \mathscr{C}[G;\Gamma])^{-1}\|_{Y_N \to X_{N,z}} \leq C(\Gamma) N^{\beta} \|\mathscr{C}[G;\Gamma]^{-1}\|_{L^2(\Gamma)\to L^2(\Gamma)}$$

*for some $\beta > 0$.*

We assume that the norm of the inverse of the finite-dimensional operator grows at most algebraically with respect to the order of approximation. In practice, this assumption is checked a posterori and we see at most logarithmic growth. With this assumption in hand we can state the main convergence theorem.

**Theorem 8.8 (see [20]).** *For the numerical method presented in [14] with Assumption 8.1 we have that if $G-I \in H^1(\Gamma)$, $G$ satisfies the zeroth-order product condition, and $\mathscr{C}[G;\Gamma]$ is invertible on $L^2(\Gamma)$, then*

$$\begin{aligned}&\|u-u_N\|_{L^2(\Gamma)}\\ &\quad \leq C(\Gamma)\|\mathscr{C}[G;\Gamma]^{-1}\|_{L^2(\Gamma)\to L^2(\Gamma)}(1+\|G-I\|_{L^\infty(\Gamma)}\|\mathscr{C}_\Gamma^-\|_{L^2(\Gamma)\to L^2(\Gamma)})N^{2+\beta-k}\|u\|_{H^k(\Gamma)},\end{aligned}$$

*where $u$ is the solution of* (8.16) *and $u_N$ is the solution of* (8.18).

The following theorem, presented here in a simplified form, is a key piece of the puzzle.

**Theorem 8.9.** *If $G-I \in H^k(\Gamma)$, $G$ satisfies the $k$th-order product condition, and $\mathscr{C}[G;\Gamma]$ is invertible on $L^2(\Gamma)$, then $\|u\|_{H^k(\Gamma)}$ is bounded by a function of $\|\mathscr{C}[G;\Gamma]\|_{L^2(\Gamma)\to L^2(\Gamma)}$, $\|\mathscr{C}[G;\Gamma]^{-1}\|_{L^2(\Gamma)\to L^2(\Gamma)}$ and norms of derivatives of $G$ up to order $k$.*

Combining these two theorems with Assumption 8.1 we obtain

$$\|u-u_N\|_{L^2(\Gamma)} \leq D_k N^{2+\beta-k}, \tag{8.19}$$

where $D_k$ depends on $\|\mathscr{C}[G;\Gamma]\|_{L^2(\Gamma)\to L^2(\Gamma)}$, $\|\mathscr{C}[G;\Gamma]^{-1}\|_{L^2(\Gamma)\to L^2(\Gamma)}$, and norms of derivatives of $G$ up to order $k$. This bound is especially useful in the case that $G$ depends on an external parameter $x$. If we bound derivatives of $G$ and $\|\mathscr{C}[G;\Gamma]^{-1}\|_{L^2(\Gamma)\to L^2(\Gamma)}$ uniformly over a range of $x$ values, this is sufficient to ensure uniform convergence of $u_N$ to $u$ over this same range of $x$ values. This allows one to realize uniform approximation in the context of RHPs.

**Remark 65.** *An important step in the numerical solution of RHPs is the truncation of the contour $\Gamma$ to one that is finite. In practice, we truncate the contour when the jump matrix $G$ is, to machine precision, the identity matrix and ignore the error induced. This truncation can be rigorously justified; see* [30].

**Remark 66.** *Assuming sufficient decay of $G-I$ it follows that*

$$\lim_{k\to\infty} k(\Phi - I) = -\frac{1}{2\pi i}\int_\Gamma u(k)dk.$$

*In the case that* $\Gamma$ *is bounded,* (8.19) *implies*

$$\left| \int_\Gamma (u_N(k) - u(k)) dk \right| \leq |\Gamma|^{1/2} D_k N^{2+\beta-k},$$

*where* $|\Gamma|$ *denotes the arclength of* $\Gamma$.

### Linear Equations

In this section we discuss a numerical method for computing solutions of linear, constant-coefficient PDEs. In much the same way as the UTM, studying the linear case in detail gives great insight into how to compute solutions of nonlinear PDEs. Consider solving

$$\begin{aligned} q_t - \omega(-i\partial_x)q = 0, \\ q(x,0) = q_0(x) \in \mathcal{S}(\mathbb{R}) \end{aligned} \tag{8.20}$$

for a polynomial $\omega(k)$. The solution of this initial value problem is given by

$$q(x,t) = \frac{1}{2\pi} \int_{-\infty}^{\infty} e^{ikx+\omega(k)} \hat{q}_0(k) dk, \quad \hat{q}_0(k) = \int_{-\infty}^{\infty} e^{-ikx} q_0(x) dx.$$

For what follows, we assume that $\lim_{|x|\to\infty} e^{\gamma|x|}|q_0(x)| < \infty$ for some $\gamma > 0$. This is important to induce analyticity in $\hat{q}_0(k)$, the Fourier transform of $q_0(x)$. We first consider computing $\hat{q}_0(k)$ and then move to computing $q(x,t)$.

### Computing Fourier Transforms

The method we present for computing the Fourier transform of a smooth function is based around the Lax pair for (8.20). In the numerical analysis community, the method we present is well known as a *Levin-type collocation method* [21]. This particular application is also described in [22, 28].

Independent of $\omega(k)$, one equation in the Lax pair is

$$\mu_x + ik\mu = q_0(x).$$

Following the technique used in the inverse scattering transform we look for two solutions:

$$\begin{aligned} \mu_+(x;k) \sim 0 \quad \text{as} \quad x \to \infty, \\ \mu_-(x;k) \sim 0 \quad \text{as} \quad x \to -\infty. \end{aligned} \tag{8.21}$$

Note that $\mu_-(0;k) - \mu_+(0;k) = \hat{q}_0(k)$. A replacement is necessary. Consider

$$\begin{aligned} (\tilde{\mu}_\pm)_x + ik\tilde{\mu}_\pm = q_0(x), \\ \tilde{\mu}_\pm(\pm L;k) = 0, \quad L > 0, \end{aligned} \tag{8.22}$$

for $L$ sufficiently large. The constant $L$ is chosen so that $|q_0(x)|$ is less than machine precision for $|x| > L$. The method described in Section 8.1.2 is well suited to compute approximations of $\tilde{\mu}_\pm$. Once an approximation is computed, $\tilde{\mu}_-(0;k) - \tilde{\mu}_+(0;k)$ is an approximation of $\hat{q}_0(k)$. The procedure is as follows: choose $k \in \mathbb{C}$ such that $|\operatorname{Im} k| < \gamma$ and solve (8.22) (see the discussion of what $\gamma$ should be in Section 8.1.2). This allows us to compute $\hat{q}_0(k)$ in a strip that contains the real axis.

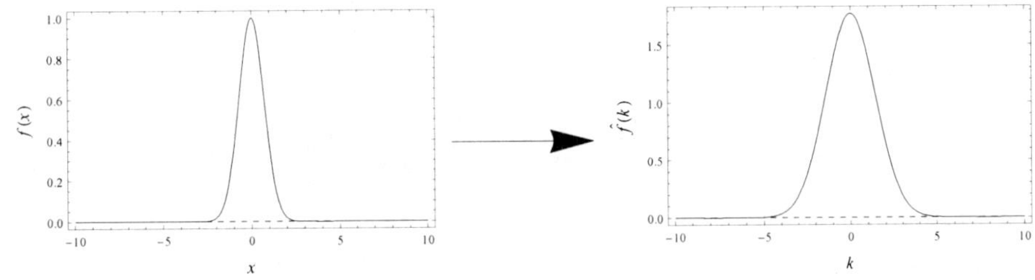


(a)

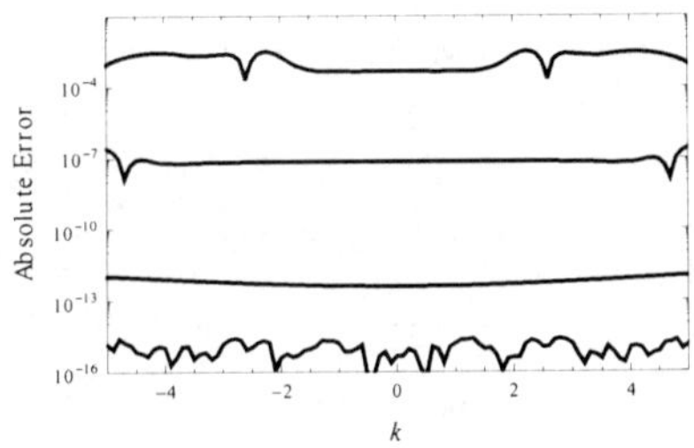


(b)

**Figure 8.4.** (a) *The Fourier transform* $\hat{f}(k) = \sqrt{\pi}e^{-k^2/4}$ *of* $f(x) = e^{-x^2}$. (b) *A plot of the error in the approximation of* $\hat{f}(k)$ *with* $N = 15,\ 30,\ 45,\ 60$.

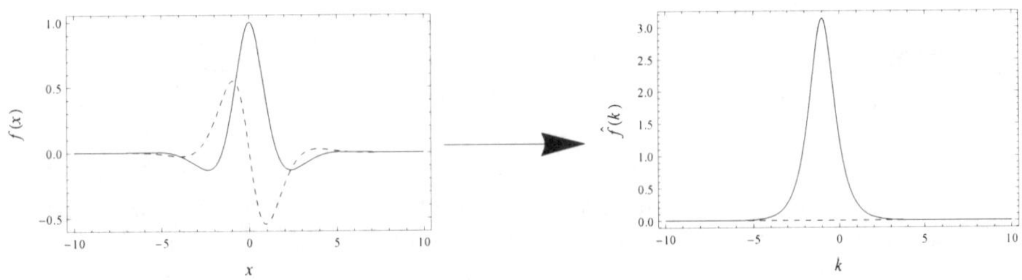


**Figure 8.5.** *The Fourier transform of* $f(x) = \operatorname{sech}(x)e^{-ix}$.

## Examples

As a prototypical test case, we consider computing the Fourier transform of $f(x) = e^{-x^2}$. Note that $\hat{f}(k) = \sqrt{\pi}e^{-k^2/4}$. We use $N$ to refer to the number of collocation points used in the numerical approximation of $\tilde{\mu}_\pm$. In Figure 8.4 we demonstrate the uniform spectral convergence of the method. We present another example in Figures 8.5 and 8.6.

**Remark 67.** *In all the examples here, we compute with smooth, exponentially decaying initial data. In practice, it is noticed that the condition number of the linear system constructed to solve* (8.22) *increases with respect to* $|k|$. *The robust numerical linear algebra routines included in software such as* Mathematica *and MATLAB allow this linear system to be solved accurately and we see no reduction in error for large* $|k|$. *This appears to occur due to smoothness and decay.*

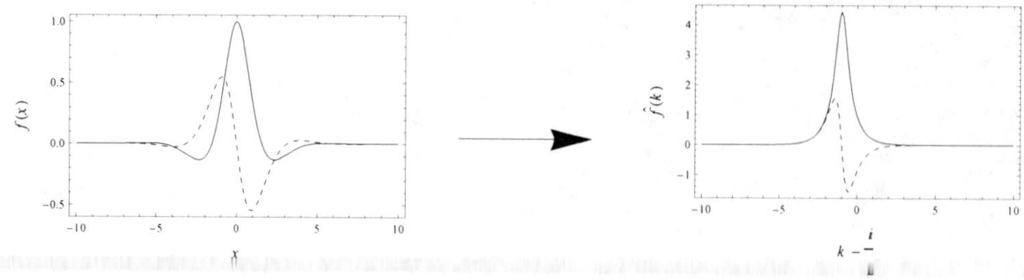


**Figure 8.6.** *The Fourier transform of* $f(x)=\operatorname{sech}(x)e^{-ix}$ *in the complex plane.*

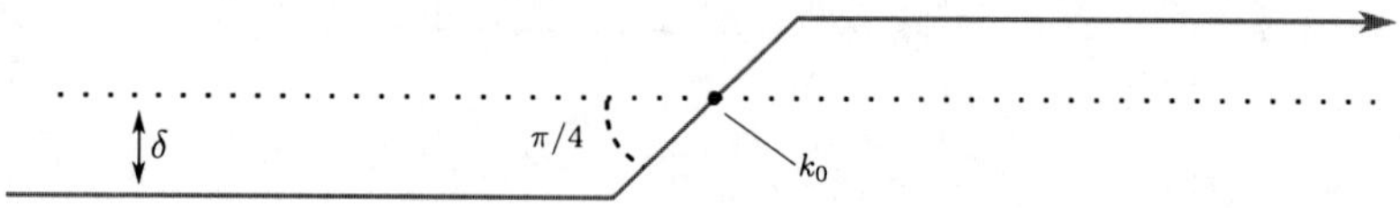


**Figure 8.7.** *The contour* $C_\delta + k_0$.

### Specifics: Linear Schrödinger

We set $\omega(k)=ik^2$ and compute solutions of (8.3) via the integral representation

$$q(x,t)=\frac{1}{2\pi}\int_{-\infty}^{\infty} e^{ikx+ik^2t}\hat{q}_0(k)dk. \tag{8.23}$$

Because we have assumed exponential decay of $q_0(x)$ ($\lim_{|x|\to\infty} e^{\gamma|x|}|q_0(x)|<\infty$) we know that $\hat{q}_0(k)$ must be analytic in the strip $|\operatorname{Im} k|<\gamma$. Contour deformations are used, motivated by the method of steepest descent for integrals [23]. We look for the stationary point $k_0$ such that the first derivative of $\Theta(k)=ikx+ik^2t$ vanishes, or $k_0=-x/(2t)$. Expand $\Theta$ around the stationary point, $\Theta(k)=ix^2/(4t)+it(k-k_0)^2$. To find the path of steepest descent we set $k-k_0=re^{i\theta}$, thus $\Theta(k)=ix^2/(4t)+itr^2e^{2i\theta}$. If $\theta=\pi/4$, then $\Theta(k)=ix^2/(4t)-tr^2$: the imaginary part is constant, while the real part is negative and decreasing away from the stationary point.

To perform contour deformations we use the fact that $\sup_{|\operatorname{Im} k|<\delta}|\hat{q}_0(k)|<\infty$ for $0<\delta<\gamma$. This is true for the class of initial conditions we consider. Define (see Figure 8.7)

$$C_\delta=(-\infty-i\delta,-\delta-i\delta]\cup\{re^{i\pi/4}:-\sqrt{2}\delta<r<\sqrt{2}\delta\}\cup[\delta+i\delta,\infty-i\delta).$$

Consider the integral, for $s<k_0$,

$$\begin{aligned} I_s &= \left|\int_{s-i\delta}^{s} e^{ikx+ik^2t}\hat{q}_0(k)dk\right| \\ &\le \left|\int_0^{\delta} e^{i(s-ir)x+i(s-ir)^2t}\hat{q}_0(s-ir)dr\right| \le \sup_{|\operatorname{Im} k|<\delta}|\hat{q}_0(k)|\int_0^{\delta} e^{rx+2srt}dr. \end{aligned} \tag{8.24}$$

Define $c=2t(k_0-s)>0$ and we see that

$$I_s \le \sup_{|\operatorname{Im} k|<\delta}|\hat{q}_0(k)|\int_0^{\delta} e^{-rc}dr = \frac{\sup_{|\operatorname{Im} k|<\delta}|\hat{q}_0(k)|}{\delta c}.$$

Since $c \to \infty$ as $s \to -\infty$ it follows that $\lim_{s\to-\infty} I_s = 0$. A similar argument applies to a contour for $s > k_0$ extending from $s$ to $s + i\delta$ for $s > k_0$. We find

$$q(x,t) = \frac{1}{2\pi}\int_{C_\delta + k_0} e^{ikx+ik^2t}\hat{q}_0(k)dk. \tag{8.25}$$

Quadrature routines are now applied to this integral. For small time $t < T_c$ we integrate over all of $C_\delta + k_0$ with truncation near $\infty$, removing pieces of the contour where the integrand is below machine precision.

A subtlety is present. As $t$ increases the integrand decays to zero away from $k_0$. On the surface this appears enough to make quadrature routines (i.e., the Clenshaw–Curtis quadrature) perform well. We examine a component of (8.25) more closely. Consider

$$\int_{k_0-\delta(1+i)}^{k_0+\delta(1+i)} e^{ikx+ik^2t}\hat{q}_0(k)dk = e^{ix^2/(4t)+i\pi/4}\int_{-\sqrt{2}\delta}^{\sqrt{2}\delta} e^{-tr^2}\hat{q}_0(k_0 + re^{i\pi/4})dr.$$

It is easy to see that the derivatives of the integrand increase as $t$ increases. This implies that if straightforward quadrature is used on this integral, the computational cost to achieve a given tolerance could increase as $t$ increases. This complication is remedied by breaking up the integral. First, for $c > 0$ and $t$ sufficiently large (so that $c/\sqrt{t} < \sqrt{2}\delta$)

$$\int_{-c/\sqrt{t}}^{c/\sqrt{t}} e^{-tr^2}\hat{q}_0(k_0 + re^{i\pi/4})dr \overset{r\mapsto r/\sqrt{t}}{=} \frac{1}{\sqrt{t}}\int_{-c}^{c} e^{-r^2}\hat{q}_0(k_0 + r/\sqrt{t}e^{i\pi/4})dr. \tag{8.26}$$

It is easy to see that all derivatives of this integrand with respect to $r$ are uniformly bounded with respect to $t \geq c > 0$ and $r \in \mathbb{R}$. As $t$ increases, $c$ may also increase while maintaining $c/\sqrt{t} < \sqrt{2}\delta$. We choose $c$ sufficiently large so that the contribution from

$$\int_{c/\sqrt{t}}^{\sqrt{2}\delta} e^{-tr^2}\hat{q}_0(k_0 + re^{i\pi/4})dr \overset{r\mapsto r/\sqrt{t}}{=} \frac{1}{\sqrt{t}}\int_{c}^{\sqrt{2t}\delta} e^{-r^2}\hat{q}_0(k_0 + r/\sqrt{t}e^{i\pi/4})dr$$

is negligible. This can be satisfied provided that $t$ is sufficiently large and this determines $T_c$. At this point, the integral on $\{re^{i\pi/4} : -\sqrt{2}\delta < r < \sqrt{2}\delta\}$ is approximated by only computing (8.26). This shows the value of contour truncation. The integrand along horizontal components of $C_\delta + k_0$ decays uniformly (and in $L^1$) to zero as $t \to \infty$. This implies that Clenshaw-Curtis applied to (8.25) with this change of variables near $k_0$ converges uniformly in $x$ and $t$, with a negligible (in the sense of machine precision) truncation error.

**Examples**

For the first example solution we take $q_0(x) = e^{-x^2}$. This produces a solution of (8.3) that is always even. See Figure 8.8 for a plot of the solution at various times. With $q_0(x) = \mathrm{sech}(x)e^{-ix}$ we plot the solution in Figure 8.9.

### 8.1.3 ▪ Nonlinear Equations

We move to the discussion of the numerical solution of nonlinear PDEs by solving RHPs numerically. The results in this section were originally published in [22]. For this we use the inverse scattering transform, which can be realized as a special case of the UTM.

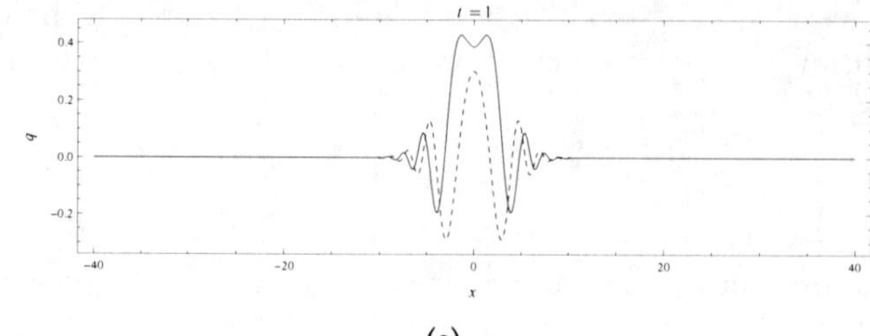


(a)

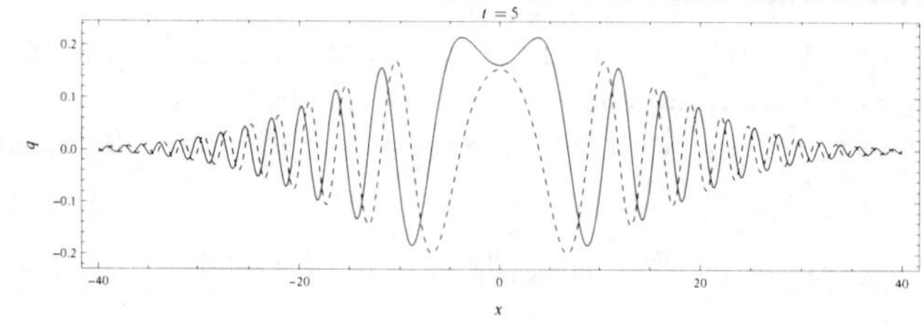


(b)

**Figure 8.8.** *Plots of the solution of* (8.3) *with* $q_0(x) = e^{-x^2}$ *(real part: solid; imaginary part: dashed). The dispersive radiation is clearly visible.* (a) $t = 1$ *and* (b) $t = 5$.

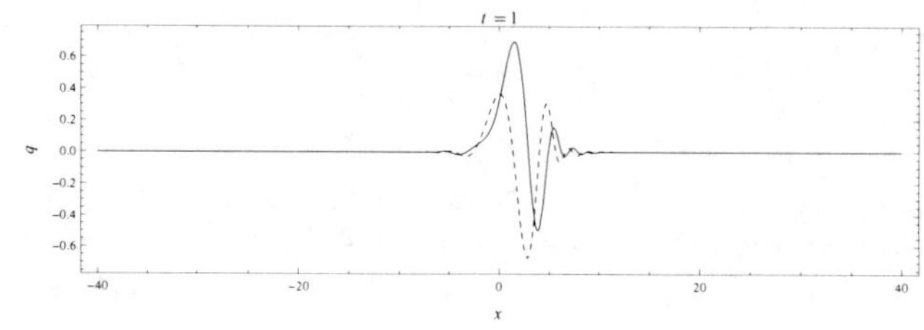


(a)

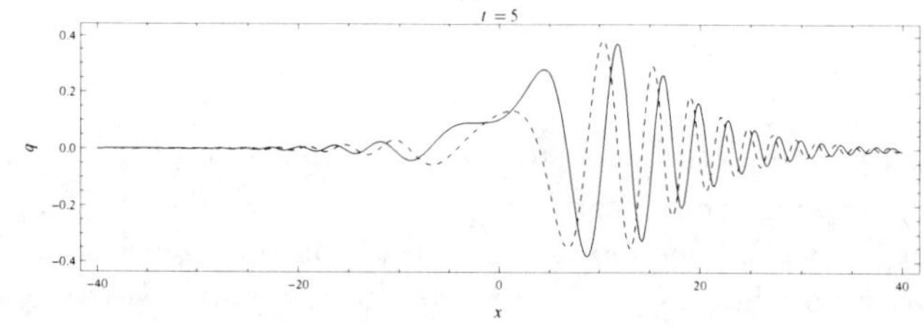


(b)

**Figure 8.9.** *Plots of the solution of* (8.3) *with* $q_0(x) = \operatorname{sech}(x)e^{-ix}$ *(real part: solid; imaginary part: dashed). The dispersive radiation is clearly visible.* (a) $t = 1$ *and* (b) $t = 5$.

We concentrate on the focusing and defocusing nonlinear Schrödinger (NLS) equations. These are given by

$$\begin{aligned} -iq_t + q_{xx} + 2\lambda|q|^2 q = 0, \quad \lambda = \pm 1, \\ q(x,0) = q_0(x) \in \mathcal{S}_\delta(\mathbb{R}). \end{aligned} \tag{8.27}$$

When $\lambda = 1$ we obtain the focusing NLS equation, and for $\lambda = -1$ we obtain the defocusing NLS equation.

The presence of an oscillatory dispersive tail is seen for both the focusing and defocusing NLS equations. This is the linear behavior that is also seen in Section 8.1.2. Examination of the linear dispersion relationship for the NLS equations produces the heuristic that small amplitude waves will travel at a speed proportional to their wavenumber. Furthermore, for the focusing NLS equation solitons are present in the solution. These solitary waves never separate from the dispersive radiation. These factors make the use of traditional numerical methods inefficient even for moderate time. The computational cost to compute the solution using time-stepping methods grows rapidly in time. This is the same complication is the effect of domain truncation that is shown in Figure 8.1. The main benefit of the method presented here is that the computational cost required to compute the solution at a given $x$ and $t$ value is seen to be independent of $x$ and $t$, just like the method in Section 8.1.2.

The inverse scattering transform can be thought of as a map from a completely integrable PDE, with initial condition, to an RHP. Once the RHP is solved, the solution of the PDE is extracted from the solution of the RHP. Since the numerical solution of RHPs has already been discussed we have two tasks at hand. The first is the construction of the relevant RHP and the second is the deformation. The construction of the RHP involves computing the reflection coefficient: a function that is analogous to the Fourier transform of the initial condition. The second proceeds in analogy with Section 8.1.2 so that oscillation is turned to exponential decay.

### Specifics: Nonlinear Schrödinger

The RHP associated with the NLS equations is

$$\begin{aligned} &\Phi^+(k) = \Phi^-(k) \begin{bmatrix} 1 + \lambda\rho(k)\overline{\rho(\bar{k})} & \lambda\overline{\rho(\bar{k})}e^{-\theta(k)} \\ \rho(k)e^{\theta(k)} & 1 \end{bmatrix}, \quad k \in \mathbb{R}, \\ &\Phi(\infty) = I, \qquad \theta(k) = 2i(xk + 2tk^2), \end{aligned} \tag{8.28}$$

where $\rho(k)$ is the reflection coefficient and is obtained from the initial condition $q_0(x)$. The residue conditions are included when $\lambda = 1$,

$$\operatorname{Res}\{\Phi(k), k = x_j\} = \lim_{k \to x_j} \Phi(k) \begin{bmatrix} 0 & 0 \\ C_j e^{\theta(x_j)} & 0 \end{bmatrix},$$

$$\operatorname{Res}\{\Phi(k), k = x_j\} = \lim_{k \to x_j} \Phi(k) \begin{bmatrix} 0 & -\bar{C}_j e^{-\theta(\bar{x}_j)} \\ 0 & 0 \end{bmatrix}.$$

Here $x_j$ and $C_j$ are also obtained from the initial condition. Once the RHP is solved the solution of the NLS equation is found with the formula

$$q(x,t) = -2i \lim_{k \to \infty} k\Phi_{21}(k) = \frac{1}{\pi} \int (\Phi_{21}^+ - \Phi_{21}^-) dk.$$

Note that once $\Phi^\pm$ are computed, the problem is reduced to contour integration. This further emphasizes that RHPs are an appropriate generalization of contour integration: one solves a linear system for the integrand of the contour integral.

### Direct Scattering: Computing $\rho$

The process we describe for computing $\rho(k)$ is in direct analogy with the method for computing $\hat{q}_0(k)$ in Section 8.1.2. Before the method is discussed, the definition of $\rho(k)$ is presented. The nonlinear Schrödinger equations have the following Lax pair [2]:

$$\mu_x + ik[\sigma_3, \mu] = Q\mu, \tag{8.29}$$
$$\mu_t + 2ik^2[\sigma_3, \mu] = \tilde{Q}\mu, \tag{8.30}$$

where

$$Q(x,t) = \begin{bmatrix} 0 & -\lambda\bar{q}(x,t) \\ q(x,t) & 0 \end{bmatrix}, \quad \tilde{Q}(x,t) = 2kQ - iQ_x\sigma_3 + i\lambda|q|^2\sigma_3, \quad \sigma_3 = \begin{bmatrix} 1 & 0 \\ 0 & -1 \end{bmatrix},$$

and $[\cdot,\cdot]$ is the standard matrix commutator.

It can be shown that the Lax pair is equivalent to the differential form

$$d\left(e^{i(kx-2k^2t)\hat{\sigma}_3}\mu(x,t,k)\right) = e^{i(kx+2k^2t)\hat{\sigma}_3}(Q\mu dx + \tilde{Q}\mu dt), \quad e^{\alpha\hat{\sigma}_3}A = e^{\alpha\sigma_3}Ae^{-\alpha\sigma_3}. \tag{8.31}$$

Integrating this differential form over various contours in the $(x,t)$-plane yields different integral equations and different solutions $\mu$. We follow [2] and define $\mu_1$, $\mu_2$ to be the solutions of

$$\begin{aligned} \mu_1(x,t,k) &= I + \int_{(-\infty,t)}^{(x,t)} e^{-ik(x-\xi)\hat{\sigma}_3}[Q(\xi,t)\mu(\xi,\tau,k)]d\xi, \\ \mu_2(x,t,k) &= I + \int_{(\infty,t)}^{(x,t)} e^{-ik(x-\xi)\hat{\sigma}_3}[Q(\xi,t)\mu(\xi,\tau,k)]d\xi. \end{aligned} \tag{8.32}$$

Define $M_i = \mu_i e^{-i(kx+2k^2t)\sigma_3}$ for $i = 1,2$. It follows that both $M_1$ and $M_2$ solve

$$M_x + ik\sigma_3 M = QM,$$
$$M_t + 2ik^2\sigma_3 M = \tilde{Q}M,$$

and hence $M_1(x,t,k) = M_2(x,t,k)S(k)$ for some matrix $S$. With this in mind, $\rho(k) = b(k)/a(k)$, where $a(k) = S_{22}(k)$ and $b(k) = S_{12}(k)$ and subscripts denote components of the matrix. Looking for boundary conditions we find

$$\mu_1(-\infty,t,k) = I = \mu_2(\infty,t,k). \tag{8.33}$$

Since $\rho(k)$ has no $t$ dependence we solve (8.29) at $t = 0$ to compute $\rho(k)$. Once approximations of $\mu_1$ and $\mu_2$ are known at $t = 0$, $M_1(0,0,k) = M_2(0,0,k)S(k)$ is easily solved for $S(k)$ and hence $\rho(k)$. To compute approximations of $\mu_1$ and $\mu_2$ we apply the method in Section 8.1.2 to (8.29) with (8.33) replaced with

$$\mu_1(-L,t,k) = I = \mu_2(L,t,k), \quad L > 0.$$

Here $L$ is chosen large enough so that $|q_0(x)|$ is less that machine precision for $|x| > L$.

### Direct Scattering: Computing $x_j$ and $C_j$

We consider (8.29) with $\lambda = 1$

$$\mu_x - Q\mu = -ik\sigma_3\mu, \quad \Rightarrow \quad i\sigma_3\mu_x - i\sigma_3 Q\mu = k\mu. \tag{8.34}$$

It follows that $\{x_j\}_{j=1}^n$ are the $L^2(\mathbb{R})$ eigenvalues of this equation that lie in the upper-half plane [22]. Making the change of variables $x \mapsto \tan(s/2)$, $U(s) = \mu(\tan(s/2))$, $H(s) = Q(\tan(s/2))$ we obtain

$$2i\cos^2(s/2)\sigma_3 U_s(s) - i\sigma_3 H(s)U(s) = kU(s).$$

We use Hill's method [3] to compute the eigenvalues of the operator

$$2i\cos^2(s/2)\sigma_3\frac{d}{ds} - i\sigma_3 H(s) \tag{8.35}$$

in the space $L^2([-\pi,\pi])$. Using the convergence of Hill's method, even for non-self-adjoint operators [24, 25], it can be shown that the only eigenvalues we obtain off the real line are those associated with (8.34). This allows us to compute the discrete spectrum $\{x_j\}_{j=1}^n$ with spectral accuracy. We may test to make sure all eigenvalues are captured by computing the inverse scattering transform at $t = 0$ and ensuring that we recover the initial condition. There is in principle no limit on the size of $\{x_j\}$ but the computational cost grows with the size of this set.

Furthermore, from the analyticity properties of $a$ we have [26, Page 75]

$$a(k) - 1 = \frac{1}{2\pi i}\int_{-\infty}^{\infty}\frac{a(s)-1}{s-k}ds, \qquad a'(k) = \frac{1}{2\pi i}\int_{-\infty}^{\infty}\frac{a'(s)}{s-k}ds.$$

Thus knowing $a(k)$ on the real line and $b(k)$ when $a(k) = 0$ allows us to compute $C_j = b(x_j)/a'(x_j)$. Also, $a'(k)$ can be obtained accurately using spectral differentiation, mapping the real line to the unit circle.

**Remark 68.** *The notation $b(x_j)$ can be misleading. In general, $b(k)$ is not analytically extendible up to $x_j$. In this case, $b(k)$ is defined within a strip of analyticity about the real axis and at each point $x_j$.*

We follow the standard procedure (see, e.g., [27, 28]) to turn the residue conditions into jump conditions. Fix $\epsilon > 0$ so that the circles $A_j^+ = \{k \in \mathbb{C} : |k - x_j| = \epsilon\}$ and $A_j^- = \{k \in \mathbb{C} : |k - \bar{x}_j| = \epsilon\}$ do not intersect each other or the real axis. We define $\hat{\Phi}$ by

$$\hat{\Phi}(k;x,t) = \begin{cases} \Phi(k;x,t)\begin{bmatrix} 1 & 0 \\ -C_j e^{\theta(x_j)}/(k-x_j) & 1 \end{bmatrix} & \text{if } |k-x_j| < \epsilon,\ j=1,\ldots,n, \\ \Phi(k;x,t)\begin{bmatrix} 1 & 0 \\ \bar{C}_j e^{-\theta(\bar{x}_j)}/(k-\bar{x}_j) & 1 \end{bmatrix} & \text{if } |k-\bar{x}_j| < \epsilon,\ j=1,\ldots,n, \\ \Phi(k;x,t) & \text{otherwise.} \end{cases} \tag{8.36}$$

It is straightforward to show that $\hat{\Phi}$ solves the RHP:

$$\hat{\Phi}^+(k) = \begin{cases} \hat{\Phi}^-(k)G(k) & \text{if } k \in \mathbb{R}, \\ \hat{\Phi}^-(k)\begin{bmatrix} 1 & 0 \\ -C_j e^{\theta(x_j)}/(k-x_j) & 1 \end{bmatrix} & \text{if } k \in A_j^+, \\ \hat{\Phi}^-(k)\begin{bmatrix} 1 & -\bar{C}_j e^{-\theta(\bar{x}_j)}/(k-\bar{x}_j) \\ 0 & 1 \end{bmatrix} & \text{if } k \in A_j^-, \end{cases} \qquad \hat{\Phi}(\infty) = I,$$

where $A_j^-$ ($A_j^+$) has a clockwise (counterclockwise) orientation.

### Linear Reduction

We further examine the relationship between the equations

$$-iq_t + q_{xx} + 2\lambda|q|^2 q = 0,$$
$$-iq_t + q_{xx} = 0$$

through their corresponding inverse problems. It is clear that setting $\lambda = 0$, which is not a priori a valid choice for $\lambda$, reduces the nonlinear problem to the linear one. It is important that setting $\lambda = 0$ in the Lax pair for the nonlinear problem reduces it to the Lax pair for the linear problem. It is a good exercise to set $\lambda = 0$ and show that $\rho(k) = \hat{q}_0(2k)$ and $a(k) = 1$ using the techniques from Section 8.1.2.

This reduction extends to the RHP. When $\lambda = 0$, we know that $a(k)$ has no zeros. Therefore we have the RHP

$$\begin{aligned} \Phi^+(k) &= \Phi^-(k)\begin{bmatrix} 1 & 0 \\ \hat{q}_0(2k)e^{\theta(k)} & 1 \end{bmatrix}, \quad k \in \mathbb{R}, \\ \Phi(\infty) &= I, \qquad \theta(k) = 2i(xk + 2tk^2). \end{aligned} \tag{8.37}$$

It is shown in [23, 30] that

$$\Phi(k) = \begin{bmatrix} 1 & 0 \\ \mathscr{C}_{\mathbb{R}}[\hat{q}_0(2\cdot)e^{\theta(\cdot)}](k) & 1 \end{bmatrix}.$$

Therefore

$$-2i \lim_{k\to\infty} \Phi_{21}(k) = \frac{1}{\pi}\int_{-\infty}^{\infty} \hat{q}_0(2k)e^{\theta(k)}dk,$$

which is precisely (8.23) after a scaling of the $k$ variable.

### Example RHPs

We continue with our prototypical example of $q_0(x) = e^{-x^2}$. We compute $\rho(k)$ for the defocusing ($\lambda = -1$) NLS equation in Figure 8.10. In Figure 8.11 we show $\rho(k)$ in the case of the focusing NLS equation ($\lambda = 1$). We also examine $a(k)$ in the upper-half plane to look for zeros. It is seen that the zero found with Hill's method is in direct correspondence with the zeros of $a(k)$. See Figure 8.11 for a demonstration of this.

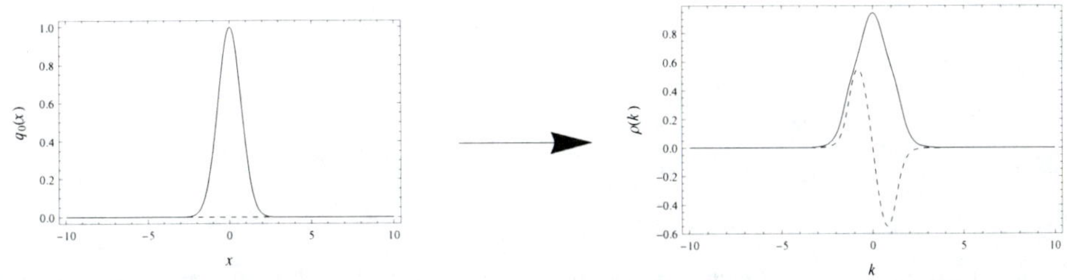

Figure 8.10. *The reflection coefficient for the defocusing NLS equation with* $q_0(x) = e^{-x^2}$.

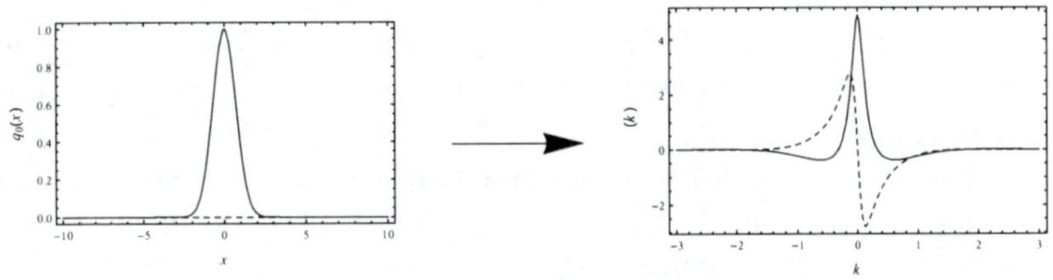

(a)

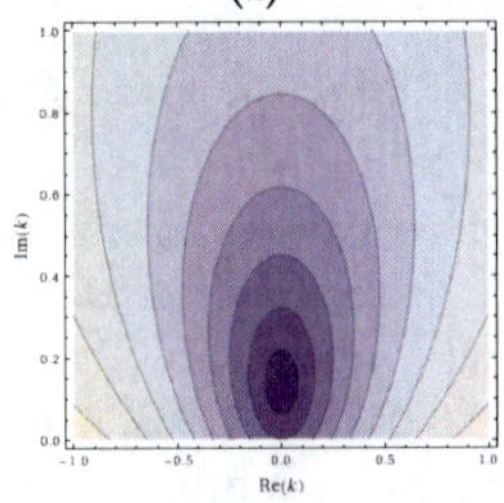

(b)

Figure 8.11. (a) *The reflection coefficient for the focusing NLS equation with* $q_0(x) = e^{-x^2}$. (b) *A contour plot of* $|a(k)|^2$ *in the upper-half plane showing the zero* ($k = 0.133316i$) *of the function.*

## Inverse Scattering: Deforming and Solving the RHP

When both $x$ and $t$ are small, the RHP does not suffer from high oscillation and it does not need to be deformed. When $t$ is small and $x$ is large, the RHP does need to be deformed. In general, the deformation of a RHP involves the factorization of the jump matrix into analytic factors. We introduce some factorizations of the jump matrix. Define

$$\begin{aligned}
G(k) &= \begin{bmatrix} 1 + \lambda\overline{\rho(\bar{k})}\rho(k) & \lambda\overline{\rho(\bar{k})}e^{-\theta(k)} \\ \rho(k)e^{\theta(k)} & 1 \end{bmatrix}, \\
P(k) &= \begin{bmatrix} 1 & 0 \\ \rho(k)e^{\theta(k)} & 1 \end{bmatrix}, \qquad M(k) = \begin{bmatrix} 1 & \lambda\overline{\rho(\bar{k})}e^{-\theta(k)} \\ 0 & 1 \end{bmatrix}, \\
L(k) &= \begin{bmatrix} 1 & 0 \\ \frac{\rho(k)}{\tau(k)}e^{\theta(k)} & 1 \end{bmatrix}, \qquad D(k) = \begin{bmatrix} \tau(k) & 0 \\ 0 & 1/\tau(k) \end{bmatrix}, \\
U(k) &= \begin{bmatrix} 1 & \frac{\lambda\overline{\rho(\bar{k})}}{\tau(k)}e^{-\theta(k)} \\ 0 & 1 \end{bmatrix}, \qquad \tau(k) = 1 + \lambda\rho(k)\overline{\rho(\bar{k})}.
\end{aligned} \tag{8.38}$$

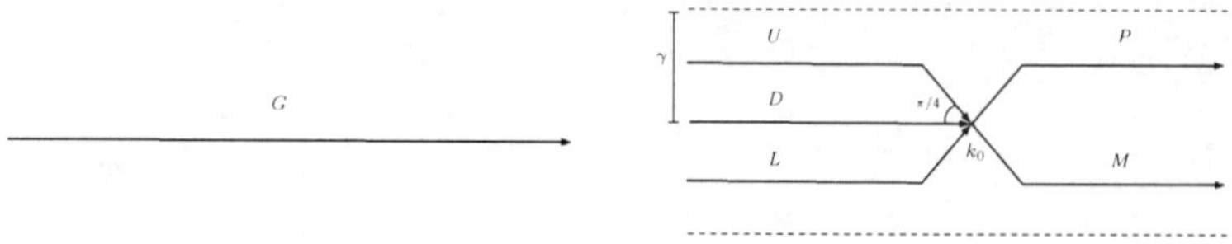


(a) (b)

**Figure 8.12.** (a) *Jump contour for the initial RHP for* $\Phi$. (b) *Jump contours after lensing for the RHP for* $\Phi_1$ [30, 22].

It follows that $G(k) = L(k)D(k)U(k) = M(k)P(k)$. For simplicity, we assume the sets $\{x_j\}_{j=1}^n$ and $\{C_j\}_{j=1}^n$ are empty. We introduce modifications to incorporate the extra contours required later.

The contours of (8.28) are deformed off the real line so that oscillations are turned to exponential decay in the same way as in Section 8.1.2. The matrix $G$ contains the two factors $\exp(\pm\theta(k))$ and if one decays the other must grow: $\exp(\pm\theta(k))$ cannot share the same path of steepest descent. This motivates separating these factors using the above factorizations.

We assume $\lim_{|x|\to\infty} e^{2\gamma|x|}|q_0(x)| < \infty$ and we know that $\rho$ is analytic in the strip $S_\delta = \{k \in \mathbb{C} : |\mathrm{Im}\, k| < \delta\}$ for $\gamma > \delta > 0$ [30]. This can be seen by considering the Volterra integral equations for the eigenfunctions $\mu$. The factors in (8.38) allow *lensing*. The process of lensing allows a right factor (i.e., $P$ in $G = MP$) to be moved to a contour that lies above the original contour. See [28, 29, 30] for a discussion of this process.

Looking for critical points of the oscillator, $\theta'(k) = 0$, we find $k = k_0 = -x/(4t)$. To ensure that the lensing process moves $e^{\pm\theta(k)}$ to the correct regions where its path of steepest descent lies, we use the $LDU$ factorization for $k < k_0$ and $MP$ for $k > k_0$. See Figure 8.12 for this deformation. In this figure, the contours are deformed along the path of steepest descent in a neighborhood of $k_0$ and therefore the jump matrix tends to the identity matrix rapidly along these contours. We denote the solution of this lensed RHP by $\Phi_1$.

This RHP can be solved numerically provided $t$ is not too large. As $t$ increases, the solution on the contour $(-\infty, k_0)$ is increasingly oscillatory, implying growth in its derivatives. Therefore, this solution will not be effectively resolved with Chebyshev polynomials.

The next deformation leads us to a method that is accurate for arbitrarily large $x$ and $t$. We modify the RHP so as to remove the contour $D$. After this removal, all jumps decay to the identity matrix away from $k_0$ as $x$ and $t$ become large. A heuristic reason can be found by considering the change of variables in (8.26). Increasing exponential decay in the integrand (or RHP) allows a change of variables to a fixed domain such that the integrand (or jump matrix) has bounded derivatives.

The function

$$\Delta(k;k_0) = \begin{bmatrix} \delta(k;k_0) & 0 \\ 0 & \delta^{-1}(k;k_0) \end{bmatrix}, \qquad \delta(k;k_0) = \exp\left(\mathscr{C}_{(-\infty,k_0)} \log \tau(k)\right)$$

satisfies

$$\Delta^+(k;k_0) = \Delta^-(k;k_0)D(k), \ \ k \in (-\infty, k_0), \qquad \Delta(\infty;k_0) = I.$$

This follows from the Plemelj lemma.

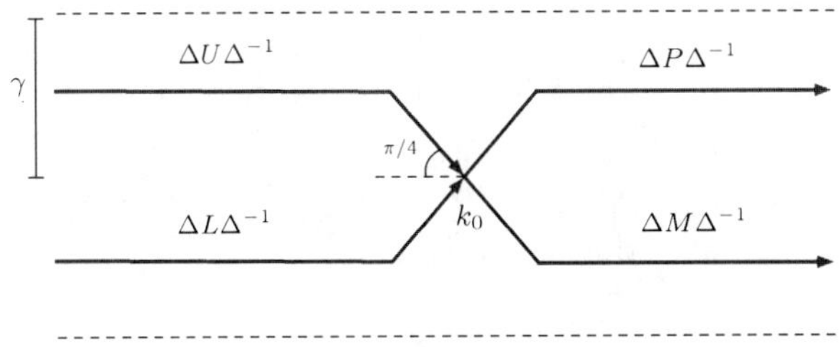


**Figure 8.13.** *Removal of the jump on* $(-\infty, k_0)$ [30, 22].

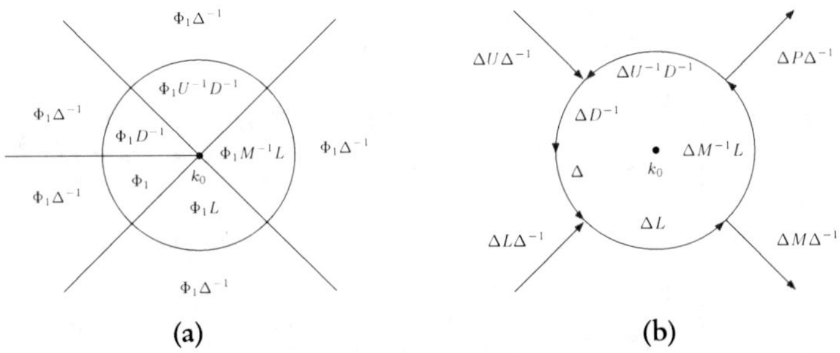

**Figure 8.14.** (a) *The piecewise definition of* $\Phi_3$. (b) *Jump contours and jump matrices for the RHP for* $\Phi_3$ [30, 22].

We multiply the solution of the RHP in Figure 8.12 by $\Delta^{-1}$ to remove the jump. To compute the solution of the new RHP we use conjugation

$$\Phi_1^+ = \Phi_1^- J \Leftrightarrow \Phi_1^+ \Delta_+^{-1} = \Phi_1^- J \Delta_+^{-1} \Leftrightarrow \Phi_1^+ \Delta_+^{-1} = \Phi_1^- \Delta_-^{-1} \Delta_- J \Delta_+^{-1}.$$

If $J = D$, $\Delta_- J \Delta_+^{-1} = I$ and $\Phi\Delta^{-1}$ has no jump. Define $\Phi_2 = \Phi_1 \Delta^{-1}$. See Figure 8.13 for a schematic overview of the new jumps. This deformation is not sufficient for a numerical solution as $\Delta(k; k_0)$ has a singularity at $k = k_0$ and so does the solution of the RHP.

We perform one last deformation in a neighborhood of $k = k_0$ to bound contours away from this singularity. We use a piecewise definition of a function $\Phi_3$ and compute the jumps (Figure 8.14). This is the final RHP and is used to compute solutions of the NLS equations for arbitrarily large time. To condition the RHP one must also truncate the contour (see Remark 65) and scale the neighborhood of the redefinition in Figure 8.14(a). The scaling of the RHP is discussed in more detail below.

The deformation in Figure 8.12 is used for small time and the deformation in Figure 8.14 is used for large time to obtain an approximation to focusing or defocusing NLS when no solitons are present in the solution. The addition of solitons for the case of focusing NLS is now addressed. There are additional jumps of the form

$$\Phi^+(k) = \begin{cases} \Phi^-(k)\begin{bmatrix} 1 & 0 \\ -C_j e^{\theta(x_j)}/(k - x_j) & 1 \end{bmatrix} & \text{if } k \in A_j^+, \\ \Phi^-(k)\begin{bmatrix} 1 & -\bar{C}_j e^{-\theta(\bar{x}_j)}/(k - \bar{x}_j) \\ 0 & 1 \end{bmatrix} & \text{if } k \in A_j^-. \end{cases}$$

We assume that $\text{imag}(x_j) > \gamma/2$.

For small $x$ and $t$, $|e^{\theta(x_j)}| = |e^{-\theta(\bar{x}_j)}|$ is close to unity and the contours and jumps need to be added to one of the deformations discussed above. This will not be effective for all $x$ and $t$. When $|C_j e^{\theta(x_j)}| > 1$ we invert this factor through a deformation that involves a redefinition of the solution inside the contours $A_j^{\pm}$. Define the index set $K_{x,t} = \{j : |C_j e^{\theta(x_j)}| > 1\}$. With this in mind, define

$$v(k) = \prod_{j \in K_{x,t}} \frac{k - x_j}{k - \bar{x}_j} \quad \text{and} \quad V(k) = \begin{bmatrix} v(k) & 0 \\ 0 & 1/v(k) \end{bmatrix}.$$

The redefinition of $\Phi$ is given by the function $\hat{\Phi}$:

$$\hat{\Phi}(k) = \Phi(k) \begin{cases} \begin{bmatrix} 1 & -(k - x_j)/(C_j e^{\theta(x_j)}) \\ C_j e^{\theta(x_j)}/(k - x_j) & 0 \end{bmatrix} V(k) & \text{if } |k - x_j| < \epsilon, \\ \begin{bmatrix} 0 & -\bar{C}_j e^{-\theta(\bar{x}_j)}/(k - \bar{x}_j) \\ (k - \bar{x}_j)/(\bar{C}_j e^{-\theta(\bar{x}_j)}) & 1 \end{bmatrix} V(k) & \text{if } |k - \bar{x}_j| < \epsilon, \\ V(k) & \text{otherwise.} \end{cases}$$

A straightforward calculation shows that the jumps that $\hat{\Phi}$ satisfies are

$$\hat{\Phi}^+(k) = \hat{\Phi}^-(k) \begin{cases} V^{-1}(k) \begin{bmatrix} 1 & -(k - x_j)/(C_j e^{\theta(x_j)}) \\ 0 & 1 \end{bmatrix} V(k) & \text{if } k \in A_j^+, \\ V^{-1}(k) \begin{bmatrix} 1 & 0 \\ -(k - \bar{x}_j)/(\bar{C}_j e^{-\theta(\bar{x}_j)}) & 1 \end{bmatrix} V(k) & \text{if } k \in A_j^- \end{cases}$$

for $j \in K_{x,t}$. This inverts growth of the exponential to force the jump matrix to decay to the identity matrix. For the sake of notation, define

$$T_{j,+}(k) = \begin{bmatrix} 1 & 0 \\ -C_j e^{\theta(x_j)}/(k - x_j) & 1 \end{bmatrix}, \quad T_{j,-}(k) = \begin{bmatrix} 1 & -\bar{C}_j e^{-\theta(\bar{x}_j)}/(k - \bar{x}_j) \\ 0 & 1 \end{bmatrix}, \tag{8.39}$$

$$S_{j,+}(k) = \begin{bmatrix} 1 & -(k - x_j)/(C_j e^{\theta(x_j)}) \\ 0 & 1 \end{bmatrix}, \quad S_{j,-}(k) = \begin{bmatrix} 1 & 0 \\ -(k - \bar{x}_j)/(\bar{C}_j e^{-\theta(\bar{x}_j)}) & 1 \end{bmatrix}. \tag{8.40}$$

In Figure 8.15 the full small-time and long-time RHPs are presented. Letting $V = T_{j\pm} = S_{j\pm} = I$ reduces this RHP to the no-soliton case.

### Example Solutions

In this section we plot some solutions of the NLS equations. First, we plot solutions of both the focusing and defocusing NLS equations with $q_0(x) = e^{-x^2}$. See Figure 8.16 for the solution of the defocusing NLS equation and Figure 8.18 for the solution of the focusing NLS equation. The solution of the focusing NLS equation exhibits a soliton at $x = 0$ with zero velocity.

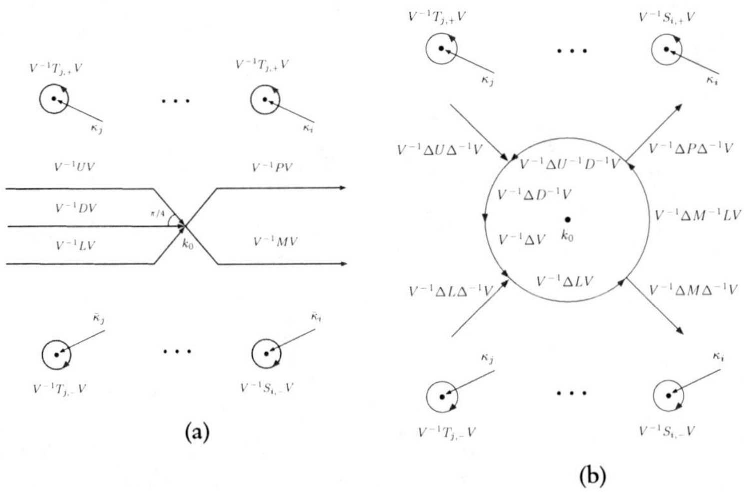


**Figure 8.15.** (a) *The jump contours and jump matrices for the final deformation for small time. In this schematic* $|C_j e^{\theta(x_j)}| \leq 1$ *and* $|C_i e^{\theta(x_i)}| > 1$. (b) *The jump contours and jump matrices for the final deformation for large time. Again, in this schematic* $|C_j e^{\theta(x_j)}| \leq 1$ *and* $|C_i e^{\theta(x_i)}| > 1$ [30, 22].

We also consider the numerical solution of both the focusing and defocusing NLS equations with $q_0(x) = \text{sech}(x)e^{-ix}$. In the case of the focusing NLS equation, the exact solution is given by $q(x,t) = \text{sech}(x-2t)e^{-ix}$ and this is captured in Figure 8.19. In this case $x_1 = 1/2 + i/2$ and $C_1 = -i$, while $\rho(k) = 0$ for $k \in \mathbb{R}$. These quantities are all accurately computed with the method presented above. This initial condition is evolved according to the defocusing NLS equation in Figure 8.20.

**Remark 69.** *For the Korteweg–de Vries equation, solutions that are a nonlinear superposition of dispersive and finite-genus (quasi-periodic) solutions can be computed numerically by solving RHPs* [31]. *Indeed, the same approach works in this case for the defocusing NLS equation. These computations will be included in a future publication.*

### Uniform Approximation

We consider the numerical solution of the RHP shown in Figure 8.14 for $t \geq c > 0$. For simplicity we concentrate only on $\lambda =$ 1 and the details of what follows can be found in [22]. The radius of the circle in Figure 8.14 and the length of the contours extending from the circle are chosen to be proportional to $1/\sqrt{t}$. For sufficiently large $t$ we truncate the RHP, following Remark 65, to precisely the RHP shown in Figure 8.14. This is possible due to exponential decay of the jump matrix to the identity matrix away from $k_0$. Any of the horizontal contours that are shown in Figure 8.15 or 8.13 are removed. We perform a conformal (affine) change of variables on the RHP, in much the same way as in Section 8.1.2. Let $k = k_0 + z/\sqrt{t}$ for $z \in \Gamma_0$ and we see that

$$\begin{aligned} \theta(k) &= \frac{ix^2}{4t} + 4it(k-k_0)^2, \\ \theta(k_0 + z/\sqrt{t}) &= \frac{ix^2}{4t} + 4iz^2. \end{aligned} \tag{8.41}$$

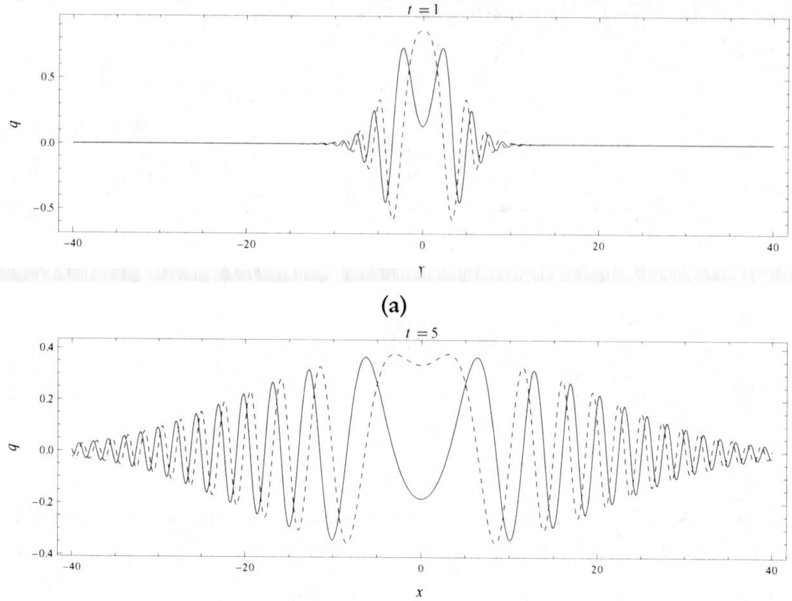


(a)

(b)

**Figure 8.16.** *Plots of the solution of the defocusing NLS equation with* $q_0(x) = e^{-x^2}$ *(real part: solid; imaginary part: dashed).* (a) $t = 1$, (b) $t = 5$.

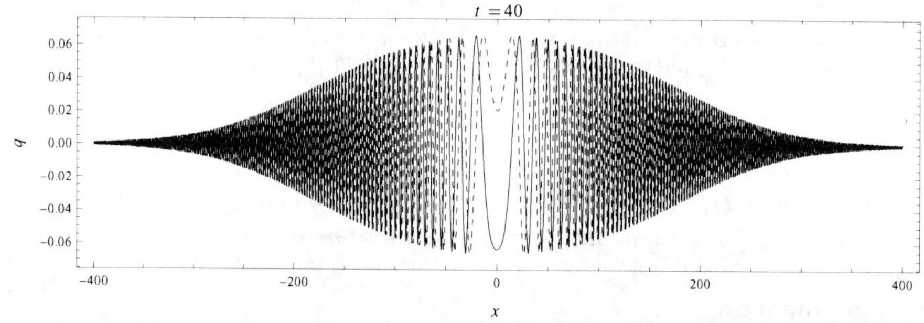


**Figure 8.17.** *Plots of the solution of the defocusing NLS equation with* $q_0(x) = e^{-x^2}$ *(real part: solid; imaginary part: dashed) when* $t = 20$. *This shows the extreme spatial extent and oscillation of the solution at even moderate times.*

It follows that $\Gamma_0$ is a fixed, $(x,t)$-independent contour. We now show that any derivatives, with respect to $z$, of the jump matrices shown in Figure 8.14 can be bounded by a constant that is independent of $x$ and $t$. By (8.41) we see that $\exp^{\pm\theta(k_0+z/\sqrt{t})}$ does not cause growth in derivatives. Additionally, it can be shown that each derivative of $\rho(k)$ can be bounded uniformly in the strip $|\mathrm{Im}\, k| < \delta < \gamma/2$ implying that derivatives (with respect to $z$) of $\rho(k_0 + z/\sqrt{t})$ are uniformly bounded in $x$ and $t$ over the entire strip.

The last factor to consider is $\delta(k_0 + z/\sqrt{t}; k_0)$. Since $\sup_{k\in\mathbb{R}} |\rho(k)| < 1$ [32] and $\rho(k)$ is rapidly decaying, it follows that $f(v) = \log(1 - |\rho(k_0 + v/\sqrt{t})|^2)$ is smooth and rapidly

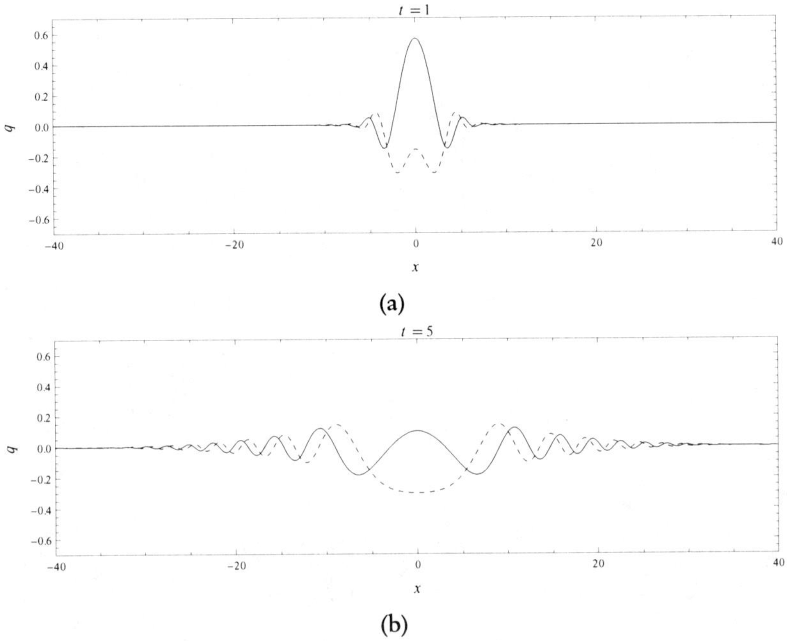


**Figure 8.18.** *Plots of the solution of the focusing NLS equation with* $q_0(x) = e^{-x^2}$ *(real part: solid; imaginary part: dashed).* (a) $t = 1$ *and* (b) $t = 5$.

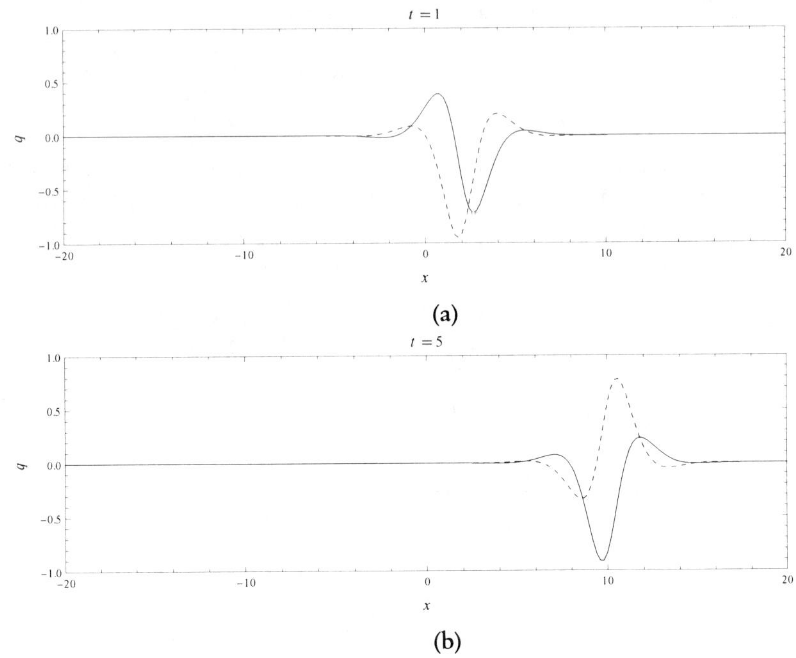


**Figure 8.19.** *Plots of the solution of the focusing NLS equation with* $q_0(x) = \text{sech}(x)e^{-ix}$ *(real part: solid; imaginary part: dashed).* (a) $t = 1$ *and* (b) $t = 5$.

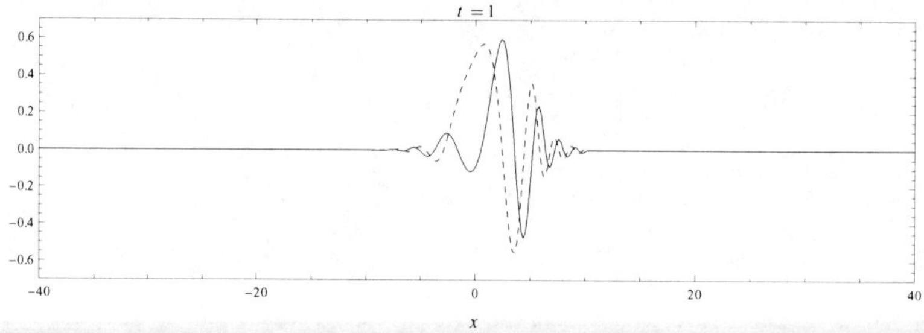


(a)

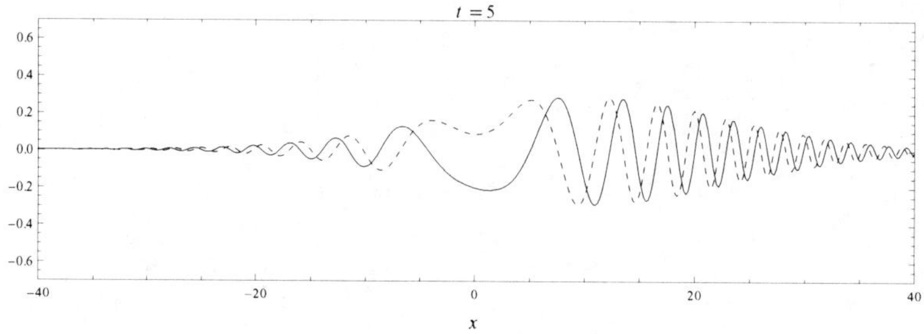


(b)

**Figure 8.20.** *Plots of the solution of the defocusing NLS equation with* $q_0(x) = \operatorname{sech}(x)e^{-ix}$ *(real part: solid, imaginary part: dashed).* (a) $t = 1$ *and* (b) $t = 5$.

decaying. We proceed to bound derivatives of

$$B(z) = \frac{1}{2\pi i}\int_{-\infty}^{k_0} \frac{\log(1-|\rho(s)|^2)}{s-k_0-z/\sqrt{t}}ds.$$

It can be shown that since $f(v)$ is real valued, $B(z)$ can be expressed in the form $B(z) = I(z) + H(z)$, where $I(z)$ is purely imaginary in a neighborhood of $z = 0$ and $\lim_{z\to 0} H(z) = 0$ uniformly. This implies that $\delta(k_0 + z/\sqrt{t}; k_0)$ is bounded near $z = 0$. Indeed, $\delta$ is uniformly bounded on $\mathbb{C}$ [32]. Let $s = k_0 + v/\sqrt{t}$ so that

$$B(z) = \frac{1}{2\pi i}\int_{-\infty}^{0} \frac{\log(1-|\rho(k_0+v/\sqrt{t})|^2)}{v-z}dv.$$

After integrating by parts, we find

$$\begin{aligned}
B'(z) &= \frac{1}{2\pi i}\int_{-\infty}^{0} \frac{f(v)}{(v-z)^2}dv \\
&= \frac{1}{2\pi i}\frac{f(0)}{z} + \frac{1}{2\pi i}\int_{-\infty}^{0}\frac{f'(v)}{v-z}dv, \\
B''(z) &= -\frac{1}{2\pi i}\frac{f(0)}{z^2} + \frac{1}{2\pi i}\frac{f'(0)}{z} + \frac{1}{2\pi i}\int_{-\infty}^{0}\frac{f''(v)}{v-z}dv.
\end{aligned}$$

Continuing this process we see that $B^{(k)}(z)$ is uniformly bounded for all $x$ and $t$ if $z \in \Gamma_0$. These pieces imply that all derivatives of the jump matrices in Figure 8.14 after the change of variables $k = k_0 + z/\sqrt{t}$ are uniformly bounded with respect to $x, t \geq c > 0$ and $z \in \Gamma_0$.

In view of (8.19) it remains to address an estimate of $||\mathscr{C}[G;\Gamma]^{-1}||_{L^2(\Gamma)\to L^2(\Gamma)}$ (an estimate of $||\mathscr{C}[G;\Gamma]^{-1}||_{L^2(\Gamma)\to L^2(\Gamma)}$ follows from the boundedness of $G$ [22]). We refer the reader to [30, 22] for a full discussion of this. The idea is as follows. In the asymptotic analysis of the RHP that is performed in [32], one explicitly constructs an approximate solution of the RHP in Figure 8.14. From this approximate solution, an approximation of $\mathscr{C}[G;\Gamma]^{-1}$ can be constructed. Since the error in the approximation tends to zero as $t\to\infty$, bounds on the approximation are used to bound the norm of $\mathscr{C}[G;\Gamma]^{-1}$.

From Remark 66 we arrive at the following theorem. Let $q_N(x,t)$ be the approximation of the solution of the defocusing NLS equation from the numerical solution of the RHP in Figure 8.14 determined by integrating $u_N$:

$$q_N(x,t)=\frac{1}{\pi}\int u_n(k)dk\approx\frac{1}{\pi}\int u(k)dk=-2i\lim_{k\to\infty}\Phi_{21}(k)=q(x,t).$$

**Theorem 8.10.** *With Assumption* 8.1 *there exists* $\epsilon>0$ *such that for* $0<c\le t$

$$|q_N(x,t)-q(x,t)|\le|\Gamma|^{1/2}D_kN^{2+\beta-k}+\epsilon,$$

*where* $D_k$ *depends only on* $k$ *and* $q_0(x)$. *Here* $\epsilon$ *is generally small and is introduced from the truncation of the RHP.*

This theorem provides us with uniform approximation.

# Bibliography


[1] A. S. Fokas, A unified transform method for solving linear and certain nonlinear PDEs, Proc. Roy. Soc. Lond. Ser. A **452** (1997), 1411–1443.

[2] A. S. Fokas, A Unified Approach to Boundary Value Problems, SIAM (2008).

[3] B. Deconinck and J. N. Kutz, Computing spectra of linear operators using the Floquet-Fourier-Hill method, J. Comput. Phys. **219** (2006), 296–321.

[4] B. Deconinck, T. Trogdon and V. Vasan, The method of Fokas for solving linear partial differential equations, SIAM Rev. **56** (2014), 59–186.

[5] A. S. Fokas and E. A. Spence, Synthesis, as opposed to separation, of variables, SIAM Rev. **54** (2012), 291–324.

[6] A. S. Fokas and N. Flyer, A hybrid analtytical-numerical method for solving evolution partial differential equations. I: The half-line, Proc R. Soc. Lond. Ser. A **464** (2008), 1823–1849.

[7] B. Fornberg and N. Flyer, A numerical implementation of Fokas boundary integral approach: Laplace's equation on a polygonal domain, Proc. R. Soc. Lond. Ser. A **2134** (2011), 2983–3003.

[8] J. Lenells, Boundary Value Problems for the Stationary Axisymmetric Einstein Equations: A Disk Rotating Around a Black Hole, Comm. Math. Phys., **304** (2011), 585–635.

[9] A. S. Fokas and I. M. Gelfand, Integrability of Linear and Nonlinear Evolution Equations and the Associated Nonlinear Fourier Transforms, Lett. Math. Phys. **32** (1994), 189–210.

[10] K. Atkinson and W. Han, Theoretical Numerical Analysis, Springer, New York (2009).

[11] F. W. J. Olver, D. W. Lozier, R. F. Boisvert, and C. W. Clark, NIST Handbook of Mathematical Functions, Cambridge University Press, Cambridge, UK (2010).

[12] J. P. Boyd, Chebyshev and Fourier Spectral Methods, Dover, New York (2001).

[13] L. N. Trefethen, Spectral Methods in MATLAB, SIAM (2000).

[14] S. Olver, A general framework for solving Riemann-Hilbert problems numerically, Numer. Math. **122** (2012), 305–340.

[15] Y. Meyer and R. R. Coifman, Wavelets: Calderón-Zygmund and Multilinear Operators, Cambridge University Press, Cambrdige, UK (1997).

[16] S. Olver, Numerical solution of Riemann–Hilbert problems: Painlevé II, Found. Comput. Math. **11** (1020), 153–179.

[17] P. Duren, Theory of $H^p$ Spaces, Academic Press, New York (1970).

[18] L. N. Trefethen, Is Gauss quadrature better than Clenshaw–Curtis?, SIAM Rev. **50** (2008), 67–87.

[19] S. Olver, RHPackage, http://www.maths.usyd.edu.au/u/olver/projects/RHPackage.html (2010).

[20] S. Olver and T. Trogdon, Nonlinear steepest descent and numerical solution of Riemann-Hilbert problems, Comm. Pure Appl. Math. **67** (2014), 1353–1389.

[21] D. Levin, Analysis of a collocation method for integrating rapidly oscillatory functions, J. Comput. Appl. Math. **78** (1997), 131–138.

[22] T. Trogdon and S. Olver, Numerical inverse scattering for the focusing and defocusing nonlinear Schrodinger equations, Proc. R. Soc. Lond. Ser. A **469** (2012), 20120330.

[23] M. J. Ablowitz and A. S. Fokas, Complex Variables: Introduction and Applications, Cambridge University Press, Cambridge, UK (2003).

[24] C. W. Curtis and B. Deconinck, On the convergence of Hill's method, Math. Comp. **269** (2010), 169–187.

[25] M. A. Johnson and K. Zumbrun, Convergence of Hill's method for nonselfadjoint operators, SIAM J. Numer. Anal. **50** (2012), 64–78.

[26] M. J. Ablowitz and P. A. Clarkson, Solitons, Nonlinear Evolution Equations and Inverse Scattering, Cambridge University Press, Cambridge, UK (1991).

[27] G. Teschl, Jacobi Operators and Completely Integrable Nonlinear Lattices, AMS, Providence, RI (2000).

[28] T. Trogdon, S. Olver and B. Deconinck, Numerical inverse scattering for the Korteweg–de Vries and modified Korteweg–de Vries equations, Phys. D **241** (2012), 1003–1025,

[29] P. Deift, Orthogonal Polynomials and Random Matrices: A Riemann-Hilbert Approach, AMS, Providence, RI (2008).

[30] T. Trogdon, Riemann-Hilbert Problems, Their Numerical Solution and the Computation of Nonlinear Special Functions, PhD thesis, University of Washington, Seattle, WA (2013).

[31] T. Trogdon and B. Deconinck, Dispersive and soliton perturbations of finite-genus solutions of the KdV equation: Computational results, Phys. Lett. A **378** (2014), 617–622.

[32] P. Deift and X. Zhou, Long-Time Behavior of the Non-Focusing Nonlinear Schröedinger Equation: A Case Study, Lectures Math. Sci. New Series **1** (1994).

# Index